Supersymmetry Beyond Minimality

Supersymmetry Beyond Minimality: From Theory to Experiment

Shaaban Khalil
Center for Fundamental Physics, Zewail City of Science
and Technology, Giza, Egypt

Stefano Moretti
School of Physics & Astronomy, University of Southampton, UK,
Particle Physics Department, STFC Rutherford Appleton
Laboratory, UK

CRC Press
Taylor & Francis Group
Boca Raton London New York

CRC Press is an imprint of the
Taylor & Francis Group, an **informa** business

CRC Press
Taylor & Francis Group
6000 Broken Sound Parkway NW, Suite 300
Boca Raton, FL 33487-2742

First issued in paperback 2019

ISBN-13: 978-1-4987-5673-0 (hbk)
ISBN-13: 978-0-367-87662-3 (pbk)

Library of Congress Cataloging-in-Publication Data

Names: Khalil, Shaaban, author. | Moretti, Stefano, 1966- author.
Title: Supersymmetry beyond minimality : from theory to experiment / Shaaban Khalil and Stefano Moretti.
Description: Boca Raton, FL : CRC Press, Taylor & Francis Group, [2017] | Includes bibliographical references and index.
Identifiers: LCCN 2016059352| ISBN 9781498756730 (hardback ; alk. paper) | ISBN 1498756735 (hardback ; alk. paper) | ISBN 9781315367903 (e-book) | ISBN 1315367904 (e-book)
Subjects: LCSH: Supersymmetry. | Symmetry (Physics)
Classification: LCC QC174.17.S9 K42 2017 | DDC 539.7/25--dc23
LC record available at https://lccn.loc.gov/2016059352

Visit the Taylor & Francis Web site at
http://www.taylorandfrancis.com

and the CRC Press Web site at
http://www.crcpress.com

To our families
For the memory of those who have left
For the support from those who are here

Contents

Section III SUSY CP and Flavour

SECTION IV MSSM Extensions

List of Figures

List of Tables

Preamble

If beauty and simplicity are to be taken as guidance in attempting to describe Nature, then supersymmetry can boast very few rivals. However, Science demands one to have proof of it. With this book, we aim at conveying to the reader both perspectives. On the one hand, we exalt its theoretical elegance. On the other hand, we stress its experimental accessibility. We could have started by almost claiming its inevitability, as it would cure at once almost all shortcomings of the current description of Nature. We might have ended by having to call for its dismissal instead, after having unsuccessfully scrutinised all its possible manifestations.

A balance is however found when one realises that other, more subtle manifestations of supersymmetry may occur than the most obvious, minimal one. A somewhat more complex than expected dynamics of the known interactions may take place, thereby calling for an extended gauge group. A somewhat more complex than expected dynamics of mass generation may take place, thereby calling for an extended Higgs sector. With this book, we therefore also aim at spurring interest and investigation of either or both of the latter solutions amongst the high energy particle physics community, at a time when many of the tenets of supersymmetry are apparently being questioned by data. While we acknowledge an irrecoverable debt to several excellent books already existing on the subject, we aim at surpassing them thanks to this leap into alternative realisations of supersymmetry.

The forthcoming Table of Contents illustrates the threads we will develop in the quest to motivate these non-minimal scenarios and enable their construction, in a bottom-up approach that interconnects experimental phenomena in both the fermionic and bosonic sectors, many of which post-date the inception of supersymmetry, like the discovery of neutrino masses and the evidence of a Higgs boson, or simply were ignored by previous textbooks, like Charge/Parity and Lepton Flavour Violation.

Preface

The paradigm that particle physics is minimalist in Nature, at least as far as Electro-Weak Symmetry Breaking (EWSB) goes, may simply be the result of early appearance if one realises that the Standard Model (SM), as we know it, appeared to be so initially, but then it revealed itself rather more articulated than we thought (or hoped). As for its interactions, there was first the photon, uncharged and massless. We later discovered its massive weak companions, Z and $W^{\pm}$, the latter being charged too. Even the gluon is one of eight, actually. Concerning matter, the story started with one generation of quarks and leptons/neutrinos, which was sufficient to keep our world stable. Then somebody ordered the muon[1] and all apparently fell apart. With it also came its neutrino (not that we saw it at the time or even now). Strangely yet charmingly, the quarks were no less zealous, producing their own second generation of two offsprings within a few decades. Discovering the third generations of fermions was even more upsetting, so bulky in comparison to the preceding two.

We may then have to dismiss a minimalist attitude also for the Higgs sector, eventually, so that the 125 GeV Higgs boson discovered at the Large Hadron Collider (LHC) at the Conseil Européen pour la Recherche Nucléaire (CERN) is nothing but the first of many companion Higgs states. We ought to be prepared this time for a non-minimal Higgs sector. Supersymmetry (SUSY), for example, calls for it. We are rather fond of this ultimate (potential) symmetry of Nature, in fact, as for the first time a theory would probably solve more problems than it could create. In particular, SUSY, whichever shape or form of it is actually realised in Nature, wants a light Higgs boson, with mass similar to that of the weak gauge bosons. (No such claim can be made by the SM instead.) Alas, SUSY has not been seen, yet. While disturbing per se, this fact may actually be a consequence of (yet again) a flawed approach, which assumes that SUSY is also minimal. Just like the SM actually is not for most of its parts (and we claim it to be for none), SUSY needs not be so either. Unsurprisingly, if one dismisses minimalism in SUSY, one may find at the same time an explanation for the absence of its manifestations at present (SUSY particle signals may be more complicated in non-minimal SUSY, thus escaping usual SUSY searches) as well as a hint of where Higgs companions might be.

"Among competing hypotheses, the one with the fewest assumptions should be selected" is an accepted English version of Ockham's razor argument [36]: *Numquam ponenda est pluralitas sine necessitate* (literally, plurality must never be posited without necessity) [37]. "Everything should be made as simple as possible, but not

---

[1]I.I. Rabi, "Who ordered that?" (a quip in 1957, verbal).

simpler" is our favoured paraphrase of Einstein's razor argument [38]. Contrasting the two, one may well conclude that to salvage the experimental situation emerged from current experimental data, which revealed no hints of SUSY signals, it suffices to migrate from Occam's to Einstein's philosophical conception, in our words, to abandon minimalism but not SUSY.

In fact, this book starts from the dual assumption that Nature is non-minimalistic and that SUSY exists. By combining the two instances, we are then drawn to motivate the importance of studying non-minimal SUSY models. The book builds this up gently, starting from what ought to be common knowledge of minimal SUSY, yet eventually constructing sophisticated implementations of it that can lend themselves to explain both theoretical and experimental flaws of the so-called Minimal Supersymmetric Standard Model (MSSM). Examples of these scenarios treated here in some depth are the Next-to-MSSM (NMSSM), the new Minimally-extended Supersymmetric Standard Model (nMSSM), the $B-L$ inspired Supersymmetric Standard Model (BLSSM) and the Exceptional Supersymmetric Standard Model (ESSM).

In attempting to achieve this, the book has been divided into eighteen Chapters, including four Appendices, grouped in six parts, other than the Preamble, Foreword, Preface and Epilogue. Part I is core material (essentially constructing the MSSM) and is necessary to develop any of the following parts, which are in turn blocks largely independent one from the other. The former could be the material for a semester course whereas any of the latter can be taught separately in more advanced and shorter modules. Part II describes the phenomenological implication of the MSSM, highlighting some of its potential successes and certain shortcomings. Part III then starts looking at how to realise non-minimal constructions of SUSY, based on well-known Charge and Parity (CP) violating phenomena that cannot be easily explained in the MSSM and thus ought to be used as guidance in building models that surpass it. Part IV develops this to the full, by starting from the realisation that the most significant evidence of Beyond the SM (BSM) physics to date, that neutrinos have masses, naturally leads to non-minimal SUSY constructs in the gauge sector, like the BLSSM, just like the absence of sparticle signals combines with the presumption of an extended Higgs sector to offer us the NMSSM/nMSSM. Assuming both, extended gauge and Higgs sectors, eventually leads one to the ESSM as well. Finally, Part V addresses the fact that SUSY cannot after all be an exact symmetry of Nature and thus must be violated, by dwelling on various possible mechanisms of SUSY breaking dynamics.

This book is based largely on our personal research with various collaborators, whom we thank here collectively. We cannot in fact acknowledge them one by one, as the list would be far too long. However, we would like to single out here those who can be more closely associated with this book, for one reason or another. We thus thank in particular, in alphabetical order, M. Abbas, W. Abdallah, S. Abel, R.C. Aggleton, A. Awad, D. Bailin, D. Barducci, L. Basso, A. Belyaev, N. Chamoun, S. De Curtis, D. Delepine, L. Delle Rose, U. Ellwanger, A. Elsayed, W. Emam, J. Fiaschi, E. Gabrielli, A. Hammad, K. Huitu, S.F. King, S.J.D. King, T. Kobayashi, E. Kou, S. Kulkarni, O. Lebedev, E. Ma, C. Marzo, A. Masiero, D.J. Miller, A.

Moursy, S. Munir, C. Munoz, A. Nassar, R. Nevzorov, H. Okada, W. Porod. P. Poulose, G.M. Pruna, L. Roszkowski, Q. Shafi, C.H. Shepherd-Themistocleous, A. Sil, F. Staub and C.S. Un. Finally, in addition to our own institutions, we would like to also thank CERN, where many parts of this book were worked upon.

Shaaban Khalil
Stefano Moretti

Foreword

Supersymmetry (SUSY) is a remarkable idea about the theory that physicists search for to describe the physical world. Essentially all other such ideas arose in response to data or puzzles, but SUSY emerged in the early 1970s from theorists studying theories, particularly, the ones in fewer than four space-time dimensions. After a few years theorists began to realise that a supersymmetric extension of the Standard Model (SM) of particle physics could solve or explain major problems and offered some opportunities that were unrecognised before the late 1970s.

The supersymmetric idea was that the underlying theory (Lagrangian) should be one that was unchanged if bosons and fermions were interchanged. That was surprising, and thought unlikely, but by 1974 Wess and Zumino and others had shown realistic four-dimensional Quantum Field Theories (QFTs) where such a symmetry could be constructed.

Two problem solutions, or opportunities, were particularly important. One had to do with the domain of validity of the SM. Quantum Electro-Dynamics (QED) had been tested to high accuracy, but was a badly behaved theory at high energies. Quantum Chromo-Dynamics (QCD) and the unified Electro-Weak (EW) theory were asymptotically free and well-behaved at high energies (and the latter contained QED). Their force strengths had the same form and ran towards one another at high energies, so one could begin to imagine a theory with a unified description of the forces. When superpartners (i.e., SUSY companion states to the SM ones) around the TeV scale or lower were incorporated in the description, the merging of the couplings was surprisingly precise. The boundaries of physics changed and began to include a unified supersymmetric underlying theory. At the same time, it was being recognised that the EW theory had a conceptual flaw, the hierarchy problem. Basically, in any QFT, virtual particle contributions brought energy scales together. Then, having a theory that included the widely separated EW and Planck scales, with scalar particles such as Higgs bosons that were sensitive to such corrections, appeared to be unworkable. SUSY stabilised the hierarchy because the contributions of bosons (fermions) and their fermionic (bosonic) superpartners cancelled automatically.

The second was that the Higgs mechanism had been shown to allow incorporating masses for quarks and leptons and gauge bosons into the SM, but it involved arbitrary assumptions about the Higgs sector potential. But because the above separation of scales allowed imagining formulating the underlying theory at a high scale, in the early 1980s, people wrote a high scale supersymmetric theory containing Higgs fields but with a normal potential and found, amazingly, that the Higgs potential needed for the SM emerged as the Lagrangian was run down to the

EW scale. There was one condition required for that result to hold, namely, that there should be a heavy (compared to the $W^{\pm}$ boson mass) quark, much heavier than all the quarks known then. Virtual particle corrections to the Z mass measured at the Large Electron-Positron (LEP) machine and Stanford Linear Collider (SLC), and later direct detection of the top quark at the Fermi National Accelerator Laboratory (FNAL) proton-antiproton collider (Tevatron), showed it indeed to be appropriately heavy. For many physicists this success in deriving the Higgs mechanism, with the associated successful prediction of the heavy top quark, was compelling evidence that Nature would be supersymmetric.

There were many additional attractive features of SUSY. The lightest superpartner was normally stable in SUSY and could give the observed relic density of Dark Matter (DM). For a Higgs boson mass of 125 GeV, the SM Higgs potential became unbounded from below at energies above about 10^{10} GeV, in the physical region below the Planck scale, an unacceptable result, while the supersymmetric generalisation was always bounded from below. With unified couplings the EW parameter $\sin^2 \theta_W$ calculated in the theory agreed with percent accuracy with the data. In a supersymmetric theory there were opportunities to understand the matter-antimatter asymmetry. The see-saw mechanism for neutrino masses required a high scale for the right-handed neutrinos, so having a way to stabilise scales was needed. The Minimal Supersymmetric Standard Model (MSSM) predicted that the Higgs boson mass was less than 135 GeV, as was indeed finally observed below this value. And a locally gauge invariant SUSY has promising connections to a theory of gravity. All of these are described in this book by Shaaban Khalil and Stefano Moretti, which crucially also includes alternative formulations of SUSY, in non-minimal form, that have the potential to reconcile this theory with all available data possibly better than the MSSM.

The Lagrangian for softly broken SUSY includes terms that determine the Higgs potential and therefore its mass and decay branching ratios. The Higgs properties observed at the LHC correspond to a particular well known limit where the Higgs sector parameters are large at the high scale and satisfy the EW Symmetry Breaking (EWSB) conditions. For some people, that is more evidence that superpartners will be found at the LHC, since part of the softly broken Lagrangian already has confirmed predictions.

Finally, there is a huge bonus if the EW scale is indeed stabilised by SUSY. The apparent successful unification of the couplings is a perturbative one, so it implies that connecting Planck scale ideas with EW data is meaningful and that Planck scale theories can be tested with EW scale data. Since compactified string/M-theories are naturally formulated near the Planck scale, that is essential for making progress towards an underlying comprehensive theory. Of course, that does not guarantee that Nature is indeed supersymmetric, but it suggests that learning about supersymmetric theories may be valuable.

This book joins a few other useful SUSY texts. It has a nice mix of phenomenological and more formal theory as well as some valuable pedagogical emphasis, guiding the reader in understanding how to make SUSY predictions. Before a few years ago no theory allowed predicting superpartner masses. An argument called 'naturalness' was used. It basically said that, if superpartners were not too heavy compared to the SM gauge bosons and top quark, then SUSY would solve the hierarchy problem and other problems rather 'naturally', so the superpartners should be light. The CERN LEP collider and the FNAL Tevatron both found no evidence for superpartners. Such an evidence should have appeared if they existed in the naturalness range. Naturalness was an argument to be used in the absence of having a theory.

In recent years, increased understanding of compactified string/M-theories has emerged, with some predictions of the superpartner spectrum. Generically scalar superpartners are rather heavy, perhaps a few tens of TeV, and sometimes gauginos (partners of gauge bosons) are much lighter. Essentially none have masses in the naturalness range, instead solving the hierarchy problem by other interesting and clever mechanisms. So, from the point of view of such theories, superpartners should not have yet been seen, but the LHC should be starting in 2017 to probe their mass range and possibly discover some. Perhaps, since the high scale of the theory is set by the Planck mass, that is not surprising. This book contains helpful introductions to supergravity and to string phenomenology, both likely to be increasingly active areas if superpartners are indeed discovered at the LHC, which will allow readers to make progress in these areas, and a broad range of topics presented both in detail and pedagogically. Other topics covered are relevant for data that will emerge from b-quark factories, such as a dedicated b-quark factory in Japan, Belle-II, and the fixed target LHCb experiment, which will reach new levels of sensitivity for studying rare b-quark decays and CP violation.

The indirect evidence that SUSY is part of our description of Nature at the physics scales where there is data is strong and has increased. Even though so far no direct evidence for superpartners has been found, there are over 20,000 papers on SUSY. If superpartners are indeed found at the LHC in the next few years, this book will be very valuable in helping us learn to interpret the data.

Gordon Kane

Acronyms

For the reader's convenience, we introduce here the acronyms that we will use in the remainder of the book. They are listed in order of appearance.

SM	Standard Model
EM	Electro-Magnetic
EW	Electro-Weak
CP	Charge/Parity
BSM	Beyond the Standard Model
DM	Dark Matter
GUT	Grand Unification Theory
UV	Ultra-Violet
QCD	Quantum Chromo-Dynamic
QED	Quantum Electro-Dynamics
SUSY	Supersymmetry
EWSB	EW Symmetry Breaking
CM	Coleman-Mandula
LSP	Lightest Supersymmetric Particle
WIMP	Weakly Interacting Massive Particle
SUGRA	Supergravity
QFT	Quantum Field Theory
VEV	Vacuum Expectation Value
CM	Centre-of-Mass
CPT	Charge/Parity/Time

WZ	Wess-Zumino
4D	4-Dimensional
SQED	Supersymmetric QED
FI	Fayet-Iliopoulos
MSSM	Minimal Supersymmetric Standard Model
cMSSM	constrained MSSM
RGE	Renormalisation Group Equation
CKM	Cabibbo-Kobayashi-Maskawa
mSUGRA	minimal SUGRA
MIA	Mass Insertion Approximation
FCNC	Flavour Changing Neutral Current
VBF	Vector Boson Fusion
HS	Higgs-Strahlung
LIPS	Lorentz-Invariant Phase Space
LHC	Large Hadron Collider
CERN	Conseil Européen pour la Recherce Nucléaire
LEP	Large Electron Positron
CMS	Compact Muon Solenoid
ATLAS	A Toroidal LHC Apparatus
ALICE	A Large Ion Collider Experiment
BR	Branching Ratio
LO	Leading Order
CMB	Cosmic Microwave Background
WMAP	Wilkinson Microwave Anisotropy Probe
PQ	Peccei-Quinn
NFW	Navarro-Frenk-White
EDM	Electric Dipole Moment
NDA	Naive Dimensional Analysis

NLSP	Next-to-LSP
B	Baryon
L	Lepton
OPE	Operator Product Expansion
NLO	Next-to-LO
GIM	Glashow-Iliopoulos-Maiani
QCDF	QCD Factorisation
RH	Right-Handed
BRPV	Bilinear R-Parity Violation
SS	Supersymmetric Seesaw
MNS	Maki-Nakagawa-Sakata
LFV	Lepton Flavour Violation
LH	Left-Handed
BLSSM	$B - L$ Supersymmetric Standard Model
BLSSM-I	BLSSM of Type-I
BLSSM-IS	BLSSM with Inverse Seesaw
IS	Inverse Seesaw
DY	Drell-Yan
SSM	Sequential Standard Model
MC	Monte Carlo
pNGB	pseudo-Nambu-Goldstone Boson
NMSSM	Next-to-MSSM
MNSSM	Minimal Non-minimal Supersymmetric SM
nMSSM	new Minimally-extended Supersymmetric SM
UMSSM	$U(1)'$ extension of the MSSM
ESSM	Exceptional Supersymmetric SM
DR	Dimensional Regularisation
AMSB	Anomaly Mediated SUSY Breaking

GMSB	Gauge Mediated SUSY Breaking
mGMSB	minimal GMSB
RNS	Ramond-Neveu-Schwarz
GSO	Gliozzi-Scherck-Olive
R	Ramond
NS	Neveu-Schwarz
CY	Calabi-Yau
CSM	Complex Structure Moduli
GKP	Giddings-Kachru-Polchinski
KKLT	Kachru-Kallosh-Linde-Trivedi

I

MSSM Construction

I

MSSM Construction

CHAPTER 1

Introduction to Supersymmetry

1.1 MOTIVATIONS FOR SUSY

The SM of elementary particle physics is a renormalisable gauge theory of the strong, EM and weak interactions, which has been confirmed experimentally. It incorporates both bosons, as mediators of such interactions, and fermions, as matter states upon which these act. Crucially, it also embeds the so-called Higgs mechanism, which enables the unification of the last two types of interactions into a single one, the EW force, generating as a by-product the masses of all such bosonic (except the photon) and fermionic (except the neutrinos) states. The SM makes genuine theoretical predictions in both such sectors, which have been eventually confirmed by experiment. On the one hand, it predicted the existence of the massive weak gauge bosons, the $W^{\pm}$ and Z states, and their mass and coupling relations. On the other hand, it predicted the existence of a third generation of leptons and quarks to account for the observed CP violation of EW interactions, which was confirmed by the discovery of the τ (tau) lepton plus b (bottom) and t (top) quarks. Its most crucial component, the self-interacting Higgs boson, which is spinless (*i.e.*, a scalar) unlike the gauge bosons which have spin one (*i.e.*, they are vectors), emerging from the spontaneous breaking of an initial EW symmetry, also seems to have been found.

However, the SM fails to explain observational evidence that has been gathered from present and past experiments. In fact, there are at least three firm measurements that necessarily imply new physics BSM. Firstly, in the SM neutrinos are massless, yet recent experiments indicate that this is not true and neutrinos have small masses. Secondly, the SM cannot describe the DM that accounts for 23% of the mass/energy density of the observable universe, as its only plausible candidate, the said neutrino, provides too modest a level of it. Thirdly, the SM cannot explain the fact that the universe is made of matter and not antimatter, as it is now established that the strength of CP violation in the SM, which is the essential ingredient for explaining the cosmological baryon asymmetry of the universe, is not sufficient.

Furthermore, from a theoretical point of view, it is also clear that the SM can-

Figure 1.1 One-loop radiative corrections to Higgs mass in the SM.

not be a valid realisation of the ultimate theory, valid to high energies (*e.g.*, up to the Planck scale, $M_{\text{Pl}} \sim 1.2 \times 10^{19}$ GeV, where gravity is expected to become strong and possibly unify with the other interactions). In fact, there are theoretical inconsistencies leading to instabilities of the Higgs self-interactions that systematically grow with the energy at which the SM is being probed, ultimately leading to either its breakdown or a solution perceived as highly *ad-hoc*, hence unnatural. This conceptual problem is known as the 'hierarchy problem'. It is associated with the absence of a symmetry protecting the Higgs mass, which is necessarily generated at the EW scale, when the interactions of the Higgs field are instead generated where the natural cutoff scale is or indeed above it, *i.e.*, the energy at which a GUT (also including gravity) will inevitably have to be formulated, $M_X \simeq 3 \times 10^{16}$ GeV. This problem can be seen explicitly from the one-loop radiative corrections to the Higgs mass (see Fig. 1.1), which lead to the following expression:

$$m_h^2 = (m_h^2)^0 + \mathcal{O}(\frac{\alpha}{4\pi})\Lambda^2, \tag{1.1}$$

where $(m_h^2)^0$ is the tree-level Higgs boson mass, given in terms of the parameters of the fundamental theory, like the EM coupling constant α, and where Λ is an UV momentum cutoff used to regulate the divergences of the loop integral. It is key to notice here that, unlike the electron mass, which is protected by a chiral symmetry and hence its radiative corrections are only logarithmically divergent, $\Delta m_e \propto \frac{\alpha}{4\pi} \ln(\Lambda)$, the absence of a similar symmetry in the case of the Higgs boson leads to the corresponding mass corrections being quadratic. Therefore, if Λ is of order of the GUT scale, then the Higgs mass will no longer be of the order of the EW scale, $\mathcal{O}(100$ GeV). While one could always technically remedy this by postulating a bare mass for the Higgs boson such that renormalisation could eventually ensure that its physical mass remains at the EW scale (indeed to match the aforementioned latest experimental results, which place it at approximately 125 GeV), this appears as a scapegoat (leading to fine-tuned cancellations, the more so the closer the energy of the interaction approaches M_{Pl}), which is not necessary in the case of other fundamental parameters of the SM. So, the strong hierarchy between the EW and GUT (or Planck) scales is directly manifest in the SM by spoiling its ability to make predictions over the entire energy range over which it can realistically be probed.

There are also a number of phenomenological issues that the SM fails to address, which may (or may not) find answers in extensions of it. For instance, can the forces of Nature be unified so that we do not have the initial four independent ones, rather that they are all interlinked as different manifestations of a sole fundamental force? Why is the symmetry group of the SM formally characterised at current energies by the structure $SU(3)_C \times SU(2)_L \times U(1)_Y$, where the $SU(3)_C$ group describes the strong interactions (through the definition of a discrete Colour degree of freedom that onsets QCD in the same way as the EM charge does in the case of QED with $SU(2)_L \times U(1)_Y$ exemplifying the SM unifying mechanism of EM and weak interactions (through the definition of two discrete degrees of freedom, of isospin (L) and hypercharge (Y))? Why are there three generations of quarks and leptons, when the aforementioned CP violation could still accommodate more, without contradicting experimental results? Why do the quarks and leptons have the masses they do, in particular, they grow from generation to generation within each family (of quarks and leptons)? Also, what is the origin of the SM distinction between bosons (gauge and Higgs particles) and fermions (matter particles)?

Thus, the SM cannot be considered as a fundamental theory and we need to search for new physics constructions beyond it, which can nonetheless reproduce the SM in the phenomenological situations where it has been probed to a phenomenal degree of accuracy. One of the best candidates for BSM physics is SUSY. In fact, SUSY theories now stand as the most promising scenarios for a unified BSM theory. Despite the absence of experimental verifications of SUSY, relevant theoretical arguments (which will be discussed in detail in the next chapters) can be given in favour of SUSY, as follows.

1. SUSY is a new symmetry which relates bosons and fermions, thereby addressing one of the questions we asked, that implies a new kind of unification between particles of different spin. In this respect, the Higgs boson is no longer a unique particle as it stands in the SM, as a SM enriched with SUSY would naturally be full of scalars (squarks, sleptons and Higgs bosons) related through SUSY to their fermionic partners (quarks, leptons, gauginos and Higgsinos), also incorporating the counterparts of the SM gauge bosons. (We ask the reader to be patient, as we will shortly introduce him or her to this nomenclature.)

2. SUSY ensures the stability of the hierarchy between the EW and Planck scales. As mentioned, in the SM, the quantum corrections to the Higgs mass are proportional to the natural highest scale, M_X or M_{Pl}. However, since these mass terms contribute to the Higgs potential, they should be of the order of the EW scale (100–1000 GeV), in order to avoid any type of fine-tuning after EWSB has taken place. This problem of stabilising the scalar masses against quantum corrections is solved in SUSY theories. As we will show, SUSY requires the scalar masses and the masses of their superpartners be related, therefore the dangerous contributions of SM particles to quantum corrections are cancelled against new ones which are present owing to the existence of additional superpartners obeying different spin statistics.

3. Through the same mechanism of unification that relates bosons and fermions, SUSY relates the mass (or equivalently the self-coupling) of a Higgs state, which is a free parameter in the SM, to the gauge boson couplings intervening in the SM interactions. As the latter are small, this provides a characteristic hallmark of SUSY, that the lightest Higgs boson (as there are more than one in any model realisation of SUSY) mass is naturally at the EW scale. So, even the mentioned experimental evidence of Higgs boson signals at 125 GeV, which looks like an accident in the SM, is a phenomenological hint in favour of an underlying Higgs mechanism subject to SUSY dynamics. In fact, on the same note, it should further be appreciated that EWSB occurs in a SUSY model radiatively, meaning that the negativity of the Higgs mass term in the Lagrangian, which is enforced by hand in the SM, can be a natural consequence of the SUSY model evolution from the GUT or Planck scale down to the EW regime.

4. SUSY facilitates the convergence of the three gauge coupling constants of the SM, when run with energy, according to renormalisation, to a single unification scale, by only taking into account that SUSY agrees with current experimental results and allowing for the masses of the new SUSY states to be in the TeV range. In turn, then, this gives us as a by-product another formidable prediction, that the new particle states predicted by SUSY are within the reach of current experiments.

5. The LSP, in the presence of a rather intuitive request that SUSY objects have some discrete quantum numbers that prevent them from decaying into SM objects only, can be a natural candidate for a WIMP , *i.e.*, a phenomenologically acceptable explanation of the DM puzzle.

6. The local version of SUSY leads to a partial unification of the ensuing BSM scenario with gravity, the so-called 'SUGRA', which is the low energy limit of a superstring theory. Conversely, string theory needs to be supersymmetric (hence superstring theory) in order to avoid the existence of tachyons in the generated model spectrum, thus leading, after compactification of any extra dimensions, to an effective SUGRA scenario.

The price one has to pay for such a beautiful theory is the abundance of new particles that must be included into any model realisation of it, since, in order to satisfy the requirements of SUSY, each particle must have a superpartner (a boson for a fermion and vice versa). Further, recall that, if ordinary particles and their superpartners only differed in spin, then SUSY would imply that they have the same masses. Therefore, there would be a superpartner for, *e.g.*, the electron (called selectron) with the same mass as the electron (≈ 0.5 MeV). However, there is no experimental evidence for such new particles, mass degenerate with the SM ones. Thus, if SUSY is meant to play a realistic role in particle physics, it must be broken, so that the superpartners are heavier than the known elementary particles. Although our theoretical ideas are still insufficient to predict the superpartner masses, there are several arguments, connected to the need of, on the one hand, ensuring EWSB, and, on the other hand, providing gauge coupling unification, that leads to the conclusion that a typical superpartner mass should be in the range $\mathcal{O}(100 - 1000 \text{ GeV})$.

In short, the proliferation of parameters imposed by a SUSY theory is somewhat counterbalanced by the phenomenological requirement that the majority of these are measurable by current or upcoming measurements. In fact, an even more intriguing condition would be the one assuming not only that SUSY classifies new and old particle states and their couplings into categories (such as bosons and fermions), but also imposes that such parameters are all identical at the GUT scale or above, as natural in fact in a mSUGRA scenario, thereby reducing the number of additional SUSY parameters (with respect to the SM count) to a handful. Hence the scene is set for current and upcoming experimental facilities to either confirm or disprove EW-scale SUSY. To be able to do so would amount to a giant step in the history of particle physics.

In the next sections we briefly introduce SUSY and show that it is an elegant extension of the Poincaré symmetry. We also discuss the SUSY algebra, superspace and superfields. There are several interesting books and review articles that provide a detailed introduction to SUSY, like [39–46] and [47–58]. Our goal in this part is to review the essential materials that will be used in the other parts of the book.

1.2 WHAT IS SUSY?

SUSY is a symmetry that transforms bosons into fermions and vice versa. The operator Q that generates such transformations must be an anti-commuting spinor[1], with

$$Q|\text{Boson}\rangle = |\text{Fermion}\rangle, \qquad Q|\text{Fermion}\rangle = |\text{Boson}\rangle. \qquad (1.2)$$

Since Q is a complex spinor, its Hermitian conjugate (h.c.) $\bar{Q}$ is also a symmetry generator. These fermionic operators carry spin 1/2, therefore SUSY must be a space-time symmetry. In this respect, SUSY is nothing but an extension of the space-time symmetry reflected in the Poincaré group.

As is known, in a QFT , symmetry is one of the basic tools for studying elementary particle processes. The known invariances are classified into two categories: (*i*) external (or space-time) symmetries; (*ii*) internal symmetries. The space-time symmetries correspond to transformations carried out directly on the space-time coordinates, i.e., $x_\mu \to x'_\mu$. The translation, $x_\mu \to x_\mu + a_\mu$, and generic Lorentz transformations, $x_\mu \to \Lambda_\mu^\nu x_\nu$, are examples of space-time symmetries. The internal symmetries are given by transformations on the fields: $\phi^a(x) \to M^a_b \phi^b(x)$, where a and b refer to field components. The EM $U(1)$ and the flavour $SU(3)$ gauges are examples of internal symmetries. If M^a_b is independent of the space-time coordinates,

---

[1]In general, the SUSY generators are given by Q^I, with $I = 1, 2, \ldots, N$. Here we focus on the case of $N = 1$ SUSY, which provides the only phenomenologically viable SUSY extension of the SM.

x_μ, then the symmetry is called *global*, whereas if it is space-time dependent, the symmetry is called *local*.

In particle physics, symmetries play crucial roles. According to Noether's theorem, every conserved physical quantity, like mass, spin, electric charge, colour, *etc.*, corresponds to a space-time or internal symmetry. Moreover, symmetries determine possible interactions among the particle through the gauge principle. Also, in the SM, we learned that symmetry can be spontaneously broken and the non-zero VEV of the field inducing EWSB is a natural way of introducing an energy scale, which is then related to the masses of the emerging particle state(s).

There were several attempts to combine internal with external symmetries in a bigger symmetry group. However, in 1967, Sidney Coleman and Jeffrey Mandula showed that it is impossible to achieve a non-trivial combination of internal and external symmetries [59]. The CM theorem is a no-go theorem, which states that for every QFT that has non-trivial interactions and its S-matrix satisfy the following conditions:

1. a symmetry group G of the S-matrix has a subgroup locally isomorphic to the Poincaré group;

2. for any mass M there is a finite number of particles with masses less than M and all particles correspond to positive-energy representations of the Poincaré group;

3. the amplitudes of elastic scattering are analytical functions of the CM energy s and momentum transfer t, u in some neighbourhood of the physical region;

4. non-trivial scattering is demanded, *i.e.*, the scattering process should have scattering angles other than $0°$ and $180°$;

one can show that G is necessarily locally isomorphic to the direct product of an internal symmetry group and the Poincaré group, *i.e.*, no mixing between these two is possible.

The proof of the CM theorem was based on the fact that, apart from the vector momentum operator P_μ, whose eigenvalues are conserved 4-momenta, p_μ, and the antisymmetric tensor angular momentum operator $M_{\mu\nu}$, any other conserved operator must be a Lorenz scalar (*i.e.*, a quantity with no Lorentz indices). To see how this proof works, let us consider two-to-two body scattering of spinless particles, $1 + 2 \rightarrow 3 + 4$, and imagine that there exists a conserved two-index symmetric tensor governing it, $\Sigma_{\mu\nu}$. By Lorentz invariance, its diagonal matrix elements between single-particle states $|P_1\rangle$ take the following general form:

$$\langle P_1 \mid \Sigma_{\mu\nu} \mid P_1 \rangle = a P_\mu^1 P_\nu^1 + b g_{\mu\nu}, \tag{1.3}$$

where a and b are constants. Then, in the scattering process, we have

$$P_\mu^1 P_\nu^1 + P_\mu^2 P_\nu^2 = P_\mu^3 P_\nu^3 + P_\mu^4 P_\nu^4, \tag{1.4}$$

$$P_\mu^1 + P_\mu^2 = P_\mu^3 + P_\mu^4. \tag{1.5}$$

Therefore, the only possible solution is $P_\mu^1 = P_\mu^3$ and $P_\mu^2 = P_\mu^4$, *i.e.*, there is no scattering at all. Thus, one can conclude that any conserved (bosonic) operator with non-trivial Lorentz transformation (not Lorentz scalar) is ruled out. In this respect, the most general algebra of the S-matrix symmetries consists of the energy momentum operator, P_μ, the angular momentum operator, $M_{\mu\nu}$, and a finite number of Lorentz scalar operators, B_l, that satisfy the following commutation relations:

$$[P_\mu, B_l] = 0, \qquad [M_{\mu\nu}, B_l] = 0, \qquad [B_l, B_m] = iC_{lm}^n B_n, \qquad (1.6)$$

where C_{lm}^n are the structure constants of the Lie algebra of the compact internal symmetry group.

SUSY is considered as a possible loophole of this theorem, since, as we will see, it contains additional generators that are not scalars but rather spinors. The spinorial operator Q_α, where the subscript α refers to the spinor component and $Q_\alpha |j\rangle = |j \pm \frac{1}{2}\rangle$, has no diagonal matrix element: $\langle P_1 | Q_\alpha | P_1 \rangle$. Such an operator does not contribute to a matrix element for a two-to-two particle elastic scattering process (in which the particle spins remain the same). Thus, the CM theorem cannot be applied. In 1975, Haag, Lopuszanski and Sohinus showed that the symmetry group of a consistent 4-dimensional QFT includes a SUSY algebra which involves, in addition to the usual commutators, anti-commutators, as a non-trivial extension of the Poincaré algebra [60].

1.3 SUSY ALGEBRA

Before we start with the SUSY algebra, let us review the Poincaré algebra first. The Poincaré transformation

$$x_\mu \to x_\mu' = \Lambda_\mu^{\ \nu} x_\nu + a_\mu \qquad (1.7)$$

is the basic symmetry of special relativity. It contains, in addition to Lorentz transformations that leave the metric tensor $\eta_{\mu\nu} \equiv \text{diag}(1, -1, -1, -1)$ invariant, $\Lambda^T \eta \Lambda = \eta$, the translations. The generators of the Poincaré group, $SO(3,1)$, are P^μ and $M^{\nu\lambda}$ with algebra:

$$[P_\mu, P_\nu] = 0, \qquad (1.8)$$
$$[M_{\mu\nu}, P_\rho] = i(\eta_{\rho\nu} P_\mu - \eta_{\rho\mu} P_\nu), \qquad (1.9)$$
$$[M_{\mu\nu}, M_{\rho\sigma}] = i(\eta_{\nu\rho} M_{\mu\sigma} - \eta_{\mu\rho} M_{\nu\sigma} - \eta_{\nu\sigma} M_{\mu\rho} + \eta_{\mu\sigma} M_{\nu\rho}). \qquad (1.10)$$

One can define Lorentz generators as $J_i = \frac{1}{2}\varepsilon_{ijk} M_{jk}$, which represent the proper rotation generators, and $K_i = M_{0i}$, as Lorentz boost generators. In this case, one finds that the linear combinations

$$J_j^\pm = \frac{1}{2}(J_j \pm iK_j) \qquad (1.11)$$

satisfy the $SU(2)$ commutation relations

$$\left[J_i^\pm, J_j^\pm\right] = i\varepsilon_{ijk} J_k^\pm, \qquad \left[J_i^\pm, J_j^\mp\right] = 0, \qquad (1.12)$$

where ε_{ijk} is the totally antisymmetric Levi-Civita symbol with $\varepsilon_{123} = 1$. This indicates that the Lorentz group is homeomorphic to $SU(2) \times SU(2)$, *i.e.*, $SO(3,1) \cong SU(2) \times SU(2)$. In addition, one can show that $SO(3,1)$ is homeomorphic to $SL(2,C)$ [57]. This can be seen by introducing the four 2×2 matrices σ_μ, with σ_0 as the identity matrix and σ_i, $i = 1,2,3$, as the three Pauli matrices. Then, for every 4-vector x_μ, the 2×2 matrix $x^\mu \sigma_\mu$ is Hermitian and has determinant equal to $x^\mu x_\mu$, which is Lorentz invariant. Therefore, the determinant and the Hermiticity of $x^\mu \sigma_\mu$ is preserved under Lorentz transformations. Hence, it must act as $x^\mu \sigma_\mu \to A x^\mu \sigma_\mu A^\dagger$ with $|\det A| = 1$, which means that A is a complex 2×2 matrix of unit determinant, *i.e.*, $A \in SL(2,C)$.

This establishes the mapping between the Lorentz and $SL(2,C)$ groups. In this respect, $SL(2,C)$ is a universal cover of the Lorentz group $SO(3,1)$, as $SU(2)$ is a universal cover of $SO(3)$. We provide a brief review on the spinor representations of the Lorentz group in Appendix A.

The first extension of the Poincaré algebra with fermionic charges was made by Golfand and Likhtman in 1971 [61]. The complete list of SUSY generators involves those of the Poincaré group (even part of the algebra) as well as the two anticommuting spinor generators Q and $\bar{Q}$ acting on the fields. The spinor generators satisfy the following algebra:

$$\{Q_\alpha, Q_\beta\} = 0, \qquad \{\bar{Q}_{\dot\alpha}, \bar{Q}_{\dot\beta}\} = 0, \tag{1.13}$$

$$\{Q_\alpha, \bar{Q}_{\dot\beta}\} = 2\sigma^\mu_{\alpha\dot\beta} P_\mu, \tag{1.14}$$

where $\sigma^\mu \equiv (1, \sigma^i)$, $\bar{\sigma}^\mu \equiv (1, -\sigma^i)$ and $\alpha, \beta, \dot\alpha, \dot\beta = 1, 2$. The commutation relations with generators of the Poincaré group are given by[2]

$$[P^\mu, Q_\alpha] = [P^\mu, \bar{Q}_{\dot\alpha}] = 0, \tag{1.15}$$

$$[M^{\mu\nu}, Q_\alpha] = -i(\sigma^{\mu\nu})^\beta_\alpha Q_\beta, \tag{1.16}$$

$$[M^{\mu\nu}, \bar{Q}^{\dot\alpha}] = -i(\bar{\sigma}^{\mu\nu})^{\dot\alpha}_{\dot\beta} \bar{Q}^{\dot\beta}, \tag{1.17}$$

where $\sigma^{\mu\nu}$ is defined as

$$(\sigma^{\mu\nu})_\alpha^{\ \beta} = \frac{1}{4}\left(\sigma^\mu_{\alpha\dot\gamma}\bar{\sigma}^{\nu\dot\gamma\beta} - \sigma^\nu_{\alpha\dot\gamma}\bar{\sigma}^{\mu\dot\gamma\beta}\right), \tag{1.18}$$

$$(\bar{\sigma}^{\mu\nu})^{\dot\alpha}_{\ \dot\beta} = \frac{1}{4}\left(\bar{\sigma}^{\mu\dot\alpha\gamma}\sigma^\nu_{\gamma\dot\beta} - \bar{\sigma}^{\nu\dot\alpha\gamma}\sigma^\mu_{\gamma\dot\beta}\right). \tag{1.19}$$

The commutators of Q_α and $\bar{Q}_{\dot\alpha}$ with internal symmetry generators T_i usually vanish. From these commutations one can easily show that, since P_μ commutes with the SUSY generators, the mass operator $P^2 = P_\mu P^\mu$ is a Casimir operator (*i.e.*, an invariant of the SUSY group):

$$[Q_\alpha, P^2] = 0. \tag{1.20}$$

---

[2]For a more detailed analysis, see [41].

Therefore, if $|F\rangle$ is a fermion state, with mass m_F, obtained from the bosonic state $|B\rangle$, with mass m_B, through the SUSY generator Q_α, $|F\rangle = Q_\alpha|B\rangle$, then

$$m_F|F\rangle = P^2|F\rangle = P^2 Q_\alpha|B\rangle = Q_\alpha P^2|B\rangle = m_B Q_\alpha|B\rangle = m_B|F\rangle. \tag{1.21}$$

Thus,

$$m_F = m_B. \tag{1.22}$$

Hence, the particle content of the irreducible representations of the SUSY algebra, $i.e.$, within one supermultiplet, has the same mass.

The square of the Pauli-Ljubanski vector, $W^2 = W_\mu W^\mu$, where W^μ is given by

$$W^\mu = -\frac{i}{2}\varepsilon^{\mu\nu\rho\sigma}M^{\nu\rho}P^\sigma, \tag{1.23}$$

which is a Casimir operator of the Poincaré algebra, is now not a Casimir operator of the super-Poincaré algebra,

$$[Q_\alpha, W^2] \neq 0. \tag{1.24}$$

Thus, the irreducible multiplets will have particles of different spin. Nevertheless, one can prove that the operator $C^2 = C_{\mu\nu}C^{\mu\nu}$, where

$$C_{\mu\nu} = B_\mu P_\nu - B_\nu P_\mu \tag{1.25}$$

and

$$B_\mu = W_\mu + \frac{1}{8}\bar{Q}_M\gamma_\mu\gamma_5 Q_M, \tag{1.26}$$

is a Casimir operator, $i.e.$,

$$[C^2, P_\mu] = [C^2, M_{\mu\nu}] = [C^2, Q_\alpha] = [C^2, \bar{Q}_{\dot\alpha}] = 0. \tag{1.27}$$

Note that, in the definition of B_μ, the 4-component Majorana spinor charge Q_M is used, where

$$Q_M = \begin{pmatrix} Q_\alpha \\ \bar{Q}^{\dot\alpha} \end{pmatrix}. \tag{1.28}$$

Therefore, the eigenvalues of the Casimir operator C^2 label the irreducible representation. It is interesting to note that C^2 can be written as

$$C^2 = 2m^4 J_\mu J^\mu, \tag{1.29}$$

where

$$J_\mu = S_\mu - \frac{1}{4m}(\bar{Q}\bar{\sigma}_\mu Q) \tag{1.30}$$

and

$$[J_\mu, J_\nu] = i\varepsilon_{\mu\nu\rho}J_\rho. \tag{1.31}$$

Thus, J^2 is an invariant operator with eigenvalues of the form $j(j+1)$ as in the case of ordinary angular momentum. Accordingly, the irreducible representations of the SUSY algebra are specified by the values of m^2 and $j(j+1)$, the eigenvalues of the Casimir operators. These states are not in general eigenstates of the spin-squared operator, S^2.

Furthermore, as the SUSY algebra shows, the generators Q_α and $\bar{Q}_{\dot\alpha}$ commute with the generators of the internal symmetry (for instance, the gauge transformation). Therefore, the particle content of the supermultiplets must have the same quantum numbers (electric charge, weak isospin, *etc.*).

Finally, we can introduce the operator $(-1)^F$, defined as

$$(-1)^F|B\rangle = |B\rangle, \qquad (-1)^F|F\rangle = -|F\rangle, \qquad (1.32)$$

i.e.,

$$F|B\rangle = 0, \qquad F|F\rangle = 1. \qquad (1.33)$$

This operator anticommutes with Q_α since

$$(-1)^F Q_\alpha |F\rangle = (-1)^F |B\rangle = |B\rangle \qquad (1.34)$$

and

$$Q_\alpha (-1)^F |F\rangle = -Q_\alpha |F\rangle = -|B\rangle. \qquad (1.35)$$

Therefore,

$$\{(-1)^F, Q_\alpha\} = 0. \qquad (1.36)$$

Also, from

$$\begin{aligned}
\mathrm{Tr}\left[(-1)^F \{Q_\alpha, \bar{Q}_{\dot\beta}\}\right] &= \mathrm{Tr}\left[(-1)^F Q_\alpha \bar{Q}_{\dot\beta} + (-1)^F \bar{Q}_{\dot\beta} Q_\alpha\right] \\
&= \mathrm{Tr}\left[-Q_\alpha (-1)^F \bar{Q}_{\dot\beta} + Q_\alpha (-1)^F \bar{Q}_{\dot\beta}\right] = 0, \quad (1.37)
\end{aligned}$$

where cyclic permutation of the trace has been used in the second term, one finds

$$\mathrm{Tr}\left[(-1)^F 2\sigma^\mu_{\alpha\dot\alpha} P_\mu\right] = 0 \quad \Rightarrow \quad \mathrm{Tr}[(-1)^F] = 0, \qquad (1.38)$$

for fixed non-zero P_μ. Therefore, the number of bosonic and fermionic states are equal in the supermultiplet.

1.4   REPRESENTATION OF THE SUSY ALGEBRA

We are now in the position to construct the representation of the complete SUSY algebra [62]. We consider the massless and massive cases separately.

Massless supermultiplet

For $m = 0$, one can choose $P_\mu = (E, 0, 0, E)$. Therefore, the Casimir operators P^2 and C^2 are zero. In this case, the anticommutation relation for Q_α and $\bar{Q}_{\dot\alpha}$ is given by

$$\{Q_\alpha, \bar{Q}_{\dot\beta}\} = 2\sigma^\mu_{\alpha\dot\beta} P_\mu = 2E(\sigma^0 + \sigma^3)_{\alpha\dot\beta} = 4E\begin{pmatrix} 1 & 0 \\ 0 & 0 \end{pmatrix}_{\alpha\dot\beta}. \qquad (1.39)$$

Therefore, one finds

$$\{Q_1, \bar{Q}_{\dot{1}}\} = 4E, \tag{1.40}$$
$$\{Q_2, \bar{Q}_{\dot{2}}\} = 0. \tag{1.41}$$

The second relation implies that

$$\langle\phi|\{Q_2, \bar{Q}_{\dot{2}}\}|\phi\rangle = 0 \;\Rightarrow\; \|\,Q_2|\phi\rangle\,\|^2 + \|\,\bar{Q}_{\dot{2}}|\phi\rangle\,\|^2 = 0 \text{ for all } \phi. \tag{1.42}$$

Thus,

$$Q_2 = \bar{Q}_{\dot{2}} = 0. \tag{1.43}$$

In this case, we are left with only Q_1 and $\bar{Q}_{\dot{1}}$. Let us define

$$a = \frac{Q_1}{2\sqrt{E}}, \qquad a^\dagger = \frac{\bar{Q}_{\dot{1}}}{2\sqrt{E}}, \tag{1.44}$$

so that

$$\{a, a^\dagger\} = 1, \qquad \{a, a\} = \{a^\dagger, a^\dagger\} = 0. \tag{1.45}$$

Then, one chooses a 'vacuum state' as the state annihilated by the operator a. The vacuum state can be characterised by a helicity λ and is denoted as $|E, \lambda\rangle$. From the commutator one has

$$[a, J_3] = \frac{1}{2}(\sigma^3)_{11}a = \frac{1}{2}a, \tag{1.46}$$

where $J_3 = M_{12}$. Thus, for any state $|\lambda\rangle$, one finds

$$J_3\,(a|\lambda\rangle) = \left(aJ_3 - \frac{a}{2}\right)|\lambda\rangle = (\lambda - \frac{1}{2})a|\lambda\rangle, \tag{1.47}$$

i.e., $a|\lambda\rangle$ has helicity $\lambda - \frac{1}{2}$. Similarly, the helicity of $a^\dagger|\lambda\rangle$ is $\lambda + \frac{1}{2}$. Therefore, the supermultiplet is of the form:

$$\Omega \equiv |E, \lambda\rangle, \qquad a^\dagger\Omega = |E, \lambda + \frac{1}{2}\rangle. \tag{1.48}$$

This supermultiplet can be denoted by $(\lambda, \lambda + \frac{1}{2})$. To satisfy CPT invariance, one must add the CPT conjugate states with opposite helicities and opposite quantum numbers. In this respect, one finds the following massless $N = 1$ supermultiplets [57].

1. *Chiral superfield*: consists of $(0, 1/2)$ and its CPT conjugate $(-1/2, 0)$, corresponding to a Weyl fermion and a complex scalar.

2. *Vector superfield*: consists of $(1/2, 1)$ and $(-1, -1/2)$, corresponding to a gauge boson (massless vector) and a Weyl fermion.

3. *Gravitino superfield*: consists of $(1, 3/2)$ and $(-3/2, -1)$, corresponding to a gravitino and a gauge boson.

4. *Graviton superfield:* consists of $(3/2, 2)$ and $(-2, -3/2)$, corresponding to a graviton and a gravitino.

Massive supermultiplet

In this case, one can choose a rest frame where $P_\mu = (m, 0, 0, 0)$. Thus, the Casimir operators are given by $P^2 = m^2$ and $C^2 = 2m^4 J^2$. Therefore, if irreducible representations are labelled by $|m, j, j_3\rangle$, where the eigenvalues of J^2 are given by $j(j+1)$ and j_3 is the third spin component, from the anticommutation relation of Q_α and $\bar{Q}_{\dot\alpha}$, one obtains

$$\{Q_\alpha, \bar{Q}_{\dot\beta}\} = 2\sigma^\mu_{\alpha\dot\beta} P_\mu = 2m\sigma^0_{\alpha\dot\beta} = 2m\delta_{\alpha\dot\beta}. \tag{1.49}$$

We define

$$a_{1,2} = \frac{Q_{1,2}}{\sqrt{2m}}, \qquad a^\dagger_{1,2} = \frac{\bar{Q}_{\dot1,\dot2}}{\sqrt{2m}}, \tag{1.50}$$

so that

$$\{a_p, a^\dagger_q\} = \delta_{pq}, \qquad \{a_p, a_q\} = \{a^\dagger_p, a^\dagger_q\} = 0. \tag{1.51}$$

This algebra is similar to the algebra of spin $1/2$ creation and annihilation operators. The vacuum state $|\Omega\rangle = |m, j, j_3\rangle$ is defined such that

$$a_1|\Omega\rangle = a_2|\Omega\rangle = 0. \tag{1.52}$$

In this case, the following four states are obtained:

$$|\Omega\rangle, \quad a^\dagger_1|\Omega\rangle, \quad a^\dagger_2|\Omega\rangle \text{ and } a^\dagger_1 a^\dagger_2|\Omega\rangle. \tag{1.53}$$

If $J_3|\Omega\rangle = j_3|\Omega\rangle$, then one can show that the values of J_3 for the above four states are given by $j_3, j_3 + \frac{1}{2}, j_3 - \frac{1}{2}, j_3$, respectively, *i.e.*, there are two bosons and two fermions. Therefore, a massive chiral supermultiplet ($j = 0$) consists of a Weyl spinor and a complex scalar field, while a massive vector supermultiplet ($j = 1/2$) consists of a scalar field, two Weyl spinors and a gauge boson.

1.5   SUPERSPACE AND SUPERFIELDS

To construct supersymmetric models, one would like to have a formalism in which SUSY is manifest. At the beginning, the way of constructing an action which is invariant under SUSY transformations was one of trial and error. As we will show in the next chapter, one starts with the first (lowest) component, the scalar field ϕ, in such a way that the SUSY transformation leads to $\phi \to \psi$. However, one cannot close the algebra with just ϕ and ψ, hence one alters the transformation law for ψ by introducing an additional bosonic field F. When superfields were introduced by Salam and Strathdee [63], a systematic way of constructing SUSY invariant actions became available.

One first introduces a superspace with coordinates defined as $(x^\mu, \theta_\alpha, \bar{\theta}^{\dot\alpha})$, where

x^μ are ordinary space-time and θ_α, $\bar{\theta}^{\dot{\alpha}}$ are two component anticommuting Grassmann variables, so that

$$\{\theta_\alpha, \theta_\beta\} = \{\bar{\theta}_{\dot{\alpha}}, \bar{\theta}_{\dot{\beta}}\} = \{\theta_\alpha, \bar{\theta}_{\dot{\beta}}\} = 0. \tag{1.54}$$

Therefore, $\theta_\alpha^2 = \bar{\theta}_{\dot{\alpha}}^2 = 0$, where $\alpha = 1, 2$ and $\dot{\alpha} = \dot{1}, \dot{2}$. Also, a generic function $f(\theta_\alpha)$ can be expanded as follows:

$$f(\theta) = a_0 + a_\alpha \theta^\alpha + a_3 \theta^2, \tag{1.55}$$

where $\theta^2 = \theta_\alpha \theta^\alpha = \varepsilon_{\alpha\beta}\theta^\beta \theta^\alpha = 2\theta^1\theta^2$. While an ordinary field is a function of x^μ only, a superfield is also a function of θ_α and $\bar{\theta}^{\dot{\alpha}}$. Since the product of more than two θ's or more than two $\bar{\theta}$'s vanishes, a superfield $\Phi(x, \theta, \bar{\theta})$ can always be expanded as follows [63]:

$$\begin{aligned}
\Phi(x, \theta, \bar{\theta}) &= \phi(x) + \theta\psi(x) + \bar{\theta}\bar{\chi} + \theta\theta\, m(x) + \bar{\theta}\bar{\theta}\, n(x) + \theta\sigma^\mu\bar{\theta}\, v_\mu(x) \\
&\quad + \theta\theta\bar{\theta}\, \bar{\lambda}(x) + \bar{\theta}\bar{\theta}\theta\, \eta(x) + \theta\theta\bar{\theta}\bar{\theta}\, d(x),
\end{aligned} \tag{1.56}$$

where $\phi(x), m(x), n(x)$ and $d(x)$ are scalars, $\psi(x), \chi(x), \lambda(x)$ and $\eta(x)$ are spinors, whereas $v_\mu(x)$ is a vector. Thus, the superfield $\Phi(x, \theta, \bar{\theta})$ has sixteen component fields of x (since all the fields under consideration are here complex), eight of them carry tensor indices and the other eight carry spinor indices.

With the anticommuting Grassmann parameters, one can transform the graded Lie algebra into a regular Lie algebra by writing the elements of the spinor sector in terms of $\theta^\alpha Q_\alpha$ and $\bar{\theta}_{\dot{\alpha}}\bar{Q}^{\dot{\alpha}}$ to obtain:

$$[\theta Q, \bar{\theta}\bar{Q}] = 2(\theta\sigma^\mu\bar{\theta})P_\mu, \qquad [\theta Q, \theta'Q] = 0, \qquad [\bar{\theta}\bar{Q}, \bar{\theta}'\bar{Q}] = 0. \tag{1.57}$$

In this case, a finite element of the corresponding group is

$$G(x^\mu, \theta, \bar{\theta}) = e^{i(-x^\mu P_\mu + \theta Q + \bar{Q}\bar{\theta})}. \tag{1.58}$$

Using the Housdorff's formula $e^A e^B = e^{A+B+\frac{1}{2}[A,B]+\cdots}$, one finds

$$G(x^\mu, \theta, \bar{\theta})G(a^\mu, \xi, \bar{\xi}) = G(x^\mu + a^\mu + i\theta\sigma^\mu\bar{\xi} - i\xi\sigma^\mu\bar{\theta}, \theta + \xi, \bar{\theta} + \bar{\xi}). \tag{1.59}$$

Therefore, the translations of the arguments x^μ, θ_α, and $\bar{\theta}^{\dot{\alpha}}$ are given by

$$x^\mu \rightarrow x^\mu + a^\mu + i\theta\sigma^\mu\bar{\xi} - i\xi\sigma^\mu\bar{\theta}, \tag{1.60}$$

$$\theta \rightarrow \theta + \xi, \tag{1.61}$$

$$\bar{\theta} \rightarrow \bar{\theta} + \bar{\xi}. \tag{1.62}$$

Under these transformations, a superfield $\Phi(x, \theta, \bar{\theta})$ takes the form

$$\begin{aligned}
\Phi(x^\mu + a^\mu + i\theta\sigma^\mu\bar{\xi} - i\xi\sigma^\mu\bar{\theta}, \theta + \xi, \bar{\theta} + \bar{\xi}) &= \Phi(x^\mu, \theta, \bar{\theta}) \\
+ (a^\mu + i\theta\sigma^\mu\bar{\xi} - i\xi\sigma^\mu\bar{\theta})\frac{\partial\Phi}{\partial x^\mu} &+ \xi^\alpha\frac{\partial\Phi}{\partial\theta^\alpha} + \bar{\xi}_{\dot{\alpha}}\frac{\partial\Phi}{\partial\bar{\theta}_{\dot{\alpha}}}.
\end{aligned} \tag{1.63}$$

This implies that the superfield transforms under the super-Poincaré algebra as

$$\Phi(x, \theta, \bar{\theta}) \rightarrow e^{i(-a^\mu P_\mu + \xi Q + \bar{\xi}\bar{Q})} \Phi(x, \theta, \bar{\theta}), \tag{1.64}$$

with the following differential forms for the generators:

$$P_\mu = i\partial_\mu, \tag{1.65}$$

$$Q_\alpha = i(-\partial_\alpha + i\sigma^\mu_{\alpha\dot{\alpha}}\bar{\theta}^{\dot{\alpha}}\partial_\mu), \tag{1.66}$$

$$\bar{Q}_{\dot{\alpha}} = i(\bar{\partial}_{\dot{\alpha}} - i\theta^\alpha \sigma^\mu_{\alpha\dot{\alpha}}\partial_\mu), \tag{1.67}$$

where $\partial_\mu = \frac{\partial}{\partial x^\mu}$, $\partial_\alpha = \frac{\partial}{\partial \theta^\alpha}$ and $\bar{\partial}_{\dot{\alpha}} = \frac{\partial}{\partial \bar{\theta}^{\dot{\alpha}}}$. We can define covariant derivatives, D_α and $\bar{D}^{\dot{\alpha}}$, which are useful in constructing SUSY-invariant Lagrangians, as follows:

$$D_\alpha = \partial_\alpha + i\sigma^\mu_{\alpha\dot{\alpha}}\bar{\theta}^{\dot{\alpha}}\partial_\mu, \tag{1.68}$$

$$\bar{D}_{\dot{\alpha}} = -\bar{\partial}_{\dot{\alpha}} - i\theta^\alpha \sigma^\mu_{\alpha\dot{\alpha}}\partial_\mu. \tag{1.69}$$

One can show that $[D_\alpha, \delta_{\text{SUSY}}] = [\bar{D}^{\dot{\alpha}}, \delta_{\text{SUSY}}] = 0$, where δ_{SUSY} is a SUSY transformation defined by $\delta_{\text{SUSY}} = \xi^\alpha Q_\alpha + \bar{\xi}_{\dot{\alpha}}\bar{Q}^{\dot{\alpha}}$. Also, one can show that D_α and $\bar{D}_{\dot{\alpha}}$ satisfy the following anticommutation relations:

$$\{D_\alpha, D_\beta\} = 0, \qquad \{\bar{D}_{\dot{\alpha}}, \bar{D}_{\dot{\beta}}\} = 0, \tag{1.70}$$

$$\{D_\alpha, \bar{D}_{\dot{\beta}}\} = 2i\sigma^\mu_{\alpha\dot{\beta}}\partial_\mu. \tag{1.71}$$

1.6 CONSTRAINED SUPERFIELDS

To obtain irreducible representations of the SUSY algebra, one can impose appropriate constraints on $\Phi(x, \theta, \bar{\theta})$ which are invariant under SUSY and eliminate the extra components. Here we study three such constraints leading to chiral, vector and linear superfields.

1.6.1 Chiral superfields

The superfields Φ and $\Phi^\dagger$ satisfying the conditions $\bar{D}_{\dot{\alpha}}\Phi = 0$ and $D_\alpha \Phi^\dagger = 0$ are called chiral and antichiral, respectively [62, 64, 65]. This constraint leads to a substantial reduction of the field components. Let us define new bosonic coordinates y^μ in superspace such that

$$y^\mu = x^\mu + i\theta\sigma^\mu\bar{\theta}. \tag{1.72}$$

Thus, we have

$$\begin{aligned} \bar{D}_{\dot{\alpha}} y^\mu &= \left(-\bar{\partial}_{\dot{\alpha}} - i\theta^\alpha \sigma^\nu_{\alpha\dot{\alpha}}\partial_\nu\right)\left(x^\mu + i\theta\sigma^\mu\bar{\theta}\right) \\ &= -i\theta^\alpha\sigma^\mu_{\alpha\dot{\alpha}} + i\theta^\beta\sigma^\mu_{\beta\dot{\alpha}} = 0. \end{aligned} \tag{1.73}$$

Also, $\bar{D}_{\dot{\alpha}}\theta^\alpha = 0$ and $D_\alpha\bar{\theta}^{\dot{\alpha}} = 0$. Hence, the superfield $\Phi = \Phi(y, \theta)$, function of y and θ, automatically satisfies the condition $\bar{D}_{\dot{\alpha}}\Phi = 0$ and thus it is a chiral superfield. The most general chiral superfield can be written as

$$\Phi(y, \theta) = \phi(y) + \sqrt{2}\theta\psi(y) + \theta\theta F(y), \tag{1.74}$$

where $\phi(y)$ is a complex scalar field, $\psi(y)$ is a spinor field and $F(y)$, as we will show, is an auxiliary field. Using a Taylor expansion, one can write the above chiral superfield in terms of x, θ, and $\bar{\theta}$ as follows:

$$\begin{aligned} \Phi(x, \theta, \bar{\theta}) &= \phi(x) + \sqrt{2}\theta\psi(x) + \theta\theta F(x) + i\theta\sigma^\mu\bar{\theta}\partial_\mu\phi(x) \\ &\quad - \frac{i}{\sqrt{2}}\theta\theta\sigma^\mu\bar{\theta}\partial_\mu\psi(x) + \frac{1}{4}\theta\theta\bar{\theta}\bar{\theta}\partial_\mu\partial^\mu\phi(x). \end{aligned} \tag{1.75}$$

This expression can easily be obtained by using the superfield transformation, as given in Eq. (1.64), with $a_\mu = \theta\sigma^\mu\bar{\theta}$, and $\xi = \bar{\xi} = 0$, i.e.,

$$\Phi(x, \theta, \bar{\theta}) = e^{\theta\sigma^\mu\bar{\theta}\partial_\mu}\left[\phi(x) + \sqrt{2}\theta\psi(x) + \theta\theta F(x)\right]. \tag{1.76}$$

The antichiral superfield $\Phi^\dagger$, which satisfies the condition $D_\alpha\Phi^\dagger = 0$, has the following expansion

$$\Phi^\dagger(\bar{y}, \bar{\theta}) = \phi^*(\bar{y}) + \sqrt{2}\bar{\theta}\bar{\psi}(\bar{y}) + \bar{\theta}\bar{\theta}F^*(\bar{y}), \tag{1.77}$$

where $\bar{y}^\mu = x^\mu - i\theta\sigma^\mu\bar{\theta}$. In terms of x, θ, and $\bar{\theta}$, the antichiral superfied $\Phi^\dagger$ has the following component field expansion

$$\begin{aligned} \Phi^\dagger(x, \theta, \bar{\theta}) &= \phi^*(x) + \sqrt{2}\bar{\theta}\bar{\psi}(x) + \bar{\theta}\bar{\theta}F^*(x) - i\theta\sigma^\mu\bar{\theta}\partial_\mu\phi^*(x) \\ &\quad + \frac{i}{\sqrt{2}}\bar{\theta}\bar{\theta}\sigma^\mu\theta\partial_\mu\bar{\psi}(x) + \frac{1}{4}\theta\theta\bar{\theta}\bar{\theta}\partial_\mu\partial^\mu\phi^*(x). \end{aligned} \tag{1.78}$$

It is remarkable that, since D_α and $\bar{D}_{\dot{\alpha}}$ obey the chain rule, any product of chiral superfields is also a chiral superfield, while any product of antichiral superfields is also an antichiral superfield. The products of chiral times antichiral superfields are neither chiral nor antichiral superfields, but just generic superfields. For example, the product of the chiral superfields Φ_1 and Φ_2 is given by

$$\begin{aligned} \Phi_1\Phi_2 &= \phi_1(y)\phi_2(y) + \sqrt{2}\theta\left[\psi_1(y)\phi_2(y) + \phi_1(y)\psi_2(y)\right] \\ &\quad + \theta\theta\left[\phi_1(y)F_2(y) + \phi_2(y)F_1(y) - \psi_1(y)\psi_2(y)\right], \end{aligned} \tag{1.79}$$

while the product of $\Phi_1^\dagger\Phi_2$ is given by

$$\begin{aligned} \Phi_1^\dagger\Phi_2 &= \phi_1^*(y)\phi_2(y) + \sqrt{2}\theta\psi_2(y)\phi_1^*(y) + \sqrt{2}\bar{\theta}\bar{\psi}_1(y)\phi_2(y) + 2\bar{\theta}\bar{\psi}_1\theta\psi_2 \\ &\quad + \theta\theta\phi_1^*(y)F_2(y) + \bar{\theta}\bar{\theta}\phi_2(y)F_1^*(y) + \sqrt{2}\theta\theta\bar{\theta}\bar{\psi}_1(y)F_2(y) \\ &\quad + \sqrt{2}\bar{\theta}\bar{\theta}\theta\psi_2(y)F_1^*(y) + \bar{\theta}\bar{\theta}\theta\theta F_1^*(y)F_2(y), \end{aligned} \tag{1.80}$$

which is not a chiral superfield. In terms of the variables y, θ and $\bar{\theta}$, we have

$$\begin{aligned} Q_\alpha &= \partial_\alpha, \tag{1.81} \\ \bar{Q}_{\dot{\alpha}} &= \bar{\partial}_{\dot{\alpha}} + 2i\theta^\alpha\sigma^\mu_{\alpha\dot{\alpha}}\partial_\mu. \tag{1.82} \end{aligned}$$

Therefore, the infinitesimal SUSY transformation of chiral superfield yields $\Phi \to \Phi + \delta\Phi$ with $\delta\Phi = i(\xi Q + \bar{\xi}\bar{Q})\Phi$ and implies

$$\delta_\xi \phi = \sqrt{2}\xi\psi, \qquad (1.83)$$

$$\delta_\xi \psi = \sqrt{2}\xi F - \sqrt{2}i\sigma^\mu \bar{\xi}\partial_\mu \phi, \qquad (1.84)$$

$$\delta_\xi F = \sqrt{2}i\psi\sigma^\mu \bar{\xi}\partial_\mu. \qquad (1.85)$$

As can be seen, δF is a total derivative. Thus, if a Lagrangian is made out of the highest component of a superfield, it is SUSY invariant.

1.6.2 Vector superfields

The chiral superfields, introduced above, can describe spin 0-bosons and spin-1/2 fermions. For describing the spin-1 gauge bosons of the SM, one introduces vector superfields V. They are defined from the general superfield by imposing a covariant reality constraint [64–66]:

$$V(x, \theta, \bar{\theta}) = V^\dagger(x, \theta, \bar{\theta}). \qquad (1.86)$$

In this regard, the vector superfield has the following expansion:

$$
\begin{aligned}
V(x, \theta, \bar{\theta}) &= C(x) + i\theta\chi(x) - i\bar{\theta}\bar{\chi}(x) + \theta\sigma^\mu\bar{\theta}v_\mu + \frac{i}{2}\theta\theta\left[M(x) + iN(x)\right] \\
&\quad - \frac{i}{2}\bar{\theta}\bar{\theta}\left[M(x) - iN(x)\right] + \theta\theta\bar{\theta}\left[\bar{\lambda}(x) + \frac{i}{2}\bar{\sigma}^\mu\partial_\mu\chi(x)\right] \\
&\quad + \bar{\theta}\bar{\theta}\theta\left[\lambda(x) - \frac{i}{2}\sigma^\mu\partial_\mu\bar{\chi}(x)\right] + \frac{1}{2}\theta\theta\bar{\theta}\bar{\theta}\left[D(x) - \frac{1}{2}\partial_\mu\partial^\mu C(x)\right], \quad (1.87)
\end{aligned}
$$

where the component fields C, D, M, N and v_μ must all be real. The presence of a real vector field in the vector superfield suggests that we use vector superfields to construct SUSY gauge theories. An example of a vector superfield is the product of an antichiral superfield and a chiral superfield $\Phi^\dagger\Phi$. Also, the sum of chiral superfield and antichiral superfield, $\Phi + \Phi^\dagger$, is another example of vector superfield.

As mentioned, the general vector superfield in Eq. (1.87) has eight bosonic and eight fermionic components, which are far too many to describe a single supermultiplet. To reduce their number, we introduce a generalisation of the usual concept of gauge transformations of spinor and gauge fields to the case of chiral and vector superfields:

$$V \to V + \Lambda + \Lambda^\dagger, \qquad (1.88)$$

where Λ is a chiral superfield. Note that, under this transformation, the real vector field v_μ transforms as

$$v_\mu(x) \to v_\mu + i\partial_\mu\left[\alpha(x) - \alpha^*(x)\right], \qquad (1.89)$$

where $\alpha(x)$ is the scalar component of the chiral superfield Λ, which corresponds to an Abelian gauge transformation. The SUSY gauge transformations allow for the possibility of gauging away the unphysical fields. In this physical gauge, the WZ gauge, the vector supermultiplet V is given by

$$V(x, \theta, \bar{\theta}) = \theta \sigma^\mu \bar{\theta} \, v_\mu(x) + \theta^2 \bar{\theta}_{\dot{\alpha}} \, \bar{\lambda}^{\dot{\alpha}}(x) + \bar{\theta}^2 \theta^\alpha \, \lambda_\alpha(x) + \frac{1}{2} \theta^2 \bar{\theta}^2 D(x), \qquad (1.90)$$

where $D(x)$ is a non-propagating auxiliary field as it was F in the chiral superfield. Notice that, analogously to F, D also transforms under a SUSY transformation into a total derivative.

1.6.3 Linear superfields

In $N = 1$ SUSY, we can also define a real linear superfield which is a vector superfield such that its second SUSY covariant derivative vanishes [67],

$$DDL = \bar{D}\bar{D}L = 0, \qquad (1.91)$$

where SUSY covariant derivatives were defined above. Solving these constraints leads to a component expansion of the form

$$\begin{aligned} L = {} & C + i\theta\chi - i\bar{\theta}\bar{\chi} + \theta\sigma^\mu\bar{\theta} \, v_\mu(x) - \frac{1}{2}\theta\theta\bar{\theta}(\partial_\mu\chi\sigma^\mu) \\ & - \frac{1}{2}\bar{\theta}\bar{\theta}\theta(\sigma^\mu\partial_\mu\bar{\chi}) - \frac{1}{4}\theta\theta\bar{\theta}\bar{\theta}\partial_\mu\partial^\mu C. \end{aligned} \qquad (1.92)$$

Furthermore, the constraint in Eq. (1.91) implies that the vector field v_μ has a vanishing divergence $\partial_\mu v^\mu = 0$. Therefore, v_μ is given by

$$v_\mu = \frac{1}{\sqrt{2}}\varepsilon_{\mu\nu\sigma\rho}\partial^\nu b^{\sigma\rho}, \qquad (1.93)$$

which is invariant under the gauge transformation

$$b_{\mu\nu} \rightarrow b_{\mu\nu} + \partial_\mu\Lambda_\nu + \partial_\nu\Lambda_\mu. \qquad (1.94)$$

It is interesting to note that the linear superfield can be expressed in terms of a chiral spinor superfield Φ_α as

$$L = i(D^\alpha\Phi_\alpha + \bar{D}_{\dot{\alpha}}\bar{\Phi}^{\dot{\alpha}}). \qquad (1.95)$$

From this definition it is easy to show that L is real and satisfies the two above-mentioned covariant constraints. Further, the expansion in Eq. (1.92) shows that the linear superfield contains a real scalar field C, a Majorana spinor χ and an antisymmetric tensor $b_{\mu\nu}$. The importance of the linear superfield is due to the fact that the gravity sector of superstrings contains an antisymmetric tensor $b_{\mu\nu}$ and a real scalar, the dilaton, along with the Majorana spinor partner. This is precisely the particle content of a linear superfield.

CHAPTER 2

SUSY Lagrangians

We now have the recipe to construct the most general renormalisable Lagrangian of a SUSY field theory. The simplest example of an interacting 4D quantum field within SUSY was considered by WZ in 1974 [68].

2.1 THE WZ MODEL

The WZ model consists of a scalar field ϕ and a spinor field ψ. The free Lagrangian density of these fields is given by

$$\mathcal{L} = \partial_\mu \phi^\dagger \partial^\mu \phi + \chi^\dagger i \bar{\sigma}^\mu \partial_\mu \chi. \tag{2.1}$$

The corresponding equation of motion for ϕ is

$$\partial_\mu \partial^\mu \phi = 0 \tag{2.2}$$

while for χ is

$$i \bar{\sigma}^\mu \partial_\mu \chi = 0. \tag{2.3}$$

However, under the SUSY transformations given in Eqs. (1.83)–(1.85), one finds that this Lagrangian is not invariant, as a term which depends on the auxiliary field, $F^\dagger F$, must be added. Thus the free WZ Lagrangian is given by

$$\mathcal{L} = \partial_\mu \phi^\dagger \partial^\mu \phi + \chi^\dagger i \bar{\sigma}^\mu \partial_\mu \chi + F^\dagger F. \tag{2.4}$$

Now one can easily show that, under a SUSY transformation, the above Lagrangian is invariant up to a total derivative, hence, the corresponding action is SUSY invariant.

So far, all fields are massless and there are no interactions. The next step is to consider mass and interaction terms that preserve SUSY. The most general renormalisable interaction terms of the fields, ϕ_i, χ_i and F_i, can be written as [68]:

$$\mathcal{L}_{\text{int}} = W_i(\phi, \phi^\dagger) F_i - \frac{1}{2} W_{ij}(\phi, \phi^\dagger) \chi_i \chi_j + \text{h.c.,} \tag{2.5}$$

where W_i and W_{ij} are quadratic and linear functions of ϕ_i and $\phi_i^\dagger$, respectively. It is clear that, due to dimensionality constraints, W_i and W_{ij} cannot depend on χ_i nor on F_i. Also note that, if we add to $\mathcal{L}_{\text{int}}$ a quartic interaction term like $G(\phi, \phi^\dagger)$, one can show that $\mathcal{L}_{\text{int}}$ cannot be SUSY invariant unless $G(\phi, \phi^\dagger) = 0$.

Since the free part of the WZ action is SUSY invariant, the interaction part must also be invariant. From $\delta_\xi \mathcal{L}_{\text{int}}$, one obtains the following four spinors terms:

$$
\begin{aligned}
\delta_\xi \mathcal{L}_{\text{int}} \supset \ & -\frac{1}{2} \frac{\partial W_{ij}}{\partial \phi_k} (\delta_\xi \phi_k)(\chi_i \chi_j) - \frac{1}{2} \frac{\partial W_{ij}}{\partial \phi_k^\dagger} (\delta_\xi \phi_k^\dagger)(\chi_i \chi_j) \\
= \ & -\frac{1}{2} \frac{\partial W_{ij}}{\partial \phi_k} (\xi \chi_k)(\chi_i \chi_j) - \frac{1}{2} \frac{\partial W_{ij}}{\partial \phi_k^\dagger} (\xi^\dagger \chi_k^\dagger)(\chi_i \chi_j).
\end{aligned}
\tag{2.6}
$$

If $\partial W_{ij}/\partial \phi_k$ is symmetric in i, j and k, then the first term vanishes, using the Fierz identity:

$$
(\xi \chi_k)(\chi_i \chi_j) + (\xi \chi_i)(\chi_j \chi_k) + (\xi \chi_j)(\chi_k \chi_i) = 0.
\tag{2.7}
$$

The second term of Eq. (2.6) vanishes only if $\partial W_{ij}/\partial \phi_k^\dagger = 0$, i.e., W_{ij} is an analytic (holomorphic) function of the complex field ϕ. In this case, one can write W_{ij} as

$$
W_{ij} = M_{ij} + Y_{ijk} \phi_k,
\tag{2.8}
$$

where M_{ij} is a dimensionful symmetric matrix while Y_{ijk} is a dimensionless coupling. Also, W_{ij} can be written as $W_{ij} = \partial^2 W / \partial \phi_i \partial \phi_j$, where W is given by

$$
W = \frac{1}{2} M_{ij} \phi_i \phi_j + \frac{1}{6} Y_{ijk} \phi_i \phi_j \phi_k.
\tag{2.9}
$$

W is called 'superpotential'.

The terms of $\delta_\xi \mathcal{L}_{\text{int}}$ that involve one derivative are given by

$$
\delta_\xi \mathcal{L}_{\text{int}} \supset -\frac{1}{2} W_{ij} \delta_\xi (\chi_i \chi_j) + W_i \delta_\xi F_i + \text{h.c.}
\tag{2.10}
$$

$$
\supset -i\sqrt{2} W_{ij} \partial_\mu \phi_j \chi_j \sigma^\mu \xi^\dagger - i\sqrt{2} W_i \partial_\mu \chi_i \sigma^\mu \xi^\dagger + \text{h.c.},
\tag{2.11}
$$

where $\delta(\chi_i \chi_j) = 2\sqrt{2} \left(i \partial_\mu \phi_i \chi_j \bar{\sigma}^\mu \xi^\dagger + F_i(\xi \chi_j) \right)$ has been used. Since $W_{ij} \partial_\mu \phi_j = \partial_\mu (\partial W/\partial \phi_i)$, the above $\delta_\xi \mathcal{L}_{\text{int}}$ can be expressed as a total derivative only if W_i is given by $\partial W/\partial \phi_i$. With these conditions on W_i and W_{ij}, one can show that the remaining terms in $\delta_\xi \mathcal{L}_{\text{int}}$ are identically cancelled. Hence, the interacting WZ action is SUSY invariant. It is important to note that the invariance of the above action under SUSY transformations does not depend on a precise form of the superpotential, W. It relies only on the fact that it is an analytic function. A general holomorphic W gives a non-renormalisable SUSY theory.

The equation of motion of $F_i^\dagger$ implies $F_i = -W_i$. Therefore, the WZ Lagrangian can be written as

$$
\mathcal{L}_{\text{WZ}} = \partial_\mu \phi_i^\dagger \partial^\mu \phi_i + i \chi_i^\dagger \bar{\sigma}^\mu \partial_\mu \chi_i - \left| \frac{\partial W}{\partial \phi_i} \right|^2 - \frac{1}{2} \left(\frac{\partial W}{\partial \phi_i \partial \phi_j} \chi_i \chi_j + \text{h.c.} \right).
\tag{2.12}
$$

For a renormalisable W, one finds

$$
\begin{aligned}
\mathcal{L}_{\mathrm{WZ}} &= \partial_\mu \phi_i^\dagger \partial^\mu \phi_i + i\chi_j^\dagger \bar{\sigma}^\mu \partial_\mu \chi_j - \left| M_{ij}\phi_i + \frac{1}{2}Y_{ijk}\phi_j\phi_k \right|^2 \\
&\quad - \frac{1}{2}\left(M_{ij}\chi_i\chi_j + Y_{ijk}\phi_i\chi_j\chi_k + \mathrm{h.c.} \right).
\end{aligned}
\tag{2.13}
$$

From this expression, it is clear that the potential $V(\phi, \phi^\dagger)$ of this model is given by

$$
\begin{aligned}
V(\phi, \phi^\dagger) &= \sum_i |F_i|^2 = \sum_i \left| \frac{\partial W}{\partial \phi_i} \right|^2 = M_{jk}^* M^{kl}\phi_j^*\phi_l + \left(\frac{1}{2}M^{jl}Y_{klm}^*\phi_i\phi^{*k}\phi^{*m} + \mathrm{h.c.} \right) \\
&\quad + \frac{1}{4}Y^{jkl}Y_{mnl}^*\phi_j\phi_k\phi^{*m}\phi^{*n}.
\end{aligned}
\tag{2.14}
$$

2.2 CHIRAL SUPERFIELD LAGRANGIAN

As emphasised in the previous chapter, both the F-component of a chiral superfield and the D-component of a vector superfield transform into a total derivative under SUSY transformations, hence, their space-time integrals are SUSY invariant. Therefore, to construct a SUSY Lagrangian of superfields, one should select the highest order component of the superfields, which can be obtained by the integration with respect to the Grassmann parameters θ and $\bar{\theta}$. In this respect, the superfield formalism provides an elegant way of constructing SUSY Lagrangians.

The integrations with respect to θ_α, $\alpha = 1, 2$, are defined as follows:

$$
\int d\theta_\alpha = 0, \qquad \int d\theta_\alpha \theta_\beta = \delta_{\alpha\beta}.
\tag{2.15}
$$

The measure $d^2\theta$ is defined as $d^2\theta = d\theta_1 d\theta_2$. Therefore,

$$
\int d^2\theta = 0, \qquad \int d^2\theta\, \theta_\alpha = 0, \qquad \frac{1}{2}\int d^2\theta\, \theta^2 = 1.
\tag{2.16}
$$

Also, we have

$$
\frac{1}{2}\int d^2\bar{\theta}\, \bar{\theta}^2 = 1
\tag{2.17}
$$

and

$$
\frac{1}{4}\int d^4\theta \equiv \frac{1}{4}\int d^2\bar{\theta}d^2\theta\, \theta^2\bar{\theta}^2 = 1.
\tag{2.18}
$$

In this case, the integration of a chiral superfield $\Phi(x, \theta)$, $\int d^2\theta \Phi(x, \theta)$, gives the coefficient of the θ^2 term in this chiral superfield, which is usually called F-term and denoted by $\Phi(x, \theta)\big|_{\theta\theta}$. Also, the integration of a general superfield $S(x, \theta)$, $\int d^2\bar{\theta}d^2\theta S(x, \theta, \bar{\theta})$, leads to the coefficient of the $\theta^2\bar{\theta}^2$ term in this superfield, which is usually called D-term and denoted by $S(x, \theta, \bar{\theta})\big|_{\theta\theta\bar{\theta}\bar{\theta}}$.

In this respect, it is now clear that the most general SUSY and renormalisable Lagrangian for chiral superfields has the form

$$\mathcal{L} = \Phi_i^\dagger \Phi^i|_{\theta\theta\bar\theta\bar\theta} + W(\Phi_i)|_{\theta\theta} + \bar{W}(\Phi_i^\dagger)|_{\bar\theta\bar\theta}, \tag{2.19}$$

where the superpotential $W(\Phi_i)$ is given by

$$W(\Phi_i) = \frac{1}{2} M_{ij} \Phi^i \Phi^j + \frac{1}{3} Y_{ijk} \Phi^i \Phi^j \Phi^k, \tag{2.20}$$

where M_{ij} and Y_{ijk} are totally symmetric terms. We stop at the product of three superfields to ensure renormalisability by simple power counting. As mentioned in Chapter 1, the chiral superfield can be written as

$$\begin{aligned}
\Phi(x,\theta,\bar\theta) &= \phi(x) + \sqrt{2}\theta\psi(x) + \theta\theta F(x) + i\theta\sigma^\mu\bar\theta\partial_\mu\phi(x) \\
&- \frac{i}{\sqrt{2}}\theta\theta\sigma^\mu\bar\theta\partial_\mu\psi(x) + \frac{1}{4}\theta\theta\bar\theta\bar\theta\partial_\mu\partial^\mu\phi(x).
\end{aligned} \tag{2.21}$$

Therefore, $\Phi_i^\dagger \Phi^i|_{\theta\theta\bar\theta\bar\theta} = \int d^4\theta\, \Phi_i^\dagger \Phi^i$ is given by

$$\Phi_i^\dagger \Phi^i|_{\theta\theta\bar\theta\bar\theta} = \partial_\mu\phi_i^\dagger \partial^\mu\phi_i + i\bar\psi_i\bar\sigma^\mu\partial_\mu\psi_i + F_i^* F_i. \tag{2.22}$$

Similarly, one can easily find the component form of $W|_{\theta\theta}$ and show that, in terms of the component fields, $\mathcal{L}$ takes the form

$$\begin{aligned}
\mathcal{L} &= \partial_\mu\phi_i^\dagger \partial^\mu\phi_i + i\bar\psi_i\bar\sigma^\mu\partial_\mu\psi_i + F_i^* F_i + \Big[M_{ij}\Big(\phi_i F_j - \frac{1}{2}\psi_i\psi_j\Big) \\
&+ Y_{ijk}\left(\phi_i\phi_j F_k - \psi_i\psi_j\phi_k\right) + \text{h.c.}\Big].
\end{aligned} \tag{2.23}$$

From this Lagrangian one notices that there is no dependence on the time derivative of the auxiliary field F_i. Thus, it does not propagate and it can be eliminated by using its field equation as

$$F_i^* = -M_{ij}\phi_i - Y_{ijk}\phi_i\phi_k \equiv -\frac{\partial W(\phi)}{\partial\phi_i}, \tag{2.24}$$

$$F_i = -M_{ij}^*\phi_i^* - Y_{ijk}^*\phi_i^*\phi_k^* \equiv -\frac{\partial \bar{W}(\phi^*)}{\partial\phi_i^*}. \tag{2.25}$$

Using these expressions, the Lagrangian (2.23) becomes

$$\begin{aligned}
\mathcal{L} &= \partial_\mu\phi_i^\dagger \partial^\mu\phi_i + i\bar\psi_i\bar\sigma^\mu\partial_\mu\psi_i - |M_{ij}\phi_j + Y_{ijk}\phi_j\phi_k|^2 \\
&- \Big(\frac{1}{2}M_{ij}\psi_i\psi_j + Y_{ijk}\psi_i\psi_j\phi_k + \text{h.c.}\Big).
\end{aligned} \tag{2.26}$$

Thus, one concludes that the tree level effective potential V is given by

$$V = \sum_i |F_i|^2 = \sum_i \left|\frac{\partial W}{\partial\phi}\right|^2. \tag{2.27}$$

This potential is positive definite with absolute minima at $F_i = 0$.

For the most general non-renormalisable SUSY interactions, the superpotential W is extended to include higher powers of Φ and the kinetic interaction term takes the form

$$\int d^4\theta K(\Phi^\dagger, \Phi), \tag{2.28}$$

where K (the Kähler potential) is a real function of Φ and $\Phi^\dagger$.

2.3 SUSY ABELIAN GAUGE THEORY

In the previous section we have constructed Lagrangians from chiral superfields which include spin-0 and spin-1/2 particles. In this section we consider the Lagrangian of a free supersymmetric Abelian gauge theory [64, 66, 69]. To construct kinetic terms for the vector field v_μ in the vector superfield $V(x, \theta, \bar\theta)$, a SUSY generalisation of the field strength is necessary. This generalisation is obtained as follows. First, define

$$W_\alpha = -\frac{1}{4}\bar{D}\bar{D}D_\alpha V(x, \theta, \bar\theta), \tag{2.29}$$

$$\bar{W}_{\dot\alpha} = -\frac{1}{4}DD\bar{D}_{\dot\alpha}V(x, \theta, \bar\theta). \tag{2.30}$$

Since $D^3 = \bar{D}^3 = 0$, W_α is a chiral superfield and $\bar{W}_{\dot\alpha}$ is an antichiral superfield. Also, they are invariant under the SUSY generalisation of a gauge transformation, $V \to V + \Lambda + \Lambda^\dagger$, where Λ and $\Lambda^\dagger$ are chiral and antichiral superfields, respectively, since

$$W_\alpha \to W_\alpha - \frac{1}{4}\bar{D}\bar{D}D_\alpha(\Lambda + \Lambda^\dagger) = W_\alpha - \frac{1}{4}\bar{D}\bar{D}D_\alpha\Lambda$$

$$= W_\alpha - \frac{1}{4}\bar{D}_{\dot\alpha}\{\bar{D}^{\dot\alpha}, D_\alpha\}\Lambda = W_\alpha - \frac{i}{2}(\sigma^\mu)_\alpha{}^{\dot\alpha}\partial_\mu\bar{D}_{\dot\alpha}\Lambda = W_\alpha. \tag{2.31}$$

It is more convenient to use the WZ gauge, Eq. (1.90), in computing the components of W_α. Also, since this is a chiral superfield, the coordinates (y, θ) simplify the computation, where $y^\mu = x^\mu + i\theta\sigma^\mu\bar\theta$. In this case, V is given by

$$V_{\text{WZ}} = \theta\sigma^\mu\bar\theta v_\mu(y) + i\theta\theta\bar\theta\bar\lambda(y) - i\bar\theta\bar\theta\theta\lambda(y) + \frac{1}{2}\theta\theta\bar\theta\bar\theta D(y). \tag{2.32}$$

In the WZ gauge, all powers V^n_{WZ} with $n \geq 3$ vanish, since they will involve at least θ^3. The non-zero V^2_{WZ} is given by

$$V^2_{\text{WZ}} = \theta\sigma^\mu\bar\theta\theta\sigma^\nu\bar\theta v_\mu v_\nu = \frac{1}{2}\theta\theta\bar\theta\bar\theta v_\mu v^\mu. \tag{2.33}$$

Also, the D_α and $\bar{D}_{\dot\alpha}$ covariant derivatives become

$$D_\alpha = \partial_\alpha + 2i\sigma^\mu_{\alpha\dot\alpha}\bar\theta^{\dot\alpha}\partial_\mu, \quad \bar{D}_{\dot\alpha} = \bar\partial_{\dot\alpha}. \tag{2.34}$$

Therefore, one finds

$$W_\alpha = -i\lambda_\alpha(y) + \left[\delta_\alpha^\beta D(y) - \frac{i}{2}(\sigma^\mu\bar{\sigma}^\nu)_\alpha{}^\beta v_{\mu\nu}(y)\right]\theta_\beta + \theta\theta\sigma^\mu_{\alpha\dot{\alpha}}\partial_\mu\bar{\lambda}^{\dot{\alpha}}(y), \qquad (2.35)$$

where $v_{\mu\nu} = \partial_\mu v_\nu - \partial_\nu v_\mu$. The SUSY transformations of the W_α components can easily be obtained from $\left[-i\xi^\alpha Q_\alpha - i\bar{\xi}_{\dot{\alpha}}\bar{Q}^{\dot{\alpha}}\right]W_\alpha$ as follows:

$$\delta_\xi v_{\mu\nu} = \left[\xi\sigma_\nu\partial_\mu\bar{\lambda} - \bar{\xi}\bar{\sigma}_\nu\partial_\mu\lambda\right] - (\mu \leftrightarrow \nu), \qquad (2.36)$$

$$\delta_\xi\lambda_\alpha = -\xi_\alpha D - \frac{i}{2}(\sigma^\mu\bar{\sigma}^\nu)_\alpha^\beta\xi_\beta v_{\mu\nu}, \qquad (2.37)$$

$$\delta_\xi D = i\xi\sigma^\mu\partial_\mu\bar{\lambda} + i\bar{\xi}\sigma^\mu\partial_\mu\lambda. \qquad (2.38)$$

From Eq. (2.35), one notices that the F-component of $W^\alpha W_\alpha$ contains the Abelian field strength associated with v_μ. Since W_α is a chiral superfield, the F-term of $W^\alpha W_\alpha$ ($\equiv \int d^2\theta W^\alpha W_\alpha$) is a SUSY invariant. Using the relation

$$\sigma^\mu\bar{\sigma}^\nu = 2\sigma^{\mu\nu} + \eta^{\mu\nu}, \qquad (2.39)$$

it is straightforward to find

$$W^\alpha W_\alpha|_{\theta\theta} = -2i\lambda\,\sigma^\mu\partial_\mu\bar{\lambda} + D^2 - \frac{1}{2}(\sigma^{\mu\nu})^{\alpha\beta}(\sigma^{\rho\sigma})_{\alpha\beta}v_{\mu\nu}v_{\rho\sigma}. \qquad (2.40)$$

Also, using the relation

$$(\sigma^{\mu\nu})^{\alpha\beta}(\sigma^{\rho\sigma})_{\alpha\beta} = \frac{1}{2}\left(g^{\mu\rho}g^{\nu\sigma} - g^{\mu\sigma}g^{\nu\rho}\right) - \frac{i}{2}\varepsilon^{\mu\nu\rho\sigma}, \qquad (2.41)$$

one finds

$$W^\alpha W_\alpha|_{\theta\theta} = -2i\lambda\,\sigma^\mu\partial_\mu\bar{\lambda} + D^2 - \frac{1}{2}v^{\mu\nu}v_{\mu\nu} + \frac{i}{4}\varepsilon^{\mu\nu\rho\sigma}v_{\mu\nu}v_{\rho\sigma}. \qquad (2.42)$$

The last term is a total divergence, so it will not affect the equations of motion. Therefore, the SUSY gauge invariant generalisation of the Lagrangian for a free vector field is given by

$$\int d^4x\mathcal{L} = \int d^4x d^2\theta W^\alpha W_\alpha = \int d^4x\left[\frac{1}{2}D^2 - \frac{1}{4}v^{\mu\nu}v_{\mu\nu} - i\lambda\,\sigma^\mu\partial_\mu\bar{\lambda}\right]. \qquad (2.43)$$

The auxiliary field $D(x)$ can be eliminated by using the equation of motion as we did with the field F in the previous section. Also, SUSY leads to a massless fermionic partner $\lambda(x)$ of the massless gauge boson $v_\mu(x)$, which is called a gaugino. Finally, one finds that the tree level effective potential V is given by

$$V = \frac{1}{2}D^2. \qquad (2.44)$$

Also, this potential is positive definite with absolute minima at $D = 0$.

2.4 SUSY NON-ABELIAN GAUGE THEORY

In this section we discuss gauge invariant interactions of chiral and vector superfields [62, 66]. The chiral superfield Lagrangian in Eq. (2.23) is invariant under global transformations,

$$\Phi' = e^{i\Lambda}\Phi, \tag{2.45}$$

where $\Lambda = \Lambda_a T^a$ are constant chiral superfields and T^a are the generators of the gauge group G that constitute the representation of G to which the chiral superfields belong. In the adjoint representation, the generators are normalised as

$$\text{Tr}[T^a T^b] = \frac{1}{2}\delta^{ab}. \tag{2.46}$$

They satisfy the relation

$$[T^a, T^b] = if^{abc}T^c, \tag{2.47}$$

where the f^{abc} are the totally antisymmetric constants of G. The kinetic terms $\Phi_i^\dagger \Phi^i|_{\theta\theta\bar\theta\bar\theta}$ are naturally invariant under this transformation, while the requirement of invariance imposes constraints on the superpotential, so each term of $W(\Phi^i)$ must be a group invariant.

When we go from global to local invariance, the transformation (2.45) is consistent with SUSY only if one allows the parameters Λ_a to be chiral superfields, since $\bar{D}_{\dot\alpha}\Phi' = (\bar{D}_{\dot\alpha}\Lambda)\Phi$, *i.e.*, Φ' is a chiral superfield only if Λ is also a chiral superfield, $\bar{D}_{\dot\alpha}\Lambda = 0$. However, if Λ_a is promoted to be a chiral superfield, $\Lambda_a^\dagger \neq \Lambda_a$ and the kinetic terms are no longer invariant. To restore the invariance, one needs to introduce a vector multiplet $V = V^a T_a$ with the transformation

$$e^V \to e^{i\Lambda^\dagger} e^V e^{-i\Lambda}. \tag{2.48}$$

The first order approximation leads to the vector superfied transformation

$$V \to V + i(\Lambda - \Lambda^\dagger). \tag{2.49}$$

The new kinetic term

$$\mathcal{L}_{\text{kin}} = \Phi^\dagger e^V \Phi|_{\theta\theta\bar\theta\bar\theta} \tag{2.50}$$

is invariant under local transformations. This introduction of the gauge vector multiplet is completely analogous to the introduction of gauge fields in non-supersymmetric gauge theories. The superfields W_α are now defined as

$$W_\alpha = -\frac{1}{4}\bar{D}\bar{D}\left(e^{-V}D_\alpha e^V\right), \tag{2.51}$$

$$\bar{W}_{\dot\alpha} = \frac{1}{4}DD\left(e^V \bar{D}_{\dot\alpha} e^{-V}\right), \tag{2.52}$$

which, at first order in V, reduce to the usual definition of the Abelian vector field in Eq. (2.30). Under the non-Abelian transformation (2.48), one finds

$$
\begin{aligned}
W_\alpha \;\rightarrow\; & -\frac{1}{4}\bar{D}\bar{D}\left[\left(e^{i\Lambda}e^{-V}e^{-i\Lambda^\dagger}\right)D_\alpha\left(e^{i\Lambda^\dagger}e^{V}e^{-i\Lambda}\right)\right] \\
= & -\frac{1}{4}\bar{D}\bar{D}\left[e^{i\Lambda}e^{-V}\left(D_\alpha e^{V}\right)e^{-i\Lambda}+e^{i\Lambda}D_\alpha e^{-i\Lambda}\right] \\
= & -\frac{1}{4}e^{i\Lambda}\bar{D}\bar{D}\left(e^{-V}D_\alpha e^{V}\right)e^{-i\Lambda}=e^{i\Lambda}W_\alpha e^{-i\Lambda}.
\end{aligned}
\tag{2.53}
$$

Similarly, one finds $\bar{W}_{\dot{\alpha}}\rightarrow e^{i\Lambda^\dagger}\bar{W}_{\dot{\alpha}}\,e^{-i\Lambda^\dagger}$. Therefore, $\mathrm{Tr}[W^\alpha W_\alpha]$ and $\mathrm{Tr}[\bar{W}_{\dot{\alpha}}\bar{W}^{\dot{\alpha}}]$ are gauge invariant. In the WZ gauge, e^{V} is simplified to $e^{V}=1+V+\frac{1}{2}V^2$. Therefore, W_α becomes

$$
W_\alpha=-\frac{1}{4}\bar{D}\bar{D}D_\alpha V+\frac{1}{8}\bar{D}\bar{D}\left[V,D_\alpha V\right],
\tag{2.54}
$$

which implies

$$
W_\alpha=-i\lambda_\alpha(y)+\theta_\alpha D(y)+i(\sigma^{\mu\nu}\theta)_\alpha v_{\mu\nu}(y)+\theta\theta(\sigma^\mu D_\mu\bar{\lambda}(y))_\alpha,
\tag{2.55}
$$

where

$$
v_{\mu\nu}=\partial_\mu v_\nu-\partial_\nu v_\mu-\frac{i}{2}\left[v_\mu,v_\nu\right]
\tag{2.56}
$$

and

$$
D_\mu\bar{\lambda}=\partial_\mu\bar{\lambda}-\frac{i}{2}\left[v_\mu,\bar{\lambda}\right].
\tag{2.57}
$$

Then, the full SUSY Lagrangian is

$$
\mathcal{L}=\Phi_i^\dagger(e^{V})^i_j\Phi^j|_{\theta\theta\bar{\theta}\bar{\theta}}+\frac{1}{4g^2}\mathrm{Tr}[W^\alpha W_\alpha]_{\theta\theta}+[W(\Phi^i)]_{\theta\theta}+\text{h.c.}
\tag{2.58}
$$

The normalisation of the gauge field kinetic term is chosen such that the component action is canonically normalised after scaling $V\rightarrow 2gV$. In the case of a $U(1)$ gauge field one can show that the kinetic term in component fields is given by

$$
\begin{aligned}
\Phi^\dagger(e^{2gV})\Phi|_{\theta^2\bar{\theta}^2}=\; & \partial_\mu\phi^\dagger\partial^\mu\phi+i\bar{\chi}\bar{\sigma}^\mu\partial_\mu\chi+FF^\dagger+g^2\phi^\dagger\phi v^\mu v_\mu-g\phi^\dagger\phi D \\
& -\left(\sqrt{2}g\chi.\lambda\phi^\dagger+igv^\mu\phi^\dagger\partial_\mu\phi+\text{h.c.}\right)+g\chi\sigma^\mu\bar{\chi}v_\mu.
\end{aligned}
\tag{2.59}
$$

If we define

$$
D_\mu=\partial_\mu+igv_\mu,
\tag{2.60}
$$

the kinetic term takes the form

$$
\begin{aligned}
\Phi^\dagger(e^{2gV})\Phi|_{\theta^2\bar{\theta}^2}=\; & (D_\mu\phi)^\dagger(D^\mu\phi)+i\bar{\chi}\bar{\sigma}^\mu D_\mu\chi+FF^\dagger \\
& -g\phi^\dagger\phi D-\left(\sqrt{2}g\chi\lambda\phi^\dagger+\text{h.c.}\right).
\end{aligned}
\tag{2.61}
$$

Similarly, one can write the most general renormalisable non-Abelian gauge invariant Lagrangian, in terms of the superfields components, as follows:

$$
\begin{aligned}
\mathcal{L} &= \sum_i \left(|D\phi_i|^2 + i\psi_i\sigma^\mu D_\mu\psi_i^* + |F_i|^2 \right) \\
&- \sum_a \frac{1}{4g_a^2} \left[(v_{\mu\nu}^a)^2 - i\lambda^a \not{D}\lambda^{a*} - \frac{1}{2}(D^a)^2 \right] \\
&+ \left(i\sqrt{2} \sum_{ia} g^a \psi_i T^a \lambda^a \phi_i^* + \text{h.c.} \right) \\
&+ \sum_{ij} \frac{1}{2} \frac{\partial^2 W}{\partial\phi_i\partial\phi_j} \psi^i\psi^j,
\end{aligned}
\tag{2.62}
$$

where $\not{D} = \sigma^\mu D_\mu$. Eliminating the D^a and F^i fields gives rise to a scalar potential of the form

$$
V_{\text{SUSY}} = \frac{1}{2}|D^a|^2 + |F^i|^2,
\tag{2.63}
$$

with $F^i = \partial W/\partial\phi_i$ and $D^a = g^a \sum_i \phi_i^* T^a \phi_i$. The first two lines in Eq. (2.62) are just the gauge invariant kinetic terms for the various fields as well as potential terms for the scalars. The third line corresponds to Yukawa interactions of the gaugino and matter fields, with strength controlled by the gauge couplings. The last line yields fermion mass terms and Yukawa couplings among the various chiral superfields.

2.5 SUSY QED

As an example of a SUSY Abelian gauge theory, we consider SQED. The latter should include the following superfields: a vector superfield containing a photon field (and of course its spin 1/2-partner, the photino λ, and an auxiliary field D), a chiral superfield Φ_+ containing the electron field (as well as its scalar partner, the selectron $\tilde{e}^-$, and an auxiliary field F_+) plus a chiral superfield Φ_- containing the positron field (and its scalar partner, the spositron $\tilde{e}^+$, and an auxiliary field F_-). The chiral superfields $\Phi_\pm$ transform under local $U(1)$ as

$$
\Phi_+ \to \Phi_+' = e^{-ie\Lambda}\Phi_+, \qquad \Phi_- \to \Phi_-' = e^{ie\Lambda}\Phi_-,
\tag{2.64}
$$

where e refers to the electron charge. In this case, the $U(1)$ gauge and SUSY invariant Lagrangian is given by

$$
\begin{aligned}
\mathcal{L}_{\text{SQED}} &= \frac{1}{4}\left(W^\alpha W_\alpha\big|_{\theta\theta} + \bar{W}_{\dot{\alpha}}\bar{W}^{\dot{\alpha}}\big|_{\bar{\theta}\bar{\theta}} \right) + \Phi_+^\dagger e^{eV}\Phi_+\big|_{\theta\theta\bar{\theta}\bar{\theta}} + \Phi_-^\dagger e^{eV}\Phi_-\big|_{\theta\theta\bar{\theta}\bar{\theta}} \\
&+ m\left(\Phi_+\Phi_-\big|_{\theta\theta} + \Phi_+^\dagger\Phi_-^\dagger\big|_{\bar{\theta}\bar{\theta}} \right).
\end{aligned}
\tag{2.65}
$$

In components, one finds

$$
\begin{aligned}
\mathcal{L}_{\text{SQED}} &= \frac{1}{2}D^2 - \frac{1}{4}v_{\mu\nu}v^{\mu\nu} - i\lambda\sigma^\mu\partial_\mu\bar{\lambda} + |F_+|^2 + |F_-|^2 + \phi_+^*\Box\phi_+ + \phi_-^*\Box\phi_- \\
&+ i\left(\partial_\mu\bar{\psi}_+\bar{\sigma}^\mu\psi_+ + \partial_\mu\bar{\psi}_-\bar{\sigma}^\mu\psi_-\right) + ev_\mu\left[\frac{1}{2}\bar{\psi}_+\bar{\sigma}^\mu\psi_+ - \frac{1}{2}\bar{\psi}_-\bar{\sigma}^\mu\psi_-\right. \\
&+ \frac{i}{2}\phi_+^*\partial^\mu\phi_+ - \frac{i}{2}\partial^\mu\phi_+^*\phi_+ - \frac{i}{2}\phi_-^*\partial^\mu\phi_- + \left.\frac{i}{2}\partial^\mu\phi_-^*\phi_-\right] \\
&- \frac{ie}{\sqrt{2}}\left(\phi_+\bar{\psi}_+\bar{\lambda} - \phi_+^*\psi_+\lambda - \phi_-\bar{\psi}_-\bar{\lambda} + \phi_-^*\psi_-\lambda\right) \\
&+ eD\left[\phi_+^*\phi_+ - \phi_-^*\phi_-\right] - \frac{1}{4}e^2 v_\mu v^\mu\left[\phi_+^*\phi_+ + \phi_-^*\phi_-\right] \\
&+ m\left[\phi_+F_- + \phi_-F_+ - \psi_+\psi_- - \bar{\psi}_+\bar{\psi}_- + \phi_+^*F_-^* + \phi_-^*F_+^*\right].
\end{aligned}
\tag{2.66}
$$

We can eliminate the auxiliary fields F_+, F_- and D using their equations of motion as follows:

$$
F_+ = m\phi_-^*, \qquad F_- = m\phi_+^*, \tag{2.67}
$$
$$
D = -e\left[|\phi_+|^2 - |\phi_-|^2\right]. \tag{2.68}
$$

If we define the covariant derivative $D_\mu = \partial_\mu + iqv_\mu$, with $q = -e$ for Φ_+ and $q = +e$ for Φ_-, one can write $\mathcal{L}_{\text{SQED}}$ as

$$
\begin{aligned}
\mathcal{L}_{\text{SQED}} &= -\frac{1}{4}v_{\mu\nu}v^{\mu\nu} - i\lambda\sigma^\mu\partial_\mu\bar{\lambda} + (D_\mu\phi_+)^\dagger(D^\mu\phi_+) + (D_\mu\phi_-)^\dagger(D^\mu\phi_-) \\
&+ i\bar{\psi}_+\bar{\sigma}_\mu\psi_+ + i\bar{\psi}_-\bar{\sigma}_\mu\psi_- \frac{ie}{\sqrt{2}}\left(\phi_+\bar{\psi}_+\bar{\lambda} - \phi_-\bar{\psi}_-\bar{\lambda} + \text{h.c.}\right) + m^2|\phi_+|^2 \\
&+ m^2|\phi_-|^2 - m\psi_+\psi_- - \bar{\psi}_+\bar{\psi}_- + \frac{e^2}{2}\left[|\phi_+|^2 - |\phi_-|^2\right]^2.
\end{aligned}
\tag{2.69}
$$

Also, the two Weyl spinors ψ_+, ψ_- combine to form one massive Dirac spinor, the electron: $\psi = \left((\psi_+)_\alpha, (\bar{\psi}_-)^{\dot{\alpha}}\right)$.

From the above Lagrangian, one can notice that the scalar partners $\phi_\pm$ have the same mass as the electron. This is a consequence of exact SUSY, which leads to ruling out this model as a viable phenomenological scenario, since no scalar particle with the electron mass (half a MeV) has been observed. As we discuss in the next chapter, this problem can be overcome through SUSY breaking.

2.6 THE $U(1)$ FAYET-ILIOPOULOS MODEL

We now consider SUSY models with gauge group defined by a $U(1)$ with gauge coupling g [70]. Let V be the vector superfields of the Abelian gauge symmetry. Therefore,

$$
V \to V + i\left(\Lambda - \Lambda^\dagger\right), \tag{2.70}
$$

where Λ is a chiral superfield. As mentioned above, under this transformation, the D-term (the term proportional to $\theta^2\bar{\theta}^2$) transforms as $D \to D + \partial_\mu\partial^\mu(\dots)$, i.e., as a total derivative. Also, the D-term transforms under SUSY as a total derivative. Thus, the following term, which is known as the FI term, is a SUSY and gauge invariant contribution and can be part of the SUSY $U(1)$ Lagrangian

$$\mathcal{L}_{\text{FI}} = \xi \int d^2\theta d^2\bar{\theta} V = \frac{1}{2}\xi D, \tag{2.71}$$

where ξ is a constant with dimension of mass squared. Therefore, the complete SUSY Lagrangian of a $U(1)$ gauge symmetry is given by $\mathcal{L} = \mathcal{L}_{\text{gauge}} + \mathcal{L}_{\text{matter}} + \mathcal{L}_{\text{FI}}$, namely,

$$\begin{aligned}
\mathcal{L} &= \frac{1}{4}\left(WW\Big|_{\theta\theta} + \bar{W}\bar{W}\Big|_{\bar{\theta}\bar{\theta}}\right) + \xi g V\Big|_{\theta\theta\bar{\theta}\bar{\theta}} + \Phi^\dagger e^{gV}\Phi\Big|_{\theta\theta\bar{\theta}\bar{\theta}} \\
&+ W(\Phi)\Big|_{\theta\theta} + \bar{W}(\Phi^\dagger)\Big|_{\bar{\theta}\bar{\theta}}.
\end{aligned} \tag{2.72}$$

In this model, one finds that the auxiliary fields are given by

$$F^i = \frac{\partial \bar{W}(\phi^*)}{\partial \phi_i^*}, \tag{2.73}$$

$$D = -g\phi^* Y\phi - g\xi, \tag{2.74}$$

where Y is the charge of the ϕ field under $U(1)$. Thus, the scalar potential $V(\phi, \phi^*)$ is given by

$$V(\phi, \phi^*) = \sum_i |F^i|^2 + \frac{1}{2}|D|^2 = \sum_i \left|\frac{\partial \bar{W}}{\partial \phi_i}\right|^2 + \frac{1}{2}|g\phi^* Y\phi + g\xi|^2. \tag{2.75}$$

It is remarkable that a term like the FI one cannot be added to the Lagrangian in case of non-Abelian gauge theory, since in this case the D-terms are not gauge invariant. Also, if the VEV of the D-term is non-vanishing, $\langle D \rangle = -g\xi \neq 0$, then the scalar potential $V(\phi, \phi^*) > 0$, which means SUSY is spontaneously broken, as we will discuss in the next chapter.

2.7 NON-RENORMALISATION THEOREM

In this section we analyse the quantum corrections to the superpotential [71], W, which, as shown above, is fully determined by the structure of SUSY theories.

In SUSY theories, like WZ or SQED, one can show that the quadratic divergences of the scalar field masses cancel to all orders in perturbation theory. In addition, the only renormalisation required is a common wavefunction renormalisation of all fields, which involves only logarithmic divergences. Thus, the masses and coupling constants do not get renormalised at all to any order in perturbation theory. The non-renormalisation theorem demonstrates precisely this.

A proof for this theorem, using a diagrammatic approach, has been considered by Seiberg [72]. This proof is based on the fact that superpotentials are holomorphic while any radiative correction to the effective action, in supergraph perturbation theory, is given in terms of superspace integration $\int d^4\theta$, *i.e.*, non-holomorphic. Therefore, the superpotential does not receive any renormalisation. In contrast, the Kähler potential, K, is renormalised.

In order to understand this theorem, let us consider the following superpotential at some scale μ

$$W_{\text{tree}} = \frac{m}{2}\Phi^2 + \frac{\lambda}{3}\Phi^3. \tag{2.76}$$

If we assign a charge $R = 1$ to the superfield Φ under a $U(1)_R$ symmetry that implies $R[W] = 2$, then we should have $R[m] = 0$ and $R[\lambda] = -1$. Also let us assume that Φ has a unit charge under a $U(1)$ symmetry of W. In this case, the $U(1)$ charges for m and λ are -2 and -3, respectively. Here the mass and the coupling are treated as background (so-called) spurion fields[1].

Now we consider the effective superpotential generated by integrating out modes from the scale μ down to some scale Λ. The holomorphic nature of the effective superpotential implies that one can write W_{eff} in the following form

$$W_{\text{eff}} = m\Phi^2 f\left(\frac{\lambda\Phi}{m}\right) = \sum_n a_n \lambda^n m^{1-n} \Phi^{n+2}, \tag{2.77}$$

for some function f. The weak limit $\lambda \to 0$ leads to $n \geq 0$ and the massless limit $m \to 0$ implies that $n \leq 1$. Thus, W_{eff} is given by

$$W_{\text{eff}} = \frac{m}{2}\Phi^2 + \frac{\lambda}{3}\Phi^3 = W_{\text{tree}}. \tag{2.78}$$

This confirms that the superpotential is not renormalised. The non-renormalisation of the superpotential is one of the most important results of supersymmetric field theories.

---

[1]That is, fictitious, auxiliary fields merely used to parameterise any symmetry breaking and determine all operators invariant under the symmetry.

CHAPTER 3

Supersymmetry Breaking

In the previous two chapters, we have studied the basic techniques in the construction of a supersymmetric Lagrangian. SUSY cannot, however, be an exact symmetry of Nature. In fact, if it were, it would imply the existence of, *e.g.*, a selectron with the same mass as the electron, ~ 0.5 MeV, and squarks with the same mass of quarks, for which there is no experimental evidence. Thus, in order for SUSY to play a role in particle physics, it must be a broken symmetry at energies at least of the order of the EW scale. As any other symmetry, SUSY can be broken either spontaneously, dynamically or explicitly.

3.1 SPONTANEOUS SUSY BREAKING

The SUSY algebra contains the following anti-commutation relations among the SUSY generators Q_α and $\bar{Q}_{\dot\alpha}$:

$$\{Q_\alpha, \bar{Q}_{\dot\alpha}\} = 2\sigma^\mu_{\alpha\dot\alpha}P_\mu. \tag{3.1}$$

Upon multiplying from the right-hand side by $(\bar{\sigma})^{\dot\beta\alpha}$, we find

$$\{Q_\alpha, \bar{Q}_{\dot\alpha}\}(\bar{\sigma})^{\dot\beta\alpha} = 2\sigma^\mu_{\alpha\dot\alpha}(\bar{\sigma})^{\dot\beta\alpha}P_\mu = 2\mathrm{Tr}[\sigma^\mu\bar{\sigma}^\nu]P_\mu = 4\eta^{\mu\nu}P_\mu. \tag{3.2}$$

For $\nu = 0$ with $\bar{\sigma}^0 = \mathbb{1}_{2\times2}$, one finds that the Hamiltonian H can be written as [73]

$$H = \frac{1}{4}(\bar{Q}_1 Q_1 + Q_1 \bar{Q}_1 + \bar{Q}_2 Q_2 + Q_2 \bar{Q}_2) \geq 0, \tag{3.3}$$

which implies that the spectrum of the Hamiltonian is semi-positive definite. Thus, if a vacuum state $|0\rangle$ is supersymmetric, *i.e.*, $Q_\alpha|0\rangle = \bar{Q}_{\dot\alpha}|0\rangle = 0$, then a zero vacuum energy is obtained:

$$E_{\text{vacuum}} = \langle 0|H|0\rangle = 0. \tag{3.4}$$

In contrast, if the vacuum state is not supersymmetric, *i.e.*, at least one SUSY generator does not annihilate the vacuum, then Eq. (3.3) implies $E_{\text{vacuum}} > 0$. Clearly, a non-zero vacuum energy is an indication of spontaneous SUSY breaking.

Furthermore, as emphasised in a previous chapter, in SUSY theories the scalar potential is given by Eq. (2.63), namely,

$$V(\phi, \phi^\dagger) = \sum_i |F^i|^2 + \frac{1}{2} \sum_a |D^a|^2. \tag{3.5}$$

Therefore, the energy of the vacuum will be non-vanishing and, hence, SUSY is spontaneously broken when one of the auxiliary fields has a non-vanishing VEV,

$$\langle 0|F_i|0\rangle \neq 0, \qquad \text{or} \qquad \langle 0|D|0\rangle \neq 0. \tag{3.6}$$

This conclusion is consistent with the fact that spontaneous SUSY breaking means that the variation of some field under SUSY transformations has a non-zero VEV, i.e., $\langle 0|\delta_\xi(\text{field})|0\rangle \neq 0$. As for a chiral superfield, only the spinor variation can be non-zero, $\langle 0|\delta_\xi\psi(x)|0\rangle \propto \langle 0|F|0\rangle \neq 0$, while, for a vector superfield, only the gaugino variation can be non-vanishing, $\langle 0|\delta_\xi\lambda(x)|0\rangle \propto \langle 0|D|0\rangle \neq 0$.

3.2 GOLDSTONE THEOREM FOR SUSY

Let us recall that the Goldstone's theorem states that, when a continuous global symmetry is spontaneously broken, a massless particle, called a 'Goldstone boson', must appear in the spectrum, for each broken generator of such an invariance. In gauge theories, the Goldstone bosons are eaten by the gauge bosons, which become massive, and their new longitudinal polarisation is indeed provided by the Goldstone boson itself. A straightforward generalisation of this theorem implies that, if global SUSY is spontaneously broken, a corresponding massless fermion, the so-called 'Goldstino', appears in the spectrum [74]. This can be seen as follows, a VEV (i.e., $\partial V/\partial \phi^i = 0$) that breaks SUSY is given by $\langle F^i \rangle \neq 0$ or $\langle D^a \rangle \neq 0$. In a general SUSY model with both gauge and chiral supermultiplets, one finds

$$\frac{\partial V}{\partial \phi^i} = F^j \frac{\partial^2 W}{\partial \phi^i \partial \phi^j} - g^a D^a \phi_j^\dagger (T^a)_i^j, \tag{3.7}$$

which must have a vanishing VEV. Also, from the requirement that the superpotential is gauge invariant, i.e., $W[\Phi] = W[(1 + i\Lambda^a T^a)\Phi]$, one finds

$$\delta_{\text{gauge}}^a W = \frac{\partial W}{\partial \phi^i} \delta_{\text{gauge}}^a \phi^i = F_i^\dagger (T^a)_i^j \phi^j = 0. \tag{3.8}$$

Therefore, we obtain

$$M \begin{pmatrix} \langle F^j \rangle \\ \langle D^a \rangle \end{pmatrix} = 0, \tag{3.9}$$

where the matrix M is given by

$$M = \begin{pmatrix} \langle \frac{\partial^2 W}{\partial \phi^i \partial \phi^j} \rangle & -g^a \langle \phi_l^\dagger \rangle (T^a)_i^l \\ -g^a \langle \phi_l^\dagger \rangle (T^a)_j^l & 0 \end{pmatrix}. \tag{3.10}$$

One can easily show that M has a zero eigenvalue. In contrast, as can be seen from Eq. (2.62), the fermionic fields, consisting of gauginos (λ^a) and chiral fermions (ψ_i), have, in the basis (ψ_i, λ^a), the matrix M as their mass matrix. Thus, there is a massless Goldstino in the spectrum.

In order to understand the decomposition of the Goldstino, one considers the SUSY current:

$$J_\alpha^\mu \sim \sum \frac{\delta\mathcal{L}}{\delta(\partial_\mu\phi)}(\delta\phi)_\alpha = \sum_i \frac{\delta\mathcal{L}}{\delta(\partial_\mu\psi_{i\alpha})}\langle F_i\rangle + \frac{1}{\sqrt{2}}\sum_a \frac{\delta\mathcal{L}}{\delta(\partial_\mu\lambda_\alpha^a)}\langle D^a\rangle, \qquad (3.11)$$

so that the Goldstino, which is linear in the fields, is given by

$$\psi_\mu^G \sim \sum_i \langle F_i\rangle\psi_i + \sum_a \langle D^a\rangle\lambda^a. \qquad (3.12)$$

Hence, the Goldstino is a combination of the fermions, which are associated with auxiliary fields acquiring non-zero VEVs.

3.3 F-TERM INDUCED SUSY BREAKING

Let us consider a simple example of a single chiral superfield Φ, with the following superpotential:

$$W = \frac{1}{2}m\Phi^2 + \frac{1}{6}\lambda\Phi^3. \qquad (3.13)$$

Therefore, $F^\dagger$ is given by

$$F^\dagger = -\frac{\partial W}{\partial\phi} = -(m\phi + \frac{1}{2}\lambda\phi^2). \qquad (3.14)$$

It is clear that $F = 0$ at $\phi = 0$. Thus, in this example, SUSY is not spontaneously broken.

The common example of SUSY breaking through a non-vanishing F-term is the model of O'Raifeartaigh [75], which is a SUSY field theory constructed from three chiral superfields, $\Phi_i, i = 1, 2, 3$, with the superpotential

$$W = m\Phi_1\Phi_2 + \lambda(M^2 - \Phi_1^2)\Phi_3, \qquad (3.15)$$

where m and λ are real parameters. The auxiliary fields F_i are given by

$$F_1^* = -\frac{\partial W}{\partial\phi_1} = -(m\phi_2 - 2\lambda\phi_1\phi_3), \qquad (3.16)$$

$$F_2^* = -\frac{\partial W}{\partial\phi_2} = -m\phi_1, \qquad (3.17)$$

$$F_3^* = -\frac{\partial W}{\partial\phi_3} = -\lambda(M^2 - \phi_1^2). \qquad (3.18)$$

From these equations, it is clear that there is no field configuration for which all the three F_i's vanish. If $F_3 = 0$, then $F_2 \neq 0$, whereas $F_2 = 0$ leads to $F_3 \neq 0$, hence SUSY must be broken spontaneously. The potential is the sum of the absolute squares of the F terms

$$V = \sum_i |F_i|^2 = |m\phi_2 - 2\lambda\phi_1\phi_3|^2 + m^2|\phi_1|^2 + \lambda^2|M^2 - \phi_1^2|^2. \qquad (3.19)$$

Differentiating this potential with respect to ϕ_i, one finds the following minimisation conditions:

$$m\phi_2 - 2\lambda\phi_1\phi_3 = 0, \qquad (3.20)$$
$$m^2\phi_1 + 2\lambda^2\phi_1^*(\phi_1^2 - M^2) = 0. \qquad (3.21)$$

Note that two independent conditions only are obtained, in addition to their complex conjugates. Thus, two of the scalars ϕ_i can be determined, while the third one remains unfixed. This is a general feature of F-term induced SUSY breaking, where the potential has, at tree level, a continuous minimum along that field, which is called flat direction. Also, Eq. (3.21) implies that ϕ_1 must be real. This can be seen by writing $\phi_1 = a_1 + ib_1$ and checking the possible solution for real and imaginary parts. One finds that these equations are satisfied if and only if $b_1 = 0$.

The solutions to Eqs. (3.20) and (3.21) depend on the value $M^2 - m^2/(2\lambda)$. If $M^2 - m^2/(2\lambda^2) < 0$, one finds the following solution:

$$\phi_1 = 0, \qquad \phi_2 = 0, \qquad \phi_3 \text{ is undetermined.} \qquad (3.22)$$

Therefore, ϕ_3 is a flat direction of the scalar potential and

$$F_1 = 0, \qquad F_2 = 0, \qquad F_3 = -\lambda M^2, \qquad V_{\min} = \lambda^2 M^4. \qquad (3.23)$$

For $M^2 - m^2/(2\lambda^2) > 0$, one gets the following solution:

$$\phi_1 = \sqrt{M^2 - \frac{m^2}{2\lambda^2}}, \qquad \phi_2 = \frac{2\lambda}{m}\sqrt{M^2 - \frac{m^2}{2\lambda^2}}\,\phi_3, \qquad \phi_3 \text{ is undetermined.} \qquad (3.24)$$

Again, ϕ_3 is a flat direction of the potential with

$$F_1 = 0, \quad F_2 = -m\sqrt{M^2 - \frac{m^2}{2\lambda^2}}, \quad F_3 = -\frac{m^2}{2\lambda}, \quad V_{\min} = m^2(M^2 - \frac{m^2}{4\lambda^2}). \qquad (3.25)$$

In both cases, we see that SUSY is broken, since one of the F-terms does not vanish. To study the mass spectrum of the model, as usual, we shift the fields by their VEVs and then rewrite the scalar potential in terms of the shifted fields. Let us consider the first solution:

$$\phi_1' = \phi_1, \qquad \phi_2' = \phi_2, \qquad \phi_3' = \phi_3 - \langle\phi_3\rangle, \qquad (3.26)$$

where $\langle\phi_3\rangle$ is the undetermined VEV of ϕ_3. From the potential (3.19), one notices that there is no bilinear term for the field ϕ_3': therefore, it must be massless. In the basis of the real and imaginary parts of $\phi'_{1,2}$, one extracts the following masses squared (for simplicity, let us set $\langle \phi_3 \rangle = 0$):

$$m^2, \quad m^2 - 2\lambda^2 M^2, \quad m^2 + 2\lambda^2 M^2. \tag{3.27}$$

Thus,

$$\sum_{\text{bosons}} \mathcal{M}^2 = 4m^2. \tag{3.28}$$

It is remarkable that the effect of F-term SUSY breaking induces a split into the mass squared of scalars by equal and opposite amounts. The fermion masses are obtained from the second derivatives of the superpotential W at the VEVs of the scalar fields. In our case, the only non-vanishing mass term is $m\bar{\psi}_1\psi_2$, which describes a Dirac fermion of mass m. The massless ψ_3 will be identified as the Goldstino, which is the superpartner of the Goldstone boson ϕ_3. The spectrum of the three Weyl fermions is

$$0, \ m, \ m \Rightarrow \sum_{\text{fermions}} \mathcal{M}^2 = 2m^2. \tag{3.29}$$

One can verify that the resultant spectrum satisfies the following SuperTrace (STr) relation:

$$\text{STr} \, \mathcal{M}^2 = \sum_J (-1)^{2J} (2J+1) m_J^2 = 0, \tag{3.30}$$

where the sum is over all particles in the theory with spin J and mass m_J. This supertrace relation remains intact if one considers the second solution in Eq. (3.24). In this respect, a salient feature of the O'Raifeartaigh model and any model with F-terms inducing SUSY breaking is that it leads to a scalar partner lighter than the corresponding fermion, $i.e.$, one would expect one selectron lighter than the electron, which is in contradiction with experiments. Therefore, it is not possible to construct a realistic SUSY model with this mechanism of SUSY breaking.

3.4 D-TERM INDUCED SUSY BREAKING

SUSY breaking with non-zero D-terms can be achieved through the FI mechanism [70]. As mentioned, if the gauge symmetry includes a $U(1)$ factor, then one can introduce a linear term in the corresponding auxiliary field of the gauge supermultiplet:

$$\mathcal{L}_{\text{FI}} = g\xi D, \tag{3.31}$$

where ξ is a constant parameter with dimension of mass squared and g is the $U(1)$ gauge coupling. In a simple model with one chiral superfield coupled to a $U(1)_{\text{FI}}$ gauge field, as shown in Eq. (2.75), the scalar potential is given by

$$V = \frac{1}{2} D^2 = \frac{1}{2} g^2 (\phi^* \phi + \xi)^2. \tag{3.32}$$

Here, we assume that the auxiliary field of the chiral superfield Φ vanishes and the charge of the scalar field ϕ under $U(1)_{FI}$ is unity. The minimisation of this potential leads to

$$\langle \phi^* \phi \rangle = 0, \quad \text{if} \quad \xi > 0 \quad \text{or} \quad \langle \phi^* \phi \rangle = -\xi, \quad \text{if} \quad \xi < 0. \tag{3.33}$$

In the first case, the minimum of the potential is given by

$$V_{\min} = \frac{1}{2} g^2 \xi^2. \tag{3.34}$$

Thus, SUSY is spontaneously broken through non-vanishing D-terms, while the $U(1)_{FI}$ gauge symmetry remains exact. In the second case, the $U(1)_{FI}$ is spontaneously broken, however, SUSY remains exact.

With respect to the associated mass spectrum, one notes that, in the first case, where SUSY is broken by the VEV of the D-term, $\langle D \rangle = g\xi$, the scalar field ϕ acquires a mass term of order $g\sqrt{\xi}$ whereas its fermionic partner remains massless. Also, the gauge boson and gaugino of the exact $U(1)_{FI}$ remain massless. Therefore, one can show that the supertrace mass relation is now given by

$$\text{STr } \mathcal{M}^2 = 2g^2 \xi. \tag{3.35}$$

Thus, this model, unlike the O'Raifeartaigh realisation, does not lead to a scalar lighter than the fermion. Also, the massless gaugino of $U(1)_{FI}$ represents the Goldstino associated to the spontaneous SUSY breaking.

In fact, in D-term induced SUSY breaking, the mass squared sum rules is generally modified as

$$\text{STr } \mathcal{M}^2 = 2\langle D \rangle \text{ Tr}[Q], \tag{3.36}$$

where Q is the charge matrix of the chiral superfields and $\langle D \rangle$ is the VEV of the auxiliary gauge field. In the SM, the trace of the hypercharge $U(1)$ generator is zero, so the mass squared sum rules do not change. If one adds an additional $U(1)$ gauge group factor, this could raise all sfermion masses. However, this $U(1)$ factor leads to gauge anomalies. Actually, the cancellation of all anomalies requires $\text{Tr}[Q] = 0$. Thus, the mass squared sum rule problem is not avoided. Such a sum rule is a consequence of the cancellation of quadratic divergences in SUSY. This can be seen from the one-loop effective scalar potential, which is given by

$$V(\phi) = \frac{\Lambda^2}{32\pi^2} \text{STr } M^2(\phi) + \frac{1}{64\pi^2} \text{STr} \left(M^2(\phi) \left[\ln \frac{M^2(\phi)}{\Lambda^2} - 1/2 \right] \right). \tag{3.37}$$

Therefore, in models with spontaneous SUSY breaking, no quadratic divergences are generated in any Green function, due to the vanishing of the STr $M^2(\phi)$. However, unfortunately, these sum rules lead again to the same phenomenological problem discussed in the previous section.

3.5 SOFT SUSY BREAKING

So far, we have seen that the mechanisms of spontaneous SUSY breaking do not lead to a phenomenological viable model. Therefore, the only remaining possibility is to break SUSY explicitly by adding to the SUSY invariant Lagrangian a set of terms that violate SUSY and do not introduce quadratic divergences. These terms are called 'soft SUSY breaking terms'. We will show in the last three chapters that these soft terms can naturally be generated through SUSY breaking in some hidden sectors of an underling theory and then the information of such breaking is conveyed to the observable sector (ordinary particles plus their superpartners) through gauge or gravitational messengers. In this way the low energy SUSY Lagrangian of the observable sector looks as follows: a SUSY invariant Lagrangian plus a set of terms which break SUSY explicitly in a soft manner. In this way the 'supertrace mass problem' can be overcome.

In order not to introduce quadratic divergences and consequently spoil the SUSY solution to the gauge hierarchy problem, the set of soft SUSY breaking terms contains only mass terms and couplings with positive mass dimension. This fact can be verified in a simple case, like the WZ model, where one can explicitly show that the cancellation of the quadratic divergences is not spoiled if, for example, the scalar boson masses are not exactly equal to the fermion masses. However, it is important to note that not all SUSY breaking terms are soft. This type of restricted terms have been catalogued by Girardello and Grisaru as follows [76]:

1. Masses for the scalars: $\tilde{m}_{ij}^2 \phi_i^* \phi_j$.

2. Masses for the gauginos: $\frac{1}{2} M_a \lambda^a \lambda^a$.

3. Bilinear scalar interactions: $\frac{1}{2} B_{ij} \phi_i \phi_j + \text{h.c.}$

4. Trilinear scalar interactions: $\frac{1}{3!} \phi_i \phi_j \phi_k + \text{h.c.}$

The soft terms are very important since they determine the SUSY spectrum and contribute to the Higgs potential generating the radiative breakdown of the EW symmetry. From the above list, it seems that the number of soft SUSY breaking terms is enormous. However, it is important to note that the general form of soft SUSY breaking terms is a parameterisation of our ignorance of the SUSY breaking mechanism. An understanding of SUSY breaking within an underlying theory will lead to a full determination of the soft SUSY terms as expressions of a few fundamental parameters.

CHAPTER 4

Minimal Supersymmetric Standard Model

The MSSM is a straightforward supersymmetrisation of the SM with the minimal number of possible new parameters [35,77]. It is consequently the most widely studied SUSY model. In this chapter we provide the reader with a detailed construction of it.

4.1 MSSM STRUCTURE

The MSSM is based on the same gauge group of the SM, *i.e.*, $SU(3)_C \times SU(2)_L \times U(1)_Y$, with the following particle content.

1. Three chiral superfields of $SU(2)_L$ doublet quarks $Q_i = (\tilde{q}_L, q_L)_i^T$ that describe the left-handed quarks $q_{L_i} = (u_L, d_L)_i^T$ and their superpartners (left-handed squarks) $\tilde{q}_{L_i} = (\tilde{u}_L, \tilde{d}_L)_i^T$, with i running over the three generations.

   The quantum numbers of Q_i under the SM gauge group are $(3, 2, 1/6)$, where the hypercharge Y is defined in terms of the electric charge Q and isospin I_3 as follows:
   $$Y = Q - I_3. \tag{4.1}$$

2. Three chiral superfields of $SU(2)_L$ singlet up-type quarks $U_i^c = (\tilde{u}_L^c, u_L^c)_i^T$ that describe the up-type right-handed quarks $u_{L_i}^c$ and their superpartners (right-handed up-squarks) $\tilde{u}_{L_i}^c$.

   The quantum numbers of U_i^c under the SM gauge group are $(\bar{3}, 1, -2/3)$. Note that here we describe the right-handed chiral particle by the left-handed chiral anti-particle.

3. Three chiral superfields of $SU(2)_L$ singlet down-type quarks $D_i^c = (\tilde{d}_L^c, d_L^c)_i^T$ that describe the down-type right-handed quarks $d_{L_i}^c$ and their superpartners (right-handed down-squarks) $\tilde{d}_{L_i}^c$. The quantum numbers of D_i^c under the SM gauge group are $(\bar{3}, 1, 1/3)$.

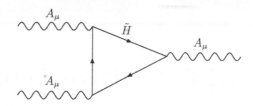

Figure 4.1 Higgsino contribution to the triangle anomaly.

4. Three chiral superfields of $SU(2)_L$ doublet leptons $L_i = (\tilde{\ell}_L, \ell_L)_i^T$ that describe the left-handed leptons $\ell_{L_i} = (\nu_{eL}, e_L)_i^T$ and their superpartners (left-handed sleptons) $\tilde{\ell}_{L_i} = (\tilde{\nu}_{eL}, \tilde{e}_L)_i^T$. The quantum numbers of L_i under the SM gauge group are $(1, 2, -1/2)$.

5. Three chiral superfields of $SU(2)_L$ singlet leptons $E_i^c = (\tilde{e}_L^c, e_L^c)_i^T$ that describe the right-handed leptons $e_{L_i}^c$ and their superpartners (right-handed sleptons) $\tilde{e}_{L_i}^c$. The quantum numbers of E_i^c under the SM gauge group are $(1, 1, 1)$.

6. Two chiral superfields of $SU(2)_L$ doublet Higgses $H_u = (H_u, \tilde{H}_u)^T$ and $H_d = (H_d, \tilde{H}_d)^T$ that describe the Higgs particles $H_{u,d}$ and their superpartners (Higgsinos) $\tilde{H}_{u,d}$, where $H_u = (H_u^+, H_u^0)^T$ and $\tilde{H}_u = (\tilde{H}_u^+, \tilde{H}_u^0)^T$, while $H_d = (H_d^0, H_d^-)^T$ and $\tilde{H}_d = (\tilde{H}_d^0, \tilde{H}_d^-)^T$.

The quantum numbers of H_u under the SM gauge group are $(1, 2, +1/2)$ and those of H_d are $(1, 2, -1/2)$. Note that the same symbols $H_{u,d}$ are usually assumed for the superfields and their scalar partners.

Also, it is worth noting that in the SM one Higgs doublet, H, is used only to generate masses for both up and down quarks after EWSB, through the Yukawa interactions $Y_d q_L H d_L^c + Y_u q_L \tilde{H} u_L^c + \text{h.c.}$, where $\tilde{H} = i\sigma_2 H^*$. In the MSSM, where the superpotential is an analytic function of only chiral superfields, the anti-chiral H^* cannot be included in the superpotential and a new Higgs doublet with opposite hypercharge should instead be introduced.

Another reason for the necessity of adding another Higgs doublet in the MSSM is to cancel the triangle anomaly generated by the fermionic partner of the Higgs superfield. This anomaly is cancelled in the SM due to the vanishing of $\text{Tr}[Y^3]$ and $\text{Tr}[T_3^2 Y]$, where T_3 stands for the $SU(2)_L$ third generator. In a SUSY model with just one Higgs doublet, the fermionic partner of this Higgs (Higgsino) contributes to the triangle anomaly, as shown in Fig. 4.1. This contribution would remain not cancelled. Therefore, a second Higgs doublet superfield, with opposite hypercharge, must be added in order to cancel this contribution.

7. Three gauge (vector) superfields corresponding to the SM gauge groups

$(B_\mu, \tilde{B})$ for $U(1)_Y$, $(W_\mu^a, \tilde{W}^a)$ for $SU(2)_L$ with $a = 1, 2, 3$ and $(G_\mu^a, \tilde{g}^a)$ for $SU(3)_C$ with $a = 1, \ldots 8$.

As a way of efficiently cataloguing the MSSM states, by relating particles and sparticles to one another, we finish this chapter by presenting Tab. 4.1, which can also work as a quick reference guide to return to in the remainder of this book while tackling MSSM phenomenology.

Supermultiplet	SM	SUSY	$SU(3)_C \times SU(2)_L \times U(1)_Y$
Q_L	quarks $q = (u_L, d_L)^T$ (spin $\frac{1}{2}$)	squarks $\tilde{q} = (\tilde{u}_L, \tilde{d}_L)^T$ (spin 0)	$(3, 2, 1/6)$
U_L^c	quarks u_L^c (spin $\frac{1}{2}$)	squarks $\tilde{u}_L^c$ (spin 0)	$(\bar{3}, 1, -2/3)$
D_L^c	quarks d_L^c (spin $\frac{1}{2}$)	squark $\tilde{d}_L^c$ (spin 0)	$(\bar{3}, 1, 1/3)$
L_L	leptons $l = (\nu_L, e_L)^T$ (spin $\frac{1}{2}$)	sleptons $(\tilde{l}) = (\tilde{\nu}_L, \tilde{e}_L)^T$ (spin 0)	$(1, 2, 1/2)$
E_L^c	leptons e_L^c (spin $\frac{1}{2}$)	sleptons $\tilde{e}_L^c$ (spin 0)	$(1, 1, 1)$
H_u	Higgs $H_u = (H_u^0, H_u^+)^T$ (spin 0)	Higgsino $\tilde{H}_u = (\tilde{H}_u^0, \tilde{H}_u^+)$ (spin $\frac{1}{2}$)	$(1, 2, 1/2)$
H_d	Higgs $H_d = (H_d^-, H_d^0)^T$ (spin 0)	Higgsino $\tilde{H}_d = (\tilde{H}_d^-, \tilde{H}_d^0)^T$ (spin $\frac{1}{2}$)	$(1, 2, 1/2)$

Table 4.1 SM and SUSY particle states in the MSSM. The family indices are implicit.

With these superfields, the MSSM Lagrangian can be written as

$$\mathcal{L}_{\text{MSSM}} = \mathcal{L}_{\text{gauge}} + \mathcal{L}_{\text{matter}} + W + \mathcal{L}_{\text{soft}}, \tag{4.2}$$

where the gauge Lagrangian $\mathcal{L}_{\text{gauge}}$ includes all the gauge interactions in the MSSM. It is given in component notation as

$$\mathcal{L}_{\text{gauge}} = -\frac{1}{4} F_G^{a\,\mu\nu} F_{G\,\mu\nu}^a + i \bar{\lambda}_G^a \bar{\sigma}^\mu D_\mu \lambda_G^a + \frac{1}{2} D^a D_a, \tag{4.3}$$

where the index G labels the colour, weak isospin and hypercharge factors in the SM gauge group, the index a refers to the adjoint representations of the non-Abelian subgroups while λ_G is the associated gaugino field. The auxiliary scalar field D^a is not a propagating field, as shown in Chapter 1, and can be eliminated from the Lagrangian by solving the equation of motion for D^a. The field strength tensors $F_{G\,\mu\nu}^a$ are given by

$$F_{G\,\mu\nu}^a = \partial_\mu A_\nu^a - \partial_\nu A_\mu^a - g_G f^{abc} A_\mu^b A_\nu^c, \tag{4.4}$$

where A_μ^a stands for the gauge fields B_μ, W_μ^a and G_μ^a while the covariant derivative D_μ is defined as

$$D_\mu = \partial_\mu + i g_G T^a A_\mu^a, \tag{4.5}$$

where T^a are the gauge group generators.

The terms obtained from the kinetic Lagrangian of chiral superfields are given by

$$
\begin{aligned}
\mathcal{L}_{\text{matter}} &= (D^\mu \phi_i)^\dagger (D_\mu \phi_i) + i \bar\psi_i \gamma^\mu D_\mu \psi_i + F_i^* F_i \\
&+ i g_a \sqrt{2} (\phi^* T^a \lambda^a \psi + \text{h.c.}) - \frac{1}{2} g_a^2 (\phi_i^* T^a \phi_i)^2.
\end{aligned} \tag{4.6}
$$

Here, $D_\mu = \partial_\mu + i g_1 Y B_\mu + i g_2 \frac{\sigma^a}{2} W_\mu^a + i g_3 \frac{\lambda^a}{2} G_\mu^a$ and ϕ, ψ refer to the MSSM scalars and fermions, respectively. The last term in Eq. (4.6) is obtained by eliminating D^a from the Lagrangian as

$$D^a = \frac{g^a}{\sqrt{2}} (\phi_i^* T^a \phi_i). \tag{4.7}$$

Moreover, F_i is an auxiliary scalar field, similar to D^a. The SM Yukawa interactions are included in the MSSM superpotential, which describes the interactions between Higgs bosons and matter superfields. This superpotential can be written as

$$W = Y_u Q U^c H_u + Y_d Q D^c H_d + Y_e L E^c H_d + \mu H_d H_u. \tag{4.8}$$

Here, the summation over generational indices is implicit. In general, the Yukawa coupling constants Y_u, Y_d and Y_e are non-diagonal 3×3 matrices in flavour space leading to the usual masses. The last term can be written as $\mu (H_u)_\alpha (H_d)_\beta \varepsilon^{\alpha\beta}$, with $\alpha, \beta = 1, 2$. Similarly, the first (other) term can be explicitly written as $\bar{u}^{ia} (Y_u)_i^j Q_{j\alpha a} (H_u)_\beta \varepsilon^{\alpha\beta}$, with the colour index $a = 1, 2, 3$ and the family indices $i, j = 1, 2, 3$. The parameter μ has mass dimension one and gives supersymmetric masses to both fermionic and bosonic components of the chiral superfields H_u and H_d. As an explicit example, let us consider the superpotential of the top superfield

$$W = Y_t Q_t H_u t_L^c. \tag{4.9}$$

As shown in Chapter 3, the interaction Lagrangian associated with this superpotential is given by

$$
\begin{aligned}
\mathcal{L}_{\text{int}} &= -\frac{1}{2} W^{ij} \psi_i \psi_j - \frac{1}{2} W^{ij*} \bar\psi_i \bar\psi_j - W^i W_i^* \\
&= -\frac{Y_t}{2} \Big(t_L H_u t_L^c + \tilde{t}_L \tilde{H}_u t_L^c + t_L \tilde{H}_u \tilde{t}_L^c + \text{h.c.} \Big) - Y_t^2 \big(|H_u \tilde{t}_L^c|^2 + |H_u \tilde{t}_L|^2 + |\tilde{t}_L \tilde{t}_L^c|^2 \big).
\end{aligned} \tag{4.10}
$$

Thus, this term in the superpotential leads to four Feynman rules, given in Fig. 4.2. Furthermore, due to the fact that Higgs and lepton doublet superfields have the

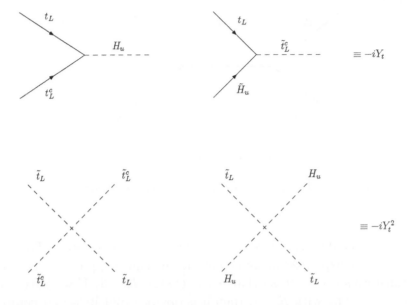

$$\equiv -iY_t$$

$$\equiv -iY_t^2$$

Figure 4.2 Feynman diagrams of top and stop interaction vertices.

same $SU(3)_C \times SU(2)_L \times U(1)_Y$ quantum numbers, we have additional terms that can be written as

$$W' = \lambda_{ijk} L_i L_j E_k^c + \lambda'_{ijk} L_i Q_j D_k^c + \lambda''_{ijk} D_i^c D_j^c U_k^c + \mu'_i L_i H_u. \qquad (4.11)$$

These terms violate baryon and lepton numbers explicitly and lead to proton decay at unacceptable rates. This can be seen from the fact that the terms $\lambda''_{211} D_2^c D_1^c U_1^c$ and $\lambda'_{112} L_1 Q_1 D_2^c$ form the four-fermion operator (through the exchange of $\tilde{s}$)

$$\frac{\lambda''_{211} \lambda'_{112}}{m_{\tilde{s}}^2} (u_L^c d_L^c)(u_L e_L). \qquad (4.12)$$

This operator would contribute to the proton decay process $p \to e^+ \pi^0$, as shown in Fig. 4.3, at a rate $\propto \Gamma \sim \lambda'_{112} \lambda''_{211} m_p^5 / m_{\tilde{s}}^4$. Therefore, the predicted lifetime of the proton is of order

$$\tau_p \sim 6 \times 10^{-13} \sec \left(\frac{m_{\tilde{s}}}{1 \text{ TeV}} \right)^4 (\lambda'_{112} \lambda''_{211})^{-2}. \qquad (4.13)$$

From the experimental limit on the proton lifetime, $\tau_p > 1.6 \times 10^{33}$ years, one finds the following extremely small upper bounds on λ'_{112} and λ''_{211}:

$$\lambda'_{112} \lambda''_{211} < 3 \times 10^{-26}. \qquad (4.14)$$

In the SM, there is no such a problem, as the requirements of gauge symmetry and renormalisability automatically lead to baryon and lepton number conservation.

To avoid this problem of too rapid a proton decay, a new symmetry (called

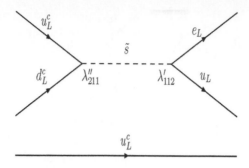

Figure 4.3 Proton decay to e^+ and π^0 through R-parity violating terms.

R-symmetry) is commonly introduced in order to remove these terms. The concept of R-symmetry can be explained in superspace formalism. Suppose we have a symmetry that transforms θ to $e^{i\alpha}\theta$ so that θ has charge $R = 1$. If we have a chiral superfield Φ transforming with $R = 1$, then it is obvious that its scalar component transforms as $\phi \rightarrow e^{i\alpha}\phi$ with $R = 1$ while the fermion component has $R(\psi) = 0$. The F-component of this superfield should have $R(F) = -1$. The vector superfield is real and consequently it has charge $R = 0$, hence $R(A_\mu) = 0$ and $R(\lambda) = 1$, *i.e.*, the gauginos transform non trivially under the R-symmetry. Let us now go back to the superpotential: there is an R-symmetry which leaves W in Eq. (4.8) as the most general superpotential. In other words this means that, if we drop the terms in Eq. (4.11), a continuous global R-symmetry appears. To forbid these terms in Eq. (4.11), in principle, a smaller symmetry, like R-parity [78],

$$R_P = (-1)^{3B+L+2S}, \qquad (4.15)$$

where B and L are baryon and lepton number and S is the spin, would be sufficient, but in general a continuous R-symmetry occurs as long as one considers only the $N = 1$ SUSY invariant Lagrangian. There are two remarkable phenomenological implications of the presence of R-parity: (*i*) SUSY particles are produced or destroyed only in pairs; (*ii*) the LSP is absolutely stable and, hence, it might constitute a possible candidate for DM.

In addition to the above interactions, one should add the soft SUSY breaking terms to the Lagrangian. Following the general classification of the soft SUSY breaking terms, we can write $\mathcal{L}_{\text{soft}}$ as follows [76]:

$$
\begin{aligned}
\mathcal{L}_{\text{soft}} = {} & -\frac{1}{2}M_a\lambda^a\lambda^a - m^2_{\tilde{q}_{ij}}\tilde{q}^*_i\tilde{q}_j - m^2_{\tilde{u}_{ij}}\tilde{u}^*_i\tilde{u}_j - m^2_{\tilde{d}_{ij}}\tilde{d}^*_i\tilde{d}_j - m^2_{\tilde{\ell}_{ij}}\tilde{\ell}^*_i\tilde{\ell}_j \\
& - m^2_{\tilde{e}_{ij}}\tilde{e}^*_i\tilde{e}_j - m^2_{H_u}|H_u|^2 - m^2_{H_d}|H_d|^2 - \Big[Y^A_{uij}\tilde{q}_i\tilde{u}_jH_u \\
& + Y^A_{dij}\tilde{q}_i\tilde{d}_jH_d + Y^A_{eij}\tilde{\ell}_i\tilde{e}_jH_d - B\mu H_uH_d + \text{h.c.}\Big].
\end{aligned}
\qquad (4.16)
$$

Note that the soft terms $m^2_{\tilde{q}}$, $m^2_{\tilde{u}}$, $m^2_{\tilde{d}}$, $m^2_{\tilde{\ell}}$ and $m^2_{\tilde{e}}$ are Hermitian 3×3 matrices in flavour space, while the trilinear couplings in most cases are given by

$Y_{fij}^A \equiv (A_f)_{ij}(Y_f)_{ij}$, with $f = u, d, e$, as complex 3×3 matrices. Also the gaugino masses M_a and the bilinear coupling B of mass dimension one are generally complex numbers. The soft SUSY terms induce about 100 free parameters which reduce the predictivity of the MSSM. In what is called cMSSM, a kind of universality among the soft SUSY breaking terms at the GUT scale $M_{GUT} = 3 \times 10^{16}$ GeV is assumed. In this case, this large number of soft SUSY breaking terms is reduced to the following four parameters:

$$
\begin{aligned}
m_{\tilde{q}_{ij}}^2 &= m_{\tilde{u}_{ij}}^2 = m_{\tilde{d}_{ij}}^2 = m_{\tilde{\ell}_{ij}}^2 = m_{\tilde{e}_{ij}}^2 = m_0^2 \delta_{ij}, \\
m_{H_u}^2 &= m_{H_d}^2 = m_0^2, \\
A_u^{ij} &= A_d^{ij} = A_e^{ij} = A_0 \delta^{ij}, \\
M_1 &= M_2 = M_3 = m_{1/2}.
\end{aligned}
\tag{4.17}
$$

The parameter m_0 is called 'universal scalar mass', A_0 is called 'universal trilinear coupling' and $m_{1/2}$ is called 'universal gaugino mass'. This class of models is motivated by mSUGRA, where SUSY breaking is mediated by gravity interactions with minimal Kähler potential and minimal gauge kinetic function, as will be shown in Chapter 16.

4.2 GAUGE COUPLING UNIFICATION IN THE MSSM

In this section, we show that SUSY in the TeV range allows for the unification of the $SU(3)_C \times SU(2)_L \times U(1)_Y$ gauge interactions [79–84]. In a QFT, the gauge couplings of these interactions are functions of the energy at which they are measured and their RGEs are given by

$$
\frac{d\alpha_i(t)}{dt} = \frac{b_i}{2\pi} \alpha_i^2(t) + \frac{1}{8\pi^2} \sum_j b_{ij} \alpha_j^2(t), \qquad i = 1, 2, 3,
\tag{4.18}
$$

where $t = \ln(Q/M_X)$ with Q the running scale and M_X the mass scale of whatever underlying GUT. The couplings α_i are defined as $\alpha_i(t) = g_i^2(t)/4\pi$. Note that the $U(1)_Y$ coupling α_1 is normalised such that

$$
g_1^2 \to \frac{5}{3} g_1^2.
\tag{4.19}
$$

The one-loop coefficients b_i of the β functions for the gauge couplings in the SM are given by

$$
\begin{aligned}
b_1 &= \frac{4}{3} N_g + \frac{N_H}{10}, \\
b_2 &= -\frac{22}{3} + \frac{4}{3} N_g + \frac{N_H}{6}, \\
b_3 &= -11 + \frac{4}{3} N_g,
\end{aligned}
\tag{4.20}
$$

where N_g is the number of generations, $N_g = 3$, and N_H is the number of the Higgs doublets, $N_H = 1$. Therefore, in the SM one finds

$$b_1 = \frac{41}{10}, \qquad b_2 = -\frac{19}{16}, \qquad b_3 = -7. \tag{4.21}$$

In the MSSM the one-loop coefficients are given by

$$
\begin{aligned}
b_1 &= 2N_g + \frac{3}{10}N_H, \\
b_2 &= -6 + 2N_g + \frac{N_H}{2}, \\
b_3 &= -9 + 2N_g.
\end{aligned}
\tag{4.22}
$$

As mentioned, the MSSM contains two Higgs superfields, $i.e.$, $N_H = 2$. Thus the b_i's are given by

$$b_1 = \frac{33}{5}, \qquad b_2 = 1, \qquad b_3 = -3. \tag{4.23}$$

The two-loop coefficient of the β-function b_{ij} for the SM are given by

$$
b_{ij} = \begin{pmatrix}
\frac{199}{50} & \frac{27}{10} & \frac{44}{5} \\
\frac{9}{10} & \frac{35}{6} & 12 \\
\frac{11}{10} & \frac{9}{2} & -26
\end{pmatrix}
\tag{4.24}
$$

whereas in the MSSM they are

$$
b_{ij} = \begin{pmatrix}
\frac{199}{25} & \frac{27}{5} & \frac{88}{5} \\
\frac{9}{5} & 25 & 24 \\
\frac{11}{5} & 9 & -14
\end{pmatrix}.
\tag{4.25}
$$

The RGEs in Eq. (4.18) can be solved to obtain $\alpha_i'(\mu')$ at a scale μ' for a given $\alpha_i(\mu)$,

$$\alpha_i'(\mu')^{-1} = \alpha_i(\mu)^{-1} + \beta_0 \ln(\frac{\mu'}{\mu}) + \frac{\beta_1}{\beta_0} \ln \left(\frac{\alpha_i'(\mu')^{-1} + \frac{\beta_1}{\beta_0}}{\alpha_i(\mu)^{-1} + \frac{\beta_1}{\beta_0}} \right) \tag{4.26}$$

with

$$\beta_0 = -\frac{1}{2\pi} \left(b_i + \frac{b_{ij}}{4\pi}\alpha_j(\mu) + \frac{b_{ij}}{4\pi}\alpha_k(\mu) \right) \tag{4.27}$$

and

$$\beta_1 = -2\frac{b_{ij}}{(4\pi)^2}. \tag{4.28}$$

Eq. (4.26) can then be solved iteratively to obtain the coupling constants at any arbitrary energy, knowing their values at a given energy. Using the experimental values of the coupling constants and the particle content of the SM, we find that the three coupling constants do not meet at the same point. In contrast, if we repeat this extrapolation within the MSSM and assume that the masses of all the SUSY particles are around 1 TeV, the coupling constants meet at a single point at a scale of order 10^{16} GeV, as shown in Fig. 4.4. This leads to the suggestion that, at a scale close to the Planck scale, all matter states and all force fields may unify into a single matter and a single force, respectively, leading to a SUSY based GUT.

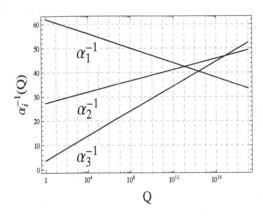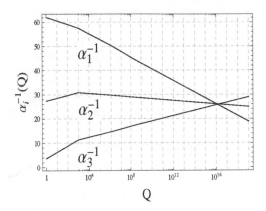

Figure 4.4 Gauge coupling evolution in the SM (left panel) and MSSM (right panel) with the energy scale Q assumed in GeV.

4.3 RADIATIVE EWSB

Let us now discuss how the EW symmetry may be broken in the MSSM. We have to study the scalar potential for the Higgs fields H_u and H_d and check if there is a minimum for which $\langle H_{u,d}^0 \rangle \neq 0$. The complete scalar potential is given by

$$V = V_F + V_D + V_{\text{soft}}, \tag{4.29}$$

where V_F, derived from the F-term, $\sum_i \left| \frac{\partial W}{\partial \phi_i} \right|^2$, is given by

$$
\begin{aligned}
V_F =\ & |Y_u \tilde{u}^c \tilde{q} + \mu H_d|^2 + |Y_d \tilde{d}^c \tilde{q} + Y_e \tilde{e}^c \tilde{\ell} + \mu H_u|^2 + |Y_u \tilde{q} H_u|^2 \\
& + |Y_d \tilde{q} H_d + Y_e \tilde{\ell} H_d|^2 + |Y_u \tilde{u}^c H_u + Y_d \tilde{d}^c H_d|^2 + |Y_e \tilde{e}^c H_d|^2,
\end{aligned} \tag{4.30}
$$

and V_D, derived from the D-term, $D^a = g^a \sum_i \phi_i^* T^a \phi_i$, is given by

$$
\begin{aligned}
V_D =\ & \frac{g_1^2}{2} \left[\frac{1}{6}|\tilde{q}|^2 - \frac{2}{3}|\tilde{u}^c|^2 + \frac{1}{3}|\tilde{d}^c|^2 - \frac{1}{2}|\tilde{\ell}|^2 + |\tilde{e}^c|^2 + \frac{1}{2}|H_u|^2 - \frac{1}{2}|H_d|^2 \right]^2 \\
& + \frac{g_3^2}{8} \left(\tilde{q}^\dagger \lambda^a \tilde{q} - \tilde{d}^{c^\dagger} \lambda^a \tilde{d}^c - \tilde{u}^{c^\dagger} \lambda^a \tilde{u}^c \right)^2 \\
& + \frac{g_2^2}{8} \left(\tilde{q}^\dagger \sigma^a \tilde{q} + \tilde{\ell}^\dagger \sigma^a \tilde{\ell} + H_u^\dagger \sigma^a H_u + H_d^\dagger \sigma^a H_d \right)^2,
\end{aligned} \tag{4.31}
$$

where g_1, g_2 and g_3 are the $U(1)_Y$, $SU(2)_L$ and $SU(3)_C$ gauge coupling constants, respectively. Here, σ^a ($a = 1, 2, 3$) are the $SU(2)_L$ generators (Pauli matrices) and λ^a ($a = 1, \ldots, 8$) are the $SU(3)_C$ Gell-Mann matrices. Finally, the scalar part of $\mathcal{L}_{\text{soft}}$ shown in Eq. (4.16) gives V_{soft}. Thus, one can write the scalar Higgs potential as

$$
\begin{aligned}
V(H_u, H_d) &= \left(|\mu|^2 + m_{H_u}^2\right)\left(|H_u^+|^2 + |H_u^0|^2\right) + \left(|\mu|^2 + m_{H_d}^2\right)\left(|H_d^-|^2 + |H_d^0|^2\right) \\
&\quad + \left[B\mu\left(H_u^+ H_d^- - H_u^0 H_d^0\right) + \text{h.c.}\right] + \frac{g_1^2 + g_2^2}{8}\left[|H_u^+|^2 + |H_u^0|^2\right. \\
&\quad \left. - |H_d^0|^2 - |H_d^-|^2\right]^2 + \frac{g_2^2}{2}\left|H_u^+ H_d^{0*} + H_u^0 H_d^{-*}\right|^2.
\end{aligned}
\tag{4.32}
$$

It is always possible, using a $SU(2)_L$ transformation, to rotate the Higgs boson fields such that their VEVs take the form

$$
\langle H_u \rangle = \begin{pmatrix} 0 \\ v_u \end{pmatrix}, \qquad \langle H_d \rangle = \begin{pmatrix} v_d \\ 0 \end{pmatrix}.
\tag{4.33}
$$

Since the electrically neutral components acquire non-vanishing VEVs, the corresponding minimum breaks the EW symmetry down to the EM one: $SU(2)_L \times U(1)_Y \to U(1)_{\text{EM}}$. Thus, the potential for the neutral Higgs fields can be written as

$$
\begin{aligned}
V(H_u^0, H_d^0) &= \left(|\mu|^2 + m_{H_u}^2\right)|H_u^0|^2 + \left(|\mu|^2 + m_{H_d}^2\right)|H_d^0|^2 - \left(B\mu H_u^0 H_d^0 + \text{h.c.}\right) \\
&\quad + \frac{g_1^2 + g_2^2}{8}\left(|H_u^0|^2 - |H_d^0|^2\right)^2.
\end{aligned}
\tag{4.34}
$$

It is clear that only the $B\mu$ term, in the above potential, can be complex. However, the phase of this term can be absorbed by a redefinition of the phases of H_u^0 and H_d^0. Therefore, all obtained VEVs are real. Consequently, CP violation is not spontaneously induced by a SUSY two-Higgs doublet scalar potential. In this respect, the scalar potential can be written as

$$
V(v_u, v_d) = m_2^2 v_u^2 + m_1^2 v_d^2 - 2m_3^2 v_u v_d + \frac{g_1^2 + g_2^2}{8}\left(v_u^2 - v_d^2\right)^2,
\tag{4.35}
$$

where

$$
m_1^2 = m_{H_d}^2 + |\mu|^2, \quad m_2^2 = m_{H_u}^2 + |\mu|^2, \quad m_3^2 = B\mu.
\tag{4.36}
$$

The potential in Eq. (4.35) is the SUSY version of the Higgs potential which induces $SU(2)_L \times U(1)_Y$ breaking in the SM, where the usual self-coupling constant is replaced by the squared gauge couplings. In order to avoid unboundness from below for this potential along the direction $v_u = v_d$, one must require

$$
m_1^2 + m_2^2 > 2m_3^2.
\tag{4.37}
$$

Now, we should closely examine the condition for EWSB. This happens when the origin $v_u = v_d = 0$ is not a local minimum. A minimum with non-vanishing VEVs may be obtained if there is a negative squared mass eigenvalue in the Higgs mass matrix

$$
M_{H^0}^2 = \begin{pmatrix} m_1^2 & -m_3^2 \\ -m_3^2 & m_2^2 \end{pmatrix}.
\tag{4.38}
$$

Therefore, $\det(M^2_{H^0}) < 0$ implies that

$$m^2_1 m^2_2 < m^4_3. \tag{4.39}$$

However, if $m^4_1 = m^4_2 < m^4_3$, then the condition in Eq. (4.37) is not satisfied and the scalar potential is unbounded from below in the direction $\langle H^0_{u,d} \rangle \to \infty$. This problem is solved by noting that the boundary conditions in Eq. (4.36) are valid only at the GUT scale. As we will show in the next section, after the running from the GUT scale down to M_W, one finds that m_1 and m_2 get renormalised differently since H_d and H_u couple with different strength to fermions.

Let us define the ratio of these VEVs as $\tan\beta = v_u/v_d$. As in the SM, the masses of the W and Z bosons can be used to fix one combination of the VEVs. As usual, the W and Z boson masses are generated after EWSB from the kinetic terms of H_u and H_d

$$(D_\mu H_u)^\dagger (D^\mu H_u) + (D_\mu H_d)^\dagger (D^\mu H_d), \tag{4.40}$$

where

$$D_\mu = \partial_\mu + ig_1 Y B_\mu + ig_2 \frac{\sigma^a}{2} W^a_\mu. \tag{4.41}$$

Then, after inserting the VEVs and defining

$$Z_\mu = \frac{-g_1 B_\mu + g_2 W^3_\mu}{\sqrt{g^2_1 + g^2_2}}, \tag{4.42}$$

one finds

$$\begin{aligned}
M^2_Z &= \frac{1}{2}(g^2_1 + g^2_2)(v^2_u + v^2_d), \\
M^2_W &= \frac{1}{2}g^2_2(v^2_u + v^2_d).
\end{aligned} \tag{4.43}$$

Therefore, one eventually gets

$$v = \sqrt{v^2_u + v^2_d} = \sqrt{\frac{2M^2_W}{g^2_2}} \simeq 246 \text{ GeV}. \tag{4.44}$$

The minimisation conditions $\partial V / \partial H^0_u = \partial V / \partial H^0_d = 0$ lead to the following equations:

$$(|\mu|^2 + m^2_{H_u})v_u = B\mu v_d + \frac{1}{4}(g^2_1 + g^2_2)(v^2_d - v^2_u)v_u, \tag{4.45}$$

$$(|\mu|^2 + m^2_{H_d})v_d = B\mu v_u - \frac{1}{4}(g^2_1 + g^2_2)(v^2_d - v^2_u)v_d, \tag{4.46}$$

which can be written as

$$|\mu|^2 + m^2_{H_u} = m^2_3 \cot\beta + \frac{M^2_Z}{2}\cos 2\beta, \tag{4.47}$$

$$|\mu|^2 + m^2_{H_d} = m^2_3 \tan\beta - \frac{M^2_Z}{2}\cos 2\beta. \tag{4.48}$$

This implies that

$$|\mu|^2 = \frac{m_{H_d}^2 - m_{H_u}^2 \tan^2 \beta}{\tan^2 \beta - 1} - \frac{M_Z^2}{2}, \tag{4.49}$$

$$\sin 2\beta = \frac{2B\mu}{m_{H_d}^2 + m_{H_u}^2 + 2|\mu|^2}. \tag{4.50}$$

These two equations are known as the EWSB conditions. Thus, two parameters (usually $|\mu|$ and B) from the free parameters of the MSSM (m_0, $m_{1/2}$, A_0, B, μ, and $\tan \beta$) can be fixed.

It is important to note that the tree-level effective potential and the corresponding tree VEVs are strongly scale dependent. The one-loop radiative corrections to the effective potential are crucial to make the potential stable against variations of the scale. The one-loop radiative correction is given by [85–87]

$$\Delta V(Q) = \frac{1}{64\pi^2} \text{STr} \left(M^4 \left[\ln \left(\frac{M^2}{Q^2} \right) - \frac{3}{2} \right] \right), \tag{4.51}$$

where M^2 is the field dependent squared mass matrix of the model and Q is the energy scale. The supertrace STr $f(M^2)$ is defined as

$$\text{STr } f(M^2) = \sum_i C_i(-1)^{2J_i}(2J_i + 1)f(m_i^2), \tag{4.52}$$

where m_i^2 is the eigenvalue of M^2 of the ith particle and its spin is J_i while C_i is its colour degree of freedom. Thus,

$$\Delta V(Q) = \frac{1}{64\pi^2} \sum_i C_i(-1)^{2J_i}(2J_i + 1)m_i^4 \left[\ln \left(\frac{m_i^2}{Q^2} \right) - \frac{3}{2} \right]. \tag{4.53}$$

Minimisation of the one-loop effective potential $V_1 = V_0 + \Delta V$ makes the one-loop VEVs much more stable against variation of the scale Q and leads to modified EWSB conditions:

$$|\mu|^2 = \frac{\bar{m}_{H_d}^2 - \bar{m}_{H_u}^2 \tan^2 \beta}{\tan^2 \beta - 1} - \frac{M_Z^2}{2}, \tag{4.54}$$

$$\sin 2\beta = \frac{2B\mu}{\bar{m}_{H_d}^2 + \bar{m}_{H_u}^2 + 2|\mu|^2}, \tag{4.55}$$

where

$$\bar{m}_{H_{u,d}}^2 = m_{H_{u,d}}^2 + \frac{\partial(\Delta V)}{\partial v_{u,d}^2}. \tag{4.56}$$

Since radiative corrections are generically expected to be small, with the exception of the contribution from the top-stop system (see below), a reasonable approximation of (4.53) leads to

$$\Delta V(Q) = \frac{3}{32\pi^2} \sum_{i=1,2} m_i^4 \left[\ln \left(\frac{m_i^2}{Q^2} \right) - \frac{3}{2} \right] - \frac{3}{16\pi^2} m_t^4 \left[\ln \left(\frac{m_t^2}{Q^2} \right) - \frac{3}{2} \right], \tag{4.57}$$

where we keep only the $\tilde{t}$, $\tilde{t}^c$ and t contributions and $m_t = Y_t v_u$. Thus, one finds

$$\frac{\partial(\Delta V)}{\partial v_d^2} = \frac{3}{16\pi^2} \sum_{i=1,2} m_i^2 \frac{\partial m_i^2}{\partial v_d^2} \left[\ln\left(\frac{m_i^2}{Q^2}\right) - 1 \right] \tag{4.58}$$

and

$$\frac{\partial(\Delta V)}{\partial v_u^2} = \frac{3}{16\pi^2} \sum_{i=1,2} m_i^2 \frac{\partial m_i^2}{\partial v_u^2} \left[\ln\left(\frac{m_i^2}{Q^2}\right) - 1 \right] - \frac{3}{8\pi^2} \frac{m_t^4}{v^2 \sin^2\beta} \left[\ln\left(\frac{m_t}{Q^2}\right) - 1 \right]. \tag{4.59}$$

Furthermore, one must impose constraints on the parameters to avoid charge and colour breaking minima in the scalar potential. Necessary conditions were deduced in the literature [88–92]. In the minimal scheme with universal soft breaking dynamics, these constraints are given by

$$|A_t|^2 < 3(m_{\tilde{t}_L}^2 + m_{\tilde{t}_L^c}^2 + m_2^2), \tag{4.60}$$

$$|A_b|^2 < 3(m_{\tilde{b}_L}^2 + m_{\tilde{b}_L^c}^2 + m_1^2), \tag{4.61}$$

$$|A_\tau|^2 < 3(m_{\tilde{\tau}_L}^2 + m_{\tilde{\tau}_L^c}^2 + m_1^2), \tag{4.62}$$

where all the quantities are evaluated at the EW scale. In general, no necessary and sufficient condition can be derived analytically and the absence of stable colour and charge breaking vacua has to be checked numerically point by point in the parameter space once the stability condition (4.37) and the breaking condition (4.39) are satisfied.

4.4   RGE ANALYSIS

In this section we analyse the running of the parameters which are involved in the $SU(2)_L \times U(1)_Y$ radiative symmetry breaking, in particular, the running of the Higgs masses [93]. We will assume that all Yukawa couplings can be neglected with respect to the top Yukawa coupling Y_t. In this approximation, the relevant RGEs are given by:

$$\frac{d\tilde{Y}_t}{dt} = \tilde{Y}_t \left(\frac{16}{3}\tilde{\alpha}_3 + 3\tilde{\alpha}_2 + \frac{13}{9}\tilde{\alpha}_1 \right) - 6\tilde{Y}_t^2, \tag{4.63}$$

$$\frac{dA_t}{dt} = \frac{16}{3}\tilde{\alpha}_3 M_3 + 3\tilde{\alpha}_2 M_2 + \frac{13}{9}\tilde{\alpha}_1 M_1 - 6\tilde{Y}_t A_t, \tag{4.64}$$

$$\frac{dm_{H_d}^2}{dt} = 3\tilde{\alpha}_2 M_2^2 + \tilde{\alpha}_1 M_1^2 + \frac{1}{2}\tilde{\alpha}_1 S, \tag{4.65}$$

$$\frac{dm_{H_u}^2}{dt} = 3\tilde{\alpha}_2 M_2^2 + \tilde{\alpha}_1 M_1^2 - \frac{1}{2}\tilde{\alpha}_1 S - 3\tilde{Y}_t \left(m_{\tilde{q}}^2 + m_{\tilde{u}}^2 + A_t^2 + m_{H_u}^2 \right), \tag{4.66}$$

$$\frac{dm_{\tilde{q}}^2}{dt} = \frac{16}{3}\tilde{\alpha}_3 M_3^2 + 3\tilde{\alpha}_2 M_2^2 + \frac{1}{9}\tilde{\alpha}_1 M_1^2 - \frac{1}{6}\tilde{\alpha}_1 S$$
$$- 3\tilde{Y}_t \left(m_{\tilde{q}}^2 + m_{\tilde{u}}^2 + A_t^2 + m_{H_u}^2 \right), \tag{4.67}$$

$$\frac{dm_{\tilde{u}}^2}{dt} = \frac{16}{3}\tilde{\alpha}_3 M_3^2 + \frac{16}{9}\tilde{\alpha}_1 M_1^2 + \frac{2}{3}\tilde{\alpha}_1 S - 2\tilde{Y}_t \left(m_{\tilde{q}}^2 + m_{\tilde{u}}^2 + A_t^2 + m_{H_u}^2 \right), \tag{4.68}$$

where the scale variable t is defined as $t = \ln M_X^2/Q^2$, with M_X the GUT scale and Q the renormalisation scale. Here we consider the limit Y_b, $Y_\tau \ll Y_t$, which simplifies the RGEs significantly, so that one may derive analytic expressions for the evolution of the coupling and mass parameters. The complete set of RGEs, which is relevant for analysing flavour violation in the MSSM, is given in Appendix B. Note that $\tilde{Y}_t$ is defined as $\tilde{Y}_t^2 = Y_t^2/(4\pi)^2$ and $\tilde{\alpha}_i = \alpha_i/4\pi = g_i^2/(4\pi)^2$. The analytic expression of $\tilde{Y}_t(t)$ is

$$\tilde{Y}_t(t) = \frac{\tilde{Y}_t(0)E(t)}{1 + 6\tilde{Y}_t(0)F(t)}, \tag{4.69}$$

where

$$E(t) = (1 + \beta_3 t)^{\frac{16}{3b_3}} (1 + \beta_2 t)^{\frac{3}{b_2}} (1 + \beta_1 t)^{\frac{13}{9b_1}} \tag{4.70}$$

and

$$F(t) = \int_0^t E(t')dt'. \tag{4.71}$$

The β_a functions are defined as $\beta_a = b_a \alpha_a(0)/4\pi$, where the b_a's for the SM gauge group are given by $b_a = 11, 1, -3$ for $a = 1, 2, 3$, respectively. The function $S(t)$ in Eqs. (4.65) and (4.66) is defined as

$$S(t) = \sum_{\text{gen}} \left(m_{\tilde{q}}^2 - 2m_{\tilde{u}^c}^2 + m_{\tilde{d}^c}^2 - m_{\tilde{\ell}}^2 + m_{\tilde{e}^c}^2 \right) - m_{H_d}^2 + m_{H_u}^2, \tag{4.72}$$

with

$$S(t) = \frac{S(0)}{1 + \beta_1 t}. \tag{4.73}$$

Therefore, in the case of a universal soft scalar mass, one has $S(t) = 0$. From Eq. (4.64), one can easily check that $A_t(t)$ is given by

$$A_t(t) = \frac{A_0}{1 + 6\tilde{Y}_t(0)F(t)} + \left(H_2(t) - \frac{6\tilde{Y}_t(0)H_3(t)}{1 + 6\tilde{Y}_t(0)F(t)} \right), \tag{4.74}$$

where the functions $H_2(t)$ and $H_3(t)$ are given by

$$H_2(t) = \tilde{\alpha}_G\, m_{1/2} \left(\frac{16}{3} \frac{t}{1 + \beta_3 t} + \frac{3\,t}{1 + \beta_2 t} + \frac{13}{9} \frac{t}{1 + \beta_1 t} \right), \tag{4.75}$$

$$H_3(t) = \int_0^t dt'\, E(t')H_2(t'), \tag{4.76}$$

where $M_i(0) = m_{1/2}$ and $\tilde{\alpha}_i(0) = \tilde{\alpha}_G$ with $\alpha_G = 1/25$ are assumed. From Eq. (4.65), one finds that $m_{H_d}^2$ at the scale t is given by

$$m_{H_d}^2(t) = m_0^2 + \frac{1}{2}\tilde{\alpha}_G\, m_{1/2}^2 \left(3f_2(t) + f_1(t) \right), \tag{4.77}$$

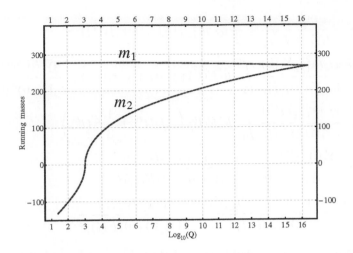

Figure 4.5 Downward evolution of $m^2_{H_{u,d}}$ from the GUT to the EW scale.

where $f_i(t)$ is given by

$$f_i(t) = \frac{1}{\beta_i} \left(1 - \frac{1}{(1 + \beta_i t)^2} \right).$$ (4.78)

In order to find the evolution of $m^2_{H_u}$, we have to solve the RGE (4.66) simultaneously with the REGs of $m^2_{\tilde{q}}$ and $m^2_{\tilde{u}}$, Eqs.(4.67) and (4.68), respectively. We can show that the linear combination of these masses

$$m^2_4 = m^2_{H_u} - m^2_{\tilde{q}} - m^2_{\tilde{u}}$$ (4.79)

has a vanishing Yukawa contribution to its anomalous dimensions. The solution of the $m^2_4(t)$ evolution equation is

$$m^2_4(t) = -m^2_0 - \frac{4}{3}\tilde{\alpha}_G \, m^2_{1/2} \left(4f_3(t) + \frac{1}{3}f_1(t) \right).$$ (4.80)

Thus, one can write the RGE in Eq. (4.66) as

$$\frac{dm^2_{H_u}}{dt} + 6Y_t m^2_{H_u} = 3\tilde{\alpha}_2 M^2_2 + \tilde{\alpha}_1 M^2_1 - \frac{1}{2}\tilde{\alpha}_1 S(t) - 3\tilde{Y}_t(t)(A^2_t - m^2_4),$$ (4.81)

which leads to the following solution:

$$
\begin{aligned}
m_{H_u}^2(t) = {} & \frac{m_0^2}{1 + 6\tilde{Y}_t(0)F(t)} + \frac{1}{2}\tilde{\alpha}_G \, m_{1/2}^2 \left(3f_2(t) + f_1(t)\right) \\
& - \frac{3\tilde{\alpha}_G \, m_{1/2}^2 \tilde{Y}_t(0)}{1 + 6\tilde{Y}_t(0)F(t)} \int_0^t dt' E(t') \left(3f_2(t') + f_1(t')\right) + \frac{3\tilde{Y}_t(0)F(t)m_4^2(0)}{1 + 6\tilde{Y}_t(0)F(t)} \\
& - \frac{4\tilde{\alpha}_G \, m_{1/2}^2 \tilde{Y}_t(0)}{1 + 6\tilde{Y}_t(0)F(t)} \int_0^t dt' E(t') \left(4f_3(t') + \frac{1}{3}f_1(t')\right) \\
& - \frac{3\tilde{Y}_t(0)A_0^2 F(t)}{(1 + 6\tilde{Y}_t(0)F(t))^2} - \frac{6\tilde{Y}_t(0)A_0 H_3(t)}{(1 + 6\tilde{Y}_t(0)F(t))^2} \\
& - \frac{3\tilde{Y}_t(0)\int_0^t dt' E(t')H_2^2(t')}{1 + 6\tilde{Y}_t(0)F(t)} + \frac{18\tilde{Y}_t^2(0)H_3^2}{(1 + 6\tilde{Y}_t(0)F(t))^2}.
\end{aligned}
\tag{4.82}
$$

From these equations, it is remarkable to find that H_u couples to a t (s)quark with a large Yukawa coupling, unlike H_d, which couples to a b (s)quark and a τ (s)lepton. The Yukawa coupling gives a negative contribution to the squared masses $m_{H_{u,d}}^2$. Therefore, the running from M_X down to the EW scale, as shown in Fig. 4.5, reduces the squared Higgs masses until, eventually, condition (4.50) is satisfied and the gauge symmetry is broken. In fact, this is an appealing feature in SUSY models that generally explains the mechanism of EWSB dynamically. In the SM with a single Higgs doublet, H, the assumption that its mass squared is negative is an ad hoc one. In constrast, from the previous discussion, we have shown that a Higgs boson mass in the MSSM becomes negative around the EW scale due to the effect of the large top Yukawa coupling.

CHAPTER 5

MSSM Mass Spectrum and Interactions

In this chapter we present the MSSM particle mass spectrum and interactions. The determination of sparticle masses and couplings requires the evolution of the relevant soft SUSY breaking terms from the GUT scale down to the EW scale. In addition, one should consider that EWSB may lead to mixing among different particles, hence different mass and interaction eigenstates can be obtained. It is worth mentioning that, in this chapter, we are considering general (non-universal) soft SUSY breaking terms, unless specified otherwise, to analyse the most general SUSY spectrum.

5.1 SUPERSYMMETRIC PARTICLES IN THE MSSM

We start with determining the masses of the new supersymmetric particle states [48]. In the MSSM, the SUSY physical spectrum consists of gluinos, charginos and neutralinos (fermionic sparticles), in addition to squarks and sleptons (bosonic sparticles).

5.1.1 Gluinos

The fermionic partner of the gluon, gluino $\tilde{g}$, acquires mass due to the soft SUSY breaking gaugino mass term

$$\mathcal{L}_{\text{soft}} \supset -\frac{1}{2}M_3\tilde{g}^a\tilde{g}^a. \tag{5.1}$$

Therefore, the tree-level gluino mass is given by $m_{\tilde{g}} = |M_3(Q)|$. Since the gluino is a colour octet fermion, it cannot mix with any other MSSM particle.

The values of the low energy gaugino masses, at one-loop level, are given by

$$M_i(Q) = \frac{\alpha_i(Q)}{\alpha_i(M_X)}M_i(M_X), \quad i = 1, 2, 3, \tag{5.2}$$

where $M_i(M_X)$ are the gaugino masses at the GUT scale and the couplings $\alpha_i(Q)$

are defined in Chapter 4. In the case of mSUGRA, the gaugino masses, at the GUT scale, are universal, *i.e.*,

$$M_1(M_X) = M_2(M_X) = M_3(M_X) = m_{1/2}. \tag{5.3}$$

Therefore, at $Q = M_Z$, one finds

$$M_1 \simeq 0.41\, m_{1/2}, \qquad M_2 \simeq 0.84\, m_{1/2}, \qquad M_3 \simeq 2.8\, m_{1/2}, \tag{5.4}$$

i.e.,

$$M_3 : M_2 : M_1 \simeq 7 : 2 : 1. \tag{5.5}$$

It is remarkable that the running gluino mass M_3 can be significantly different from the physical gluino mass, which can be expressed as [94]

$$m_{\tilde{g}} = M_3(Q) \left[1 + \frac{\alpha_3(Q)}{4\pi} \left(15 - 18 \ln \frac{M_3(Q)}{Q} + \sum_q B_1(M_3, m_{\tilde{q}}, m_q) \right) \right], \tag{5.6}$$

where the loop function $B_1 \sim \ln(M_1(Q)/Q)$. Therefore, the difference between $m_{\tilde{g}}$ and $M_3(M_Z)$ can be, *e.g.*, of order 30% for $m_{\tilde{q}} \simeq 1$ TeV.

5.1.2 Charginos

The mass spectrum for the charged gauginos is more complicated, because once $SU(2)_L \times U(1)_Y$ is broken, the charged winos, $\tilde{W}^\pm$, and the charged Higgsinos, $\tilde{H}_u^+$ and $\tilde{H}_d^-$, mix amongst themselves to form two mass eigenstates with charge ± 1 called charginos. In this way, the following mass matrix is obtained

$$\mathcal{L} \supset -\left(\tilde{W}^- \ \tilde{H}_d^- \right) \mathcal{M}_C \begin{pmatrix} \tilde{W}^+ \\ \tilde{H}_u^+ \end{pmatrix} + \text{h.c.}, \tag{5.7}$$

with

$$\mathcal{M}_C = \begin{pmatrix} M_2 & \sqrt{2} M_W \sin\beta \\ \sqrt{2} M_W \cos\beta & \mu \end{pmatrix}, \tag{5.8}$$

where the mass M_2 is the soft SUSY breaking mass of the gaugino partner of the $W^\pm$ gauge boson, the wino $\tilde{W}^\pm$. Since the chargino mass matrix $\mathcal{M}_C$ is non-symmetric, it can be diagonalised by two unitary matrices U and V, such that

$$\mathcal{M}_C^{\text{diag}} = U \mathcal{M}_C V^{-1}, \tag{5.9}$$

where

$$U = \mathcal{O}_-, \quad V = \begin{cases} \mathcal{O}_+ & \text{if } \det \mathcal{M}_C > 0 \\ \sigma_3 \mathcal{O}_+ & \text{if } \det \mathcal{M}_C < 0 \end{cases}, \tag{5.10}$$

where σ_3 is the diagonal Pauli matrix introduced to make the chargino masses positive and the rotation matrices $\mathcal{O}_\pm$ are given by

$$\mathcal{O}_\pm = \begin{pmatrix} \cos\phi_\pm & -\sin\phi_\pm \\ \sin\phi_\pm & \cos\phi_\pm \end{pmatrix}, \tag{5.11}$$

where

$$\tan 2\phi_- = 2\sqrt{2}M_W \frac{-\mu\sin\beta + M_2\cos\beta}{M_2^2 - \mu^2 - 2M_W^2\cos 2\beta} \qquad (5.12)$$

and

$$\tan 2\phi_+ = 2\sqrt{2}M_W \frac{-\mu\cos\beta + M_2\sin\beta}{M_2^2 - \mu^2 - 2M_W^2\cos 2\beta}. \qquad (5.13)$$

The matrix $\mathcal{M}_C$ has two eginstates, $\tilde{\chi}_1^\pm$ and $\tilde{\chi}_2^\pm$ (the charginos), with the following mass eigenvalues

$$M_{\tilde{\chi}_{1,2}^\pm}^2 = \frac{1}{2}\left[M_2^2 + \mu^2 + 2M_W^2 \right.$$

$$\left. \mp \sqrt{(M_2^2 - \mu^2)^2 + 4M_W^2(M_W^2\cos^2 2\beta + M_2^2 + \mu^2 + 2M_2\mu\sin 2\beta)} \right].$$

$$(5.14)$$

The lightest chargino mass $m_{\tilde{\chi}_1^\pm}$ is often of order M_Z and can be the lightest charged SUSY particle.

5.1.3 Neutralinos

The neutralinos χ_i ($i = 1, 2, 3, 4$) are the physical (mass) superpositions of two fermionic partners of the two neutral gauge bosons, called gauginos $\tilde{B}$ (bino) and $\tilde{W}^3$ (wino), and of the two neutral Higgs bosons, called Higgsinos $\tilde{H}_d^0$, and $\tilde{H}_u^0$. The neutralino mass matrix is given by

$$\mathcal{L} \supset -\left(\tilde{B}\ \tilde{W}^3\ \tilde{H}_d^0\ \tilde{H}_u^0\right) \mathcal{M}_N \begin{pmatrix} \tilde{B} \\ \tilde{W}^3 \\ \tilde{H}_d^0 \\ \tilde{H}_u^0 \end{pmatrix} + \text{h.c.}, \qquad (5.15)$$

with

$$\mathcal{M}_N = \begin{pmatrix} M_1 & 0 & -M_Z s_W c_\beta & M_Z s_W s_\beta \\ 0 & M_2 & M_Z c_W c_\beta & -M_Z c_W s_\beta \\ -M_Z s_W c_\beta & M_Z c_W c_\beta & 0 & -\mu \\ M_Z s_W s_\beta & -M_Z c_W s_\beta & -\mu & 0 \end{pmatrix}, \qquad (5.16)$$

where $c_W \equiv \cos\theta_W$, $s_W \equiv \sin\theta_W$, $c_\beta \equiv \cos\beta$ and $s_\beta \equiv \sin\beta$ while M_1 and M_2 are the $U(1)_Y$ and $SU(2)_L$ soft SUSY breaking gaugino masses, respectively. Note that the off-diagonal elements are generated from the Higgs-Higgsino-gaugino mixing terms. The Hermitian matrix $\mathcal{M}_N$ can be diagonalised by a unitary transformation of the neutralino fields, so that

$$N^\dagger \mathcal{M}_N N = \mathcal{M}_N^{\text{diag}}. \qquad (5.17)$$

The lightest eigenvalue of this matrix and the corresponding eigenstate, say χ, have good chance of being the LSP. The LSP will be a linear combination of the original fields

$$\chi = N_{11}\tilde{B} + N_{12}\tilde{W}^3 + N_{13}\tilde{H}_d^0 + N_{14}\tilde{H}_u^0. \tag{5.18}$$

The phenomenology of the neutralino is governed primarily by its mass and composition. A useful parameter for describing the neutralino composition is the gaugino 'purity' function $f_g = |N_{11}|^2 + |N_{12}|^2$. If $f_g > 0.5$, then the neutralino is primarily gaugino, whereas if $f_g < 0.5$, then the neutralino is primarily Higgsino. Actually if $|\mu| > |M_2| \geq M_Z$, the two lightest neutralino states will be determined by the gaugino components. Similarly, the lightest chargino will be mostly a charged wino, while, if $|\mu| < |M_2|$, the two lighter neutralinos and the lighter chargino are all mostly Higgsinos, with mass close to $|\mu|$. Finally, if $|\mu| \simeq |M_2|$, the states will be strongly mixed.

In the limit of large $|\mu|$, *i.e.*, $|\mu| \gg M_{1,2} \gg M_Z$, the mass expressions of the physical neutralino states can be approximated as follows [95]:

$$m_{\chi_1} \simeq M_1 - \frac{M_Z^2}{\mu^2}(M_1 + \mu s_{2\beta})s_W^2, \tag{5.19}$$

$$m_{\chi_2} \simeq M_2 - \frac{M_Z^2}{\mu^2}(M_2 + \mu s_{2\beta})c_W^2, \tag{5.20}$$

$$m_{\chi_{3,4}} \simeq |\mu| + \frac{1}{2}\text{sign}(\mu)\frac{M_Z^2}{\mu^2}(1 \mp s_{2\beta})\left(|\mu| \pm M_2 s_W^2 \mp M_1 c_W^2\right). \tag{5.21}$$

5.1.4 Squarks

The scalar partners of the quarks get their masses from the following contributions.

1. A contribution comes from the scalar potential once the doublets $H_{u,d}$ get VEVs, namely, one gets the following contribution from V_F:

$$V_F \supset |Y_q\tilde{q}^c\tilde{q} + \mu H_q|^2 + |Y_q\tilde{q}H_q|^2$$
$$\supset Y_q v_q \mu \tilde{q}^c\tilde{q}. \tag{5.22}$$

In addition, V_D leads to

$$V_D \supset \frac{1}{4}g_2^2\left(\tilde{u}_L^*\tilde{u}_L + H_u^{0*}H_u^0 - H_d^{0*}H_d^0 + \dots\right)^2$$
$$+ \frac{1}{2}g_1^2\left(\frac{1}{6}\tilde{u}_L^*\tilde{u}_L - \frac{2}{3}\tilde{u}_L^{c*}\tilde{u}_L^c - \frac{1}{2}H_u^{0*}H_u^0 + \frac{1}{2}H_d^{0*}H_d^0 + \dots\right)^2$$
$$\supset \frac{1}{2}(v_u^2 - v_d^2)\left(\frac{1}{2}g_2^2 - \frac{1}{6}g_1^2\right)\tilde{u}_L^*\tilde{u}_L + \frac{1}{3}g_1^2(v_u^2 - v_d^2)\tilde{u}_L^{c*}\tilde{u}_L^c. \tag{5.23}$$

2. The soft SUSY breaking scalar mass terms m_{ij}^2 are the most important contributions. Note that these terms are in general non-universal. Even if they are assumed to be universal at the GUT scale, due to their different evolution down to low energy, non-universality will be produced.

3. A contribution due to the soft breaking trilinear terms, after the Higgs doublets acquire VEVs, also emerges. In most SUSY breaking scenarios, this term is proportional to the product of the trilinear coupling A and the mass of the fermionic partner. Therefore, it is negligible for the first two generations. This mass contribution mixes left-handed with right-handed scalars. Hence, one finds that the soft SUSY breaking contribution to squarks is given by

$$M_{\text{soft}}^2 = \begin{pmatrix} m_{\tilde{q}_L}^2 & v_q Y_q A_q^* \\ v_q Y_q A_q & m_{\tilde{q}_L^c}^2 \end{pmatrix}. \tag{5.24}$$

4. The supersymmetric contribution, which is equal to the mass squared of the respective fermionic partners, is essentially negligible except for the quarks of the third generation. This contribution is obtained from the V_F term $|Y_q \tilde{q} H_q|^2 \supset m_q^2 \tilde{q}^* \tilde{q}$.

Thus, the mass term Lagrangian for the up and down squarks, in the gauge eigenstate basis, can be written as

$$\mathcal{L} \supset -(\tilde{q}_L^\dagger \ \tilde{q}_L^{c\,\dagger}) \, \mathcal{M}_{\tilde{q}}^2 \begin{pmatrix} \tilde{q}_L \\ \tilde{q}_L^c \end{pmatrix}. \tag{5.25}$$

Here we drop generation indices: q denotes u or d and $\mathcal{M}_{\tilde{q}}^2$ is given by

$$\mathcal{M}_{\tilde{q}}^2 = \begin{pmatrix} M_{LL}^2 & M_{LR}^2 \\ M_{RL}^2 & M_{RR}^2 \end{pmatrix}. \tag{5.26}$$

For $q = u$ we have

$$M_{LL}^2 = v_u^2 Y_u^* Y_u^T + m_{\tilde{q}_L}^2 + \frac{1}{6}(4M_W^2 - M_Z^2)\cos 2\beta \mathbf{1}, \tag{5.27}$$

$$M_{RR}^2 = v_u^2 Y_u^* Y_u^T + m_{\tilde{u}_L^c}^2 + \frac{2}{3}M_Z^2 \cos 2\beta \sin^2\theta_W \mathbf{1}, \tag{5.28}$$

$$M_{LR}^2 = v_u A_u^* - \mu v_d Y_u^*, \tag{5.29}$$

$$(M_{RL}^2)_u = (M_{LR}^2)_u^\dagger, \tag{5.30}$$

while for $q = d$ we have

$$M_{LL}^2 = v_d^2 Y_d^* Y_d^T + m_{\tilde{q}_L}^2 - \frac{1}{6}(2M_W^2 + M_Z^2)\cos 2\beta \mathbf{1}, \tag{5.31}$$

$$M_{RR}^2 = v_d^2 Y_d^* Y_d^T + m_{\tilde{d}_L^c}^2 - \frac{1}{3}M_Z^2 \cos 2\beta \sin^2\theta_W \mathbf{1}, \tag{5.32}$$

$$M_{LR}^2 = v_d A_d^* - \mu v_u Y_d^*, \tag{5.33}$$

$$(M_{RL}^2)_d = (M_{LR}^2)_d^\dagger, \tag{5.34}$$

where $\mathbf{1}$ stands for the 3×3 unit matrix.

It is necessary to work in the mass eigenstate basis of quarks and squarks. For the quarks, mass eigenstates are related to the gauge eigenstates through 3×3 unitary matrices $V_{L,R}^{u,d}$ as follows:

$$
\begin{aligned}
u' &= V_L^u u + V_R^u C \bar{u}^{c\,T}, \\
d' &= V_L^d d + V_R^d C \bar{d}^{c\,T}.
\end{aligned}
\tag{5.35}
$$

Thus, their diagonal 3×3 mass matrices are related to the Yukawa couplings via

$$
m_u = v \sin\beta V_R^u Y_u^T V_L^{u\,\dagger}, \qquad m_d = v \cos\beta V_R^d Y_d^T V_L^{d\,\dagger}.
\tag{5.36}
$$

The CKM matrix element (V_{CKM}) is defined as in the SM and so we have

$$
V_{\mathrm{CKM}} = V_L^u V_L^{d\,\dagger}.
\tag{5.37}
$$

In the same way, the squarks are rotated parallel to their corresponding quarks forming the so-called 'Super-CKM' basis:

$$
\tilde{U} = \begin{pmatrix} V_L^u \tilde{u}_L \\ V_R^u \tilde{u}_L^{c\,*} \end{pmatrix}, \qquad
\tilde{D} = \begin{pmatrix} V_L^d \tilde{d}_L \\ V_R^d \tilde{d}_L^{c\,*} \end{pmatrix}.
\tag{5.38}
$$

Hence, the mass matrix given in (5.26) takes the following form:

$$
\mathcal{M}_{\tilde{U}}^2 = \begin{pmatrix} (M_{\tilde{u}}^2)_{LL} + m_u^2 - \frac{\cos 2\beta}{6}(M_Z^2 - 4M_W^2) & (M_{\tilde{u}}^2)_{LR} - \cot\beta\mu m_u \\ (M_{\tilde{u}}^2)_{LR}^\dagger - \cot\beta\mu^* m_u & (M_{\tilde{u}}^2)_{RR} + m_u^2 + \frac{2\cos 2\beta}{3}M_Z^2 \sin^2\theta_W \end{pmatrix},
\tag{5.39}
$$

$$
\mathcal{M}_{\tilde{D}}^2 = \begin{pmatrix} (M_{\tilde{d}}^2)_{LL} + m_d^2 - \frac{\cos 2\beta}{6}(M_Z^2 + 2M_W^2) & (M_{\tilde{d}}^2)_{LR} - \tan\beta\mu m_d \\ (M_{\tilde{d}}^2)_{LR}^\dagger - \tan\beta\mu^* m_d & (M_{\tilde{d}}^2)_{RR} + m_d^2 - \frac{\cos 2\beta}{3}M_Z^2 \sin^2\theta_W \end{pmatrix},
\tag{5.40}
$$

where

$$
(M_{\tilde{u}}^2)_{LL} = V_L^u m_{\tilde{q}_L}^2 V_L^{u\dagger}, \quad (M_{\tilde{u}}^2)_{RR} = V_R^u m_{\tilde{u}_L^c}^{2\,T} V_R^{u\dagger}, \quad (M_{\tilde{u}}^2)_{LR} = v \sin\beta V_L^u A_u^* V_R^{u\dagger},
\tag{5.41}
$$

$$
(M_{\tilde{d}}^2)_{LL} = V_L^d m_{\tilde{q}_L}^2 V_L^{d\dagger}, \quad (M_{\tilde{d}}^2)_{RR} = V_R^d m_{\tilde{d}_L^c}^{2\,T} V_R^{d\dagger}, \quad (M_{\tilde{d}}^2)_{LR} = v \cos\beta V_L^d A_d^* V_R^{d\dagger}.
\tag{5.42}
$$

The squark mass matrices $\mathcal{M}_{\tilde{U}}^2$ and $\mathcal{M}_{\tilde{D}}^2$ can be diagonalised to real matrices by two unitary matrices Γ^u and Γ^d, respectively,

$$
\begin{aligned}
M_{\tilde{u}}^2 &= \Gamma^U \mathcal{M}_{\tilde{U}}^2 \Gamma^{U\dagger}, \tag{5.43} \\
M_{\tilde{d}}^2 &= \Gamma^D \mathcal{M}_{\tilde{D}}^2 \Gamma^{D\dagger}, \tag{5.44}
\end{aligned}
$$

where

$$M_{\tilde{u}}^2 = \mathrm{diag}(m_{\tilde{u}_1}^2, m_{\tilde{u}_2}^2, m_{\tilde{u}_3}^2, m_{\tilde{u}_4}^2, m_{\tilde{u}_5}^2, m_{\tilde{u}_6}^2), \tag{5.45}$$

$$M_{\tilde{d}}^2 = \mathrm{diag}(m_{\tilde{d}_1}^2, m_{\tilde{d}_2}^2, m_{\tilde{d}_3}^2, m_{\tilde{d}_4}^2, m_{\tilde{d}_5}^2, m_{\tilde{d}_6}^2). \tag{5.46}$$

The squark mass eigenstates $\tilde{u}$ and $\tilde{d}$ are related to $\tilde{U}$ and $\tilde{D}$, respectively, via

$$\tilde{u} = \Gamma^U \tilde{U}, \quad \tilde{d} = \Gamma^D \tilde{D}. \tag{5.47}$$

In mSUGRA, where the soft SUSY breaking terms are universal, the mass matrix of the first two generations is almost diagonal and the stop mass matrix is given by

$$\mathcal{M}_{\tilde{t}}^2 = \begin{pmatrix} m_{\tilde{q}_L}^2 + m_t^2 - \frac{\cos 2\beta}{6}(M_Z^2 - 4M_W^2) & m_t(A_t - \mu \cot \beta) \\ m_t(A_t - \mu \cot \beta) & m_{\tilde{t}_L^c}^2 + m_t^2 + \frac{2}{3}M_Z^2 \sin^2 \theta_W \cos 2\beta \end{pmatrix}. \tag{5.48}$$

Therefore, the following matrix can be used to diagonalise $\mathcal{M}_{\tilde{t}}^2$:

$$\begin{pmatrix} \tilde{t}_1 \\ \tilde{t}_2 \end{pmatrix} = \begin{pmatrix} \cos \theta_{\tilde{t}} & \sin \theta_{\tilde{t}} \\ -\sin \theta_{\tilde{t}} & \cos \theta_{\tilde{t}} \end{pmatrix} \begin{pmatrix} \tilde{t}_L \\ \tilde{t}_L^c \end{pmatrix}. \tag{5.49}$$

Here, the eigenvalues are denoted by $m_{\tilde{t}_1}^2$ and $m_{\tilde{t}_2}^2$, with $m_{\tilde{t}_1}^2 < m_{\tilde{t}_2}^2$. The mixing angle $\theta_{\tilde{t}}$ is given by

$$\sin 2\theta_{\tilde{t}} = \frac{2m_t(A_t - \mu \cot \beta)}{m_{\tilde{t}_2}^2 - m_{\tilde{t}_1}^2}. \tag{5.50}$$

The off-diagonal elements in Eq. (5.48) are proportional to the large top quark mass and will result in the mass of the lighter stop $m_{\tilde{t}_1}$ being smaller than the mass of any other squark, which is phenomenologically interesting.

For the $\tilde{b}$ sector, the squared-mass matrix takes the form

$$\mathcal{M}_{\tilde{b}}^2 = \begin{pmatrix} m_{\tilde{q}_L}^2 + m_b^2 - \frac{\cos 2\beta}{6}(M_Z^2 + 2M_W^2) & m_b(A_b - \mu \tan \beta) \\ m_b(A_b - \mu \tan \beta) & m_{\tilde{b}_L^c}^2 + m_b^2 - \frac{1}{3}m_Z^2 \sin^2 \theta_W \cos 2\beta \end{pmatrix}. \tag{5.51}$$

Clearly, the $\tilde{b}_L$-$\tilde{b}_L^c$ mixing depends on $\tan \beta$. For $\tan \beta \gg 1$, this mixing could be large and phenomenologically relevant.

5.1.5 Sleptons

In analogy to the discussion on the squarks, the mass term Lagrangian for the sleptons in the gauge eigenstate basis is given by

$$\mathcal{L} \supset -(\tilde{\ell}_L^\dagger \ \tilde{\ell}_L^{c\,\dagger}) \, \mathcal{M}_{\tilde{\ell}}^2 \begin{pmatrix} \tilde{\ell}_L \\ \tilde{\ell}_L^c \end{pmatrix}, \tag{5.52}$$

where we drop generation indices and $\mathcal{M}_{\tilde{\ell}}^2$ is given by

$$\mathcal{M}_{\tilde{\ell}}^2 = \begin{pmatrix} M_{LL}^2 & M_{LR}^2 \\ M_{RL}^2 & M_{RR}^2 \end{pmatrix}, \tag{5.53}$$

with

$$M_{LL}^2 = v_d^2 Y_\ell^* Y_\ell^T + m_{\tilde{\ell}_L}^2 + \frac{1}{2}(M_Z^2 - 2M_W^2)\cos 2\beta \mathbf{1}, \qquad (5.54)$$

$$M_{RR}^2 = v_d^2 Y_\ell^* Y_\ell^T + m_{\tilde{\ell}_L^c}^2 - M_Z^2 \cos 2\beta \sin^2\theta_W \mathbf{1}, \qquad (5.55)$$

$$M_{LR}^2 = v_d A_\ell^* - \mu v_u Y_\ell^*, \qquad (5.56)$$

$$M_{RL}^2 = M_{LR}^{2\dagger}. \qquad (5.57)$$

It is useful to work in the mass eigenstate basis for the leptons and sleptons. The mass eigenstates of the leptons are related to the gauge eigenstates through 3×3 unitary matrices $V_{L,R}^\ell$ as follows:

$$\nu' = V_L^\ell \nu, \qquad\qquad \ell' = V_L^\ell \ell + V_R^\ell C \bar{\ell}^{cT}. \qquad (5.58)$$

Therefore, the 3×3 diagonal lepton mass matrix can be related to the Yukawa couplings through

$$m_\ell = v\cos\beta V_R^\ell Y_\ell^T V_L^{\ell\,\dagger}. \qquad (5.59)$$

In the same way, we rotate the sleptons parallel to their corresponding SM partners,

$$\tilde{N} = V_L^\ell \tilde{\nu}, \qquad \tilde{L} = \begin{pmatrix} V_L^\ell \tilde{\ell}_L \\ V_R^\ell \tilde{\ell}_L^{c\,*} \end{pmatrix}. \qquad (5.60)$$

In this case, the slepton mass matrix is defined in Eq. (5.57) and the sneutrino mass matrix takes the form

$$\mathcal{M}_{\tilde{L}}^2 = \begin{pmatrix} (M_{\tilde{L}}^2)_{LL} + m_\ell^2 + \frac{\cos 2\beta}{2}(M_Z^2 - 2M_W^2) & (M_{\tilde{L}}^2)_{LR} - \tan\beta\mu m_\ell \\ (M_{\tilde{L}}^2)_{LR}^\dagger - \tan\beta\mu^* m_\ell & (M_{\tilde{L}}^2)_{RR} + m_\ell^2 - \cos 2\beta M_Z^2 \sin^2\theta_W \end{pmatrix},$$

$$\mathcal{M}_{\tilde{N}}^2 = V_L^\ell m_{\tilde{\ell}_L}^2 V_L^{\ell\dagger} + \frac{\cos 2\beta}{2} M_Z^2, \qquad (5.61)$$

where

$$(M_{\tilde{L}}^2)_{LL} = V_L^\ell m_{\tilde{\ell}_L}^2 V_L^{\ell\dagger}, \quad (M_{\tilde{L}}^2)_{RR} = V_R^\ell m_{\tilde{\ell}_L^c}^2 V_R^{\ell\dagger}, \quad (M_{\tilde{L}}^2)_{LR} = v\cos\beta V_L^\ell A_\ell^* V_R^{\ell\dagger}. \qquad (5.62)$$

The two matrices $\mathcal{M}_{\tilde{L}}^2$ and $\mathcal{M}_{\tilde{N}}^2$ can be diagonalised by two unitary matrices $\Gamma^{N,E}$

$$M_{\tilde{\nu}}^2 = \Gamma^N \mathcal{M}_{\tilde{N}}^2 \Gamma^{N\dagger},$$

$$M_{\tilde{\ell}}^2 = \Gamma^E \mathcal{M}_{\tilde{L}}^2 \Gamma^{E\dagger}. \qquad (5.63)$$

In mSUGRA, the stau mass matrix is given by

$$\mathcal{M}_{\tilde{\tau}}^2 = \begin{pmatrix} m_{\tilde{\ell}_L}^2 + m_\tau^2 + \frac{\cos 2\beta}{2}(M_Z^2 - 2M_W^2) & m_\tau(A_\tau - \mu\tan\beta) \\ m_\tau(A_\tau - \mu\tan\beta) & m_{\tilde{\ell}_L^c}^2 + m_\tau^2 - \cos 2\beta M_Z^2 \sin^2\theta_W \end{pmatrix}. \qquad (5.64)$$

As for the case of the sbottom, the $\tilde{\tau}_L$-$\tilde{\tau}_L^c$ mixing depends on $\tan\beta$. For $\tan\beta \gg 1$, this mixing could be large and phenomenologically relevant. In this case, diagonalisation of $\mathcal{M}_{\tilde{\tau}}^2$ results in a mass-squared matrix with eigenvalues denoted by $m_{\tilde{\tau}_1}^2$ and $m_{\tilde{\tau}_2}^2$, where $\tilde{\tau}_1$ will be the lightest slepton in analogy to the lightest squarks $\tilde{t}_1$ and $\tilde{b}_1$.

5.2 MSSM INTERACTION LAGRANGIAN

In this section we provide the interaction Lagrangian and the corresponding Feynman rules in the MSSM [49,96,97]. Here we focus on the strong and EW interactions. We will save the Higgs interactions for the next chapter, where a detailed discussion for the MSSM Higgs spectrum and interactions will be given.

5.2.1 Strong interactions in the MSSM

The SM QCD Lagrangian (of gluon-gluon and gluon-quark interactions) remains intact in the MSSM. Therefore, we will not mention it here and we will focus on the SQCD interactions.

Gluino-quark-squark interactions

The interaction of gluino-quark-squark can be obtained from the kinetic term of quark superfields:

$$\int d^2\theta d^2\bar{\theta} \, Q \exp\left[g_3 \frac{T^a}{2} V^a\right] Q. \tag{5.65}$$

Also, one can relate the quark current eigenstates, $q_{L,R}^I$, where $I = 1, 2, 3$, and the squark mass eigenstates, $\tilde{q}_i$, where $i = 1, \ldots, 6$ as follows:

$$q_{L,R}^I = V_{L,R}^q q, \tag{5.66}$$

$$\tilde{q}_{L,R} = \Gamma_{L,R}^{Q\dagger} \tilde{q}. \tag{5.67}$$

The $\Gamma_{L,R}^Q$ are 6×3 mixing matrices, extracted from the 6×6 matrices Γ^Q that diagonalise up and down squarks, and are as follows:

$$\Gamma^Q = \left[\Gamma_{6\times 3}^{Q_L}, \, \Gamma_{6\times 3}^{Q_R}\right]. \tag{5.68}$$

Therefore, one finds that the gluino-quark-squark interaction is given by

$$\mathcal{L} = \sqrt{2}g_3 T_{\beta\alpha}^a \overline{\tilde{g}^a} \tilde{q}_i^\beta \left(-\Gamma_{QL}^{iK} V_{qL}^{KI} P_L + \Gamma_{QR}^{iK} V_{qR}^{KI} P_R\right) q_I^\alpha + \text{h.c.,} \tag{5.69}$$

where T^a are the $SU(3)_C$ generators in the fundamental representation, normalised to $\text{Tr}(T^a T^b) = \delta^{ab}/2$. Also, the T^a's satisfy the following relation:

$$T_{\alpha\beta}^a T_{\gamma\delta}^a = \frac{1}{2}\left[\delta_{\alpha\delta}\delta_{\beta\gamma} - \frac{1}{3}\delta_{\alpha\beta}\delta_{\gamma\delta}\right]. \tag{5.70}$$

Gluon-squark-squark interactions

Similarly, one can show that the interactions of gluon-squark-squark are given by

$$\mathcal{L} = -g_3 \left[\tilde{q}_i^* \frac{T^a}{2} \overleftrightarrow{\partial^\mu} \tilde{q}_j \right] g_\mu^a + \text{h.c.} \tag{5.71}$$

Gluon-gluino-gluino interactions

The gluon-gluino-gluino interactions are given by

$$\mathcal{L} = i\frac{g_3}{2} f_{abc} \tilde{g}^a \gamma^\mu \tilde{g}^b g_\mu^c + \text{h.c.} \tag{5.72}$$

Gluon-gluon-squark-squark interactions

Finally, the gluon-gluon-squark-squark interactions are given by

$$\mathcal{L} = g_3^2 \left[\tilde{u}_i^* \frac{T^a}{2} \frac{T^b}{2} \tilde{u}_i + \tilde{d}_i^* \frac{T^a}{2} \frac{T^b}{2} \tilde{d}_i \right] g_\mu^a g^{b\mu}. \tag{5.73}$$

5.2.2   EW interactions in the MSSM

(W, γ, Z)-sfermion-sfermion

Let us start with the SM EW gauge boson interactions with sfermions. The W-sfermion-sfermion interactions can be written as

$$\mathcal{L} = -\frac{ig_2}{\sqrt{2}} \left[\Gamma_{ik}^{D_L} \Gamma_{kj}^{U_L} (\tilde{d}_i^* \overleftrightarrow{\partial^\mu} \tilde{u}_j) W_\mu^- + \Gamma_{IJ}^N \Gamma_{iI}^{L_L} (\tilde{\ell}_i^* \overleftrightarrow{\partial^\mu} \tilde{\nu}^J) W_\mu^- + \text{h.c.} \right]. \tag{5.74}$$

The photon-sfermion-sfermion interaction is given by

$$\mathcal{L} = -ie \left[\frac{2}{3} (\tilde{u}_i^* \overleftrightarrow{\partial^\mu} \tilde{u}_i) A_\mu - \frac{1}{3} (\tilde{d}_i^* \overleftrightarrow{\partial^\mu} \tilde{d}_i) A_\mu - (\tilde{\ell}_i^* \overleftrightarrow{\partial^\mu} \tilde{\ell}_i) A_\mu \right]. \tag{5.75}$$

Finally, the Z-sfermion-sfermion interaction can be written as

$$\begin{aligned}
\mathcal{L} = \frac{-ig_2}{2c_W} \Big[& (\Gamma_{iI}^{U_L *} \Gamma_{jI}^{U_L} - \frac{4}{3} s_W^2 \gamma^{ij}) (\tilde{u}_i^* \overleftrightarrow{\partial^\mu} \tilde{u}_j) Z_\mu - (\Gamma_{iI}^{D_L} \Gamma_{jI}^{D_L *} - \frac{2}{3} s_W^2 \gamma^{ij}) (\tilde{d}_i^* \overleftrightarrow{\partial^\mu} \tilde{d}_j) Z_\mu \\
& - (\Gamma_{iI}^{E_L} \Gamma_{jI}^{E_L *} - 2s_W^2 \gamma^{ij}) (\tilde{\ell}_i^* \overleftrightarrow{\partial^\mu} \tilde{\ell}_j) Z_\mu + (\tilde{\nu}^{I*} \overleftrightarrow{\partial^\mu} \tilde{\nu}^I) Z_\mu \Big].
\end{aligned} \tag{5.76}$$

$W - W -$sfermion-sfermion

These read as follows:

$$\mathcal{L} = \frac{g_2^2}{2} \left[\Gamma_{iI}^{U_L *} \Gamma_{jI}^{U_L} \tilde{u}_i^* \tilde{u}_j + \Gamma_{iI}^{D_L} \Gamma_{jI}^{D_L *} \tilde{d}_i^* \tilde{d}_j W_\mu^+ W^{-\mu} + \Gamma_{iI}^{E_L *} \Gamma_{jI}^{E_L} W_\mu^+ W^{-\mu} \tilde{\ell}_i^* \tilde{\ell}_j + \text{h.c.} \right]. \tag{5.77}$$

$W - Z(\gamma)$-sfermion-sfermion

Here, we have

$$\mathcal{L} = -\frac{g_2 e}{3\sqrt{2}c_W}\Gamma_{iI}^{D_L}\Gamma_{jJ}^{U_L *}V_{\mathrm{CKM}}^{JI *}\tilde{d}_i^*\tilde{u}_j(s_W^2 Z^\mu - c_W A^\mu)W_\mu^- + \mathrm{h.c.} \tag{5.78}$$

$(Z,\gamma) - (Z,\gamma)$-sfermion-sfermion

In this case, one has

$$\begin{aligned}
\mathcal{L} =& \frac{g_2^2}{3c_W^2}\left[(\frac{4}{3}s_W^4\gamma^{ij} + \frac{3-8s_W^2}{4}\Gamma_{iI}^{U_L *}\Gamma_{jI}^{U_L})\tilde{u}_i^*\tilde{u}_j \right.\\
&\left.+ (\frac{1}{3}s_W^4\gamma^{ij} + \frac{3-4s_W^2}{4}\Gamma_{iI}^{D_L}\Gamma_{jI}^{D_L *})\tilde{d}_i^*\tilde{d}_j\right]Z_\mu Z^\mu\\
&+ \frac{g_2 e}{3c_W}\left[2(\Gamma_{iI}^{U_L *}\Gamma_{jI}^{U_L} - \frac{4}{3}s_W^2\gamma^{ij})\tilde{u}_i^*\tilde{u}_j + (\Gamma_{iI}^{D_L}\Gamma_{jI}^{D_L *} - \frac{2}{3}s_W^2\gamma^{ij})\tilde{d}_i^*\tilde{d}_j\right]A_\mu Z^\mu\\
&+ \frac{g_2^2}{c_W^2}\left[\frac{1}{4}\tilde{\nu}^{I*}\tilde{\nu}^I + (s_W^4\gamma^{ij} + \frac{1-4s_W^2}{4}\Gamma_{iI}^{E_L *}\Gamma_{jI}^{E_L})\tilde{\ell}_i^*\tilde{\ell}_j\right]Z_\mu Z^\mu + e^2\tilde{\ell}_i^*\tilde{\ell}_i A_\mu A^\mu\\
&+ \frac{e^2}{9}(4\tilde{u}_i^*\tilde{u}_i + \tilde{d}_i^*\tilde{d}_i)A_\mu A^\mu + \frac{g_2 e}{c_W}(\Gamma_{iI}^{E_L *}\Gamma_{jI}^{D_L} - 2s_W^2\gamma^{ij})\tilde{\ell}_i^*\tilde{\ell}_j A_\mu Z^\mu.
\end{aligned} \tag{5.79}$$

Chargino-quark-squark

The chargino-quark-squark interactions can be written as

$$\begin{aligned}
\mathcal{L} =& -g_2\sum_{j=1}^{2}\left[C^{-1}\chi_j^{+T}\tilde{u}_i^\dagger(G_{IJij}^{U_L}P_L + G_{IJij}^{U_R}P_R)d^I + \overline{d}^I(G_{IJij}^{U_L *}P_R + G_{IJij}^{U_R *}P_L)\overline{\chi_j^+}^T\tilde{u}_i C\right]\\
&+ g_2\sum_{j=1}^{2}\left[\overline{\chi_j^-}\tilde{d}_i^\dagger(G_{IJij}^{D_L}P_L + G_{IJij}^{D_R}P_R)u^I + \overline{u}^I(G_{IJij}^{D_R *}P_L + G_{IJij}^{D_L *}P_R)\chi_j^-\tilde{d}_i\right], \tag{5.80}
\end{aligned}$$

where C is the charge conjugate matrix and the mixing matrices G are defined as

$$\begin{aligned}
G_{IJij}^{U_L} &= \left(\frac{m_{u_J}}{\sqrt{2}M_W\sin\beta}V_{j2}^*\Gamma_{iJ}^{U_R} - V_{j1}^*\Gamma_{iJ}^{U_L}\right)V_{\mathrm{CKM}}^{JI}, \tag{5.81}\\
G_{IJij}^{U_R} &= \frac{m_{d_I}}{\sqrt{2}M_W\cos\beta}U_{j2}\Gamma_{iJ}^{U_L}V_{\mathrm{CKM}}^{JI}, \tag{5.82}\\
G_{IJij}^{D_L} &= \left(\frac{m_{d_J}}{\sqrt{2}M_W\cos\beta}\Gamma_{iJ}^{D_R}U_{j2}^* - \Gamma_{iJ}^{D_L}U_{j1}^*\right)V_{\mathrm{CKM}}^{IJ}, \tag{5.83}\\
G_{IJij}^{D_R} &= \frac{m_{u_I}}{\sqrt{2}M_W\sin\beta}\Gamma_{iJ}^{D_L}V_{j2}V_{\mathrm{CKM}}^{IJ}. \tag{5.84}
\end{aligned}$$

Here, U and V are the chargino mixing matrices defined above.

Chargino-lepton-slepton

The chargino-lepton-slepton interactions are given by

$$
\begin{aligned}
\mathcal{L} &= g_2 \sum_{i=1}^{2} \overline{\tilde{\chi}_i^-} \tilde{\nu}_J^\dagger \left(H_{IJi}^L P_L + H_{IJi}^R P_R \right) \ell_I + g_2 \sum_{i=1}^{2} \overline{\ell}_I \left(H_{IJi}^{R*} P_L + H_{IJi}^{L*} P_R \right) \tilde{\chi}_i^- \tilde{\nu}_J \\
&+ g_2 \sum_{j=1}^{2} \overline{\tilde{\chi}_j^+} (G_{Iij}^L P_L + G_{Iij}^R P_R) \nu^I \tilde{\ell}_i^\dagger + g_2 \sum_{j=1}^{2} \overline{\nu}^I \tilde{\ell}_i (G_{Iij}^{R*} P_L + G_{Iij}^{L*} P_R) \chi_j^+, \quad (5.85)
\end{aligned}
$$

where

$$
\begin{aligned}
H_{IJi}^L &= -V_{i1}^* \Gamma_{IJ}^N, & (5.86) \\
H_{IJi}^R &= \frac{m_\ell}{\sqrt{2} M_W \cos\beta} U_{i2} \Gamma_{IJ}^N, & (5.87) \\
G_{Iij}^L &= \frac{m_\ell}{\sqrt{2} M_W \cos\beta} \Gamma_{iI}^{E_R} U_{j2}^* - g_2 \Gamma_{iI}^{E_L} U_{j1}^*, & (5.88) \\
G_{Iij}^R &= 0. & (5.89)
\end{aligned}
$$

(Z, γ)-chargino-chargino

The Z-chargino-chargino interactions are given by

$$
\mathcal{L} = -\frac{g_2}{2c_W} \sum_{i,j=1}^{2} \left[\overline{\tilde{\chi}_i^+} \gamma^\mu \left(V_{ij} P_L + U_{ij} P_R + A^{ij} \right) \tilde{\chi}_j^+ Z_\mu + \text{h.c.} \right], \quad (5.90)
$$

where

$$
V_{ij}^L = V_{i1}^* V_{j1}, \quad U_{ij}^R = U_{i1} U_{j1}^*, \quad A^{ij} = (c_W^2 - s_W^2) \delta^{ij}. \quad (5.91)
$$

The γ-chargino-chargino interactions are given by

$$
\mathcal{L} = -\sum_{i=1}^{2} e \overline{\tilde{\chi}_i^+} \gamma^\mu \tilde{\chi}_i^+ A_\mu. \quad (5.92)
$$

Neutralino-quark-squark

The neutralino-quark-squark interactions are given by

$$
\begin{aligned}
\mathcal{L} &= g_2 \sum_{k=1}^{4} \left[\overline{d}^I (G_{0Iik}^{D_R} P_R + G_{0Iik}^{D_L} P_L) \tilde{\chi}_k \tilde{d}_i + \overline{\tilde{\chi}_k} (G_{0Iik}^{*D_L} P_R + G_{0Iik}^{*D_R} P_L) \tilde{d}_i^* d^I \right] \\
&+ g_2 \sum_{k=1}^{4} \left[\overline{u}^I (H_{0Iik}^R P_R + H_{0Iik}^L P_L) \tilde{\chi}_k \tilde{u}_i + \overline{\tilde{\chi}_k} (H_{0Iik}^{*L} P_R + H_{0Iik}^{*R} P_L) \tilde{u}_i^* u^I \right],
\end{aligned}
$$

$$
(5.93)
$$

where

$$G_{0Iik}^{D_R} = \frac{m_d}{\sqrt{2}M_W\cos\beta}N_{k3}\Gamma_{iI}^{D_R*} + (N_{k2} - \frac{1}{3}\tan\theta_W N_{k1})\Gamma_{iI}^{D_L*}, \quad (5.94)$$

$$G_{0Iik}^{D_L} = \frac{m_d}{\sqrt{2}M_W\cos\beta}N_{k3}^*\Gamma_{iI}^{D_L*} - \frac{2}{3}\tan\theta_W N_{k1}^*\Gamma_{iI}^{D_R*}, \quad (5.95)$$

$$G_{0Iik}^{U_R} = \frac{-m_u}{\sqrt{2}M_W\cos\beta}N_{k4}\Gamma_{iI}^{U_R*} + (N_{k2} + \frac{1}{3}\tan\theta_W N_{k1})\Gamma_{iI}^{U_L*}, \quad (5.96)$$

$$G_{0Iik}^{U_L} = -\frac{m_u}{\sqrt{2}M_W\cos\beta}Z_{k4}^*\Gamma_{iI}^{U_L*} + \frac{4}{3}\tan\theta_W N_{k1}^*\Gamma_{iI}^{U_R*}. \quad (5.97)$$

Neutralino-lepton-slepton

Here, one has

$$\mathcal{L} = \frac{g_2}{\sqrt{2}}\bar{\ell}_I(G_{0Iik}^{L_R}P_R + G_{0Iik}^{L_L}P_L)\tilde{\chi}_k\tilde{\ell}_i + \frac{g_2}{\sqrt{2}}\overline{\tilde{\chi}_k}(G_{0Iik}^{L_L*}P_R + G_{0Iik}^{L_R*}P_L)\ell_I\tilde{\ell}_i^*$$

$$+ \frac{g_2}{\sqrt{2}}\bar{\nu}^I H_{0IJk}^R P_R\tilde{\chi}_k\tilde{\nu}^J + \frac{g_2}{\sqrt{2}}\overline{\tilde{\chi}_k}H_{0IJk}^{*R}P_L\nu^I\tilde{\nu}^{*J}, \quad (5.98)$$

where

$$G_{0Iik}^{L_R} = \frac{m_\ell}{M_W\cos\beta}N_{k3}\Gamma_{iI}^{E_R*} + (N_{k2} + \tan\theta_W N_{k1})\Gamma_{Ii}^{E_L*}, \quad (5.99)$$

$$G_{0Iik}^{L_L} = \frac{m_\ell}{M_W\cos\beta}N_{k3}\Gamma_{iI}^{E_L*} - 2\tan\theta_W N_{k1}^*\Gamma_{iI}^{E_R*}, \quad (5.100)$$

$$H_{0IJk}^R = (\tan\theta_W N_{k1} - N_{k2}^N)\Gamma_{IJ}^N, \quad (5.101)$$

$$H_{0IJk}^R = 0. \quad (5.102)$$

W-chargino-neutralino

In this case, one obtains

$$\mathcal{L} = g_2\overline{\tilde{\chi}_j^+}\gamma^\mu\left(A_{ij}^L P_L + A_{ij}^R P_R\right)\tilde{\chi}_i W_\mu^+ + \text{h.c.}, \quad (5.103)$$

where

$$A_{ij}^L = (N_{i2}V_{j1}^* - \frac{1}{\sqrt{2}}N_{i4}V_{j2}^*), \quad A_{ij}^R = (N_{i2}^*U_{j1} + \frac{1}{\sqrt{2}}N_{i3}^*U_{2j}). \quad (5.104)$$

Z-neutralino-neutralino

Finally, one obtains

$$\mathcal{L} = \frac{g_2}{4c_W}\sum_{i,j=1}^{4}\overline{\tilde{\chi}_i}\gamma^\mu(N_{ij}^L P_L - N_{ij}^R P_R)\tilde{\chi}_j Z_\mu, \quad (5.105)$$

where

$$N_{ij}^L = (N_{i4}^* N_{j4} - N_{i3}^* N_{j3}^*), \quad N_{ij}^R = (N_{i4}^* N_{j4} - N_{i3}^* N_{j3}^*). \tag{5.106}$$

5.3 MASS INSERTION APPROXIMATION

In this section we consider a method for analysing supersymmetric effects in a model independent way. This approach is known as the MIA. The MIA is a technique which was developed to include the soft SUSY breaking terms without specifying the model behind them [18,98,99]. In this approximation, one adopts a basis where the couplings of the fermions and sfermions to neutral gauginos are flavour diagonal, leaving all the sources of flavour violation inside the off-diagonal terms of the sfermion mass matrix.

As is well known, at low energy scales the quark and squark mass terms are given by

$$\mathcal{L}_{\text{mass}} = \bar{q} M_q q + \tilde{q}^\dagger \tilde{m}_q^2 \tilde{q} + \text{h.c.} \tag{5.107}$$

Also, the gluino interactions are given by

$$\mathcal{L}_{\text{gluino}} = g_3 T^a \bar{q} \tilde{q} \tilde{g}^a + \text{h.c.} \tag{5.108}$$

Therefore, in mass eigenstate basis, where $q \to U q$ and $\tilde{q} \to V \tilde{q}$, in order to diagonalise the mass matrices M_q and $\tilde{m}_q^2$, one obtains

$$\mathcal{L}_{\text{gluino}} = g_3 T^a \left(\bar{q} U^\dagger V \tilde{q} \tilde{g}^a \right) + \text{h.c.} \tag{5.109}$$

It is clear that, if $U^\dagger V \neq 1$, then tree level FCNC processes are mediated by gluinos. The experimental limits on these FCNC processes impose stringent constraints on the supersymmetric parameter space. The MIA is therefore a powerful model-independent tool for specifically analysing and constraining SUSY models from FCNC effects.

As mentioned, the super-CKM basis is obtained by performing a superfield rotation that diagonalises the quark mass matrix M_q. In this case, the gluino vertices $q_L^i \tilde{q}_L^i \tilde{g}^a$ and $q_R^i \tilde{q}_R^i \tilde{g}^a$ are flavour diagonal. However, the squark mass matrix $\tilde{m}_q^2$ remains non-diagonal, $\tilde{m}_q^2 \to U^\dagger \tilde{m}_q^2 U_q \neq (\tilde{m}_q^2)_{\text{diag}}$. In general, in the super-CKM basis, the couplings of fermions and their SUSY partners to neutral gauginos are flavour-diagonal and flavour-violating SUSY effects are encoded in the non-diagonal entries of the sfermion mass matrix. If one assumes that the off-diagonal elements of $\tilde{m}_f^2$ are smaller than the diagonal ones, one can write

$$\tilde{m}_f^2 = \tilde{m}^2 \mathbf{1} + \Delta, \tag{5.110}$$

where the Δ's are the off-diagonal terms in the sfermion mass matrices and $\mathbf{1}$ is the unit matrix. There exist four different Δ's connecting flavour i to j, namely: $(\Delta_{LL})_{ij}$, $(\Delta_{LR})_{ij}$, $(\Delta_{RL})_{ij}$ and $(\Delta_{RR})_{ij}$. The indices L and R refer to the helicity of the fermion parameters. The Hermiticity of the sfermion mass matrix implies that $(\Delta_{LL})_{ij} = (\Delta^*_{LL})_{ji}$, $(\Delta_{RR})_{ij} = (\Delta^*_{RR})_{ji}$ and $(\Delta_{LR})_{ij} = (\Delta^*_{RL})_{ji}$. Using this approximation, the sfermion propagators can be expanded as follows:

$$\langle \tilde{f}^a_A \tilde{f}^{b*}_B \rangle = i(k^2\mathbf{1} - \tilde{m}^2\mathbf{1} - \Delta^f_{AB})^{-1}_{ab} \simeq \frac{i\delta_{ab}}{k^2 - \tilde{m}^2} + \frac{i(\Delta^f_{AB})_{ab}}{(k^2 - \tilde{m}^2)^2} + \mathcal{O}(\Delta^2), \quad (5.111)$$

where $A, B = (L, R)$. The SUSY contributions are parameterised in terms of the dimensionless terms

$$(\delta^f_{AB})_{ij} = (\Delta^f_{AB})^{ij}/\tilde{m}^2. \tag{5.112}$$

In this respect, the relevant Feynman rules in MIA are given by:

Furthermore, in the Super-CKM basis, the interacting Lagrangian involving charginos is given by

$$
\begin{aligned}
\mathcal{L}_{q\tilde{q}\tilde{\chi}^+} = & -g_2 \sum_k \sum_{a,b} \Big(V_{k1} K^*_{ba} \bar{d}^a_L (\tilde{\chi}^+)^* \tilde{u}^b_L - U^*_{k2} (Y^{\mathrm{diag}}_d . K^+)_{ab} \bar{d}^a_R (\tilde{\chi}^+)^* \tilde{u}^b_L \\
& - V^*_{k2} (K . Y^{\mathrm{diag}}_u)_{ab} \bar{d}^a_L (\tilde{\chi}^+)^* \tilde{u}^b_R \Big),
\end{aligned}
\tag{5.113}
$$

where $Y^{\mathrm{diag}}_{u,d}$ are the diagonal Yukawa matrices and K is the usual CKM matrix. The indices a, b and k label flavour and chargino mass eigenstates, respectively, and V, U are the chargino mixing matrices. As one can see from Eq. (12.61), the Higgsino couplings are suppressed by the Yukawas of the light quarks and therefore they are negligible, except for the stop-bottom interaction which is directly enhanced by the top Yukawa (Y_t). The other vertex involving the down and stop could also be enhanced by Y_t, but one should pay the price of a λ^3 suppression, where λ is the Cabibbo mixing. Since here we adopt the approximation of retaining only terms proportional to λ, we will neglect the effect of this vertex. Moreover, we also set to zero the Higgsino contributions proportional to the Yukawa couplings of the light quarks with the exception of the bottom Yukawa Y_b, since its effect could be enhanced by large $\tan\beta$. The associated Feynman rules are given by:

$$-g_2 K^\dagger_{ab} V_{j1},$$

$$g_2 \left(K^\dagger \hat{Y}_U \right)_{ab} V_{j2},$$

$$g_2 \left(\hat{Y}_D K^\dagger \right)_{ab} U^*_{j2}.$$

CHAPTER 6

Higgs Bosons in the MSSM

As mentioned, the Higgs sector of the MSSM consists of two complex Higgs doublets, H_u and H_d. After EWSB, three of the eight degrees of freedom contained in H_u and H_d are acquired in the usual way by the $W^\pm$ and Z in order to become massive. The five physical degrees of freedom that remain form a neutral pseudoscalar (or CP-odd) Higgs boson, A, two neutral scalars (or CP-even), h and H, plus a charged Higgs boson pair (with mixed CP quantum numbers), $H^\pm$. This varied spectrum is to be compared with the single physical neutral scalar Higgs boson of the SM. From Eq. (4.32), the scalar potential $V(H_u, H_d)$ can be written as

$$
\begin{aligned}
V(H_u, H_d) &= m_2^2 \left(|H_u^+|^2 + |H_u^0|^2\right) + m_1^2 \left(|H_d^-|^2 + |H_d^0|^2\right) \\
&+ \left[m_3^2 \left(H_u^+ H_d^- - H_u^0 H_d^0\right) + \text{h.c.}\right] + \frac{g_1^2 + g_2^2}{8} \left[|H_u^+|^2 + |H_u^0|^2 \right. \\
&- \left. |H_d^0|^2 - |H_d^-|^2\right]^2 + \frac{g_2^2}{2} \left|H_u^+ H_d^{0*} + H_u^0 H_d^{-*}\right|^2.
\end{aligned}
$$

(6.1)

In this chapter we will identify the physical states of the MSSM Higgs bosons and compute their masses [100]. Then we will discuss the possible production (at both hadron and lepton colliders) as well as decay modes of both neutral and charged Higgs bosons of the MSSM (borrowing material from [101]).

6.1 CHARGED HIGGS BOSONS

We begin with the physical masses of the charged Higgs bosons. The relevant potential terms of $V(H_u^+, H_d^-)$, extracted from Eq. (6.1), are given by

$$
\begin{aligned}
V(H_u^+, H_d^-) &= m_2^2|H_u^+|^2 + m_1^2|H_d^-|^2 + m_3^2(H_u^+ H_d^- + \text{h.c.}) \\
&+ \frac{g_1^2 + g_2^2}{4}(|H_u^0|^2 - |H_d^0|^2)(|H_u^+|^2 - |H_d^-|^2) + \frac{g_2^2}{2}\left|H_u^+ H_d^{0*} + H_u^0 H_d^{-*}\right|^2.
\end{aligned}
$$

(6.2)

Therefore, the mass matrix $M_{H^\pm}^2$ is obtained as follows:

$$
\mathcal{L} \supset - \left(H_d^\mp \ H_u^{\pm *}\right) M_{H^\pm}^2 \begin{pmatrix} H_d^{\mp *} \\ H_u^\pm \end{pmatrix},
$$

(6.3)

with

$$M^2_{H^\pm} = \frac{1}{2} \left(\begin{array}{cc} \frac{\partial^2 V(H^+_u, H^-_d)}{\partial H^\mp_d \partial H^{\mp*}_d} & \frac{\partial^2 V(H^+_u, H^-_d)}{\partial H^\mp_d \partial H^\pm_u} \\ \frac{\partial^2 V(H^+_u, H^-_d)}{\partial H^{\pm*}_u \partial H^{\mp*}_d} & \frac{\partial^2 V(H^+_u, H^-_d)}{\partial H^{\pm*}_u \partial H^\pm_u} \end{array} \right) \Bigg|_{\langle H^0_{u,d}\rangle=v_{u,d},\ \langle H^+_u\rangle=0,\ \langle H^-_d\rangle=0}$$

$$= \left(\begin{array}{cc} m^2_1 + \frac{g^2_1+g^2_2}{4}(v^2_d - v^2_u) + \frac{g^2_2}{2}v^2_u & m^2_3 + \frac{g^2_2}{2}v_d v_u \\ m^2_3 + \frac{g^2_2}{2}v_d v_u & m^2_2 - \frac{g^2_1+g^2_2}{4}(v^2_d - v^2_u) + \frac{g^2_2}{2}v^2_d \end{array} \right).$$

(6.4)

The eigenvalues of this mass matrix are zero and[1]

$$m^2_{H^\pm} = M^2_W + m^2_1 + m^2_2.$$

(6.5)

Here, we have used the minimisation conditions of Eq. (4.46). The zero mass corresponds to the Goldstone bosons $G^\pm$ 'eaten' by the massive $W^\pm$ bosons while $m_{H^\pm}$ is the mass of the charged Higgs bosons $H^\pm$. The mixing matrix, which gives the physical states, is given by

$$\left(\begin{array}{c} G^\pm \\ H^\pm \end{array} \right) = \left(\begin{array}{cc} \cos\beta & \sin\beta \\ -\sin\beta & \cos\beta \end{array} \right) \left(\begin{array}{c} H^{\mp*}_d \\ H^\pm_u \end{array} \right).$$

(6.6)

Therefore, from Eq. (4.8) of the superpotential, one can verify that the couplings of the physical charged Higgs boson with the SM fermions are given by

$$Y_{H^+ \bar{u} d} = -\frac{i}{\sqrt{2}v} V^*_{ud} \left[m_d \tan\beta(1+\gamma_5) + m_u \cot\beta(1-\gamma_5) \right],$$

(6.7)

$$Y_{H^+ \ell^- \bar{\nu}} = -\frac{i}{\sqrt{2}v} m_\ell \tan\beta(1+\gamma_5).$$

(6.8)

Also, the couplings of physical charged Higgs bosons to SM neutral gauge bosons, derived from Eq. (4.40), are given by

$$Z_\mu H^+ H^- = -\frac{1}{2}\frac{g_2}{\cos\theta_W} \cos 2\theta_W (p+p')_\mu,$$

(6.9)

$$\gamma_\mu H^+ H^- = -ie(p+p')_\mu,$$

(6.10)

where p and p' are the incoming and outgoing momenta of the charged Higgs bosons. Notice that they have no MSSM parameter dependence.

---

[1]The eigenvalues (masses) and the eigenvectors (mixings) of any real symmetric 2×2 matrix, say A, are given by

$$\lambda_{1,2} = \frac{1}{2} \left[A_{11} + A_{22} \mp \sqrt{(A_{11}+A_{22})^2 - 4(A_{11}A_{22} - A^2_{12})} \right]$$

$$\sin 2\theta = \frac{2A_{12}}{\sqrt{(A_{11}-A_{22})^2 + 4A^2_{12}}}, \quad \cos 2\theta = \frac{A_{11} - A_{22}}{\sqrt{(A_{11}-A_{22})^2 + 4A^2_{12}}},$$

where λ_1 and λ_2 are the eigenvalues (masses) and θ is the mixing angle.

6.2 CP-ODD NEUTRAL HIGGS BOSONS

Now, we consider the masses of the neutral Higgs bosons. Upon expanding H_u^0 and H_d^0 around the vacua v_u and v_d, respectively, $i.e.$,

$$H_u^0 = v_u + \sigma_u + i\chi_u, \quad H_d^0 = v_d + \sigma_d + i\chi_d, \tag{6.11}$$

the mass matrix at tree level of the CP-odd neutral Higgs bosons (stemming from the imaginary components of the above fields), in the basis (χ_d, χ_u), is given by

$$M_{H_I^0}^2 = \frac{1}{2} \left(\begin{array}{cc} \frac{\partial^2 V(H_u^0, H_d^0)}{\partial \chi_d \partial \chi_d} & \frac{\partial^2 V(H_u^0, H_d^0)}{\partial \chi_d \partial \chi_u} \\ \frac{\partial^2 V(H_u^0, H_d^0)}{\partial \chi_u \partial \chi_d} & \frac{\partial^2 V(H_u^0, H_d^0)}{\partial \chi_u \partial \chi_u} \end{array} \right) \Bigg|_{\langle \sigma_{u,d} \rangle = 0, \ \langle \chi_{u,d} \rangle = 0}$$

$$= \left(\begin{array}{cc} m_1^2 + \frac{g_1^2 + g_2^2}{4}(v_d^2 - v_u^2) & -m_3^2 \\ -m_3^2 & m_2^2 - \frac{g_1^2 + g_2^2}{4}(v_d^2 - v_u^2) \end{array} \right). \tag{6.12}$$

The eigenvalues of this mass matrix are zero and

$$m_A^2 = m_1^2 + m_2^2. \tag{6.13}$$

The zero mass corresponds to the Goldstone boson, G^0, 'eaten' by the massive Z boson while m_A is the mass of the pseudoscalar Higgs state A. Therefore, at the tree level, one finds

$$m_{H^\pm}^2 = m_A^2 + M_W^2. \tag{6.14}$$

As usual, in unitary gauge, the Goldstone boson G^0 disappears from the Lagrangian. The mixing matrix for G^0 and the pseudoscalar Higgs boson A is given by

$$\left(\begin{array}{c} G^0 \\ A \end{array} \right) = \left(\begin{array}{cc} \cos\beta & \sin\beta \\ -\sin\beta & \cos\beta \end{array} \right) \left(\begin{array}{c} \chi_d \\ \chi_u \end{array} \right). \tag{6.15}$$

Thus, one can show that the couplings of the pseudoscalar Higgs state A to SM fermions are given by

$$Y_{Au\bar{u}} = \frac{m_u}{v} \cot\beta\gamma_5, \tag{6.16}$$

$$Y_{Ad\bar{d}} = \frac{m_d}{v} \tan\beta\gamma_5, \tag{6.17}$$

$$Y_{A\ell^+\ell^-} = \frac{m_\ell}{v} \tan\beta\gamma_5. \tag{6.18}$$

In addition, the pseudoscalar Higgs state A has the following coupling with the $W^\pm$-gauge boson and charged Higgs boson:

$$W_\mu^\pm H^\mp A \ : \ \frac{g_2}{2}(p + p')_\mu, \tag{6.19}$$

where p and p' are the incoming and outgoing momenta of $H^\mp$ and A, respectively.

6.3 CP-EVEN NEUTRAL HIGGS BOSONS

The mass matrix of the tree level masses pertaining to the CP-even neutral Higgs bosons, in the basis (σ_d, σ_u), is given by

$$M^2_{H^0_R} = \frac{1}{2} \left(\begin{array}{cc} \frac{\partial^2 V(H^0_u, H^0_d)}{\partial \sigma_d \partial \sigma_d} & \frac{\partial^2 V(H^0_u, H^0_d)}{\partial \sigma_d \partial \sigma_u} \\ \frac{\partial^2 V(H^0_u, H^0_d)}{\partial \sigma_u \partial \sigma_d} & \frac{\partial^2 V(H^0_u, H^0_d)}{\partial \sigma_u \partial \sigma_u} \end{array} \right) \Bigg|_{\langle \sigma_{u,d} \rangle = 0, \ \langle \chi_{u,d} \rangle = 0}$$

$$= \left(\begin{array}{cc} m_1^2 + \frac{g_1^2 + g_2^2}{4}(3v_d^2 - v_u^2) & -m_3^2 - \frac{g_1^2 + g_2^2}{2} v_d v_u \\ -m_3^2 - \frac{g_1^2 + g_2^2}{2} v_d v_u & m_2^2 + \frac{g_1^2 + g_2^2}{4}(3v_u^2 - v_d^2) \end{array} \right). \qquad (6.20)$$

The eigenvalues of the mass matrix $M^2_{H^0_R}$ (stemming from the real parts of the Higgs doublets) are

$$m^2_{h,H} = \frac{1}{2} \left[m_A^2 + M_Z^2 \mp \sqrt{(m_A^2 + M_Z^2)^2 - 4 \left[\frac{2 m_3^2 M_Z^2}{\sin 2\beta}(\cos^4 \beta + \sin^4 \beta) - m_3^2 M_Z^2 \sin 2\beta \right]} \right],$$

$$(6.21)$$

wherein $-(+)$ refer to $m_h(m_H)$, so that (conventionally) $m_h \leq m_H$. Using the relation $m_A^2 = 2m_3^2 / \sin 2\beta$, one finds

$$m^2_{h,H} = \frac{1}{2} \left(m_A^2 + M_Z^2 \mp \sqrt{(m_A^2 + M_Z^2)^2 - 4 m_A^2 M_Z^2 \cos^2 2\beta} \right). \qquad (6.22)$$

These physical states can be written in terms of σ_d and σ_u as

$$\left(\begin{array}{c} H \\ h \end{array} \right) = \left(\begin{array}{cc} \cos\alpha & \sin\alpha \\ -\sin\alpha & \cos\alpha \end{array} \right) \left(\begin{array}{c} \sigma_d \\ \sigma_u \end{array} \right), \qquad (6.23)$$

where the mixing angle α is given by

$$\alpha = \frac{1}{2} \tan^{-1} \left[\tan 2\beta \frac{m_A^2 + M_Z^2}{m_A^2 - M_Z^2} \right], \quad -\frac{\pi}{2} \leq \alpha \leq 0. \qquad (6.24)$$

From Eq. (6.22) one can easily derive the following upper bound on the light neutral Higgs mass

$$m_h \leq M_Z |\cos 2\beta|. \qquad (6.25)$$

This implies that the lightest CP-even Higgs state in the MSSM must be lighter than the Z boson. The origin of this strong bound can be traced back to the fact that the only Higgs self-couplings in Eq. (6.1) are EW gauge couplings. In contrast, in the non-supersymmetric SM the strength of the Higgs self-interaction is unknown. However, the scalar potential $V(H_u, H_d)$ receives loop corrections, which may be important in some ranges of the MSSM parameters. These corrections modify the MSSM tree-level prediction for m_h, as will be explicitly shown in the next section.

From the rotation matrix in Eq. (6.23), one can easily verify that the couplings of the light CP-even MSSM neutral Higgs state to the SM fermions are given by

$$Y_{hu\bar{u}} \;=\; i\frac{m_u}{v}\frac{\cos\alpha}{\sin\beta}, \tag{6.26}$$

$$Y_{hd\bar{d}} \;=\; -i\frac{m_d}{v}\frac{\sin\alpha}{\cos\beta}, \tag{6.27}$$

$$Y_{h\ell^+\ell^-} \;=\; -i\frac{m_\ell}{v}\frac{\sin\alpha}{\cos\beta}. \tag{6.28}$$

It is clear that at large $\tan\beta$ the neutral Higgs couplings with $d\bar{d}$ and $\ell^+\ell^-$ will go like $\tan\beta$ and hence they can be strongly enhanced, while $u\bar{u}$ coupling goes like $\cot\beta$ and thus can be strongly suppressed. Also, one can easily show that the couplings of the MSSM heavy CP-even neutral Higgs state to SM fermions are given by

$$Y_{Hu\bar{u}} \;=\; i\frac{m_u}{v}\frac{\sin\alpha}{\sin\beta}, \tag{6.29}$$

$$Y_{Hd\bar{d}} \;=\; i\frac{m_d}{v}\frac{\cos\alpha}{\cos\beta}, \tag{6.30}$$

$$Y_{H\ell^+\ell^-} \;=\; i\frac{m_\ell}{v}\frac{\cos\alpha}{\cos\beta}. \tag{6.31}$$

Furthermore, the couplings of the light CP-even MSSM neutral Higgs state with the SM gauge bosons are given by

$$hW_\mu^+ W_\nu^- \;=\; ig_2 M_W g_{\mu\nu}\sin(\beta-\alpha), \tag{6.32}$$

$$hZ_\mu Z_\nu \;=\; ig_2\frac{M_Z}{\cos\theta_W}g_{\mu\nu}\sin(\beta-\alpha), \tag{6.33}$$

$$Z_\mu hA \;=\; \frac{1}{2}\frac{g_2}{\cos\theta_W}\cos(\beta-\alpha)(p+p')_\mu, \tag{6.34}$$

$$W_\mu^\pm H^\mp h \;=\; \mp i\frac{g_2}{2}\cos(\beta-\alpha)(p+p')_\mu. \tag{6.35}$$

Therefore, in the so-called 'decoupling limit', when

$$m_A \gg M_Z \quad\text{and}\quad \cos(\beta-\alpha) \sim \frac{1}{2}\sin 4\beta\frac{M_Z^2}{m_A^2} \approx 0, \tag{6.36}$$

all h couplings approach their SM values (so that the lightest CP-even MSSM neutral Higgs state of the MSSM is effectively SM-like). The couplings of the heavy CP-even MSSM neutral Higgs state of the MSSM to the $W^\pm$ and Z gauge bosons can be obtained from the above expressions by changing $\sin(\beta-\alpha) \to \cos(\beta-\alpha)$. The couplings of two gauge bosons and two scalar Higgs states can be found in [100].

6.4   RADIATIVE CORRECTIONS TO THE HIGGS BOSON MASSES

The full one-loop effective Higgs potential is given by [85, 86, 102]

$$V_1(Q) = V_0(Q) + \Delta V(Q), \tag{6.37}$$

where $\Delta V(Q)$ is given in Eq. (4.53). The tree-level minimisation conditions, mentioned in the previous chapter, remain intact if $\partial(\Delta V)/\partial v_k = 0$, where

$$\frac{\partial(\Delta V)}{\partial v_k} = \frac{1}{64\pi^2} \sum_i (-1)^{2J_i}(2J_i + 1) \frac{\partial m_i^2}{\partial v_k} F(m_i^2), \qquad (k = 1, 2), \qquad (6.38)$$

with $F(m_i^2) = 2m_i^2 \left[\ln\left(m_i^2/Q^2\right) - 1\right]$. The scale $\hat{Q}$ that satisfies this condition can be determined from the following equation [86]:

$$\sum_i (-1)^{2J_i}(2J_i + 1)\, F(m_i^2) = 0. \qquad (6.39)$$

Here, we have assumed that the first derivatives of all square masses m_i^2 are equal, i.e., $\partial m_i^2/\partial v_k = \partial m_j^2/\partial v_k$, for all i, j, k. Therefore, the one-loop corrections to the CP-even neutral Higgs masses are given by

$$(\Delta M_{H^0}^2)_{k\ell} = \frac{1}{2}\frac{\partial^2(\Delta V)}{\partial v_k \partial v_\ell} = \frac{1}{64\pi^2}\sum_i (-1)^{2J_i}(2J_i + 1) \frac{\partial m_i^2}{\partial v_k}\frac{\partial m_i^2}{\partial v_\ell}\ln\frac{m_i^2}{\hat{Q}^2}. \qquad (6.40)$$

The dominant contribution to ΔV, and in turn to the Higgs mass corrections, is due to the top and stop particles flowing in the loops. As shown in the previous chapter, the stop masses are given by

$$m_{\tilde{t}_{1,2}}^2 = m_t^2 + \frac{1}{2}(m_q^2 + m_u^2) + \frac{1}{4}M_Z^2 \cos 2\beta$$
$$\pm \sqrt{\left[\frac{1}{2}(m_q^2 - m_u^2) + \frac{1}{12}(8M_W^2 - 5M_Z^2)\cos^2 2\beta\right]^2 + m_t^2(A_t + \mu \cot \beta)^2}. \qquad (6.41)$$

Note that the soft terms m_Q, m_U and A_t are not field dependent. If one neglects, for simplicity, the D-term and off-diagonal contributions in the stop mass matrix, one finds that

$$m_{\tilde{t}_1}^2 = m_t^2 + m_q^2, \qquad (6.42)$$

$$m_{\tilde{t}_2}^2 = m_t^2 + m_u^2. \qquad (6.43)$$

Therefore, in this simplified case, we have

$$\frac{\partial m_t^2}{\partial v_i} = \frac{\partial m_{\tilde{t}_1}^2}{\partial v_i} = \frac{\partial m_{\tilde{t}_2}^2}{\partial v_i}. \qquad (6.44)$$

Recalling that $m_t = Y_t v_2$, one further gets

$$\frac{\partial^2(\Delta V)}{\partial v_1^2} = 0, \qquad (6.45)$$

$$\frac{\partial^2(\Delta V)}{\partial v_2^2} = \frac{3m_t^4}{8\pi^2 v_2^2}\ln\frac{m_{\tilde{t}_1}^2 m_{\tilde{t}_2}^2}{m_t^4}, \qquad (6.46)$$

$$\frac{\partial^2(\Delta V)}{\partial v_1 \partial v_2} = 0. \qquad (6.47)$$

Note that the factor 3 in the second equation is due to the colour degrees of freedom of the top (s)quarks. In this case, the correction mass matrix takes the form

$$\Delta M_{H^0}^2 = \begin{pmatrix} 0 & 0 \\ 0 & \delta_t^2 \end{pmatrix}, \tag{6.48}$$

where

$$\delta_t^2 = \frac{3m_t^4}{16\pi^2 v_2^2} \ln \frac{m_{\tilde{t}_1}^2 m_{\tilde{t}_2}^2}{m_t^4}. \tag{6.49}$$

In the decoupling limit, $m_A \gg M_Z$, the new upper bound derived from the total CP-even neutral Higgs mass matrix is given by

$$m_h^2 \leq M_Z^2 + \delta_t^2. \tag{6.50}$$

It was found that, for a wide range of parameters, δ_t can reach 100 GeV. Therefore, the new numerical upper bound on the mass of the lightest CP-even Higgs boson of the MSSM is now

$$m_h \lesssim \sqrt{90^2 + 100^2} \text{ GeV} \sim 135 \text{ GeV}. \tag{6.51}$$

For completeness, we provide the full expressions of $\partial^2(\Delta V)/\partial v_i \partial v_j$:

$$\frac{\partial^2(\Delta V)}{\partial v_1^2} = \frac{m_t^4}{\sin^2 \beta} \left(\frac{\mu(A_t + \mu \cot \beta)}{m_{\tilde{t}_1}^2 - m_{\tilde{t}_2}^2} \right)^2 g(m_{\tilde{t}_1}^2, m_{\tilde{t}_2}^2), \tag{6.52}$$

$$\frac{\partial^2(\Delta V)}{\partial v_2^2} = \frac{m_t^4}{\sin^2 \beta} \left[\ln \frac{m_{\tilde{t}_1}^2 m_{\tilde{t}_2}^2}{m_t^4} + \frac{2A_t(A_t + \mu \cot \beta)}{m_{\tilde{t}_1}^2 - m_{\tilde{t}_2}^2} \ln \frac{m_{\tilde{t}_1}^2}{m_{\tilde{t}_2}^2} \right]$$

$$+ \frac{m_t^4}{\sin^2 \beta} \left(\frac{A_t(A_t + \mu \cot \beta)}{m_{\tilde{t}_1}^2 - m_{\tilde{t}_2}^2} \right)^2 g(m_{\tilde{t}_1}^2, m_{\tilde{t}_2}^2), \tag{6.53}$$

$$\frac{\partial^2(\Delta V)}{\partial v_1 \partial v_2} = \frac{m_t^4}{\sin^2 \beta} \frac{\mu(A_t + \mu \cot \beta)}{m_{\tilde{t}_1}^2 - m_{\tilde{t}_2}^2} \left[\ln \frac{m_{\tilde{t}_1}^2}{m_{\tilde{t}_2}^2} + \frac{A_t(A_t + \mu \cot \beta)}{m_{\tilde{t}_1}^2 - m_{\tilde{t}_2}^2} g(m_{\tilde{t}_1}^2, m_{\tilde{t}_2}^2) \right], \tag{6.54}$$

where the loop function $g(m_1^2, m_2^2)$ is given as

$$g(m_1^2, m_2^2) = 2 - \frac{m_1^2 + m_2^2}{m_1^2 - m_2^2} \ln \frac{m_1^2}{m_2^2}. \tag{6.55}$$

It turns out that the two-loop corrections reduce this upper bound by a few GeVs [103–105]. Hence, the MSSM predicts the following upper bound for the Higgs mass:

$$m_h \lesssim 130 \text{ GeV}. \tag{6.56}$$

Concerning the charged Higgs boson mass, it does not receive significant radiative corrections and the leading contribution is of $\mathcal{O}(\alpha m_t^2)$ [106–109]. A rather simple expression for the corrected charged Higgs boson mass, which gives a result that is reasonably accurate, is [110]

$$M_{H^\pm} = \sqrt{M_A^2 + M_W^2 - \epsilon_+} \quad \text{with} \quad \epsilon_+ = \frac{3G_\mu M_W^2}{4\sqrt{2}\pi^2} \left[\frac{\overline{m}_t^2}{\sin^2\beta} + \frac{\overline{m}_b^2}{\cos^2\beta} \right] \log\left(\frac{M_S^2}{m_t^2}\right). \tag{6.57}$$

However, for most practical purposes, the tree level result (*i.e.*, the above expression for $\epsilon_+ = 0$) suffices.

6.5 MSSM NEUTRAL HIGGS BOSONS PRODUCTION AT COLLIDERS

Based on Ref. [101] and the related HERWIG implementation [111], for $2 \to 2$ hard scattering subprocesses of the form $12 \to 34$, let us define the usual Mandelstam variables in terms of the partonic momenta as

$$s = (p_1 + p_2)^2, \tag{6.58}$$
$$t = (p_1 - p_3)^2, \tag{6.59}$$
$$u = (p_1 - p_4)^2. \tag{6.60}$$

As initial state partons are taken to have zero kinematic mass, it follows that $s + t + u = m_3^2 + m_4^2$, $s + t_3 + u_4 = 0$ and $ut - m_3^2 m_4^2 = s p_T^2$ where $t_3 = t - m_3^2$, $u_4 = u - m_4^2$, and p_T is the outgoing transverse momentum. The electromagnetic coupling is e, such that $\alpha = e^2/4\pi$. The QCD colour factors are $N_C = 3$ and $C_F = (N_C^2 - 1)/2N_C = 4/3$.

Colour flow directions are not stated explicitly as they are trivial in most of the processes whose matrix elements are given below. For a general treatment, see [112]. Matrix elements given here are summed over the final state colour and spin and averaged over the initial state colour and spin. As usual, this is indicated by a long overline. Statistical factors of two for identical particles in the final state are written explicitly.

Let us now consider the widths $\Gamma(q^2)$ appearing in the propagators of unstable massive particles with mass $M(q^2)$. In order to simplify the notation, let us define a function $\overline{\Gamma}(q^2)$, which is defined for positive q^2 by:

$$\sqrt{q^2}\Gamma(q^2) \equiv M(q^2)\overline{\Gamma}(q^2). \tag{6.61}$$

For negative q^2, the width is taken to be zero. The propagator is then given in general, again for positive q^2, by:

$$\frac{1}{q^2 - M^2(q^2) + i\sqrt{q^2}\Gamma(q^2)} \equiv \frac{1}{q^2 - M^2(q^2) + iM(q^2)\overline{\Gamma}(q^2)}. \tag{6.62}$$

In the remainder, we adopt the naive running width approximation and neglect the running of mass, so that:

$$\overline{\Gamma}(q^2) = q^2 \Gamma(M^2)/M^2. \tag{6.63}$$

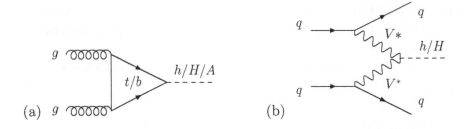

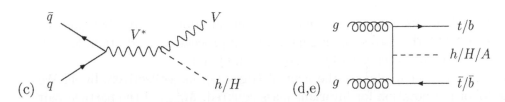

Figure 6.1 Representative Feynman diagrams for the dominant MSSM neutral Higgs production mechanisms in hadronic collisions.

The following coupling notations are used in processes involving EW interactions:

$$L_f = \frac{I_f^3 - Q_f \sin^2 \theta_W}{\cos \theta_W \sin \theta_W}, \qquad (6.64)$$

$$R_f = \frac{-Q_f \sin^2 \theta_W}{\cos \theta_W \sin \theta_W}, \qquad (6.65)$$

where Q_f and I_f^3 are the electric charge and the weak isospin for the SM fermions, respectively.

For hadronic collisions, the following five classes of production subprocesses are used in the search for MSSM neutral Higgs states [113]:

(a) gluon-gluon fusion via loops of heavy quarks $q = b$ and t and squarks $\tilde{q} = \tilde{b}$ and $\tilde{t}$,

(b) W^+W^- and ZZ fusion (also known as VBF),

(c) associated production with $W^\pm$ and Z bosons (also known as Higgs-Strahlung off heavy gauge bosons),

(d) associated production with pairs of heavy quarks $q\bar{q}$ (also known as HS off heavy quarks),

(e) associated production with pairs of heavy squarks $\tilde{q}\tilde{q}^*$ (also known as HS off heavy squarks).

As can be seen from Fig. 6.1, the scalar Higgs bosons h and H can be produced by all five subprocesses (a)–(e), whereas the pseudoscalar A is only produced via the first and the last two, as the AVV vertex, with $V = W^\pm, Z$, is absent at tree level (in a CP-conserving theory). Higgs bosons are also produced in decays of heavier (s)particles.

For leptonic collisions, the following channels afford experimental exploitation:

(f) W^+W^- and ZZ fusion,

(g) associated production with Z,

(h) hA and HA production.

Of these, (f) and (g) are expressed by formulae that are obtained by changing the Z boson couplings in the corresponding hadronic subprocesses. In addition, for (g), the colour factor $1/N_C$ is replaced by 1 ($N_C = 3$ in QCD).

In the following, let us denote the neutral Higgs bosons collectively by Φ. We define the following notation for invariant mass squared, $M^2_{[ab]}$, of the particle pair $[ab]$:

$$M^2_{[a_i b_i]} = (p_a + p_b)^2 = s, \tag{6.66}$$

$$M^2_{[a_f b_f]} = (p_a + p_b)^2, \tag{6.67}$$

$$M^2_{[a_i b_f]} = (p_a - p_b)^2, \tag{6.68}$$

where the subscripts i and f indicate particles in the initial and final states, respectively. $M^2_{[ab]}$ is negative only in the last of these three cases. Finally, notice that, for the purpose of making our formulae useful to the reader who wants to carry out explicit calculations of realistic processes, we present the completely differential cross sections (i.e., apart from a common flux factor, the scattering 'matrix elements'), unlike the case of many other textbooks (and indeed, specialised reviews), which report integrated (i.e., fully inclusive) cross sections. Their simple (analytical or numerical) integration over the LIPS of the processes presented (and also on the PDFs for the case of hadro-production), ideally in the presence of acceptance and selection cuts, offers the reader a chance to directly compare with experimental results. Finally, we limit ourselves to the lowest order. (Plenty of specialised literature exists on the calculation of higher order effects in the full MSSM, see [100] and references therein for a review.)

For the gluon-gluon fusion subprocess, class (a), the matrix elements are cast in the form seen in [114]:

$$\overline{|\mathcal{M}|^2}(gg \to h, H) = \frac{G_F \alpha_3^2 m_{h,H}^4}{36\sqrt{2}\,\pi^2(N_C^2 - 1)}$$

$$\times \left| \sum_q g_q^{h,H} A_q^{h,H}(\tau_q) + \sum_{\tilde{q}} g_{\tilde{q}}^{h,H} A_{\tilde{q}}^{h,H}(\tau_{\tilde{q}}) \right|^2, \tag{6.69}$$

Higgs boson		$g_{u,\nu}$	$g_{d,\ell}$	$g_{W^\pm, Z}$
SM	h	1	1	1
MSSM	h	$\cos\alpha/\sin\beta$	$-\sin\alpha/\cos\beta$	$\sin(\beta-\alpha)$
	H	$\sin\alpha/\sin\beta$	$\cos\alpha/\cos\beta$	$\cos(\beta-\alpha)$
	A	$1/\tan\beta$	$\tan\beta$	0

Table 6.1 Higgs coupling coefficients in the MSSM to isospin $+1/2$ and $-1/2$ fermions and to the gauge bosons $W^\pm, Z$, as first appearing in Eqs. (6.69) and (6.70).

Higgs boson		$g_{\tilde{f}_i}$
SM	h	0
MSSM	h	$m_{\tilde{f}_i}^{-2}\left[m_f^2 g_f^h + M_Z^2 \cos\theta_W \sin\theta_W (L_f Q_{Li}^2 - R_f Q_{Ri}^2)\sin(\alpha+\beta)\right.$ $\left. + (\mu m_f(\delta_{fu}\frac{\sin\alpha}{\sin\beta} - \delta_{fd}\frac{\cos\alpha}{\cos\beta}) + A_f m_f g_f^h)Q_{Li}Q_{Ri}\right]$
	H	$m_{\tilde{f}_i}^{-2}\left[m_f^2 g_f^H + M_Z^2 \cos\theta_W \sin\theta_W (R_f Q_{Ri}^2 - L_f Q_{Li}^2)\cos(\alpha+\beta)\right.$ $\left. + (\mu m_f(-\delta_{fu}\frac{\cos\alpha}{\sin\beta} - \delta_{fd}\frac{\sin\alpha}{\cos\beta}) + A_f m_f g_f^H)Q_{Li}Q_{Ri}\right]$
	A	0

Table 6.2 MSSM Higgs coupling coefficients to sfermions $\tilde{f} = \tilde{q}, \tilde{\ell}$, as first appearing in Eq. (6.69). For the leptonic case, the mixing matrices Q_{Li} and Q_{Ri} should be replaced by the corresponding slepton mixing matrices L_{Li} and L_{Ri}, respectively.

$$\overline{|\mathcal{M}|^2}(gg \to A) = \frac{G_F \alpha_3^2 m_A^4}{36\sqrt{2}\,\pi^2(N_C^2-1)}\left|\sum_q g_q^A A_q^A(\tau_q)\right|^2, \qquad (6.70)$$

for scalar and pseudoscalar production, respectively, where G_F is the Fermi coupling constant. The couplings g_q and $g_{\tilde{q}}$ are those appearing in Tabs. 6.1 and 6.2, respectively. We have defined $\tau = 4m^2/m_\Phi^2$, where m_Φ is the mass of the Higgs boson concerned and m is the mass of the particle inside the loop. The functions A are given by

$$A_q^{h,H}(\tau) = \frac{3}{2}\tau[1 + (1-\tau)f(\tau)], \qquad (6.71)$$

$$A_{\tilde{q}}^{h,H}(\tau) = -\frac{3}{4}[1 - \tau f(\tau)] \qquad (6.72)$$

for the scalars and

$$A_q^A(\tau) = \tau f(\tau) \qquad (6.73)$$

for the pseudoscalar, with the function $f(\tau)$ given by

$$
f(\tau) = \begin{cases} \arcsin^2 \dfrac{1}{\sqrt{\tau}} & \tau \geq 1, \\[2ex] -\dfrac{1}{4}\left(\log \dfrac{1 + \sqrt{1-\tau}}{1 - \sqrt{1-\tau}} - i\pi\right)^2 & \tau < 1. \end{cases} \tag{6.74}
$$

For process (b), the matrix element for the $W^{\pm}W^{\mp}$ fusion process is given by

$$
\overline{|\mathcal{M}|^2}(q_1 q_2' \to q_3'' q_4''' \Phi) = 2M_W^2 (g_{W^{\pm}}^{\Phi})^2 \left(\frac{e^2}{2\sin^2\theta_W}\right)^3
$$
$$
\times \frac{|V_{\text{CKM}}[qq'']|^2 |V_{\text{CKM}}[q'q''']|^2 M_{[12]}^2 M_{[34]}^2}{(M_{[13]}^2 - M_W^2)(M_{[24]}^2 - M_W^2)}, \tag{6.75}
$$

whereas for ZZ fusion we have:

$$
\overline{|\mathcal{M}|^2}(q_1 q_2' \to q_3 q_4' \Phi) = 4M_Z^2 (g_Z^{\Phi})^2 \left(\frac{e^2}{4\sin^2\theta_W \cos^2\theta_W}\right)^3
$$
$$
\times \frac{(L_q^2 L_{q'}^2 + R_q^2 R_{q'}^2)M_{[12]}^2 M_{[34]}^2 + (L_q^2 R_{q'}^2 + R_q^2 L_{q'}^2)M_{[14]}^2 M_{[23]}^2}{(M_{[13]}^2 - M_Z^2)(M_{[24]}^2 - M_Z^2)}. \tag{6.76}
$$

The coupling coefficients g_V^{Φ} are as given in Tab. 6.1. The formulae are identical to those for the case of leptonic collisions, with the replacement of the CKM matrix elements by unity for the $W^{\pm}W^{\mp}$ fusion subprocess and the substitution of the couplings L, R with the appropriate leptonic couplings L_e, R_e for the ZZ fusion subprocess.

For process (c), the matrix elements are given by

$$
\overline{|\mathcal{M}|^2}(u_1 \bar{d}_2 \to W_3^+ \Phi_4) = (g_{W^{\pm}}^{\Phi})^2 \frac{1}{N_C}\left(\frac{e^2}{2\sin^2\theta_W}\right)^2 \left(\frac{1}{2}\right)
$$
$$
\times |V_{\text{CKM}}[ud]|^2 \frac{sp_T^2 + 2M_W^2 s}{(s - M_W^2)^2 + M_W^2 \overline{\Gamma}_W^2}, \tag{6.77}
$$

$$
\overline{|\mathcal{M}|^2}(q_1 \bar{q}_2 \to Z_3 \Phi_4) = (g_Z^{\Phi})^2 \frac{1}{N_C}\left(\frac{e^2}{4\sin^2\theta_W \cos^2\theta_W}\right)^2
$$
$$
\times (L_q^2 + R_q^2) \frac{sp_T^2 + 2M_Z^2 s}{(s - M_Z^2)^2 + M_Z^2 \overline{\Gamma}_Z^2}. \tag{6.78}
$$

The coupling coefficients g_V^{Φ} are again as given in Tab. 6.1. The above formula, for the Z mediated subprocess, is immediately applicable to the leptonic case with the substitution of the colour factor 1 to replace $1/N_C$. The couplings L_q, R_q are replaced by L_e, R_e in this case.

For process (d), the matrix element for the $q\bar{q}$ initiated subprocess is given by

$$\overline{|\mathcal{M}|^2}(q_1\bar{q}_2 \to q_3\bar{q}_4\Phi_5) = \frac{e^2 g_3^4 (g_q^\Phi)^2}{2\sin^2\theta_W M_W^2 s} \left(\frac{C_F}{N_C}\right)\mathcal{G}\left[\frac{1}{2\mathcal{G}_5} - \frac{M^2 s\mathcal{G}}{2}\right.$$
$$+ \left. M^2(\mathcal{G}_3 p_{T3}^2 + \mathcal{G}_4 p_{T4}^2) - \mathcal{G}_5(s + M^2)p_{T5}^2\right].$$

(6.79)

Here, $M^2 = m_\Phi^2$ for A and $M^2 = m_\Phi^2 - 4m_q^2$ for h, H. The p_T's are the transverse momenta of the final state particles. These can be expressed as $p_{T_i}^2 = 4p_1 \cdot p_i \, p_2 \cdot p_i/s - m_i^2$ (where $i = 3, 4, 5$) as the incoming particles are considered massless. The propagator factors are defined by:

$$\mathcal{G}_3 = \frac{1}{M_{[35]}^2 - m_q^2}, \tag{6.80}$$

$$\mathcal{G}_4 = \frac{1}{M_{[45]}^2 - m_q^2}, \tag{6.81}$$

$$\mathcal{G} = \mathcal{G}_3 + \mathcal{G}_4, \tag{6.82}$$

$$\mathcal{G}_5 = \frac{\mathcal{G}_3 \mathcal{G}_4}{\mathcal{G}}. \tag{6.83}$$

For the case $q = b$, some care must be excercised in the calculation of the cross section in order to avoid double counting of the $2 \to 3$ subprocess with the companion $2 \to 1$ and $2 \to 2$ channels, the matrix elements of which are:

$$\overline{|\mathcal{M}|^2}(b_1\bar{b}_2 \to \Phi_3) = \frac{e^2 m_H^2 m_b^2 (g_b^\Phi)^2}{8\sin^2\theta_W M_W^2 N_C} \tag{6.84}$$

and

$$\overline{|\mathcal{M}|^2}(g_1 b_2 \to b_3 \Phi_4) = \left(\frac{g_3^2}{2N_C}\right)\left(\frac{e^2}{2\sin\theta_W^2}\right)\left(\frac{m_b^2(g_b^\Phi)^2}{2M_W^2}\right)$$
$$\times \left(\frac{-u_4^2}{st_3}\right)\left[1 + 2\frac{m_4^2 - m_3^2}{u_4}\left(1 + \frac{m_3^2}{t_3} + \frac{m_4^2}{u_4}\right)\right],$$

(6.85)

respectively. The $2 \to 3$ matrix element for $gg \to q\bar{q}\Phi$ is unfortunately not publishable because of its length.

For process (e), the matrix element for the $q\bar{q}$ initated subprocess is given by

$$\overline{|\mathcal{M}|^2}(q_1\bar{q}_2 \to \tilde{q}_3\tilde{q}_4^*\Phi_5) = \left(\frac{g_3^4 e^2 g_{\tilde{q}}^{\Phi^2} m_{\tilde{q}}^2}{\sin^2\theta_W M_W^2 s}\right)\left(\frac{C_F}{N_C}\right)$$
$$\times \left[|\mathcal{G}_3|^2 p_{T3}^2 + |\mathcal{G}_4|^2 p_{T4}^2 - 2\mathrm{Re}\,(\mathcal{G}_3^*\mathcal{G}_4)\,p_{T3}\cdot p_{T4}\right].$$

(6.86)

The propagator factors are given by

$$\mathcal{G}_3 = \frac{1}{(p_4 + p_5)^2 - m_3^2 + im_3\overline{\Gamma}_3}, \tag{6.87}$$

$$\mathcal{G}_4 = \frac{1}{(p_3 + p_5)^2 - m_4^2 + im_4\overline{\Gamma}_4}. \tag{6.88}$$

The dot product of two transverse momenta is two dimensional, *i.e.*, $p_{T3} \cdot p_{T4} = p_{x3}p_{x4} + p_{y3}p_{y4}$.

This time the gg initiated subprocess can be written in a compact form, as follows:

$$\overline{|\mathcal{M}|^2}(g_1g_2 \to \tilde{q}_3\tilde{q}_4^*\Phi_5) = \left(\frac{g_3^4 e^2 g_{\tilde{q}}^{\Phi 2} m_{\tilde{q}}^2 N_C}{4\sin^2\theta_W M_W^2 (N_C^2 - 1)}\right) \sum_{i=x,y}\sum_{j=x,y} \left[|\mathcal{A}_t(i,j)|^2 \right.$$
$$\left. + |\mathcal{A}_u(i,j)|^2 - \frac{1}{N_C^2}|\mathcal{A}_t(i,j) + \mathcal{A}_u(i,j)|^2\right]. \tag{6.89}$$

The two amplitude functions $\mathcal{A}_t$ and $\mathcal{A}_u$ are given by

$$\mathcal{A}_t(i,j) = 2\left(p_{i3}p_{j4}\mathcal{G}_{24}\mathcal{G}_{13} + p_{i3}p_{j3}\mathcal{G}_{13}\mathcal{G}_{45} + p_{i4}p_{j4}\mathcal{G}_{24}\mathcal{G}_{35}\right)$$
$$- \frac{\delta_{ij}}{\sqrt{s}}\left(p_{z3}\mathcal{G}_{45} - p_{z4}\mathcal{G}_{35}\right) - \frac{\delta_{ij}}{2}\left(\mathcal{G}_{35} + \mathcal{G}_{45}\right), \tag{6.90}$$

$$\mathcal{A}_u(i,j) = 2\left(p_{i4}p_{j3}\mathcal{G}_{14}\mathcal{G}_{23} + p_{i3}p_{j3}\mathcal{G}_{23}\mathcal{G}_{45} + p_{i4}p_{j4}\mathcal{G}_{14}\mathcal{G}_{35}\right)$$
$$+ \frac{\delta_{ij}}{\sqrt{s}}\left(p_{z3}\mathcal{G}_{45} - p_{z4}\mathcal{G}_{35}\right) - \frac{\delta_{ij}}{2}\left(\mathcal{G}_{35} + \mathcal{G}_{45}\right). \tag{6.91}$$

The propagator factors are given by

$$\mathcal{G}_{13} = 1/\left((p_1 + p_3)^2 - m_3^2\right), \tag{6.92}$$

$$\mathcal{G}_{23} = 1/\left((p_2 + p_3)^2 - m_3^2\right), \tag{6.93}$$

$$\mathcal{G}_{14} = 1/\left((p_1 + p_4)^2 - m_4^2\right), \tag{6.94}$$

$$\mathcal{G}_{24} = 1/\left((p_2 + p_4)^2 - m_4^2\right), \tag{6.95}$$

$$\mathcal{G}_{35} = 1/\left((p_3 + p_5)^2 - m_4^2 + im_4\overline{\Gamma}_4\right), \tag{6.96}$$

$$\mathcal{G}_{45} = 1/\left((p_4 + p_5)^2 - m_3^2 + im_3\overline{\Gamma}_3\right). \tag{6.97}$$

For process (h), the matrix element is given by

$$\overline{|\mathcal{M}|^2}(e_1^- e_2^+ \to h_3 A_4, H_3 A_4) = e^4 s p_T^2 \cdot \frac{L_\nu^2\left(L_e^2 + R_e^2\right)\left(\cos^2(\beta - \alpha), \sin^2(\beta - \alpha)\right)}{(s - M_Z^2)^2 + M_Z^2\overline{\Gamma}_Z^2}. \tag{6.98}$$

Note that the coupling factor $\cos^2(\beta - \alpha)$ is for hA production whereas $\sin^2(\beta - \alpha)$ is for HA production.

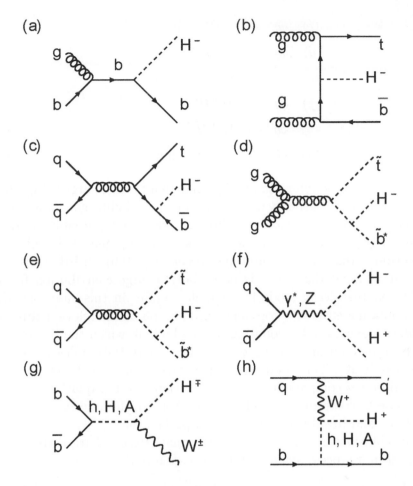

Figure 6.2 Representative Feynman diagrams for the dominant MSSM charged Higgs production mechanisms in hadronic collisions.

6.6 MSSM CHARGED HIGGS BOSON PRODUCTION AT COLLIDERS

We adopt here the same approach as in the previous section concerning the presentation of the results, that is, we evaluate scattering matrix elements at lowest order. (For reviews on higher order effects in the full MSSM, again, see [100].) For hadronic collisions, the following production modes are available [101]:

$$(a) \quad gb \quad \rightarrow \quad tH^- + \text{h.c.}, \tag{6.99}$$

$$(b) \quad gg \quad \rightarrow \quad t\bar{b}H^- + \text{h.c.}, \tag{6.100}$$

$$(c) \quad q\bar{q} \quad \to \quad t\bar{b}H^- + \text{h.c.}, \tag{6.101}$$

$$(d) \quad gg \quad \to \quad \tilde{t}\,\tilde{b}^*H^- + \text{h.c.}, \tag{6.102}$$

$$(e) \quad q\bar{q} \quad \to \quad \tilde{t}\,\tilde{b}^*H^- + \text{h.c.}, \tag{6.103}$$

$$(f) \quad q\bar{q} \quad \to \quad H^+H^-, \tag{6.104}$$

$$(g) \quad b\bar{b} \quad \to \quad W^{\pm}H^{\mp}, \tag{6.105}$$

$$(h) \quad qb \quad \to \quad q'bH^+ + \text{h.c.} \tag{6.106}$$

See Fig. 6.2 for the corresponding sample Feynman diagrams. Charged Higgs bosons are also produced through the decay of heavier objects, including gauginos and third generation sfermions. Perhaps most notably, an important potential production mode for charged Higgs bosons is through the decay of top quarks. In this vein, it should be mentioned that, when computing processes (6.100), (6.101) and (6.106), one should bear in mind that there is potentially a large contribution from top quark decay when this decay mode is available. Again, in this case, one should take care of simulating either the top quark production or the relevant three-body process in order to avoid double counting. In the limit in which the bottom quark is generated by the collinear splitting $g \to b\bar{b}$, process (6.100) is described by the two body process (6.99). This contribution becomes significant if the mass of the charged Higgs boson is comparable to or greater than the top quark mass. Again, one should choose to compute either of the two processes in order to avoid double counting (or both and to subtract the common part).

For leptonic collisions, in addition to the top quark decay, charged Higgs bosons can be produced in a process analogous to process (6.104):

$$e^-e^+ \to H^+H^-. \tag{6.107}$$

The matrix element is obtained immediately from the hadronic case by changing the appropriate photon and Z couplings and removing the colour factor $1/N_C$.

Process (6.99) is given by

$$\overline{|\mathcal{M}|^2}(g_1 b_2 \to t_3 H_4^-) = \left(\frac{g_3^2}{2N_C}\right)\left(\frac{e^2}{2\sin\theta_W^2}\right)|V_{\text{CKM}}[tb]|^2 \left(\frac{m_b^2 \tan^2\beta + m_t^2 \cot^2\beta}{2M_W^2}\right)$$
$$\times \left(\frac{-u_4^2}{st_3}\right)\left[1 + 2\frac{m_4^2 - m_3^2}{u_4}\left(1 + \frac{m_3^2}{t_3} + \frac{m_4^2}{u_4}\right)\right]. \tag{6.108}$$

Process (6.100) yields, like in the evaluation of the neutral Higgs boson production process $gg \to q\bar{q}\Phi$, an intractable analytical expression, one that we chose not to reproduce here.

Process (6.101) can be evaluated using a scattering matrix element cast in the more general formula for $q_1\bar{q}_2 \to q_3\bar{q}_4\Phi$, which is given by

$$\overline{|\mathcal{M}|^2}(q_1\bar{q}_2 \to q_3\bar{q}_4\Phi) = \left(\frac{g_3^4 C_F}{2N_C}\right)\frac{2\phi_H^2}{s}\left[(p_{T_3}^2 + p_{T_4}^2 - p_{T_5}^2) + 4p_0 \cdot p_3\, p_0 \cdot p_4|\mathcal{G}|^2\right.$$

$$+\quad 2\left(p_3 \cdot p_4 - \lambda m_3 m_4\right)\left[-s|\mathcal{G}|^2 + 2\mathrm{Re}\left(\mathcal{G}\mathcal{G}_3^* p_{T_4}^2 + \mathcal{G}\mathcal{G}_4^* p_{T_3}^2 - \mathcal{G}_3\mathcal{G}_4^* p_{T_5}^2\right)\right]$$

$$-\quad 4\mathrm{Re}\left[\mathcal{G}\mathcal{G}_3^* p_{T_4}^2 + \mathcal{G}\mathcal{G}_4^* p_{T_3}^2 + \mathcal{G}\left(\mathcal{G}_3^* p_0 \cdot p_4 + \mathcal{G}_4^* p_0 \cdot p_3\right)\frac{p_{T_3}^2 + p_{T_4}^2 - p_{T_5}^2}{2}\right]$$

$$-\quad 4\mathrm{Im}\left[\mathcal{G}\mathcal{G}_3^* p_{z_4} - \mathcal{G}\mathcal{G}_4^* p_{z_3}\right]\lambda'\sqrt{s}\left(p_{x_4}p_{y_3} - p_{x_3}p_{y_4}\right)\right], \qquad (6.109)$$

with $p_0 = p_1 + p_2$. Denoting the $q_3\bar{q}_4\Phi$ vertex as $(g_S + g_P\gamma_5)$, the various coupling coefficients appearing above are defined by:

$$\phi_H^2 = g_S^2 + g_P^2, \qquad (6.110)$$

$$\lambda = \frac{g_S^2 - g_P^2}{\phi_H^2}, \qquad (6.111)$$

$$\lambda' = \frac{2g_S g_P}{\phi_H^2}. \qquad (6.112)$$

Process (6.102) yields at matrix element level expressions which are the same as Eqs. (6.89)–(6.97), but the coupling coefficient needs to be changed. Let us consider the vertex involving an incoming H^+ and outgoing $\tilde{t}_i\bar{b}_j^*$. We have the following replacement:

$$\frac{g_{\bar{q}}m_{\bar{q}}^2}{M_W^2} \to \frac{1}{\sqrt{2}}\left[\left(s_{2\beta} - \frac{m_b^2\tan\beta + m_t^2\cot\beta}{M_W^2}\right)Q_{Li}^t Q_{Lj}^b + \frac{m_b}{M_W^2}(\mu - A_b\tan\beta)Q_{Li}^t Q_{Rj}^b\right.$$

$$\left. - \frac{m_t m_b}{M_W^2}(\tan\beta + \cot\beta)Q_{Ri}^t Q_{Rj}^b + \frac{m_t}{M_W^2}(\mu - A_t\cot\beta)Q_{Ri}^t Q_{Lj}^b\right]. \qquad (6.113)$$

The scattering matrix element for process (6.104) is given by

$$\overline{|\mathcal{M}|^2} = \frac{e^4 s p_T^2}{N_C}$$

$$\times\quad \left[\left|\frac{Q_q Q_{H^+}}{s} + \frac{L_q L_{H^+}}{s - M_Z^2 + iM_Z\overline{\Gamma}_Z}\right|^2 + \left|\frac{Q_q Q_{H^+}}{s} + \frac{R_q L_{H^+}}{s - M_Z^2 + iM_Z\overline{\Gamma}_Z}\right|^2\right], \qquad (6.114)$$

where $Q_{H^+} = +1$ and L_{H^+} is given by Eq. (6.64) with $I_f^3 = +1/2$. For the leptonic case the subscripts q should be replaced with e and the colour factor $1/N_C$ should be replaced by 1. The matrix element for process (6.105) is as given in [115]. The matrix element expression for (6.106) from [116] can be manipulated to obtain a compact expression which sets the kinematic bottom quark mass to zero. For the process $b_1 u_2 \to b_3 d_4 H_5^+$ we have:

$$\overline{|\mathcal{M}|^2}(b_1 u_2 \to b_3 d_4 H_5^+) = \frac{1}{2}\left(\frac{e^2}{2\sin^2\theta_W}\right)^3 |V_{\text{CKM}}[ud]|^2 \frac{1}{(M_{[24]}^2 - M_W^2)^2}$$

$$\times \left[|V_{\text{CKM}}[tb]|^2 \overline{|A_t|^2} + \overline{|A_\Phi|^2} + \text{Re}(V_{\text{CKM}}[tb] \cdot \overline{2A_\Phi^* A_t})\right], \quad (6.115)$$

where

$$\overline{|A_t|^2} = 2M_{[12]}^2 |\mathcal{G}_t|^2 \left[\left(\frac{m_b \tan\beta}{M_W}\right)^2 \frac{[3545]}{2} + \left(\frac{m_t}{M_W \tan\beta}\right)^2 m_t^2 M_{[34]}^2\right],$$

$$(6.116)$$

$$\overline{|A_\Phi|^2} = (\mathcal{G}_{h,H}^2 + \mathcal{G}_A^2)(-M_{[13]}^2)\frac{[2545]}{2}, \quad (6.117)$$

$$\text{Re}(\overline{2A_\Phi^* A_t}) = \frac{m_b \tan\beta}{M_W}(\mathcal{G}_{h,H} + \mathcal{G}_A)\Big[2\text{Re}(\mathcal{G}_t)(M_{[45]}^2 - m_{H^\pm}^2)(M_{[12]}^2)(-M_{[13]}^2) -$$

$$- \text{Re}(\mathcal{G}_t)M_{[45]}^2[1243] + \text{Im}(\mathcal{G}_t)M_{[45]}^2 \cdot 4\det(1234)\Big]. \quad (6.118)$$

Here, we have defined $[abcd] = 4(p_a \cdot p_b\, p_c \cdot p_d + p_a \cdot p_d\, p_b \cdot p_c - p_a \cdot p_c\, p_b \cdot p_d)$ and $\det(abcd) = \det(p_a^\mu, p_b^\nu, p_c^\rho, p_d^\sigma) = \sum_{\mu\nu\rho\sigma} \epsilon_{\mu\nu\rho\sigma} p_a^\mu p_b^\nu p_c^\rho p_d^\sigma$ where $\epsilon_{0123} = +1$.

The propagators are defined by:

$$\mathcal{G}_t = \frac{1}{M_{[35]}^2 - m_t^2 + im_t\overline{\Gamma}_t} = \frac{M_{[35]}^2 - m_t^2 - im_t\overline{\Gamma}_t}{(M_{[35]}^2 - m_t^2)^2 + m_t^2\overline{\Gamma}_t^2}, \quad (6.119)$$

$$\mathcal{G}_{h,H} = \frac{m_b \sin\alpha\cos(\beta - \alpha)}{(M_{[13]}^2 - m_h^2)M_W\cos\beta} + \frac{m_b\cos\alpha\sin(\beta-\alpha)}{(M_{[13]}^2 - m_H^2)M_W\cos\beta}, \quad (6.120)$$

$$\mathcal{G}_A = \frac{m_b\tan\beta}{M_W(M_{[13]}^2 - m_A^2)}. \quad (6.121)$$

Processes with antiquarks in the initial state can be obtained by exchanging the subscripts $(1 \leftrightarrow 3)$ for the subprocess $\bar{b}u \to \bar{b}dH^+$, $(2 \leftrightarrow 4)$ for the subprocess $b\bar{d} \to b\bar{u}H^+$ and both for the subprocess $\bar{b}\bar{d} \to \bar{b}\bar{u}H^+$. Note that the definitions of the invariant masses squared are such that these imply momentum substitutions of the form $(p_1 \leftrightarrow -p_3)$.

6.7 MSSM NEUTRAL AND CHARGED HIGGS BOSON DECAYS

In the case of Higgs boson decays, we naturally compute integrated widths, as there is no physics observable associated to differential decay widths. Again, we limit ourselves to the lowest order and refer the reader to, *e.g.*, [100] for a review of higher order effects. This section is largely based on [114].

6.7.1 Decays into lepton and heavy quark pairs

At lowest order, the leptonic decay width of neutral MSSM Higgs boson[2] decays is given by [117, 118]

$$\Gamma\left[\Phi \to \ell^+\ell^-\right] = \frac{G_F m_\Phi}{4\sqrt{2}\pi}(g_\ell^\Phi)^2 m_\ell^2 \beta^p, \tag{6.122}$$

where g_ℓ^Φ denotes the corresponding MSSM coupling, as presented in Tab. 6.1, $\beta = (1 - 4m_\ell^2/m_\Phi^2)^{1/2}$ is the velocity of the final-state leptons and $p = 3\,(1)$ the exponent for scalar (pseudoscalar) Higgs particles.

The analogous expression for the leptonic decays of the charged Higgs state reads as follows:

$$\Gamma\left[H^+ \to \nu\bar{\ell}\right] = \frac{G_F m_{H^\pm}}{4\sqrt{2}\pi} m_\ell^2 \tan^2\beta \left(1 - \frac{m_\ell^2}{m_{H^\pm}^2}\right)^3. \tag{6.123}$$

The decay widths of the MSSM neutral Higgs particles into quarks can instead be obtained as follows:

$$\Gamma[\Phi \to q\bar{q}] = \frac{3G_F m_\Phi}{4\sqrt{2}\pi}(g_q^\Phi)^2 m_q^2 \beta^p, \tag{6.124}$$

where β and p are as above.

The partial decay width of the charged Higgs particles into heavy quarks may be written as [119, 120]

$$\begin{aligned}
\Gamma\left[H^+ \to u\bar{d}\right] &= \frac{3G_F m_{H^\pm}}{4\sqrt{2}\pi}|V_{ud}|^2 \lambda^{\frac{1}{2}}\left[(1 - \mu_u - \mu_d)\left(\frac{m_u^2}{\tan^2\beta} + m_d^2 \tan^2\beta\right)\right.\\
&\quad \left. - 4m_u m_d\sqrt{\mu_u \mu_u}\right],
\end{aligned} \tag{6.125}$$

where $\mu_i = m_i^2/m_{H^\pm}^2$ and $\lambda = (1 - \mu_u - \mu_d)^2 - 4\mu_u \mu_d$ denotes the usual two-body phase-space function.

Even below the $t\bar{t}$ threshold, heavy neutral Higgs boson decays into off-shell top quarks are sizable, as shown in [121, 122], thus modifying the decay profile of these Higgs particles significantly in the threshold region $m_\Phi \approx 2m_t$. The dominant below-threshold contributions can be obtained from the SM expression

$$\frac{d\Gamma}{dx_1 dx_2}(H \to t\bar{t} \to Wtb) = (g_t^H)^2 \frac{d\Gamma}{dx_1 dx_2}(h_{SM} \to t\bar{t} \to Wtb), \tag{6.126}$$

where

$$\frac{d\Gamma}{dx_1 dx_2}(h_{SM} \to t\bar{t} \to Wtb) = \frac{3G_F^2}{32\pi^3}m_t^2 m_{h_{SM}}^3 \frac{\Gamma_0}{y_1^2 + \gamma_t \kappa_t}, \tag{6.127}$$

with the reduced energies $x_{1,2} = 2E_{t,b}/m_A$, the scaling variables $y_{1,2} = 1 - x_{1,2}$,

---

[2]Again, in the following we denote the different types of neutral Higgs particles by $\Phi = h, H, A$.

$\kappa_i = m_i^2/m_A^2$, and the reduced decay widths of the virtual particles $\gamma_i = \Gamma_i^2/m_A^2$. The squared amplitude may be written as [121, 122]

$$\Gamma_0 = y_1^2(1 - y_1 - y_2 + \kappa_W - \kappa_t) + 2\kappa_W(y_1 y_2 - \kappa_W) - \kappa_t(y_1 y_2 - 2y_1 - \kappa_W - \kappa_t). \quad (6.128)$$

The differential decay width in Eq. (6.127) has to be integrated over the x_1, x_2 region, which is bounded by

$$\left| \frac{2(1 - x_1 - x_2 + \kappa_t + \kappa_b - \kappa_W) + x_1 x_2}{\sqrt{x_1^2 - 4\kappa_t}\sqrt{x_2^2 - 4\kappa_b}} \right| \leq 1. \quad (6.129)$$

The corresponding dominant below-threshold contributions of the pseudoscalar Higgs particle are given by [121, 122]

$$\frac{d\Gamma}{dx_1 dx_2}(A \to t\bar{t} \to Wtb) = \frac{3G_F^2}{32\pi^3} m_t^2 m_A^3 (g_t^A)^2 \frac{\Gamma_0}{y_1^2 + \gamma_t \kappa_t}. \quad (6.130)$$

The differential decay width of (6.130) also has to be integrated over the x_1, x_2 region given by Eq. (6.129).

Similarly to the neutral Higgs case, below the $t\bar{b}$ threshold, off-shell decays $H^+ \to t\bar{b} \to b\bar{b}W^+$ are important. For $m_{H^\pm} < m_t + m_b - \Gamma_t$, their expression can be cast into the form [121, 122]

$$\begin{aligned}
\Gamma(H^+ \to t\bar{b} \to Wb\bar{b}) &= \frac{3G_F^2 m_t^4 m_{H^\pm}}{64\pi^3 \tan^2\beta} \left[\frac{\kappa_W^2}{\kappa_t^3}(4\kappa_W\kappa_t + 3\kappa_t - 4\kappa_W)\log\left[\frac{\kappa_W(\kappa_t - 1)}{\kappa_t - \kappa_W}\right] \right. \\
&\quad + \frac{1 - \kappa_W}{\kappa_t^2}(3\kappa_t^3 - \kappa_t\kappa_W - 2\kappa_t\kappa_W^2 + 4\kappa_W^2) + \kappa_W\left(4 - \frac{3}{2}\kappa_W\right) \\
&\quad \left. + (3\kappa_t^2 - 4\kappa_t - 3\kappa_W^2 + 1)\log\left(\frac{\kappa_t - 1}{\kappa_t - \kappa_W}\right) - \frac{5}{2} \right],
\end{aligned} \quad (6.131)$$

with the scaling variables $\kappa_i = m_i^2/m_{H^\pm}^2 \ (i = t, W)$. The b mass has been neglected in Eq. (6.131).

6.7.2 Gluonic decay modes

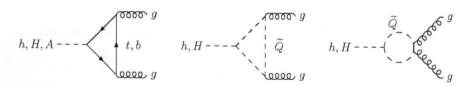

Figure 6.3 Typical diagrams contributing to $\Phi \to gg$ at lowest order.

The leading order width for $h, H \to gg$ is generated by quark and squark loops.

Φ		$H^\pm$	$\tilde{\chi}_i^\pm$
SM	h	0	0
MSSM	h	$\dfrac{M_W^2}{m_{H^\pm}^2}\left[\sin(\beta-\alpha)+\dfrac{\cos 2\beta \sin(\beta+\alpha)}{2\cos^2\theta_W}\right]$	$2\dfrac{M_W}{M_{\tilde{\chi}_i^\pm}}(S_{ii}\cos\alpha - Q_{ii}\sin\alpha)$
	H	$\dfrac{M_W^2}{m_{H^\pm}^2}\left[\cos(\beta-\alpha)-\dfrac{\cos 2\beta \cos(\beta+\alpha)}{2\cos^2\theta_W}\right]$	$2\dfrac{M_W}{M_{\tilde{\chi}_i^\pm}}(S_{ii}\sin\alpha + Q_{ii}\cos\alpha)$
	A	0	$2\dfrac{M_W}{M_{\tilde{\chi}_i^\pm}}(-S_{ii}\cos\beta - Q_{ii}\sin\beta)$

Φ		$\tilde{f}_{L,R}$
SM	h	0
MSSM	h	$\dfrac{m_f^2}{m_{\tilde{f}}^2}g_f^h \mp \dfrac{M_Z^2}{m_{\tilde{f}}^2}(I_f^3 - Q_f\sin^2\theta_W)\sin(\alpha+\beta)$
	H	$\dfrac{m_f^2}{m_{\tilde{f}}^2}g_f^H \pm \dfrac{M_Z^2}{m_{\tilde{f}}^2}(I_f^3 - Q_f\sin^2\theta_W)\cos(\alpha+\beta)$
	A	0

Table 6.3 MSSM Higgs couplings to charged Higgs bosons, charginos and sfermions relative to SM couplings. Q_{ii} and S_{ii} $(i = 1,2)$ are related to the mixing angles between the charginos $\tilde{\chi}_1^\pm$ and $\tilde{\chi}_2^\pm$, see [33–35].

The contributing diagrams are depicted in Fig. 6.3. The partial decay widths are given by [33, 34, 123, 124]

$$\Gamma(h, H \to gg) = \frac{G_F \alpha_3^2 m_{h,H}^3}{36\sqrt{2}\pi^3}\left|\sum_q g_q^{h,H} A_q^{h,H}(\tau_q) + \sum_{\tilde{q}} g_{\tilde{q}}^{h,H} A_{\tilde{q}}^{h,H}(\tau_{\tilde{q}})\right|^2, \quad (6.132)$$

where

$$A_q^{h,H}(\tau) = \frac{3}{2}\tau[1 + (1-\tau)f(\tau)],$$

$$A_{\tilde{q}}^{h,H}(\tau) = -\frac{3}{4}\tau[1 - \tau f(\tau)], \quad (6.133)$$

with $\tau_i = 4m_i^2/m_{h,H}^2$ $(i = q, \tilde{q})$. The function $f(\tau)$ is defined in Eq. (6.74) and the MSSM couplings $g_q^{h,H}$ can be found in Tab. 6.1. The squark couplings $g_{\tilde{q}}^{h,H}$ are summarised in Tab. 6.3. The amplitudes approach constant values in the limit of large loop particle masses:

$$A_q^{h,H}(\tau) \to 1 \qquad \text{for } m_{h,H}^2 \ll 4m_q^2, \quad (6.134)$$

$$A_{\tilde{q}}^{h,H}(\tau) \to \frac{1}{4} \qquad \text{for } m_{h,H}^2 \ll 4m_{\tilde{q}}^2. \quad (6.135)$$

For the pseudoscalar Higgs decays, only quark loops are contributing and we obtain

[123]

$$\Gamma\left[A \to gg\right] = \frac{G_\mathrm{F}\,\alpha_3^2\,m_A^3}{16\sqrt{2}\,\pi^3}\left|\sum_q g_q^A A_q^A(\tau_q)\right|^2,\qquad(6.136)$$

where

$$A_q^A(\tau) = \tau f(\tau),$$

with $\tau_q = 4m_q^2/m_A^2$. The MSSM couplings g_q^A can be found in Tab. 6.1. For large quark masses, the quark amplitude approaches unity.

6.7.3 Decays into photon pairs

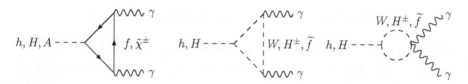

Figure 6.4 Typical diagrams contributing to $\Phi \to \gamma\gamma$ at lowest order.

The decay of scalar Higgs bosons to photon pairs is mediated by $W^\pm$ and heavy fermion (predominantly top) loops in the SM, while in the MSSM, in addition, one also has $H^\pm$, sfermion and chargino loops. In fact, in the MSSM, the b-quark loop can become competitive with the t-quark one, depending on α and β. The relevant diagrams are shown in Fig. 6.4. The partial decay widths [33, 34, 123] are given by

$$\Gamma[h, H \to \gamma\gamma] = \frac{G_\mathrm{F}\alpha^2 m_{h,H}^3}{128\sqrt{2}\pi^3}\left| \sum_f N_C Q_f^2 g_f^{h,H} A_f^{h,H}(\tau_f) + g_W^{h,H} A_W^{h,H}(\tau_W) \right.$$

$$\left. + \; g_{H^\pm}^{h,H} A_{H^\pm}^{h,H}(\tau_{H^\pm}) + \sum_{\tilde{\chi}^\pm} g_{\tilde{\chi}^\pm}^{h,H} A_{\tilde{\chi}^\pm}^{h,H}(\tau_{\tilde{\chi}^\pm}) + \sum_{\tilde{f}} N_C Q_{\tilde{f}}^2 g_{\tilde{f}}^{h,H} A_{\tilde{f}}^{h,H}(\tau_{\tilde{f}}) \right|^2,$$

$$(6.137)$$

with the form factors

$$A_{f,\tilde{\chi}^\pm}^{h,H}(\tau) = 2\tau\left[1 + (1-\tau)f(\tau)\right],\qquad(6.138)$$
$$(6.139)$$
$$A_{H^\pm,\tilde{f}}^{h,H}(\tau) = -\tau\left[1 - \tau f(\tau)\right],\qquad(6.140)$$
$$(6.141)$$
$$A_W^{h,H}(\tau) = -\left[2 + 3\tau + 3\tau(2-\tau)f(\tau)\right],\qquad(6.142)$$

where the function $f(\tau)$ is defined in Eq. (6.74). For large loop particle masses, the form factors approach constant values,

$$A_{f,\tilde{\chi}^\pm}^{h,H}(\tau) \rightarrow \frac{4}{3} \qquad \text{for } m_{h,H}^2 \ll 4m_{f,\tilde{\chi}^\pm}^2$$

$$A_{H^\pm,\tilde{f}}^{h,H}(\tau) \rightarrow \frac{1}{3} \qquad \text{for } m_{h,H}^2 \ll 4m_{H^\pm,\tilde{f}}^2$$

$$A_W^{h,H}(\tau) \rightarrow -7 \qquad \text{for } m_{h,H}^2 \ll 4M_W^2. \qquad (6.143)$$

The photonic decay mode of the pseudoscalar Higgs boson is generated by heavy charged fermion and chargino loops only, see Fig. 6.4. Bosonic loops (*e.g.*, $W^\pm$ and/or $H^\pm$ ones) are in fact forbidden by CP conservation. The partial decay width reads as [33, 34, 123]

$$\Gamma(A \rightarrow \gamma\gamma) = \frac{G_F \alpha^2 m_A^3}{32\sqrt{2}\pi^3} \left| \sum_f N_C Q_f^2 g_f^A A_f^A(\tau_f) + \sum_{\tilde{\chi}^\pm} g_{\tilde{\chi}^\pm}^A A_{\tilde{\chi}^\pm}^A(\tau_{\tilde{\chi}^\pm}) \right|^2, \qquad (6.144)$$

with the amplitudes

$$A_{f,\tilde{\chi}^\pm}^A(\tau) = \tau f(\tau). \qquad (6.145)$$

For large loop particle masses, the pseudoscalar amplitudes approach unity. The parameters $\tau_i = 4m_i^2/m_\Phi^2$ $(i = f, W, H^\pm, \tilde{\chi}^\pm, \tilde{f})$ are defined by the corresponding mass of the heavy loop particle, and the MSSM couplings $g_{f,W,H^\pm,\tilde{\chi}^\pm,\tilde{f}}^\Phi$ are summarised in Tabs. 6.1 and 6.3.

6.7.4 Decays into Z boson and photon

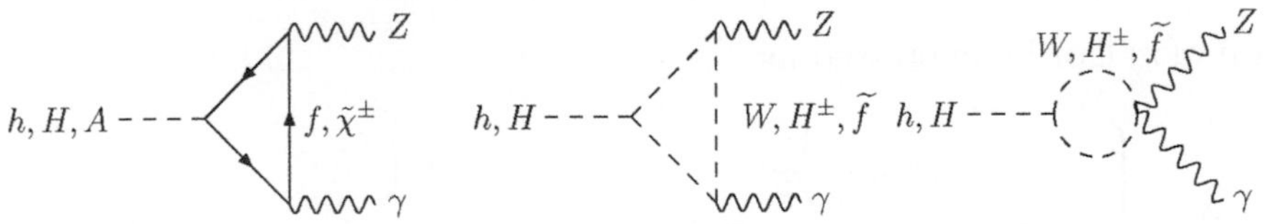

Figure 6.5 Typical diagrams contributing to $\Phi \rightarrow Z\gamma$ at lowest order.

The decays of scalar Higgs bosons into $Z\gamma$ are mediated by $W^\pm$ and heavy fermion loops as in the SM, plus, in addition in the MSSM, by $H^\pm$, sfermion and chargino loops. In this case, the contributing diagrams are found in Fig. 6.5. The partial decay widths read as [123, 125]

$$
\begin{aligned}
\Gamma\left[h, H \to Z\gamma\right] &= \frac{G_F^2 M_W^2 \, \alpha \, m_{h,H}^3}{64 \, \pi^4} \left(1 - \frac{M_Z^2}{m_{h,H}^2}\right)^3 \Bigg| \sum_f g_f^{h,H} A_f^{h,H}(\tau_f, \lambda_f) \\
&+ g_W^{h,H} A_W^{h,H}(\tau_W, \lambda_W) + g_{H^\pm}^{h,H} A_{H^\pm}^{h,H}(\tau_{H^\pm}, \lambda_{H^\pm}) \\
&+ \sum_{\tilde{\chi}_i^\pm, \tilde{\chi}_j^\mp} g_{\tilde{\chi}_i^\pm \tilde{\chi}_j^\mp}^{h,H} g_{\tilde{\chi}_i^\pm \tilde{\chi}_j^\mp}^Z A_{\tilde{\chi}_i^\pm \tilde{\chi}_j^\mp}^{h,H} + \sum_{\tilde{f}_i, \tilde{f}_j} g_{\tilde{f}_i \tilde{f}_j}^{h,H} g_{\tilde{f}_i \tilde{f}_j}^Z A_{\tilde{f}_i \tilde{f}_j}^{h,H} \Bigg|^2 ,
\end{aligned}
$$

(6.146)

with the form factors $A_f^{h,H}, A_W^{h,H}$ given by

$$
A_f^{h,H}(\tau, \lambda) = 2N_C \frac{Q_f(I_{3f} - 2Q_f \sin^2\theta_W)}{\cos\theta_W} [I_1(\tau, \lambda) - I_2(\tau, \lambda)] \quad (6.147)
$$

$$
\begin{aligned}
A_W^{h,H}(\tau, \lambda) &= \cos\theta_W \Bigg[4(3 - \tan^2\theta_W) I_2(\tau, \lambda) \\
&+ \left[\left(1 + \frac{2}{\tau}\right) \tan^2\theta_W - \left(5 + \frac{2}{\tau}\right) \right] I_1(\tau, \lambda) \Bigg].
\end{aligned}
$$

(6.148)

The functions I_1, I_2 are given by

$$
\begin{aligned}
I_1(\tau, \lambda) &= \frac{\tau\lambda}{2(\tau - \lambda)} + \frac{\tau^2\lambda^2}{2(\tau - \lambda)^2} [f(\tau) - f(\lambda)] + \frac{\tau^2\lambda}{(\tau - \lambda)^2} [g(\tau) - g(\lambda)] , \\
I_2(\tau, \lambda) &= -\frac{\tau\lambda}{2(\tau - \lambda)} [f(\tau) - f(\lambda)] ,
\end{aligned}
$$

(6.149)

where the function $g(\tau)$ can be expressed as

$$
g(\tau) = \begin{cases} \sqrt{\tau - 1} \arcsin \dfrac{1}{\sqrt{\tau}} & \tau \geq 1 \\[2mm] \dfrac{\sqrt{1 - \tau}}{2} \left(\log \dfrac{1 + \sqrt{1 - \tau}}{1 - \sqrt{1 - \tau}} - i\pi \right) & \tau < 1, \end{cases}
$$

(6.150)

and the function $f(\tau)$ is defined in Eq. (6.74).

Also,

$$
A_{H^\pm}^{h,H}(\tau, \lambda) = \frac{\cos 2\theta_W}{\cos\theta_W} I_1(\tau, \lambda),
$$

(6.151)

where the function $I_1(\tau, \lambda)$ is defined after Eq. (6.148).

The $Z\gamma$ decay mode of the pseudoscalar Higgs boson is generated by heavy charged fermion and chargino loops, see Fig. 6.5. The partial decay width is given by [125]

$$\Gamma(A \to Z\gamma) = \frac{G_F^2 M_W^2 \alpha m_A^3}{16\pi^4} \left(1 - \frac{M_Z^2}{m_A^2}\right)^3$$

$$\times \left| \sum_f g_f^A A_f^A(\tau_f, \lambda_f) + \sum_{\tilde{\chi}_i^\pm, \tilde{\chi}_j^\mp} g_{\tilde{\chi}_i^\pm \tilde{\chi}_j^\mp}^A g_{\tilde{\chi}_i^\mp \tilde{\chi}_j^\pm}^Z A_{\tilde{\chi}_i^\pm \tilde{\chi}_j^\pm}^A \right|^2, \quad (6.152)$$

with the fermion amplitudes

$$A_f^A(\tau, \lambda) = 2N_C \frac{Q_f(I_f^3 - 2Q_f \sin^2\theta_W)}{\cos\theta_W} I_2(\tau, \lambda). \quad (6.153)$$

The contributions of charginos and sfermions involve mixing terms. Their analytical expressions can be found in [125]. For large loop particle masses and small Z mass, the form factors approach the photonic amplitudes *modulo* couplings. The parameters $\tau_i = 4m_i^2/m_\Phi^2$, $\lambda_i = 4m_i^2/M_Z^2$ $(i = f, W, H^\pm, \tilde{\chi}^\pm, \tilde{f})$ are defined by the corresponding mass of the heavy loop particle and the non-mixing MSSM couplings $g_{f,W,H^\pm,\tilde{\chi}^\pm,\tilde{f}}^\Phi$ are summarised in Tabs. 6.1 and 6.3, while the mixing and Z boson couplings g_i^Z can be found in [33, 34].

6.7.5 Decays into intermediate gauge bosons

The partial widths of the scalar MSSM Higgs bosons into $W^\pm$ and Z boson pairs can be obtained from the SM Higgs decay widths by rescaling through the corresponding MSSM couplings $g_V^{h,H}$ listed in Tab. 6.1:

$$\Gamma(h, H \to V^{(*)}V^{(*)}) = (g_V^{h,H})^2 \Gamma(h_{SM} \to V^{(*)}V^{(*)}). \quad (6.154)$$

(Notice that the pseudoscalar Higgs particle does not couple to $W^\pm$ and Z bosons at tree level.)

6.7.6 Decays into Higgs particles

The heavy scalar Higgs particle can decay into pairs of light scalar as well as pseudoscalar Higgs bosons, see Fig. 6.6. The partial decay widths are given by [33, 34]

$$\Gamma(H \to hh) = \lambda_{Hhh}^2 \frac{G_F M_Z^4}{16\sqrt{2}\pi m_H} \sqrt{1 - 4\frac{m_h^2}{m_H^2}}, \quad (6.155)$$

$$\Gamma(H \to AA) = \lambda_{HAA}^2 \frac{G_F M_Z^4}{16\sqrt{2}\pi m_H} \sqrt{1 - 4\frac{M_A^2}{m_H^2}}. \quad (6.156)$$

The self-couplings λ_{Hhh} and λ_{HAA} can be derived from the effective Higgs potential [126–130].

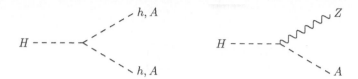

Figure 6.6 Typical diagrams contributing to Higgs decays with Higgs bosons in the final state.

The contributions of final states containing off-shell scalar or pseudoscalar Higgs bosons may be significant. Their expressions read as [121, 122]

$$
\begin{aligned}
\Gamma(H \to \phi\phi^*) &= \lambda_{H\phi\phi}^2 g_{\phi bb}^2 m_b^2 \frac{3 G_{\rm F}^2 M_Z^4}{16\pi^3 m_H} \left[(\kappa_\phi - 1)\left(2 - \frac{1}{2}\log\kappa_\phi \right) \right. \\
&\left. + \frac{1 - 5\kappa_\phi}{\sqrt{4\kappa_\phi - 1}} \left(\arctan\frac{2\kappa_\phi - 1}{\sqrt{4\kappa_\phi - 1}} - \arctan\frac{1}{\sqrt{4\kappa_\phi - 1}} \right) \right],
\end{aligned}
$$

$$(6.157)$$

where $\kappa_\phi = m_\phi^2 / m_H^2$. They slightly expand the regions where the hh, AA decay modes of the heavy scalar Higgs boson H are sizable.

Moreover, Higgs bosons can decay into a gauge and a Higgs boson, see Fig. 6.6. The various partial widths can be expressed as

$$
\Gamma(H \to AZ) = \lambda_{HAZ}^2 \frac{G_{\rm F} M_Z^4}{8\sqrt{2}\pi m_H} \sqrt{\lambda(m_A^2, M_Z^2; m_H^2)} \lambda(m_A^2, m_H^2; M_Z^2), \quad (6.158)
$$

$$
\Gamma(H \to H^\pm W^\mp) = \lambda_{HH+W}^2 \frac{G_{\rm F} M_W^4}{8\sqrt{2}\pi m_H} \sqrt{\lambda(m_{H^\pm}^2, M_W^2; m_H^2)} \lambda(m_{H^\pm}^2, m_H^2; M_W^2),
$$

$$(6.159)$$

$$
\Gamma(A \to hZ) = \lambda_{hAZ}^2 \frac{G_{\rm F} M_Z^4}{8\sqrt{2}\pi m_A} \sqrt{\lambda(m_h^2, M_Z^2; m_A^2)} \lambda(m_h^2, m_A^2; M_Z^2), \quad (6.160)
$$

$$
\Gamma(H^+ \to hW^+) = \lambda_{hH+W}^2 \frac{G_{\rm F} M_W^4}{8\sqrt{2}\pi m_{H^\pm}} \sqrt{\lambda(m_h^2, M_W^2; m_{H^\pm}^2)} \lambda(m_h^2, m_{H^\pm}^2; M_W^2),
$$

$$(6.161)$$

where the couplings λ_{ijk}^2 can be determined from the effective Higgs potential [126–130]. The functions $\lambda(x, y; z) = (1 - x/z - y/z)^2 - 4xy/z^2$ denote the usual two-body phase-space factors. The branching ratios of these decay modes may be sizable in specific regions of the MSSM parameter space.

Below-threshold decays into a Higgs particle and an off-shell gauge boson turn out to be very important for the heavy Higgs bosons of the MSSM. The individual contributions are given by [121, 122]

$$\Gamma(H \to AZ^*) = \lambda_{HAZ}^2 \delta_Z' \frac{9G_F^2 M_Z^4 m_H}{8\pi^3} G_{AZ}, \qquad (6.162)$$

$$\Gamma(H \to H^\pm W^{\mp*}) = \lambda_{HH^\pm W}^2 \frac{9G_F^2 M_W^4 m_H}{8\pi^3} G_{H^\pm W}, \qquad (6.163)$$

$$\Gamma(A \to hZ^*) = \lambda_{hAZ}^2 \delta_Z' \frac{9G_F^2 M_Z^4 m_A}{8\pi^3} G_{hZ}, \qquad (6.164)$$

$$\Gamma(H^+ \to hW^{+*}) = \lambda_{hH^\pm W}^2 \frac{9G_F^2 M_W^4 m_{H^\pm}}{8\pi^3} G_{hW}, \qquad (6.165)$$

$$\Gamma(H^+ \to AW^{+*}) = \frac{9G_F^2 M_W^4 m_{H^\pm}}{8\pi^3} G_{AW}. \qquad (6.166)$$

The generic functions G_{ij} can be written as

$$\begin{aligned}
G_{ij} = \frac{1}{4} \Bigg\{ & 2(-1 + \kappa_j - \kappa_i)\sqrt{\lambda_{ij}} \left[\frac{\pi}{2} + \arctan\left(\frac{\kappa_j(1 - \kappa_j + \kappa_i) - \lambda_{ij}}{(1 - \kappa_i)\sqrt{\lambda_{ij}}} \right) \right] \\
& + (\lambda_{ij} - 2\kappa_i)\log\kappa_i + \frac{1}{3}(1 - \kappa_i)\left[5(1 + \kappa_i) - 4\kappa_j - \frac{2}{\kappa_j}\lambda_{ij} \right] \Bigg\},
\end{aligned}$$
$$(6.167)$$

using the parameters

$$\lambda_{ij} = -1 + 2\kappa_i + 2\kappa_j - (\kappa_i - \kappa_j)^2, \qquad \kappa_i = \frac{m_i^2}{m_\phi^2}. \qquad (6.168)$$

Finally, the coefficient δ_Z' is defined as

$$\delta_Z' = \frac{7}{12} - \frac{10}{9}\sin^2\theta_W + \frac{40}{27}\sin^4\theta_W. \qquad (6.169)$$

II

Collider Phenomenology

CHAPTER 7

Search for Supersymmetric Particles at the LHC

One of the main motivations of the experiments at the LHC, started in September 2008, is to search for SUSY particles (or sparticles, for short). The LHC first operational run (2009–2013) has seen unprecedented success, including the discovery of a Higgs boson. During this phase the LHC has reached record energies of 7 and 8 TeV from pp collisions. After a 2 year shutdown, in April 2015, the LHC started its second operational run (2015–2018) at 13 TeV, which is still ongoing as we write. In this chapter, we will briefly describe the LHC project. Then we will discuss the region of parameter space of the MSSM where SUSY particles can be probed at the CERN machine [131, 132]. We will show that, if SUSY exists at the EW scale, it should be easy to find signals for it at the LHC. If the LHC does find SUSY, this would be one of the greatest achievements in the history of theoretical physics.

7.1 THE LHC PROJECT

It is believed that the LHC at the CERN is the machine that will take particle physics into a new phase of discovery. The LHC is a proton-proton collider designed to explore energies on the order of few TeV, which are about ten times the energies accessed at previous particle accelerators. The LHC is a two-ring-superconducting-hadron accelerator and collider built in the existing 26.7 km tunnel recently occupied by the LEP machine. The LHC, after the aforementioned initial stages at 7, 8 and 13 TeV, will probably have a final center of mass energy of 14 TeV and a final design luminosity of 10^{34} cm^{-2} s^{-1}, due to the collision of bunches of up to 10^{11} protons 40 million times per second. Note that the number of events per second generated at the LHC is given by $N_{\text{event}} = L \times \sigma_{\text{event}}$, where σ_{event} is the cross section of the event and L is the machine luminosity that depends on the beam parameters, such as number of particles per bunch and number of bunches per beam [133].

Unlike the electrons and positrons at LEP, the protons at the LHC are composed particles seen as bunches of constituents, *i.e.*, quarks, antiquarks and gluons, so

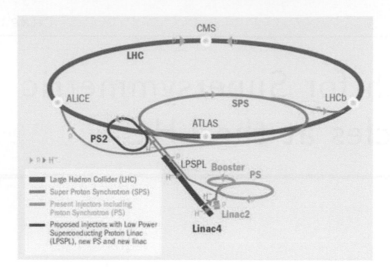

Figure 7.1 Schematic layout of the LHC at CERN.

that the explorable mass range of the produced particles is less than the energy of the interacting beams. However, a large number of events will occur due to the high intensity of the strong force. At present, four main experiments have been constructed at the LHC, as shown in Fig. 7.1. The two largest, CMS and ATLAS, are general-purpose experiments designed to detect any possible old and new physics processes. Two more special-purpose experiments have been designed for the LHC machine. ALICE is a dedicated heavy-ion detector that will exploit the physics potential of nucleus-neucleus interactions at LHC energies. At these extreme collision energies a new phase of matter, the so-called quark-gluon plasma, is generated. In addition, the LHCb detector is devoted to the study of CP violation and other rare phenomena in the decays of B-mesons (b in LHCb stands for beauty).

7.2   THE CMS AND ATLAS EXPERIMENTS AT THE LHC

The CMS detector is 21.6 m long and has a diameter of 14.6 m [134]. It has a total weight of 12500 tons. As displayed in Fig. 7.2, the CMS detector is built around a very high field solenoid magnet. It is designed in order to fulfil the following tasks in an excellent manner: (i) muon identification and momentum resolution over a wide range of momenta and angles; (ii) charged-particle momentum resolution and reconstruction efficiency in the inner tracker; (iii) EM energy resolution and di-photon and di-electron mass resolution; (iv) missing-transverse-energy and di-jet-mass resolution, which require hadron calorimeters with a large hermetic geometric coverage and with fine lateral segmentation.

The origin of the CMS coordinate system is centered at the nominal collision point inside the experiment. The y-axis is chosen to be vertically upward, while the x-axis is defined as radially inward toward the center of the LHC. In this case, the

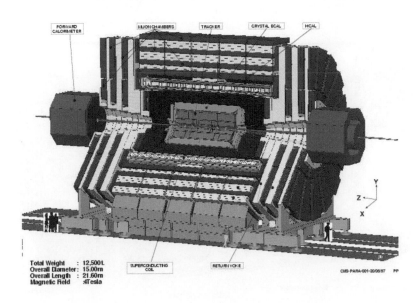

Figure 7.2 Cutaway view of the CMS detector.

z-axis points along the beam direction. The azimuthal angle ϕ is measured from the x-axis in the $x-y$ plane and the radial coordinate in this plane is denoted by r. The polar angle θ is measured from the z-axis, with respect to the counterclockwise beam direction. Pseudorapidity, η, is defined as $\eta = -\ln\tan(\theta/2)$. Thus, the momentum and energy transverse to the beam direction, denoted by p_T and E_T, respectively, are computed from the x and y components.

Like CMS, the ATLAS experiment is a general purpose detector. However, it is substantially larger and essentially relies upon an air cored toroidal magnet system for the measurement of the muons. The overall ATLAS detector layout is shown in Fig. 7.3. The ATLAS detector is about 45 meters long, more than 25 meters high, and about 7000 tons in weight [135]. It is nominally forward-backward symmetric with respect to the interaction point, therefore it is somewhat cylindrically shaped. The magnet configuration comprises a thin superconducting solenoid surrounding the inner-detector cavity, and three large superconducting toroids (one barrel and two end-caps) arranged with an eightfold azimuthal symmetry around the calorimeters. The magnets bend charged particles such that their momentum is measured.

The ATLAS detector is also divided into many components such that each component probes a particular type of particle. In fact, each type of particle has its own signature in the detector. High granularity liquid-argon EM sampling calorimeters,

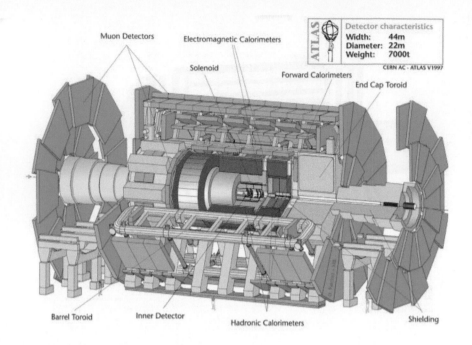

Figure 7.3 Cutaway view of the ATLAS detector.

with excellent performance in terms of energy and position resolution, cover the pseudorapidity range $|\eta| < 3.2$. The hadronic calorimeter surrounds the electromagnetic calorimeter and covers the range $|\eta| < 1.7$. It absorbs and measures the energies of hadrons, including protons, neutrons, pions and kaons. It consists of steel absorbers separated by tiles of scintillating plastic. Interactions of high energy hadrons in the plates transform the incident energy into a hadronic shower of many low energy protons, neutrons, and other hadrons. This shower, when traversing the scintillating tiles, causes them to emit light in an amount proportional to the incident energy.

7.3 THE ALICE EXPERIMENT AT THE LHC

ALICE is a general-purpose heavy ion detector at the LHC which focuses on QCD, the strong interaction sector of the SM [136]. It is designed to address the physics of strongly interacting matter and the quark-gluon plasma at extreme values of energy density and temperature in nucleus-nucleus collisions. It will allow a comprehensive study of hadrons, electrons, muons and photons produced in the collision of heavy nuclei, up to the highest multiplicities anticipated at the LHC. The physics program also includes collisions with lighter ions and at lower energy, in order to vary energy density and interaction volume, as well as dedicated proton-nucleus runs. The overall ALICE detector layout is shown in Fig. 7.4. Data taking during proton-proton runs at the top LHC energy will provide reference data for the heavy-ion

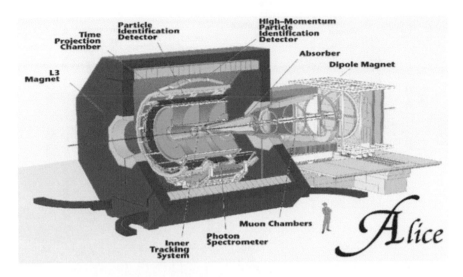

Figure 7.4   Cutaway view of the ALICE detector.

programme and address a number of specific strong-interaction topics for which ALICE is complementary to the other LHC detectors.

7.4   THE LHCb EXPERIMENT AT THE LHC

The LHCb experiment is dedicated to precision measurements of CP violation and rare decays of B hadrons at the LHC. On the one hand, the current results in heavy flavour physics obtained at the B factories and at the Tevatron are, so far, fully consistent with the CKM mechanism [137]. On the other hand, the level of CP violation in the SM weak interactions cannot explain the amount of matter in the universe. A new source of CP violation beyond the SM is therefore needed to solve this puzzle. With much improved precision, the effect of such a new source might be seen in heavy flavour physics. Many models of new physics indeed produce contributions that change the expectations of the CP violating phases: rare decay branching fractions can change significantly and they may even generate decay modes which are forbidden in the SM.

With the large $b\bar{b}$ production cross section of, $e.g.$, ~ 500 μb expected at an energy of 14 TeV, the LHC will be the most copious source of B mesons ever in the world. Also, B_c and b-baryons such as Λ_b will be produced in large quantities. With a luminosity of order 2×10^{32} cm^{-2} s^{-1} for LHCb, 10^{12} $b\bar{b}$ pairs would be produced in one year of data taking.

In order to exploit this large number of B hadrons, the LHCb detector must include an efficient trigger in order to cope with such a severe hadronic environment. The trigger must be sensitive to many different final states. In LHCb, a series of different detectors are placed one after another to measure different aspects of the particles produced in the proton-proton collisions. Also, excellent vertex

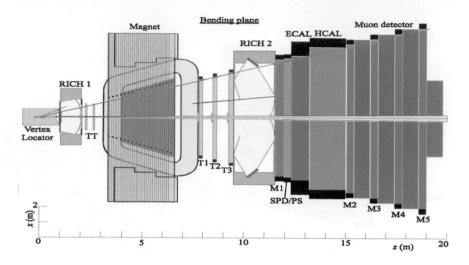

Figure 7.5 The setup of the LHCb detector.

and momentum resolution are essential prerequisites for the proper-time resolution necessary to study the rapidly oscillating $B_s - \bar{B}_s$ meson system and also for the invariant mass resolution needed to reduce combinatorial background. In addition to electron, muon, γ, π^0 and η detection, identification of protons, kaons, and pions is crucial in order to cleanly reconstruct many hadronic B meson decay final states such as $B \to \pi^+\pi^-$, $B \to DK^{(*)}$, and $B_s \to D_s^\pm K_s^\mp$. These are key channels for the physics goals of the experiment. The setup of the LHCb detector is displayed in Fig. 7.5.

7.5   SUSY SEARCHES AT THE LHC

At LHC, the total SUSY particle production cross section is largely dominated by strongly interacting sparticles. Thus, a typical high mass SUSY signal has squarks and gluinos which decay through a number of steps to quarks, gluons, charginos, neutralinos, W, Z, Higgses and, finally, to a stable $\tilde{\chi}_1^0$, as shown in Fig. 7.6.

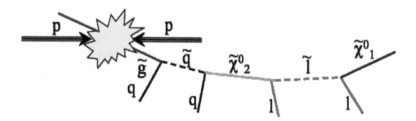

Figure 7.6 SUSY particle production at the LHC and typical decay chain.

Point	m_0 (GeV)	$m_{1/2}$ (GeV)	A_0 (GeV)	$\tan\beta$	$\mathrm{sgn}\mu$	lightest $\tilde{q}$ (GeV)	$\tilde{\chi}_1^0$ (GeV)
LM0	200	160	-400	10	$+$	207	60
LM1	60	250	0	10	$+$	410	97
LM2	185	350	0	35	$+$	582	141
LM3	330	240	0	20	$+$	446	94
LM4	210	285	0	10	$+$	483	112
LM5	230	360	0	10	$+$	603	145

Table 7.1 Possible selected points in the cMSSM considered for analysis by the CMS and ATLAS experiments.

Note that squark-squark production usually leads to two jets while gluino production typically results in higher jet multiplicities. Therefore, the search for the production and decay of SUSY particles by the multi-purpose LHC experiments (CMS and ATLAS) is described by events with two or more energetic jets and significant missing transverse energy. In addition, the two largest BRs are $\mathrm{BR}(\tilde{g} \to \tilde{\chi}_1^- t\bar{b}) \approx \mathrm{BR}(\tilde{g} \to \tilde{\chi}_2^- t\bar{b}) \approx 23\%$. Therefore, decays into both charginos and all four neutralinos with all allowed combinations of quarks are significant, which leads to many complex signatures.

It is essentially impossible to scan the entire MSSM parameter space, even in the case of the most popular constrained model, the cMSSM , which is characterised by four parameters and a sign, namely the scalar mass parameter m_0, the gaugino mass parameter $m_{1/2}$, the trilinear coupling A_0, the ratio of the Higgs vacuum expectation values, $\tan\beta$, and the sign of the SUSY Higgs(ino) mass parameter, μ. Therefore, the CMS and ATLAS collaborations at the LHC have selected specific points in the aforementioned cMSSM for their analysis. In Tab. 7.1 we list examples of these selected points, the corresponding masses of the lightest squark, $\tilde{t}_1$, and the lightest neutralino, $\tilde{\chi}_1^0$.

7.6 GLUINOS AND SQUARKS

In the cMSSM, the gluino mass is given by $m_{\tilde{g}} \approx 2.5 m_{1/2}$ and the five squark flavours $(\tilde{d}, \tilde{u}, \tilde{s}, \tilde{c}, \tilde{b})$, with their left and right chiral states, have a degenerate mass: $m_{\tilde{q}} \approx \sqrt{m_0^2 + 6m_{1/2}^2}$. The $\tilde{t}_L$ and $\tilde{t}_R$ are treated differently, since they have a large mixing which leads to two mass eigenstates, $\tilde{t}_1$ and $\tilde{t}_2$, with $\tilde{t}_1$ significantly lighter than all other $\tilde{q}$ states, which requires a separate discussion in both cases of production and decay.

At the LHC, the squarks and gluinos can be produced in pairs $\{\tilde{g}\tilde{g}, \tilde{q}\tilde{q}, \tilde{q}\tilde{g}\}$ and can decay through $\tilde{q} \to q\tilde{\chi}_1^0$ and $\tilde{g} \to q\bar{q}\tilde{\chi}_1^0$, as shown in Fig. 7.7, where $\tilde{\chi}_1^0$ is the lightest neutralino. A search for squarks and gluinos in final states containing high

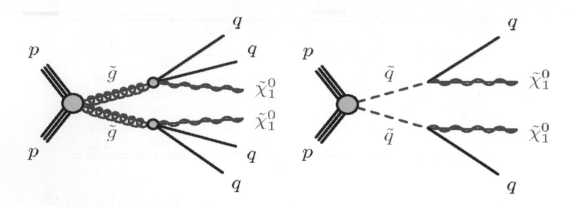

Figure 7.7 Representative diagrams for (left) gluino- and (right) squark-pair production and their direct decays: $\tilde{g} \to q\bar{q}\tilde{\chi}^0$ and $\tilde{q} \to q\tilde{\chi}^0$.

p_T jets and missing transverse momentum has been carried out in both the AT-LAS and CMS experiments, with CM energies $\sqrt{s} = 7, 8$ TeV and total integrated luminosity of order 20 fb^{-1} [1,2]. No significant excess above the SM expectations was observed. In Fig. 7.8 we display the ATLAS exclusion limits at 95% CL for the cMSSM model, with $\tan\beta = 30$, $A_0 = -2m_0$, and $\mu > 0$ presented (left) in the $(m_0 - m_{1/2})$ plane and (right) in the $(m_{\tilde{g}} - m_{\tilde{q}})$ plane [1]. CMS limits are essentially the same.

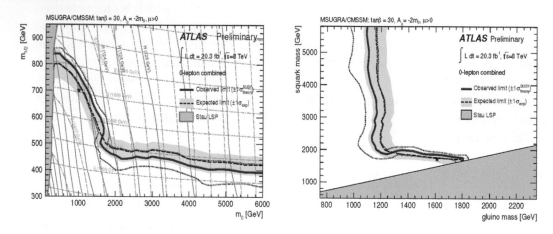

Figure 7.8 ATLAS exclusion limits, at 95% CL, for the cMSSM model with $\tan\beta = 30$, $A_0 = -2m_0$ and $\mu > 0$, projected on (left) the $(m_0 - m_{1/2})$ plane and (right) the $(m_{\tilde{g}} - m_{\tilde{q}})$ plane [1]. CMS limits are similar [2].

As can be seen from this figure, in this scenario, values of $m_{1/2} < 340$ GeV are excluded for $m_0 < 6$ TeV, which is consistent with the constraints obtained in the next chapter from the measured lightest Higgs mass, $m_h \approx 125$ GeV. At 13

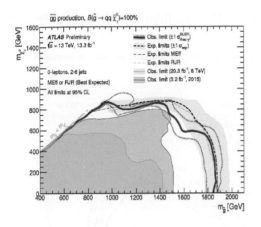
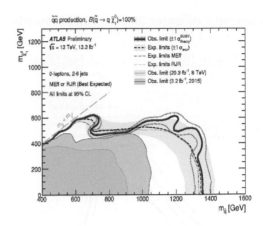

Figure 7.9 ATLAS exclusion limits for the gluino mass (left) and squark mass (right) at the CoM energy of 13 TeV [3]. CMS limits are similar [4].

TeV with integrated luminosity ~ 13 fb^{-1}, the ATLAS and CMS collaborations provided the exclusion limits given in Fig. 7.9 for gluino and squark masses [3,4]. It is important to note that in SUSY searches at the LHC, a simplified model where one production process and one decay channel with a 100% branching fraction is usually assumed. As shown from this figure, the upper limit of the excluded light-flavour squark mass region in this scenario of the simplified model is 1.35 TeV, assuming massless $\tilde{\chi}^0$ and that the gluino mass is 1.86 TeV.

7.7 STOPS

The stop plays an essential role in solving the hierarchy problem in the MSSM, through the cancellations of the quadratic divergences of top-quark loops. As mentioned, in most SUSY models, a light stop quark mass arises naturally. A search for the pair production of stop quarks has been performed by both the ATLAS and CMS experiments, using the full data set collected at $\sqrt{s} = 7, 8$ TeV and integrated

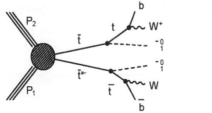
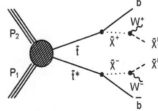

Figure 7.10 Representative diagrams for stop pair production and decay into $b\bar{b}$ $W^+W^- \tilde{\chi}_1^0 \tilde{\chi}_1^0$.

luminosity of order 20 fb^{-1}. LHC searches focus on the semileptonic decay mode, where $\tilde{t} \to t\tilde{\chi}_1^0 \to bW^{\pm}\tilde{\chi}_1^0$ and $\tilde{t} \to b\tilde{\chi}_1^{\pm} \to bW^{\pm}\tilde{\chi}_1^0$, as shown in Fig. 7.10. Then one $W^{\pm}$ decays hadronically and the other leptonically. Thus, the final state searched for is $4j + \ell +$ missing energy ($\ell = e, \mu$) [138–141].

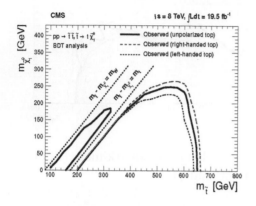
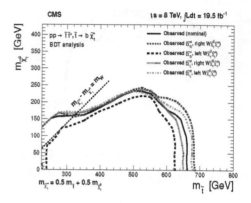

Figure 7.11 (Left) The observed 95% CL excluded regions for the $\tilde{t} \to t\tilde{\chi}_1^0$ decay. (Right) The observed 95% CL excluded regions for the $\tilde{t} \to b\tilde{\chi}_1^+$ decay. Both are from CMS data [5]. ATLAS results are similar.

In Fig. 7.11 we show the CMS exclusion limits at 95% CL in the $(m_{\tilde{t}} - m_{\tilde{\chi}_1^0})$ plane, for $\sqrt{s} = 8$ TeV. The ATLAS collaboration has also obtained similar results. One can conclude from these plots that, for small $m_{\tilde{\chi}_1^0}$, stop masses are excluded up to 650 GeV. However, this result strongly depends on a sizable mass splitting between the stop and the lightest neutralino into which it decays. This limit is drastically reduced in the region of low mass splitting, as shown in the above figure. In fact, if $m_{\tilde{t}} \sim m_t + m_{\tilde{\chi}_1^0}$, then the stop signal becomes difficult to distinguish from the $t\bar{t}$ background and the LHC data are unable to constrain the stop mass.

Recently, an analysis of ~ 13 fb^{-1} and $\sqrt{s} = 13$ TeV at ATLAS [6] and CMS [7] detectors indicated no significant excess over the SM background. Exclusion limits at 95% CL. shown in Fig. 7.12, are reported. These limits are presented in terms of stop and lightest neutralino masses. In the considered simplified model, where $BR(\tilde{t} \to t\tilde{\chi}^0) = 1$, the stop masses between ~ 310 and ~ 820 GeV are excluded for $m_{\tilde{\chi}^0} < 160$ GeV.

7.8 CHARGINOS AND NEUTRALINOS

As mentioned above, the mixing of the fermionic partners of the EW gauge and Higgs bosons, the gauginos, and Higgsinos, gives rise to the physical mass eigenstates called the charginos and neutralinos. They are among the lightest SUSY particles and therefore present a particular interest. The two lightest neutralinos and the lightest chargino ($\tilde{\chi}_1^0$, $\tilde{\chi}_2^0$, $\tilde{\chi}_1^{\pm}$) have as their largest mixing component the gauginos and their masses are determined by the common gaugino mass, $m_{1/2}$.

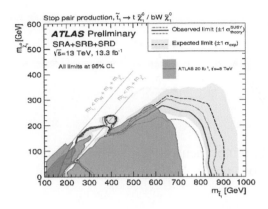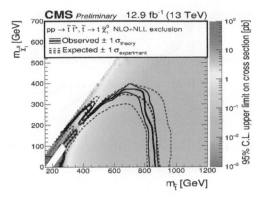

Figure 7.12 ATLAS (left) and CMS (right) exclusion limits at 95% CL for direct stop production, followed by the $\tilde{t} \to t\tilde{\chi}^0$ decay [6,7].

Within the cMSSM, we have

$$m_{\tilde{\chi}_1^0} \approx 0.45 m_{1/2}, \tag{7.1}$$

$$m_{\tilde{\chi}_2^0} \approx M_{\tilde{\chi}_1^\pm} \approx 2 m_{\tilde{\chi}_1^0}, \tag{7.2}$$

$$m_{\tilde{\chi}_3^0} \approx (0.25 - 0.35) m_{\tilde{g}}. \tag{7.3}$$

The lightest chargino ($\tilde{\chi}_1^\pm$) has several leptonic decay modes, producing an isolated lepton and missing energy, due to the undetectable neutrino and the LSP [142,143]:

$$\tilde{\chi}_1^\pm \ \to \ \tilde{\chi}_1^0 \ \ell^\pm \ \nu, \tag{7.4}$$

$$\tilde{\chi}_1^\pm \ \to \ \tilde{\ell}_L^\pm \ \nu \to \tilde{\chi}_1^0 \ \ell^\pm, \tag{7.5}$$

$$\tilde{\chi}_1^\pm \ \to \ \tilde{\nu}_L \ \ell^\pm \to \tilde{\chi}_1^0 \ \nu, \tag{7.6}$$

$$\tilde{\chi}_1^\pm \ \to \ \tilde{\chi}_1^0 \ W^\pm \to \ell^\pm \ \nu. \tag{7.7}$$

The BRs for these decay channels have been calculated in terms of the SUSY breaking terms (m_0, $m_{1/2}$). It was shown that, in different parameter space regions, these different leptonic decay modes are complementary. Three body decays are open in the region $m_{1/2} \lesssim 200$ GeV and $m_{1/2} \gtrsim 0.5 m_0$, whereas in the rest of the parameter space the two body decays are dominant. Leptonic decays of $\tilde{\chi}_2^0$, which give two isolated leptons and missing energy ($\tilde{\chi}_1^0$), have also been studied. It was emphasised that, as in the $\tilde{\chi}_1^\pm$ case, the three body decay BRs are sizable for $m_{1/2} \lesssim 200$ GeV and $m_{1/2} \gtrsim 0.5 m_0$. In hadronic collisions, the charginos and neutralinos can be produced directly via 21 different reactions or through the cascade decays of strongly interacting sparticles.

It is finally worth noting that, in the cMSSM, the bounds imposed on $m_{1/2}$ from the lightest Higgs mass lead to much stronger limits on the lightest chargino (~ 900 GeV) and lightest neutralino (~ 600 GeV) masses, which are far above the current direct limits from the LHC.

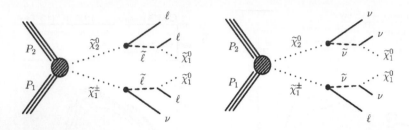

Figure 7.13 Representative diagrams for direct charginos/neutralinos production and decay mediated by sleptons (left) and sneutrinos (right), leading to leptonic final states.

The first searches for the direct EW of production supersymmetric charginos and neutralinos at $\sqrt{s} = 13$ TeV were carried out with pp collision of luminosity ~ 13 fb^{-1} in the CMS detector [8]. The signals of these searches are signatures with two light leptons of the same flavour and with three or more leptons, see Fig. 7.13. As shown in Fig. 7.14, the result is classified into categories formed according to the number, sign and flavour of the leptons. These results display no significant deviation from the SM expectations, so they are used to set limits on simplified models with large chargino-neutralino pair production cross sections. In particular, they probe chargino and neutralino masses up to 400–1000 GeV depending on the assumed model parameters.

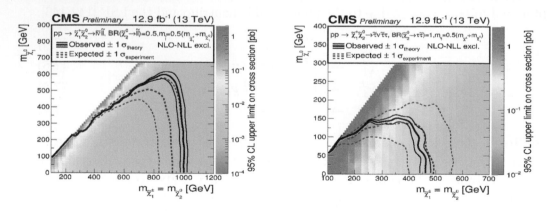

Figure 7.14 CMS results of the three-lepton search in the flavour-democratic signal model with the slepton/sneutrino mass given by $m_{\tilde{\ell}} = m_{\tilde{\nu}} = \frac{1}{2}(m_{\tilde{\chi}^+} + m_{\tilde{\chi}^0})$ [8].

7.9 SLEPTONS

The SUSY partners of ordinary leptons are scalars. Left- and right- handed charged sleptons are not mass degenerate. The slepton masses are determined by m_0 and

$m_{1/2}$:

$$m^2_{\tilde{\ell}_L} \quad = \quad m^2_0 + 0.52m^2_{1/2} - 0.5(1 - 2\sin^2\theta_W)M^2_Z\cos 2\beta, \tag{7.8}$$

$$m^2_{\tilde{\ell}_R} \quad = \quad m^2_0 + 0.15m^2_{1/2} - \sin^2\theta_W M^2_Z\cos 2\beta, \tag{7.9}$$

$$m^2_{\tilde{\nu}} \quad = \quad m^2_0 + 0.52m^2_{1/2} + 0.5M^2_Z\cos 2\beta, \tag{7.10}$$

where the coefficients are determined by the RGE running. It is clear that charged left sleptons are the heaviest. The mass dependence on $m_{1/2}$ for left sleptons is stronger than for right ones, hence the left-right slepton mass splitting increases with $m_{1/2}$. The left sleptons can decay to charginos and neutralinos through the following modes [142, 143]:

$$\tilde{\ell}^\pm_L \rightarrow \ell^\pm\ \tilde{\chi}^0_{1,2} \text{ or to } \nu_\ell\ \tilde{\chi}^\pm_1 \tag{7.11}$$

and

$$\tilde{\nu}^\pm_\ell \rightarrow \nu_\ell\ \tilde{\chi}^0_{1,2} \text{ or to } \ell^\pm\ \tilde{\chi}^\mp_1. \tag{7.12}$$

The right sleptons only decay to neutralinos and mainly to the LSP:

$$\tilde{\ell}^-_R \rightarrow \ell^-\ \tilde{\chi}^0_1. \tag{7.13}$$

The total cross section and the BR of these decays have been studied in terms of m_0 and $m_{1/2}$. It was shown that, in almost the entire space of these two parameters, slepton pair production is dominated by right sleptons.

CHAPTER 8

SUSY Higgs Prospects at the LHC

8.1 SM-LIKE HIGGS BOSON SEARCHES AT THE LHC

In July 2012 the ATLAS and CMS collaborations announced the detection of a new particle consistent with a Higgs boson. The combined measured mass is [144,145]:

$$m_h = 125.09 \pm 0.21(\text{stat.}) \pm 0.11(\text{syst.}) \text{ GeV}. \tag{8.1}$$

This discovery was dramatic, since a Higgs boson was the last undiscovered particle desperately searched for to complete experimental verification of the SM. Although a Higgs discovery is not *a priori* an indication for SUSY, still, the confirmed existence of a light Higgs boson is in favour of it (as dicussed in Chapter 6 for the MSSM, see Eq. (6.56)[1]). Furthermore, there are fundamental differences between the SM and the MSSM Higgs sector that might allow us to distinguish between the two models. Therefore, the light Higgs discovery gives hope for probing SUSY since, as we will show, SUSY models are consistent with the properties of the discovered Higgs-like boson within the accuracy of the experimental data (some are even more preferred by data in comparison to the SM).

The most recent analyses of Higgs boson properties reported by ATLAS [146] and CMS [147] are based on 4.7 fb^{-1} at 7 TeV and up to 13 fb^{-1} at 8 TeV data processed by the ATLAS collaboration plus 5.1 fb^{-1} at 7 TeV and up to 19.6 fb^{-1} at 8 TeV processed by the CMS collaboration. The results are presented for various Higgs boson production and decay channels. The studied decay modes include $H \rightarrow \gamma\gamma$, ZZ, W^+W^-, $\tau^+\tau^-$ and $b\bar{b}$. Some results have also appeared concerning invisible Higgs boson decays [148–150]. The discovery has been achieved via the decay channel $h \rightarrow \gamma\gamma$ with Higgs mass around 125 GeV, as shown in Fig. 8.1. However, the signal for the golden decay channel $h \rightarrow ZZ \rightarrow 4\ell$ is easier to recognise from the background, as shown in Fig. 8.2.

---

[1]Somewhat higher upper limits on an SM-like Higgs mass are obtained in non-minimal SUSY models, see Chapter 15.

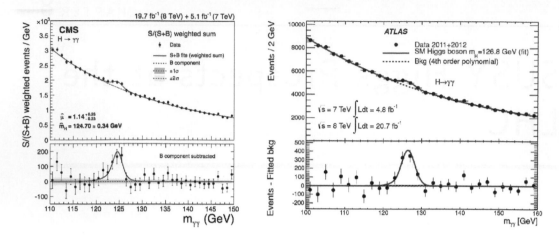

Figure 8.1 Signal of Higgs decaying to $h \to \gamma\gamma$ discovery at CMS [9] in the left panel and at ATLAS [10] in the right panel.

The magnitude of the Higgs signal is usually expressed via the "signal strength" parameters μ, defined for either the entire combination of or the individual decay modes, relative to the SM expectations. The respective results as reported by ATLAS are given by [146]

$$\mu(h \to \gamma\gamma) = 1.65 \pm 0.35, \tag{8.2}$$

$$\mu(h \to ZZ) = 1.7 \pm 0.5, \tag{8.3}$$

$$\mu(h \to WW) = 1.01 \pm 0.31, \tag{8.4}$$

$$\mu(h \to b\bar{b}) = -0.4 \pm 1.0, \tag{8.5}$$

$$\mu(h \to \tau\bar{\tau}) = 0.8 \pm 0.7 \tag{8.6}$$

while from the CMS collaboration one has [147]

$$\mu(h \to \gamma\gamma) = 0.78 \pm 0.28, \tag{8.7}$$

$$\mu(h \to ZZ) = 0.91^{+0.3}_{-0.24}, \tag{8.8}$$

$$\mu(h \to WW) = 0.67 \pm 0.21, \tag{8.9}$$

$$\mu(h \to b\bar{b}) = 1.15 \pm 0.62, \tag{8.10}$$

$$\mu(h \to \tau\bar{\tau}) = 1.10 \pm 0.41. \tag{8.11}$$

In Fig. 8.3 the signal strengths are presented as given by CMS and ATLAS [151]. One can see that these results, on the one hand, are consistent with the SM at 95% CL, while, on the other hand, there is still a significant statistical probability to accommodate deviations from the SM. One should also notice that the $h \to \gamma\gamma$ measurement reported by ATLAS is about 2σ above the SM, while the CMS result for it is about 1σ below the SM one. Thus, one can also see that there is some tension between the ATLAS and CMS results.

We will focus on the $h \to \gamma\gamma, W^+W^-, ZZ$ final states (for which the most accurate results are given) and the $gg \to h$ production channel (which is the most substantial

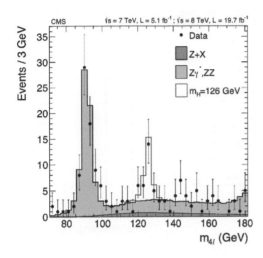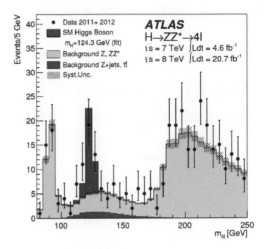

Figure 8.2 Signal of Higgs decaying to four charged leptons $h \to ZZ \to 4\ell$ discovery at CMS [11] in the left panel and at ATLAS [10] in the right panel.

one), henceforth labelled as ggh. We will, in particular, discuss deviations and possible excesses/depletions in the $\gamma\gamma$ channel in the framework of the MSSM and assess how these could affect the Higgs production or decay dynamics (or indeed both), while maintaining the W^+W^- and ZZ signal strengths broadly compatible with the experiment. In order to do so, we define μ_{XY} for a given production (X) and decay (Y) channel, in terms of production cross sections σ and decay widths Γ:

$$
\begin{aligned}
\mu_{X,Y} &= \frac{\sigma_X^{\mathrm{MSSM}}}{\sigma_X^{\mathrm{SM}}} \times \frac{BR_Y^{\mathrm{MSSM}}}{BR_Y^{\mathrm{SM}}} = \kappa_X \times \frac{\Gamma_Y^{\mathrm{MSSM}}/\Gamma_{\mathrm{tot}}^{\mathrm{MSSM}}}{\Gamma_Y^{\mathrm{SM}}/\Gamma_{\mathrm{tot}}^{\mathrm{SM}}} \\
&= \kappa_X \times \frac{\Gamma_Y^{\mathrm{MSSM}}}{\Gamma_Y^{\mathrm{SM}}} \times \frac{\Gamma_{\mathrm{tot}}^{\mathrm{SM}}}{\Gamma_{\mathrm{tot}}^{\mathrm{MSSM}}} = \kappa_X \times \kappa_Y \times \kappa_h^{-1},
\end{aligned}
\tag{8.12}
$$

where, generally, $X = ggh$ and $Y = \gamma\gamma$, W^+W^-, ZZ, $b\bar{b}$, $\tau^+\tau^-$, $etc.$ Notice that, here, κ_X and κ_Y are equal to the respective ratios of the couplings squared while κ_h is the ratio of the total Higgs boson width in the MSSM relative to the SM. For example, for $gg \to h \to \gamma\gamma$, we have

$$
\mu_{X,Y} = \mu_{ggh,\gamma\gamma} = \kappa_{ggh} \times \kappa_{\gamma\gamma} \times \kappa_h^{-1} = \frac{\sigma_{ggh^{\mathrm{MSSM}}}}{\sigma_{ggh^{\mathrm{SM}}}} \times \frac{\Gamma_{\gamma\gamma h^{\mathrm{MSSM}}}}{\Gamma_{\gamma\gamma h^{\mathrm{SM}}}} \times \kappa_h^{-1}.
\tag{8.13}
$$

Quite apart from the fact that current data are not entirely compatible with SM Higgs production and/or decay rates, the LHC measurements point to a rather light Higgs mass. While the possibility that the SM Higgs state had such a mass would be merely a coincidence (as its mass is a free parameter), in the MSSM, in contrast, the mass of the lightest Higgs boson with SM-like behaviour is naturally confined to be less than 130 GeV by SUSY itself, as discussed, which in essence relates trilinear Higgs and gauge couplings, so that the former are of the same size as the latter, in turn implying such a naturally small Higgs mass value. Therefore, to some extent, the new LHC results may be in favour of some low energy SUSY realisation, so that it is of the utmost importance to test the validity of the SUSY hypothesis through the Higgs data and the respective viable parameter space.

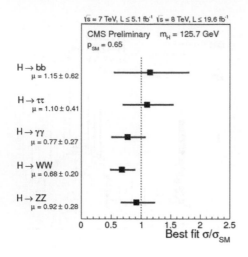
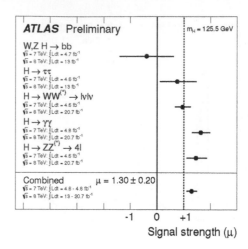

Figure 8.3   CMS and ATLAS results for the signal strength parameter $\mu \equiv \sigma/\sigma_{\rm SM}$, where σ is the production cross section times the relevant branching fractions and $\sigma_{\rm SM}$ is the SM expectation.

One should finally mention that, initially, both collaborations had observed an enhancement in the $h \to \gamma\gamma$ channel while, later on, CMS results shifted towards the SM value or even below it. In the next section we explore the compatibility of the MSSM against the aforementioned LHC data, in particular, its ability to produce enhanced (with respect to the SM) rates in the di-photon channel, by assessing the contribution of the lightest sfermions, sbottom, stop as well as stau, which can alter the effective ggh and $h\gamma\gamma$ couplings. Besides this, we explore the effects of deviations entering all other Higgs boson couplings to SM particles due to quantum corrections or mixing effects (or both). Hence, in all generality, in our approach both decay BRs and production rates for the Higgs boson will be modified.

8.2   IMPLICATIONS OF THE HIGGS BOSON DISCOVERY FOR THE MSSM

As discussed in Part 1, in the cMSSM, the whole SUSY spectrum and the Higgs boson mass are determined by the GUT scale universal soft SUSY breaking parameters m_0, $m_{1/2}$, A_0 and $\tan\beta$, in addition to sign(μ). In Fig. 8.4 we display the contour plot of the SM-like Higgs boson: $m_h \in [124, 126]$ GeV on the $(m_0, m_{1/2})$ plane for different values of A_0 and $\tan\beta$ [12].

It is remarkable that, the smaller A_0 is, the smaller $m_{1/2}$ needs to be to satisfy this value of Higgs mass. It is also clear that the scalar mass m_0 remains essentially unconstrained by the Higgs mass and can vary from a few hundred GeVs to a few TeVs. Such large values of $m_{1/2}$ seem to imply quite a heavy SUSY spectrum, much heavier than the lower bound imposed by direct searches at the LHC experiments at the CM energies of $\sqrt{s} = 7, 8$ TeV and total integrated luminosity of order 20 fb^{-1}. Furthermore, the LHC lower limit on the gluino mass, $m_{\tilde{g}} \gtrsim 1.4$ TeV [152, 153], excluded the values of $m_{1/2} < 620$ GeV that was allowed by Higgs mass constraints for $m_0 > 4$ TeV. This region is shown with dashed lines in Fig. 8.4.

In order to understand the features of the SUSY spectrum in this allowed range, we

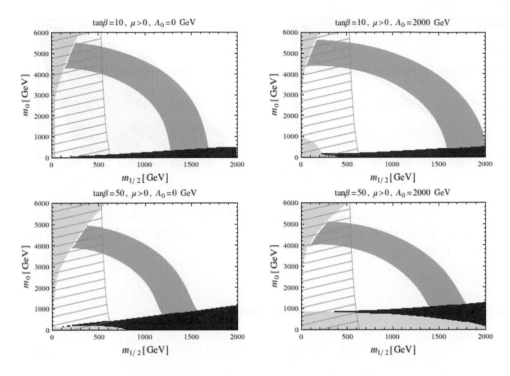

Figure 8.4 MSSM parameter space for $\tan\beta = 10$ and 50 with $A_0 = 0$ and 2 TeV [12]. Green region corresponds to $124 \lesssim m_h \lesssim 126$ GeV. Blue region is excluded because the lightest neutralino is not the LSP. Pink region is excluded by EWSB conditions. Gray shadow lines show the excluded area by $m_{\tilde{g}} < 1.4$ TeV.

plot in Fig. 8.5 the correlation between lightest chargino and lightest neutralino masses (left panel) and between light stop and light stau masses (right panel). It is noticeable that in this region both lightest chargino and lightest neutralino are gaugino-like, hence their masses satisfy the ratio of M_2/M_1. Also, the scalar masses are typically very heavy, particularly the squarks (including the light stop). However, the lightest chargino has a chance to be light and even an $m_{\tilde{\tau}}$ of order 100 GeV is allowed.

8.3 ENHANCED $H \to \gamma\gamma$ RATE IN THE MSSM

In this section, we discuss MSSM effects which alter the Higgs event rates at the LHC as compared to those of the SM. We start our discussion with MSSM Higgs boson production via the gluon-gluon fusion process which, as previously mentioned, is the dominant channel for Higgs searches at the LHC. In the SM, this mode is predominantly mediated by top quarks via a one-loop triangle diagram, while the contribution from other quarks, even the bottom one, is only at the few percent level.

In the MSSM, however, this is no longer the case, as, depending on the value of $\tan\beta$, bottom quark loops can compete with top quark ones at production level. Furthermore, strongly interacting superpartners of the SM quarks, $i.e.$, the squarks, could provide a sizable contribution to the triangle loop entering $gg \to h$. Based on the results from Chapter 6 (we

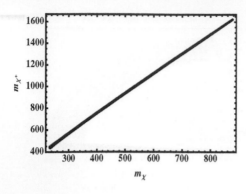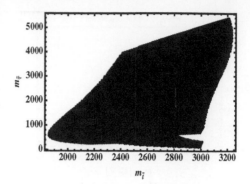

Figure 8.5 Mass correlations of lightest chargino-lightest neutralino (left panel) and light stop-light stau (right panel) in cMSSM after imposing the Higgs mass and gluino mass limits.

now use $x = \tau^{-1}$), the LO parton-level cross section can be written as

$$\hat{\sigma}_{\text{LO}}(gg \to h) = \frac{\pi^2}{8m_h}\Gamma_{\text{LO}}(h \to gg)\Delta(\hat{s} - m_h^2), \tag{8.14}$$

where $\hat{s}$ is the CM energy at the partonic level and $\Delta(\hat{s} - m_h^2)$ is the Breit-Wigner form of the Higgs boson propagator, which is given by

$$\Delta(\hat{s} - m_h^2) = \frac{1}{\pi}\frac{\hat{s}\Gamma_h/m_h}{(\hat{s} - m_h^2)^2 + (\hat{s}\Gamma_h/m_h)^2}, \tag{8.15}$$

where Γ_h is the total Higgs boson decay width while its partial decay width, $\Gamma_{\text{LO}}(h \to gg)$, is given by

$$\Gamma_{\text{LO}}(h \to gg) = \frac{\alpha_3^2 m_h^3}{512\pi^3}\left|\sum_f \frac{2Y_f}{m_f}F_{1/2}(x_f) + \sum_S \frac{g_{hSS}}{m_S^2}F_0(x_S)\right|^2, \tag{8.16}$$

where Y_f and g_{hSS} are the MSSM Higgs couplings to the respective (s)particle species for fermion (spin-1/2) and scalar (spin-0) particles, respectively, entering the triangle diagram. The loop functions $F_{1/2,0}$ are given by [154]

$$F_{1/2}(x) = 2x^2\left[x^{-1} + (x^{-1} - 1)f(x^{-1})\right], \tag{8.17}$$

$$F_0(x) = -x^2\left[x^{-1} - f(x^{-1})\right], \tag{8.18}$$

with $f(x) = \arcsin^2\sqrt{x}$. We define x_i as $4m_i^2/m_h^2$, where m_i is the mass of the particle running in the loop. Note that the functions $F_{1/2}(x)$ and $F_0(x)$ reach a plateau very quickly for $x > 1$ and their values are about 1.4 and 0.4, respectively. This fact has important consequences, which will be discussed below in connection with the Higgs decay into two photons.

Now we turn to the Higgs decay into di-photons. In the MSSM the one-loop partial decay width of the h state into two photons also gets a contribution from scalar particles represented by sfermions and charged Higgs bosons and this is given by

$$\Gamma(h \to \gamma\gamma) = \frac{\alpha^2 m_h^3}{1024\pi^3} \left| \frac{g_{hWW}}{M_W^2} F_1(x_V) + \sum_f \frac{2Y_f}{m_f} N_{c,f} Q_f^2 F_{1/2}(x_f) \right.$$

$$\left. + \sum_S \frac{g_{hSS}}{m_S^2} N_{c,S} Q_S^2 F_0(x_S) \right|^2, \tag{8.19}$$

where V, f and S stand for vector, fermion and scalar particles, respectively, entering the one-loop triangle diagram, g_{hWW} is the MSSM Higgs coupling to $W^\pm$ bosons while Y_f and g_{hSS} are the MSSM couplings of the Higgs boson to fermions and scalars, respectively.

The genuine SUSY contributions to $\Gamma(h \to \gamma\gamma)$ are mediated by charged Higgs bosons, charginos and charged sfermions. The SM-like part is dominated by $W^\pm$ bosons, for which $F_1(x_W) \simeq -8.3$, whereas the top quark loop is subdominant and enters with opposite sign, $N_{c,f} Q_f^2 F_{1/2}(x_f) \simeq 1.8$, with all other fermions contributing negligibly. It is also worth mentioning that $F_0(x_S) \sim 0.4$, which is about a factor of 20 smaller than $F_1(x_W)$ and approximately a factor of 4 smaller than $F_{1/2}(x_f)$.

In [13], it has been shown that, in the MSSM, $h \to \gamma\gamma$ may be enhanced through one of the following possibilities:

1. by the introduction of a large scalar contribution, due to the light stop and/or sbottom and/or stau, with negative coupling g_{hSS} so that it interferes constructively with the dominant $W^\pm$ contribution;

2. via charged Higgs boson contributions;

3. via chargino contributions;

4. via modification of the Yukawa couplings of top and bottom quarks in the loop, the latter also affecting the $b\bar{b}$ partial width.

It is worth mentioning that the only way to have a sizable effect from the scalar loops is to be in a scenario with large coupling g_{hSS} and light scalars. In such a scenario the scalar loop competing with the fermion loop has a larger relative contribution to $\Gamma(h \to gg)$ than to $\Gamma(h \to \gamma\gamma)$, where it would also compete with the dominant vector boson loop. At the same time, the contribution from squarks is opposite for Higgs production via gluon-gluon fusion compared to the di-photon decay: depending on the sign of g_{hSS}, they will destructively (constructively) interfere with top quarks in production loops and constructively (destructively) interfere with W boson loops in Higgs boson decays. Therefore, any squark loop which causes an increase (decrease) in $\Gamma(h \to \gamma\gamma)$ will cause a proportionally larger decrease (increase) in $\Gamma(h \to gg)$.

The total Higgs decay width in the MSSM is given, similarly to the SM, by the sum of all the Higgs partial decay widths, $i.e.$, $\Gamma_{\text{tot}} = \Gamma_{b\bar{b}} + \Gamma_{WW} + \Gamma_{ZZ} + \Gamma_{\tau\bar{\tau}} + \dots$. Other partial decay widths into SM particles are much smaller and can safely be neglected. As per decays into SUSY states, we assume that the lightest neutralino is heavy enough, so we do not have invisible decay channels with large rates.

8.3.1 Stau effects

The lightest stau may give important contributions that in particular could enhance $\kappa_{\gamma\gamma}$. For staus, $N_C Q^2 = 1$, a factor of 3 larger than sbottoms, and since the Higgs-stau coupling like the Higgs-sbottom coupling also does not depend on X_t, and hence is not constrained

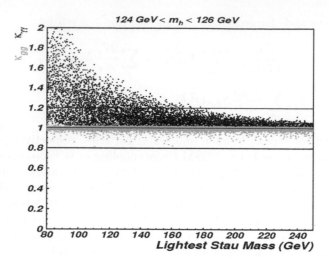

Figure 8.6 $\kappa_{\gamma\gamma}$ (black) and κ_{gg} (green) as a function of lightest stau mass for 124 GeV $< m_h <$ 126 GeV. Taken from [13].

by the Higgs mass, light stau effects on $\Gamma(h \to \gamma\gamma)$ could be more significant than sbottom effects, with the caveat of a different (running) bottom mass versus the tau mass.

The Higgs coupling to the lightest stau, normalised by $v/\sqrt{2} = M_W/g_2$, with v the SM Higgs VEV, is given by

$$\hat{g}_{h\tilde{\tau}_1\tilde{\tau}_1} = \cos 2\beta \left(-\frac{1}{2}\cos^2\theta_{\tilde{\tau}} + \sin^2\theta_W \cos 2\theta_{\tilde{\tau}} \right) + \frac{m_\tau^2}{M_Z^2} + \frac{m_\tau X_\tau}{2M_Z^2}\sin 2\theta_{\tilde{\tau}}. \tag{8.20}$$

with $X_\tau = A_\tau - \mu\tan\beta$.

For a large positive μ, with large $\tan\beta$, X_τ is large and negative and we find that

$$\hat{g}_{h\tilde{\tau}_1\tilde{\tau}_1} \simeq \frac{m_\tau X_\tau}{2M_Z^2}, \tag{8.21}$$

which is large and negative. Thus the stau contribution may enhance $\Gamma(h \to \gamma\gamma)$ in a large $\tan\beta$ scenario with large and positive μ.

This is demonstrated in Fig. 8.6, where $\kappa_{\gamma\gamma}$ (black) is plotted against the stau mass, indeed showing that we can have $\kappa_{\gamma\gamma} > 1.2$ when $m_\tau \lesssim 180$ GeV. It also shows that light staus have no effect on κ_{gg}, as expected. (The points with a slight reduction in κ_{gg} have sbottoms or stop masses ~ 300 GeV.)

8.3.2 Chargino effects

In the MSSM, the interactions of the lightest neutral MSSM Higgs boson, h, with the charginos are given by

$$\mathcal{L} = g_2 \tilde{\chi}_i^+ \left(C_{ij}^L P_L + C_{ij}^R P_R \right) \tilde{\chi}_j^+ h + \text{h.c.} \tag{8.22}$$

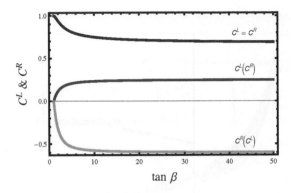

Figure 8.7 Chargino-Higgs couplings C_L and C_R versus $\tan\beta$ for $\mu = M_2 = 200$ GeV (green line), $M_2 = 2\mu = 300$ (red line for C_L and blue line for C_R), and $\mu = 2M_2 = 300$ GeV, red line for C_R and blue line for C_L (from [14]).

where the $C_{ij}^{L,R}$'s are given by

$$C_{ij}^L = \frac{1}{\sqrt{2}\sin\theta_W}\left[-\sin\alpha V_{j1}U_{i2} + \cos\alpha V_{j2}U_{i1}\right], \tag{8.23}$$

$$C_{ij}^R = \frac{1}{\sqrt{2}sin\theta_W}\left[-\sin\alpha V_{i1}U_{j2} + \cos\alpha V_{i2}U_{j1}\right]. \tag{8.24}$$

In Fig. 8.7 we display these couplings as a function of $\tan\beta$ for different values of μ and M_2. As can be seen from this plot, such couplings can reach their maximum values and become $\mathcal{O}(\pm 1)$ if $\tan\beta$ is very small, close to 1, and $\mu \simeq M_2$. It is also remarkable that, if $\mu > M_2$ ($\mu < M_2$), the coupling $C_R(C_L)$ flips its sign, which leads to destructive interferences involving chargino contributions. From this plot, it is clear that the Higgs coupling to charginos can be negative, hence the chargino can give a constructive interference with the $W^\pm$ boson, which may in turn lead to a possible enhancement for $\kappa_{\gamma\gamma}$ and $\mu_{\gamma\gamma}$. In Fig. 8.8 we display the results for $\kappa_{\gamma\gamma}$ as a function of the lightest chargino mass, $M_{\tilde\chi_1^+}$, with $m_h \simeq 125$ GeV. We scan over the following expanse of parameter space: $1.1 < \tan\beta < 5$, 100 GeV $< \mu < 300$ GeV, and 100 GeV $< M_2 < 300$ GeV. Other dimensionful SUSY parameters are fixed to be on order of a few TeV so that all other possible SUSY effects onto $H \to \gamma\gamma$ are essentially negligible. As can be seen from this figure, in order to have a significant chargino contribution to $\kappa_{\gamma\gamma}$, quite a light chargino mass ($M_{\tilde\chi_1^+} \sim 104$ GeV), approximately the LEP limit, is required [86, 155–157]. Precisely at this limiting value, one finds that the Higgs signal strength is enhanced by about 25%.

8.4 IMPLICATIONS FOR HIGGS VACUUM STABILITY

In the SM, the variation of the quartic Higgs coupling with the energy scale Q at which it is probed is described by the RGE (up to one loop)

$$\frac{d\lambda}{dt} = \frac{1}{16\pi^2}\left(24\lambda^2 - 6Y_t^4 + \frac{9}{8}g_2^4 + \frac{3}{8}g_1^4 + \frac{3}{4}g_1^2g_2^2 + 12\lambda Y_t^2 - 9\lambda g_2^2 - 3\lambda_1 g_1^2\right). \tag{8.25}$$

In Fig. 8.9, we display the running of the Higgs self-coupling in the SM with Higgs mass $m_h = 126.6$ GeV. As can be seen from this figure, the SM Higgs potential becomes unstable

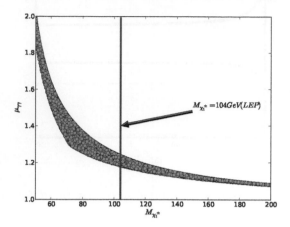

Figure 8.8 Signal strength of the di-photon channel as a function of the lightest chargino mass for $1.1 < \tan\beta < 5$, 100 GeV $< \mu <$ 500 GeV, and 100 GeV $< M_2 <$ 500 GeV (From [14]).

at a scale on the order of $\mathcal{O}(10^{10}$ GeV). This can be easily understood from the RGE structure in Eq. (8.25), where the top Yukawa coupling Y_t contributes to the evolution of λ with fourth power and negative sign. Therefore, it is natural that the scale of energy up to which the SM is valid depends on the precise value of the top quark mass. If one assumes that $\lambda \ll \lambda_t, g_1, g_2$, then the RGE takes the approximate form:

$$\frac{d\lambda}{dt} \simeq \frac{1}{16\pi^2}\left[-12\frac{m_t^4}{v^4} + \frac{3}{16}\left[2g_2^4 + (g_1^2 + g_2^2)^2\right]\right],\tag{8.26}$$

which leads to the following solution at the weak scale:

$$\lambda(Q^2) = \lambda(v^2) + \frac{1}{16\pi^2}\left[-12\frac{m_t^4}{v^4} + \frac{3}{16}\left[2g_2^4 + (g_1^2 + g_2^2)^2\right]\right]\log\frac{Q^2}{v^2}.\tag{8.27}$$

If the coupling λ is too small, the top quark contribution can be dominant and could drive λ to a negative value $\lambda(Q^2) < 0$, which implies $V(Q^2) < V(v)$; hence, the vacuum is not stable where it has no minimum. For a scalar potential to be bounded from below we should keep $\lambda(Q^2) > 0$, and therefore the Higgs boson mass should be larger than the value

$$m_h^2 > \frac{v^2}{8\pi^2}\left[-12\frac{m_t^4}{v^4} + \frac{3}{16}\left[2g_2^4 + (g_1^2 + g_2^2)^2\right]\right]\log\frac{Q^2}{v^2},\tag{8.28}$$

which provides the following constraint:

$$m_h \simeq 125 \text{ GeV} \Rightarrow \Lambda_C \lesssim 10^{10} \text{ GeV}.\tag{8.29}$$

A natural solution for this problem is to consider some possible new physics beyond the SM, like SUSY, that changes the running of the quartic coupling and prevents it from running into negative values. We should therefore analyse the Higgs vacuum stability in

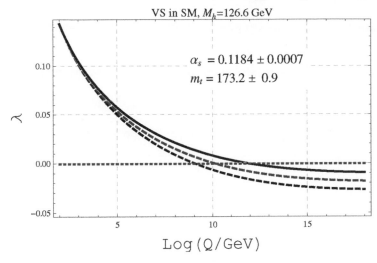

Figure 8.9 The running of scalar self coupling using a top quark mass $m_t = 173.2\pm0.9$ GeV and $\alpha_3(M_Z) = 0.1184 \pm 0.0007$.

SUSY extensions of the SM. As mentioned, the Higgs scalar potential in the MSSM is given by

$$V(H_d, H_u) = m_1^2\, H_d^2 + m_2^2\, H_u^2 - 2m_3^2\, H_d H_u + \frac{g_1^2 + g_2^2}{8} \left(H_d^2 - H_u^2\right)^2, \qquad (8.30)$$

where the masses m_i^2 are given in terms of the soft SUSY breaking terms $m_{H_{d,u}}^2$, B, and the μ parameter is introduced as follows:

$$m_{1,2}^2 = m_{H_{d,u}}^2 + |\mu|^2, \qquad m_3^2 = B\mu. \qquad (8.31)$$

This potential is the SUSY version of the Higgs potential which induces $SU(2)_L \times U(1)_Y$ breaking in the SM, where the usual self-coupling constant is replaced by the squared gauge couplings.

In order to study the stability of the MSSM Higgs potential, one should consider the following two cases: (i) flat direction, where $H_d = H_u = H$; (ii) non-flat directions. In the flat direction, the quartic terms vanish and the potential takes the simple form:

$$V(H) = (m_1^2 + m_2^2 - 2m_3^2)H^2, \qquad (8.32)$$

which is stable only if the coefficient $(m_1^2 + m_2^2 - 2m_3^2)$ is non-negative. This is the well known condition for avoiding the unboundedness of the MSSM potential from below. In contrast, along non-flat directions the quartic terms in Eq. (8.30) are non-vanishing and dominate the potential for large values of the scalar fields $H_{1,2}$. Thus, the stability is unconditionally guaranteed because the quartic coupling $(g_1^2 + g_2^2)/8$ is always positive. Therefore, one concludes that the MSSM Higgs potential is identically stable in any direction except the flat one, which requires the following condition:

$$m_1^2 + m_2^2 \geq 2m_3^2. \qquad (8.33)$$

FIGURE 4.3 The unification scales and couplings using a top quark mass $m_t = 170.2 \pm 0.9$ GeV and $\alpha_s(M_Z) = 0.184 \pm 0.0007$.

SUSY extensions of the SM. As a result of this, a clear potential in the MSSM is even

$$V(H_u, H_d) = m_1^2 |H_d|^2 + m_2^2 |H_u|^2 - m_3^2 (H_u H_d + \text{h.c.}) + \frac{g^2 + g'^2}{8}(|H_d|^2 - |H_u|^2)^2 \quad (4.30)$$

where the masses m_i are given in terms of the soft SUSY breaking terms m_{H_u}, m_{H_d} and the μ parameter is introduced as follows:

$$m_{1,2}^2 = m_{H_{d,u}}^2 + \mu^2, \qquad m_3^2 = B\mu. \quad (4.31)$$

This potential is the SUSY version of the Higgs potential, which induces $SU(2) \times U(1)_Y$ breaking in the SM, where the usual self-coupling constant is replaced by the gauge boson couplings.

In order to avoid the stability of the MSSM Higgs potential, one should consider the following cases: (i) flat direction: when $|H_u| = |H_d|$; (ii) non-flat direction: in the flat direction, the condition holds and the potential takes the simple form:

$$V(H) = (m_1^2 + m_2^2 - 2m_3^2)|H|^2 \quad (4.32)$$

which is stable only if the coefficient $m_1^2 + m_2^2 - 2m_3^2 \geq 0$. This is the well-known condition for avoiding the unboundedness of the effective potential from below. In contrast, along the flat directions, the quartic terms in Eq. (4.30) are unpublishable and dominate the potential for large values of the scalar fields Re ϕ. Thus, the stability is specifically satisfied because the quartic term vanishes if $\langle H \rangle$ is always positive. Therefore, one concludes that the MSSM Higgs potential is identically stable in any direction except the one which satisfies the following condition:

$$m_1^2 + m_2^2 \geq 2m_3^2, \quad (4.33)$$

CHAPTER 9

Supersymmetric DM

One of the great scientific mysteries still unsolved is the existence of DM. Most astronomers, cosmologists and particle physicists are convinced that at least 26.8% of the mass of the Universe is constituted by some matter of non-luminous form, called DM. Although the existence of DM was suggested 68 years ago, we still do not know its composition. Despite the fact that the nature of this DM is still unknown, its hypothetical existence is not so odd if we remember that the discovery of Neptune in 1846 by Galle was due to the suggestion of Le Verrier on the basis of the irregular motion of Uranus.

One of the most interesting candidates for DM is the lightest neutralino, which is typically the LSP in the MSSM. These neutralinos are stable and therefore may be leftovers from the Big Bang. Thus, they will cluster gravitationally with ordinary stars in the galactic halos and, in particular, they will be present in our own galaxy, the Milky Way. As a consequence, there will be a flux of these DM particles through the Earth. In this chapter we study the possibility that the LSP accounts for the observed DM relic abundance. We start with an introduction, based on the brief review on DM from Ref. [158]. We also discuss the most important experiments searching, directly or indirectly, for these relic objects.

9.1 INTRODUCTION TO DM

The first proposal for the existence of a DM component in the Universe dates back to 1933, when the astronomer Fritz Zwicky provided evidence that the mass of the luminous matter (stars) in the Coma cluster, which consists of about 1000 galaxies, was much smaller than its total mass implied by the motion of galaxies in the cluster. However, only in the 1970s the existence of DM began to be considered seriously. Its presence in spiral galaxies was the most plausible explanation for the anomalous rotation curves of these galaxies. The computation of the rotation velocity of stars or hydrogen clouds located far away from galactic centres is quite straightforward. One only needs to extrapolate Newton's gravitational law, which works outstandingly well for nearby astronomical phenomena, to galactic distances. Let us recall that, $e.g.$, for an average distance r of a planet from the center of the Sun, Newton's gravitational law implies that $v^2(r)/r = GM(r)/r^2$, where $v(r)$ is the average orbital velocity of the planet, $G = 6.67 \times 10^{-11}$ m^3 kg^{-1} s^{-2} is the Newton's constant and $M(r)$ is the total mass enclosed by the orbit. Therefore one obtains

$$v(r) = \sqrt{\frac{G\,M(r)}{r}} \ . \tag{9.1}$$

Thus, if the mass of the galaxy is concentrated in its visible part, one would expect $v(r) \sim 1/\sqrt{r}$ for distances far beyond the visible radius. Instead, astronomers, by means of the

Doppler effect, observe that the velocity rises towards a constant value about 100 to 200 km s^{-1}. Hence, for large distances, $M(r)/r$ is generically constant, as if the mass inside r increased linearly. An example of this can be seen in Fig. 9.1, where the rotation curve of M33, one of about 45 galaxies which form our small cluster, is shown. For comparison, the expected velocity from luminous disks is also shown.

This phenomenon has already been observed for about a thousand spiral galaxies, including our galaxy, the Milky Way. The most common explanation for these flat rotation curves is to assume that disk galaxies are immersed in extended DM halos. While at small distances this DM is only a small fraction of the galaxy mass, it becomes a very large amount at larger distances. Cosmologists usually express the present-day mass density av-

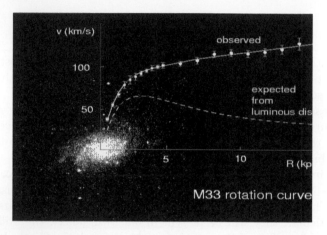

Figure 9.1 Observed rotation curve of the nearby dwarf spiral galaxy M33, superimposed on its optical image (from [15]).

eraged over the Universe, ρ, in units of the so-called critical density, $\rho_c \approx 10^{-29}$ g cm^{-3}, i.e., they define $\Omega = \rho/\rho_c$. Whereas current observations of luminous matter in galaxies determine $\Omega_{\text{lum}} \lesssim 0.01$, analyses of rotation curves imply $\Omega \approx 0.1$. The recent observations by the Planck satellite [159], which measures angular differences in temperature of the CMB radiation across the sky, have pointed out, as depicted in Fig. 9.2, that the Universe is made up of 4.9% atoms, the building blocks of stars and planets, about 26.8% DM and about 68.3% dark energy, which acts as a sort of anti-gravity. This energy, distinct from DM, is responsible for the present-day acceleration of the universal expansion.

9.2 DM CANDIDATES

The DM problem provides a potentially important interference between particle physics and cosmology, since only elementary particles can be reliable candidates for DM in the Universe. As the density of baryonic matter, formed from protons and neutrons such as, *e.g.*, gas, brown dwarfs, *etc.* is very limited and quite small to account for the whole of DM in the Universe, a non-baryonic matter is definitely required. Fortunately, particle physics offers various candidates for DM. Although the current SM of matter and interactions does not have non-baryonic particles that can account for DM, several extensions of the minimal model do have them. Indeed, detecting non-baryonic DM in the Universe would be a direct evidence of new physics beyond the SM. The most promising candidates for

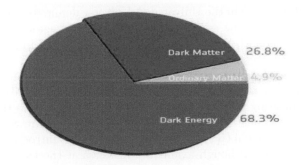

Figure 9.2 Recent results by the Planck satellite on the composition of the Universe.

DM are 'axions', 'neutrinos' and 'neutralinos' with masses of the order of 10^{-5} eV, 10 eV and 100 GeV, respectively. Although these particles are not present in the current SM of particle physics, they are well motivated by theories that attempt to unify the forces and particles of Nature, *i.e.*, by extensions of the SM.

Neutrinos are the only DM candidates which are known to exist. The SM has three families or flavours of left-handed neutrinos, ν_L. These SM neutrinos are strictly massless because there are no right-handed neutrinos ν_R that could combine with the ν_L's to form a Dirac mass term through their interaction with the Higgs doublet. However, several extensions of the SM do allow for ν_R and hence neutrino masses are generated. Moreover, the observation of solar and atmospheric neutrinos has indicated that one flavour might change to another. Remarkably, this is a quantum process (neutrino oscillations) which can only happen if the neutrino has a mass. The best evidence for neutrino mass comes from the Super Kamiokande experiment. The results of this experiment indicate a mass difference of the order of 0.05 eV between the muon neutrino and tau neutrino. These neutrinos left over from the Big Bang were in thermal equilibrium in the early Universe and decoupled when they were moving with relativistic velocities. They fill the Universe in enormous quantities and their current number density is similar to the one of photons (by entropy conservation in the adiabatic expansion of the Universe). In particular, $n_\nu = \frac{3}{11} n_\gamma$. Moreover, the number density of photons is very accurately obtained from the CMB measurements: $n_\gamma \approx 410.5$ cm^{-3}. The SM neutrino, with mass m_ν, gives a contribution to the total energy density of the Universe which is given by

$$\rho_\nu = m_{\rm tot}\, n_\nu, \tag{9.2}$$

where the total mass $m_{\rm tot} = \sum_\nu (g_\nu/2) m_\nu$ and with the number of degrees of freedom $g_\nu = 4(2)$ for Dirac (Majorana) neutrinos. Writing $\Omega_\nu = \rho_\nu/\rho_c$, where ρ_c is the critical energy density of the Universe, we have

$$\Omega_\nu h^2 \;=\; 10^{-2}\, m_{\rm tot}\ ({\rm eV}), \tag{9.3}$$

where h is the reduced Hubble constant, for which recent analyses give the favoured value: $h = 0.71\pm0.08$. Thus a neutrino of 10 eV would give $\Omega_\nu h^2 \sim \mathcal{O}(0.1)$, consistently with recent WMAP observations. However, there is now significant evidence against neutrinos being the bulk of DM. Neutrinos belong to the so-called 'hot' DM because they were moving with relativistic velocities at the time the galaxies started to form. But hot DM cannot correctly reproduce the observed structure in the Universe. A Universe dominated by neutrinos would form large structures first, and the small structures later, by fragmentation of the larger objects. Such a Universe would produce a 'top-down' cosmology, in which the galaxies were formed last and quite recently. This time scale seems incompatible with our present understanding of galaxy evolution, such as the initial enthusiasm for a neutrino dominated Universe fades away. Hence, many cosmologists now favour an alternative model, one in which the particles dominating the Universe are 'cold' (non-relativistic) rather than hot.

As already discussed, axions are spin-0 particles with zero charge associated with the spontaneous breaking of the global $U(1)$ PQ symmetry, which was introduced to dynamically solve the strong CP problem. Although axions are massless at the tree level they pick up a small mass by non-perturbative effects. The mass of the axion, m_a, and its couplings to ordinary matter, g_a, are proportional to $1/f_a$, where f_a is the (dimensionful) axion decay constant which is related to the scale of the new symmetry breaking. In particular, the coupling of an axion with two fermions of mass m_f is given by $g_a \sim m_f/f_a$. Likewise, $m_a \sim \Lambda_{\text{QCD}}^2/f_a$, i.e.,

$$m_a \sim 10^{-5} \text{ eV} \times \frac{10^{12} \text{ GeV}}{f_a}. \tag{9.4}$$

On the one hand, a lower bound on f_a can be obtained from the requirement that axion emission does not over-cool stars: the supernova SN1987 imposed the strongest bound, $f_a \gtrsim 10^9$ GeV. On the other hand, since coherent oscillations of the axion around the minimum of its potential may give an important contribution to the energy density of the Universe, the requirement $\Omega \lesssim 1$ puts a lower bound on the axion mass, implying $f_a \lesssim 10^{12}$ GeV. The combination of both constraints, astrophysical and cosmological, gives rise to the following window for the value of the axion constant:

$$10^9 \text{ GeV} \lesssim f_a \lesssim 10^{12} \text{ GeV}. \tag{9.5}$$

The lower bound implies an extremely small coupling of the axion to the ordinary matter and therefore a very large lifetime, larger than the age of the Universe by many orders of magnitude. As a consequence, the axion is a candidate for DM. Axions would then have been produced copiously in the Big Bang: they were never in thermal equilibrium and are always non-relativistic (i.e., they are cold DM). In addition, the upper bound implies that $m_a \sim 10^{-5}$ eV if the axion is to be a significant component of DM.

WIMPs are very interesting candidates for DM in the Universe. They were in thermal equilibrium with the SM particles in the early Universe and decoupled when they were non-relativistic. The process was as follows: when the temperature T of the Universe was larger than the mass of the WIMP, the number density of WIMPs and photons was roughly the same, $n_{\text{WIMP}} \propto T^3$, and the WIMP was annihilated with its own antiparticle into lighter particles and vice versa. However, shortly after the temperature dropped below the mass of the WIMP, m, its number density dropped exponentially, $n_{\text{WIMP}} \propto e^{-m/T}$, because only a small fraction of the light particles mentioned above had sufficient kinetic energy to create WIMPs. As a consequence, the WIMP annihilation rate dropped below the expansion rate of the Universe. At this point WIMPs came away, as they could not be annihilated and their density has been the same since then. Following these arguments, the relic density of WIMPs, as it will be shown in detail in the next section, can be written as

$$\Omega_{\text{WIMP}} \simeq \frac{7 \times 10^{-27} \text{ cm}^3 \text{ s}^{-1}}{\langle \sigma_A v \rangle}, \tag{9.6}$$

where σ_A is the total cross section for annihilation of a pair of WIMPs into SM particles, v

is the relative velocity between two WIMPs, $\langle .. \rangle$ denotes thermal averaging and the number in the numerator is obtained using the value of the temperature of the cosmic background radiation, Newton's constant, *etc*. As expected from the above discussion about the early Universe, the relic WIMP density decreases with increasing annihilation cross section.

Now we can understand easily why WIMPs are suitable candidates for DM. If a new particle with weak interactions exists in Nature, its cross section will be $\sigma_A \simeq \alpha^2/m_{\text{weak}}^2$, where $\alpha \simeq \mathcal{O}(10^{-2})$ is the weak coupling and $m_{\text{weak}} \simeq \mathcal{O}(100 \text{ GeV})$ is a mass of the order of one of the $W^{\pm}, Z$ gauge boson masses, which are associated to the $SU(2)$ gauge group of the SM. Thus, one may obtain $\sigma_A \approx 10^{-9} \text{ GeV}^{-2}$. Since at the freeze-out temperature v is close to the speed of light, one obtains $\langle \sigma_A v \rangle \approx 10^{-26} \text{ cm}^3 \text{ s}^{-1}$. To our surprise, this number is remarkably close to the value that we need in Eq. (9.6) in order to obtain the observed density of the Universe. This is a possible hint that new physics at the weak scale provides us with a reliable solution to the DM problem. This is a qualitative improvement with respect to the axion DM case, where a small mass for the axion, about 10^{-5} eV, has to be postulated. As we will discuss in the next sections, the lightest neutralino in SUSY theories is a natural WIMP candidate.

9.3 CALCULATION OF WIMP RELIC ABUNDANCE

In the early Universe, where the temperature was very high, the WIMPs, denoted by χ with mass m_χ, were abundant and rapidly converting to light particles, and vice versa. Shortly after T dropped below m_χ the number density of χ's diminished exponentially and the rate for the annihilation of χ's fell below the expansion rate, *i.e.*,

$$\Gamma_A < H, \tag{9.7}$$

where $H = \dot{a}/a$ is the Hubble expansion rate (or parameter), a is the scale factor of the Universe and the dot denotes the derivative with respect to time. In terms of matter density, the Hubble parameter is given by

$$H = \left(\frac{8\pi\rho}{3\tilde{M}_{\text{Pl}}^2} \right)^{1/2}, \tag{9.8}$$

where $\tilde{M}_{\text{Pl}}$ is the reduced Planck mass which is given by $\tilde{M}_{\text{Pl}} \equiv M_{\text{Pl}}/\sqrt{8\pi} \sim 2.435 \times 10^{18}$ GeV. At this point, the χ's cease to annihilate, fall out of equilibrium and a remnant remains. This simple picture is described quantitatively by the Boltzmann equation, which describes the time evolution of the number density $n_\chi(t)$ of WIMPs,

$$\frac{dn_\chi}{dt} + 3Hn_\chi = -\langle \sigma_A v \rangle \left[(n_\chi)^2 - (n_\chi^{\text{eq}})^2 \right], \tag{9.9}$$

where n_χ is the number density of χ particles in terms of the phase space density:

$$n_\chi = \frac{g_\chi}{(2\pi)^3} \int d^3p f(E). \tag{9.10}$$

The parameter g_χ is the number of internal degrees of freedom of the particle, so that $g_\chi = 4$ for Dirac fermions and $g_\chi = 2$ for Majorana fermions. Further, $f(E)$ is the Fermi-Dirac or Bose-Einstein distribution, which is given by

$$f(E) = \frac{1}{e^{E/T} \pm 1}. \tag{9.11}$$

Here the plus (minus) sign is for fermions (bosons). Finally, n_χ^{eq} is the number density of χ's in thermal equilibrium, which is given by

$$n_\chi^{\text{eq}} = \frac{g_\chi T^3}{2\pi^2} \left(\frac{m_\chi}{T}\right)^2 K_2 \left(\frac{m_\chi}{T}\right), \qquad (9.12)$$

where $K_2(m/T)$ is the modified Bessel function. Therefore, at temperatures $(T \ll m_\chi)$, one finds

$$n_\chi^{\text{eq}} = g_\chi \left(\frac{m_\chi T}{2\pi}\right)^{3/2} e^{-m_\chi/T}. \qquad (9.13)$$

This means that there are roughly as many χ particles as photons. Finally, $\langle \sigma_A v \rangle$ is the thermal averaged total cross section for annihilation of $\chi\bar{\chi}$ into lighter particles times the relative velocity v, defined as follows:

$$\langle \sigma_A v \rangle = \frac{1}{(n_\chi^{\text{eq}})^2} \int d^3 p_1 d^3 p_2 f(E_1) f(E_2) \sigma_A v, \qquad (9.14)$$

with $\sigma_A v$ given by

$$\sigma_A v = \frac{1}{4 E_1 E_2} \int \prod_i \frac{d^3 p_i}{(2\pi)^3 2 p_i^0} (2\pi)^4 \delta^4 \left(p_1 + P_2 - \sum_j p_j\right) |\mathcal{M}|^2, \qquad (9.15)$$

where p_1 and p_2 are incoming momenta. The sum and product are over the outgoing particles. Here, $\mathcal{M}$ is the annihilation amplitude of two χ-particles.

Note that the term $3 H n_\chi$ in Eq. (9.9) describes the dilution of the number density due to the Hubble expansion and, in the absence of number-changing interactions, the right-hand side would be zero. Thus, in order to determine the abundance of any stable WIMP, one needs to calculate the cross section for the annihilation of this WIMP into all lighter particles. By introducing $Y = n/s$, where $s = 2\pi g_* T^3/45$ is the entropy density and the parameter $x = m_\chi/T$, and using the conservation of entropy per comoving volume ($sa^3 =$ constant), one obtains $s\dot{Y} = \dot{n} + 3Hn$. Thus, Eq. (9.9) can be written as

$$\frac{dY}{dx} = \frac{-s\langle \sigma_A v \rangle}{Hx} \left(Y^2 - Y_{\text{eq}}^2\right). \qquad (9.16)$$

In Fig. 9.3, we display the comoving number density, Y, of a WIMP in the early Universe as a function of $x = m_\chi/T$. The solid lines stand for the equilibrium abundance and the dashed lines refer to the actual abundances. It is clear that the larger annihilation cross section is, the smaller number density is left. The Boltzmann equation (9.16) can be solved by non-relativistic expansion in x:

$$\langle \sigma_A v \rangle \approx a + \frac{b}{x}, \qquad (9.17)$$

where a and b are the s-wave and p-wave contributions of annihilation processes, respectively. One may rewrite Eq. (9.16) in terms of the variable $\Delta = Y - Y_{\text{eq}}$ as

$$\Delta' = -Y_{\text{eq}}' - f(x)\Delta(2Y_{\text{eq}} + \Delta), \qquad (9.18)$$

where prime stands for the derivative with respect to x and $f(x) = s\langle \sigma_A v \rangle/Hx$, which, in the non-relativistic approximation of $\langle \sigma_A v \rangle$, is given by

$$f(x) = \sqrt{\frac{\pi g_*}{45}} m_\chi \tilde{M}_{\text{Pl}}(a + 6b/x)x^{-2}. \qquad (9.19)$$

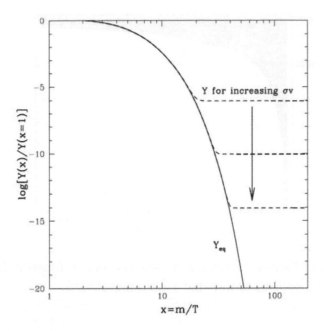

Figure 9.3 Number density of a WIMP as a function of $x = m_\chi/T$.

At early times, $x \ll x_F$, one finds $\Delta = -\frac{Y'_{\text{eq}}}{f(x)(2Y_{\text{eq}}+\Delta)}$, while at late times one has $\Delta \sim Y \gg Y^{\text{eq}}$ and $\Delta' \gg Y'_{\text{eq}}$: therefore, we get

$$\Delta^{-2}\Delta' = -f(x), \tag{9.20}$$

which can be integrated between x_F and ∞ to obtain

$$\Delta_\infty = \left[\int_{x_F}^\infty f(x)dx\right]^{-1} \simeq Y_\infty. \tag{9.21}$$

Thus, the non-relativistic approximated annihilation cross section Y_∞, which is the value of Y at long time after the freeze out, is given by

$$Y_\infty^{-1} = \frac{\pi g_*}{45}\tilde{M}_{\text{Pl}}m_\chi x_F^{-1}\left(a + \frac{3b}{x_F}\right). \tag{9.22}$$

The WIMPs freeze out at the temperature T_F, which is determined by

$$\Gamma_A(T_F) = H(T_F), \tag{9.23}$$

where $\Gamma_A = n_\chi^{\text{eq}}\langle\sigma_A v\rangle$, and, during the radiation dominated epoch, one has $H(T) = 1.66g_*^{1/2}T^2/\tilde{M}_{\text{Pl}}$ with g_* defined as

$$g_* = \sum_{i=\text{bosons}} g_i\left(\frac{T_i}{T}\right)^3 + \frac{7}{8}\sum_{i=\text{fermions}} g_i\left(\frac{T_i}{T}\right)^3. \tag{9.24}$$

Therefore, one finds

$$x_F = \ln\frac{0.74\tilde{M}_{\text{Pl}}(a + 6b/x_F)c(2+c)m_\chi}{(g_* x_F)^{1/2}}, \tag{9.25}$$

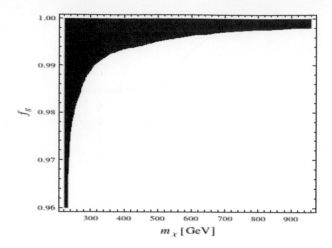

Figure 9.4 The mass of lightest neutralino versus the purity function in the region of parameter space allowed by gluino and Higgs mass limits.

where g_* is evaluated at T_F and c is a constant of order unity ($c = 1/2$ is usually assumed) determined by matching the late-time and early-time solutions. This equation can be solved by iteration. It leads to $x_F \simeq 20$.

Finally, the contribution to the relic abundance Ωh^2 from the WIMP is given by

$$\Omega h^2 = \frac{m_\chi n_\chi}{\rho_c/h^2} = \frac{Y_\infty s_0 m_\chi}{\rho_c/h^2} \simeq 2.82 \times 10^8 Y_\infty \left(\frac{m_\chi}{\text{GeV}}\right), \tag{9.26}$$

where s_0 refers to the value of entropy density today, which is given by $s_0 \simeq 4000 \text{ cm}^{-3}$, the critical density today being $\rho_c \simeq 10^{-5} h^2 \text{ GeV cm}^{-3}$, while h is the already introduced Hubble constant. Therefore, one finds

$$\Omega h^2 \simeq 0.1 \times \left(\frac{a + 3b/x_F}{10^{-9}\text{GeV}^{-2}}\right)^{-1} \left(\frac{x_F}{20}\right) \left(\frac{g_*}{90}\right)^{-1/2}, \tag{9.27}$$

which is consistent with the recent results of [160],

$$\Omega_{\text{DM}} h^2 = 0.1197 \pm 0.0022, \tag{9.28}$$

if the annihilation cross section is of order 10^{-9} GeV^{-2}.

9.4   NEUTRALINO DM

As mentioned in Sect. 5.1.3, the lightest neutralino is a linear combination of the flavour eigenstates:

$$\chi = N_{11}\tilde{B}^0 + N_{12}\tilde{W}^0 + N_{13}\tilde{H}_1^0 + N_{14}\tilde{H}_2^0. \tag{9.29}$$

The collider phenomenology and cosmology of the neutralino are governed primarily by its mass and composition, described by the gaugino purity function defined in Eq. (9.4). As shown previously, the constraints on $m_{1/2}$ from Higgs mass limit and gluino mass lower bound imply that $m_\chi \gtrsim 240$ GeV, which is larger than the limits obtained from direct searches at the LHC. Moreover, an upper bound of order 1 TeV is also obtained (from Higgs mass constraints). In this region of allowed parameter space, the LSP is essentially pure bino, as shown in Fig. 9.4. This can be easily understood from the fact that the μ parameter, determined by the radiative EWSB condition, Eq. (4.54), is typically of order m_0 and hence it is much heavier than the gaugino mass M_1.

Concerning the annihilation cross section contributing to the density of the Universe in Eq. (9.6), there are numerous final states into which the neutralinos can annihilate. The most important of these are the two body final states which occur at the tree level. In general, the neutralinos may annihilate into $f\bar{f}$ (*i.e.*, fermion-antifermion pairs), W^+W^-, ZZ, W^+H^-, ZA, ZH, Zh, H^+H^- and all other combinations of neutral Higgs states. For a bino-like LSP, *i.e.*, $N_{11} \simeq 1$ and $N_{1i} \simeq 0$, $i = 2, 3, 4$, one finds that the relevant annihilation channels are the fermion-antifermion ones, as shown in Fig. 9.5, and all other channels are instead suppressed (including h-mediation, which would be off-shell, and H-exchange, which has suppressed couplings to both χ and f states). Also, the annihilation process mediated by the Z gauge boson is suppressed due to the small $Z\chi\chi$ coupling $\propto N_{13}^2 - N_{14}^2$, except at the resonance when $m_\chi \sim M_Z/2$. Furthermore, one finds that the annihilation is predominantly into leptons through the exchanges of the three slepton families ($\tilde{\ell}_L, \tilde{\ell}_R$) ($\ell = e, \mu, \tau$). In fact, the squarks exchanges are suppressed due to their large masses.

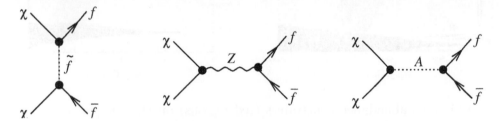

Figure 9.5 Feynman diagrams contributing to early Universe neutralino ($\tilde{\chi}_1^0$) annihilation into fermions through sfermions, Z gauge boson, and Higgs states.

From Chapter 5, one can find that the interactions of sleptons with leptons and neutralinos are given by

$$\mathcal{L}_{\ell\tilde{\ell}\chi} = \sqrt{2}g'\left[Y_L\bar{\ell}_i P_R\tilde{\ell}_{L_i} + Y_R\bar{\ell}_i P_L\tilde{\ell}_{R_i}\right]\chi + \text{h.c.} \tag{9.30}$$

where $Y_L = T_{3l} - Q_l = \frac{1}{2}$, with T_{3l} being the lepton weak isospin and Q_l the lepton electric charge. Further, $Y_R = Q_l = -1$ and $P_{L,R}$ are defined as $P_{L,R} = \frac{1}{2}(1 \mp \gamma_5)$. Note that small off-diagonal elements in $M_{\tilde{l}}^2$ are neglected. In this case the thermal average annihilation cross section is given by

$$\langle \sigma_A v \rangle = \left(1 - \frac{m_l^2}{m_\chi^2}\right)^{1/2}\frac{g_1^4}{32\pi}\left[(Y_L^2 + Y_R^2)^2\frac{m_l^2}{m_{\tilde{l}}^2 + m_\chi^2 - m_l^2}\right.$$
$$\left. + (Y_L^4 + Y_R^4)\frac{4m_\chi^2}{m_{\tilde{l}}^2 + m_\chi^2 - m_l^2}\frac{1}{x}\right]. \tag{9.31}$$

Thus, in the limit of $m_\ell/m_{\tilde{\ell}} \to 0$, we have $a = 0$ and

$$b = \frac{g'^4}{8\pi}(Y_L^4 + Y_R^4)\frac{m_\chi^2}{m_\chi^2 + m_{\tilde{l}}^2}. \tag{9.32}$$

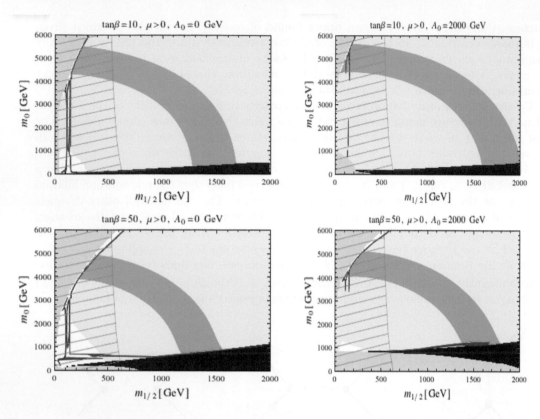

Figure 9.6 LSP relic abundance constraints (red regions) on the $(m_0, m_{1/2})$ plane. The LUX result is satisfied by the yellow region while the green region indicates the $124 \lesssim m_h \lesssim 126$ GeV constraint. The blue region is excluded because the lightest neutralino is not the LSP. The pink region is excluded due to the absence of radiative EWSB. The gray shadow lines denote exclusion because $m_{\tilde{g}} < 1.4$ TeV. Taken from [12].

In the cMSSM, the slepton mass at the EW scale is given in terms of m_0 and $m_{1/2}$ as follows:

$$m_{\tilde{l}}^2 \simeq m_0^2 + 0.5 m_{1/2}^2. \tag{9.33}$$

Therefore, the cosmological upper limit on $\Omega_\chi h^2$ leads to an upper bound on m_0 for a fixed $m_{1/2}$. In Fig. 9.6, we show the constraint from the observed limits of Ωh^2 on the $(m_0, m_{1/2})$ plane for $A_0 = 0, 2000$ GeV, $\tan \beta = 10, 50$ and $\mu > 0$. Here we used micrOMEGAs [161] to compute the complete relic abundance of the lightest neutralino, taking into account the possibility of having co-annihilation with the next-to-LSP, which is typically the lightest stau. In this figure the red regions correspond to a relic abundance within the measured limits [159]:

$$0.09 < \Omega h^2 < 0.14. \tag{9.34}$$

It is noticeable that, with low $\tan \beta$ (~ 10), this region corresponds to light $m_{1/2}$ (< 500 GeV), where a significant co-annihilation between the LSP and stau takes place.

However, this possibility is now excluded by the Higgs and gluino mass constraints [162]. At large $\tan\beta$, another region is allowed because of a possible resonance due to s-channel annihilation of the DM pair into fermion-antifermion via the pseudoscalar Higgs boson A at $M_A \simeq 2m_\chi$ [12]. For $A_0 = 0$, a very small part of this region is allowed by the Higgs mass constraint, while for large A_0 (~ 2 TeV) a slight enlargement of this part can be achieved. In Fig. 9.7, we zoom into this region to show the explicit dependence of the relic abundance on the LSP mass and large values of $\tan\beta$. As can be seen from this figure, there is no point that can satisfy the relic abundance stringent constraints with $\tan\beta < 30$.

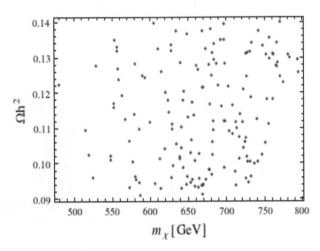

Figure 9.7 The relic abundance versus the mass of the LSP for different values of $\tan\beta$. Red points indicate $40 \leq \tan\beta \leq 50$ and blue points indicate $30 \leq \tan\beta < 40$. All points satisfy the above-mentioned constraints. Taken from [12].

9.5 NEUTRALINO DIRECT DETECTION

If neutralinos are the bulk of DM, they will form not only a background density in the Universe, but also will cluster gravitationally with ordinary stars in the galactic halos. In particular, they will be present in our own galaxy, the Milky Way. This raises the hope of detecting relic neutralinos directly, by experiments on the Earth. In 1985 Goodman and Witten showed that a direct experimental detection through elastic scattering with nuclei in a detector, as shown schematically in Fig. 9.8, is possible in principle [163].

The differential detection rate of neutralino DM, where the latter effects are induced in appropriate detectors by neutralino-nucleus elastic scattering, is given by

$$\frac{dR}{dQ} = \frac{\sigma\rho_\chi}{2m_\chi m_r^2} F^2(Q) \int_{v_{\min}}^{\infty} \frac{f_1(v)}{v} dv, \qquad (9.35)$$

where $f_1(v)$ is the distribution of speeds relative to the detector. The reduced mass is $m_r = \frac{m_\chi m_N}{m_\chi + m_N}$, where m_N is the mass of the nucleus, $v_{\min} = \left(\frac{Qm_N}{2m_r^2}\right)^{1/2}$, Q is the energy deposited in the detector and ρ_χ is the density of the neutralino near the Earth. It is common to fix the DM density as $\rho_\chi = 0.3$ GeV/cm^3.

The quantity σ is the elastic-scattering cross section of the LSP with a given nucleus.

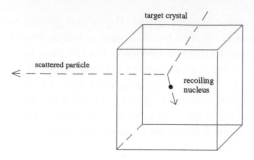

Figure 9.8 Elastic scattering of a DM particle with an atomic nucleus in a detector.

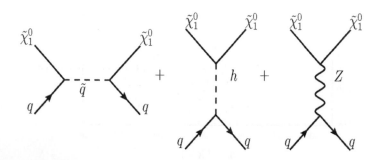

Figure 9.9 Feynman diagrams contributing to neutralino ($\tilde{\chi}_1^0$)-quark (q) cross section.

In general σ has two contributions: a spin-dependent contribution, arising from Z and $\tilde{q}$ exchange diagrams, and a spin-independent (scalar) contribution due to the Higgs and squark exchange diagrams, as shown in Fig. 9.9. The effective scalar interaction of a neutralino with a quark is given by

$$\mathcal{L} = a_q \bar{\chi}\chi \, \bar{q}q, \tag{9.36}$$

where a_q is the neutralino-quark effective coupling. The scalar cross section of the neutralino scattering with target nucleus at zero momentum transfer is given by [164]

$$\sigma_0^{\text{SI}} = \frac{4m_r^2}{\pi} \left(Z f_p + (A - Z) f_n \right)^2, \tag{9.37}$$

where Z and $A - Z$ are the number of protons and neutrons, respectively, and f_p, f_n are the neutralino couplings to protons and neutrons, respectively. The differential scalar cross section for non-zero momentum transfer q can now be written as

$$\frac{d\sigma_{\text{SI}}}{dq^2} = \frac{\sigma_0^{\text{SI}}}{4m_r^2 v^2} F^2(q^2), \ 0 < q^2 < 4m_r^2 v^2, \tag{9.38}$$

where v is the neutrino velocity and $F(q^2)$ is the form factor [164]. In Fig. 9.10, we display the MSSM prediction for a spin-independent scattering cross section of the LSP with a proton $(\sigma_{\text{SI}}^p = \int_0^{4m_r^2 v^2} \frac{d\sigma_{\text{SI}}}{dq^2}\big|_{f_n = f_p} dq^2)$ after imposing the LHC and relic abundance constraints. It is clear that our results for σ_{SI}^p are less than the recent LUX bound (blue curve) by at

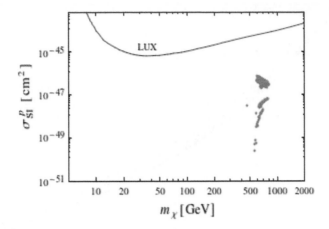

Figure 9.10 Spin-independent scattering cross section of the LSP with a proton versus the mass of the LSP within the region allowed by all constraints (from the LHC and relic abundance). Taken from [12].

least two orders of magnitude. This would explain the negative results of direct searches so far. It is important to note that the cMSSM yields bino-like DM, which provides a very low scattering cross section. However, some other models may give different DM structures such as a bino-wino mixture, which has a little bit higher cross section and also bino-Higgsino mixture or Higgsino-like DM with higher cross section. For an example of this type of different cases of DM structure, see [165].

9.6 NEUTRALINO INDIRECT DETECTION

A promising method for indirect detection of neutralinos in the halo is the observation of the energetic neutrinos from the annihilation of neutralinos that accumulate in the Sun or in the Earth. Among the annihilation products are ordinary neutrinos, which may be observable in suitable detectors. The energies of the neutrinos are about a third of the LSP mass, so they are easily distinguished from solar neutrinos or any other known background. The technique for the detection of such energetic neutrinos is through observation of upward muons produced by the charged current interactions of the neutrinos in the rock below the detector. Concentrating on the neutralino annihilation on the Sun, the flux of such muons can be written as [164]

$$
\Gamma = 2.9 \times 10^8 m^{-2} \mathrm{yr}^{-1} \tanh^2(t/\tau) \left(\frac{\rho_\chi}{0.3 \ \mathrm{GeV/cm}^3} \right) f(m_\chi) \zeta(m_\chi)
$$
$$
\times \left(\frac{m_\chi^2}{\mathrm{GeV}} \right)^2 \left(\frac{f_P}{\mathrm{GeV}^{-2}} \right)^2 .
\tag{9.39}
$$

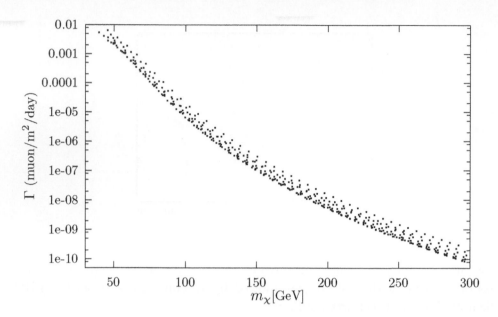

Figure 9.11 The mounic flux Γ versus m_χ for $\rho_\chi = 0.3$ and $\tan\beta \simeq 2$.

The neutralino-mass dependence of the capture rates is described by [164]

$$f(m_\chi) = \sum_i f_i \phi_i S_i(m_\chi) F_i(m_\chi) \frac{m_i^3 m_\chi}{(m_\chi + m_i)^2}, \tag{9.40}$$

where the quantities ϕ_i and f_i describe the distribution of element i in the Sun and they are listed in [164]. The quantity $S_i(m_\chi) = S(\frac{m_\chi}{m_{N_i}})$ is the kinematic suppression factor for the capture of a neutralino of mass m_χ from a nucleus i of mass m_{N_i} while $F_i(m_\chi)$ is the corresponding form factor. Finally, the functions $F_i(m_\chi)$ and $\zeta(m_\chi)$, which describes the energy spectrum from neutralino annihilation for a given mass, are given in [164]. In Fig. 9.11, we present an example of the muonic flux resulting from captured neutralinos in the Sun for $\rho_\chi = 0.3 \text{GeV/cm}^3$. We see that the predicted muonic flux lies between 10^{-2} and 10^{-9} muon/m^2/day. Clearly, large scale detectors are best suited for neutralino detection.

9.6.1 PAMELA, ATIC, FERMI and HESS Anomalies

The PAMELA experiment measured an excess of cosmic ray positrons with no indication of any excess of anti-proton flux . Therefore, if it is indeed DM that is responsible for the positron excess, it seems natural to consider a type of DM particle that annihilates predominantly into l^+l^- channels. The positron flux in the galactic halo is given in terms of the production rate of the positrons from DM annihilation, which is given by

$$Q(E, \mathbf{r}) = \frac{1}{2} \left(\frac{\rho(\mathbf{r})}{m_\chi} \right)^2 \sum_f \langle \sigma_A v \rangle_f \left(\frac{dN}{dE} \right)_f, \tag{9.41}$$

where $\langle \sigma_A v \rangle_f \equiv a_f$ refers to the averaged annihilation cross section into the final state f and $(dN/dE)_f$ is the fragmentation function, representing the number of positrons with

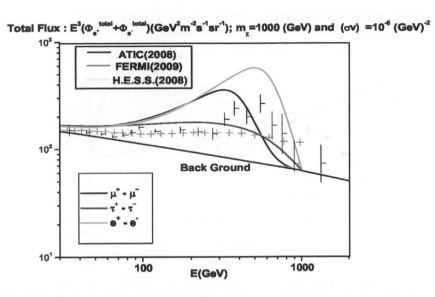

Figure 9.12 The total absolute flux in units of $\text{GeV}^2\text{m}^{-2}\text{s}^{-1}\text{sr}^{-1}$, generated by the DM annihilation into e^+e^-(green line), $\mu^+\mu^-$(black line) and $\tau^+\tau^-$(red line), as a function of positron energy for $m_\chi = 1$ TeV, for thermal averaging cross section 10^{-6} GeV^{-2}. Taken from [16].

energy E produced from the final state f. Here, $\rho(\mathbf{r})$ is a DM halo mass profile. Though there are several types of proposed halo DM density profiles, we adopt here the standard NFW profile [166].

Fig. 9.12 shows the total absolute flux, $\Phi_{e^+}^{\text{total}} + \Phi_{e^-}^{\text{total}}$, generated by DM annihilation into e^+e^-, $\mu^+\mu^-$ and $\tau^+\tau^-$ as a function of the positron energy for $m_\chi = 1$ TeV and hermal averaging cross section of the order of 10^{-6} GeV^{-2}. In our analysis, the so-called MED diffusion model [167] and NFW galactic halo with scale-length 20 kpc [168] are assumed. Also, $\Phi_{e^+}^{\text{total}}$ and $\Phi_{e^-}^{\text{total}}$ are defined as

$$\Phi_{e^+}^{\text{total}} \sim \Phi_{e^+}^{\text{DM}} + \Phi_{e^+}^{\text{sec}} \quad , \quad \Phi_{e^-}^{\text{total}} \sim \Phi_{e^-}^{\text{prim}} + \Phi_{e^-}^{\text{sec}}. \tag{9.42}$$

The flux of positrons $\Phi_{e^+}^{\text{DM}}$ is given from the number density of the positron through several steps. The fragmentation function for the direct process is almost monotonic: $\left(\frac{dN}{dE}\right)_{ee} \sim \delta(E - m_\chi)$. The functions $\left(\frac{dN}{dE}\right)_{\mu\mu}$ and $\left(\frac{dN}{dE}\right)_{\tau\tau}$ can be found in [169].

Calculated in terms of fitting functions matching the fluxes deduced via standard simulations of cosmic ray production and propagation and the astrophysical background fluxes

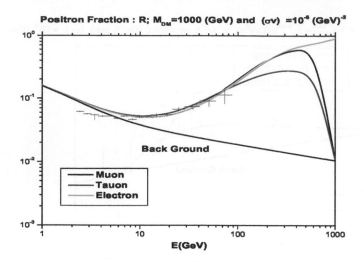

Figure 9.13 The positron fraction for DM annihilation into e^-e^+(green line), $\mu^+\mu^-$(black line) and $\tau^+\tau^-$(red line) for DM mass $m_\chi = 1$ TeV and thermal averaging cross section 10^{-6} GeV^{-2}. Taken from [16].

of positrons, Φ_{e+}^{sec}, $\Phi_{e-}^{\mathrm{prim}}$ and Φ_{e-}^{sec}, are given by the following rates:

$$\Phi_{e-}^{\mathrm{prim}}(\epsilon) = \frac{0.16\epsilon^{-1.1}}{1 + 11\epsilon^{0.9} + 3.2\epsilon^{2.15}}(\mathrm{cm}^{-2}\mathrm{s}^{-1}\mathrm{sr}^{-1}),$$

$$\Phi_{e-}^{\mathrm{sec}}(\epsilon) = \frac{0.70\epsilon^{-0.7}}{1 + 110\epsilon^{1.5} + 600\epsilon^{2.9} + 580\epsilon^{4.2}}(\mathrm{cm}^{-2}\mathrm{s}^{-1}\mathrm{sr}^{-1}),$$

$$\Phi_{e+}^{\mathrm{sec}}(\epsilon) = \frac{4.5\epsilon^{0.7}}{1 + 650\epsilon^{2.3} + 1500\epsilon^{4.2}}(\mathrm{cm}^{-2}\mathrm{s}^{-1}\mathrm{sr}^{-1}), \qquad (9.43)$$

where $\epsilon \equiv E/(1\ \mathrm{GeV})$.

We use a DM particle of mass ~ 1 TeV, consistently with the maximum allowed by FERMI γ-ray observations of DM-dominated virialised gravitational systems. As it is immediately seen, Φ_{e+}^{sec} and Φ_{e-}^{sec} are quite small at all energies and can be neglected compared to $\Phi_{e-}^{\mathrm{prim}}$. The same goes for the computed positron flux for $\langle\sigma_A v\rangle \sim 10^{-9}$ GeV^{-2}, which remains far below the background. Therefore, in such a situation, and in terms of the plotted quantities, any excess in the flux would have to be explained in terms of modifications to the astrophysical background (including the possible effects of local pulsars, supernovae remnants, etc.). Alternatively, a larger cross section can lift the computed DM flux above the background, so as to explain any apparent excess. It turns out that a $\langle\sigma_A v\rangle \sim 10^{-6}$ GeV^{-2} is compatible with PAMELA observations (without invoking any changes in the background model). In addition, for a positron energy $E > 200$ GeV, the positron flux Φ_{e+}^{DM} exceeds the electron background, which fares well in explaining the ATIC, HESS and FERMI excess at 300–800 GeV.

In Fig. 9.13, we plot the positron fraction

$$R = \frac{\Phi_{e+}^{\mathrm{DM}}(E) + \Phi_{e+}^{\mathrm{sec}}(E)}{\Phi_{e+}^{\mathrm{DM}}(E) + \Phi_{e+}^{\mathrm{sec}}(E) + \Phi_{e-}^{\mathrm{prim}}(E) + \Phi_{e-}^{\mathrm{sec}}(E)}, \qquad (9.44)$$

the quantity that PAMELA actually measured for DM annihilation into e^+e^-, $\mu^+\mu^-$ and $\tau^+\tau^-$, along with the relevant measurements. It is clear, as expected, that DM annihilation with $\langle\sigma_A v\rangle \sim 10^{-6}$ GeV^{-2} can easily account for the PAMELA measurements while employing a standard astrophysical background. These measurements explore the existence of a positron excess for energies confined between 8 and 80 GeV.

It is worth mentioning that various 'boost factors' have been proposed, including, for example, in terms of increased DM density inside the galaxy's subhalo population. Yet very little of such substructure is expected in the solar neighborhood: efficient stripping dissolves subhalos in these inner regions, thereby ensuring that most mass in subhalos is concentrated at far larger galactocentric radii. Further, positrons with energies of order 100 GeV or above are expected to emanate from DM annihilation within a few hundred pc of the solar system. This is quite a generic result, independent of the details of the diffusion model adopted for cosmic ray propagation, having its origins in such fairly tractable processes as the expected positron energy loss due to inverse Compton scattering off CMB and starlight. It is therefore quite unlikely that the required boost in DM annihilation flux can be obtained by invoking the galactic subhalo population, a result confirmed by detailed modelling of its effect.

9.7   DM SEARCHES AT THE LHC

The LHC provides an alternative and a complementary way to search for DM [170, 171]. In fact, searching for DM at colliders has some advantages compared with direct detection. It is independent of any astrophysical assumptions and also it can probe much lighter DM. The typical signatures of DM pair production at the LHC are mono-jet plus missing energy, mono-photon plus missing energy and mono-Z plus missing energy [172, 173], as shown in Fig. 9.14.

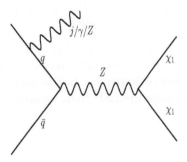

Figure 9.14   Mono-jet, mono-γ and mono-Z signatures for DM at the LHC.

In the MSSM, DM pair production can occur through the exchange of a Z boson or neutral Higgs bosons, h, H and A. However, the latter process leads to small cross section so we will focus on the vector mediator. The interaction Lagrangian of Z with the lightest neutralino is given by:

$$\mathcal{L} = \frac{g}{2\cos\theta_W}\left(|N_{14}|^2 - |N_{13}|^2\right)\bar{\chi}\gamma^\mu\gamma_5\chi Z_\mu. \tag{9.45}$$

In order to enhance this coupling, N_{13} and/or N_{14} should be of order one. Thus, the lightest neutralino has to be Higgsino-like. As explained in Chapter 4, the lightest neutralino becomes Higgsino-like when $|\mu| \ll M_{1,2}$. In this case, the LSP mass is given by

$$m_{\chi_{3,4}} \simeq |\mu| \mp \frac{1}{2} M_Z^2 (1 \mp \sin 2\beta) \left(\frac{s_W^2}{\mu \pm M_1} + \frac{c_W^2}{\mu \pm M_2} \right). \tag{9.46}$$

Light μ can be realised if $m_{H_d}^2 \approx \tan^2 \beta (M_Z^2 + m_{H_u}^2)$ and, in order to avoid large fine tuning, one should require $m_{H_d} \sim |m_{H_u}| \sim \mathcal{O}(M_Z)$. Therefore, non-universal gaugino masses are essential for this scenario to have a very large $M_3 \sim \mathcal{O}(\text{TeV})$ scale to guarantee heavy stop masses, required by the SM-like Higgs mass and also to account for a heavy gluino with $M_{1,2}$ of order of the EW scale.

However, in the limit of Higgsino-like LSP , one can show that $|N_{13}| \simeq |N_{14}| \simeq \sin \pi/4$ and degenerate LSP and NLSP. Therefore, the coupling $Z\chi_1\chi_1$ becomes very suppressed and the corresponding cross section $\sigma(pp \to Z \to \chi_1\chi_1 + j)$ is of order of the 10^{-7} pb, which is extremely small with respect to the SM backgrounds, which are dominated by the following channels: *(i)* the irreducible background $pp \to Z(\to \nu\bar{\nu}) + j$, which is the main one because it has the same topology as the signal; *(ii)* $pp \to W(\to \ell\nu) + j$ ($\ell = e, \mu, \tau$), though this process fakes the signal only when the charged lepton is outside the acceptance of the detector or close to the jet; *(iii)* $pp \to W(\to \tau\nu) + j$, as this process may fake the signal, since a secondary jet from hadronic tau decays tends to localise on the side opposite to $\not{E}_T$; *(iv)* $pp \to t\bar{t}$, given that this process may resemble the signal but also contains extra jets and leptons, which allow one to highly suppress it by applying b-jet and lepton vetoes; *(v)* the di-boson background $pp \to ZZ(\to 2\nu 2\bar{\nu}) + j$, which is generically suppressed due to its small cross section at production level but topologically mimics the signal rather well.

One may avoid the cancellation in the $Z\chi_1\chi_1$ coupling by considering the Z decay into $\chi_1\chi_2$, where its coupling is proportional to

$$\frac{g}{2\cos\theta_W} (N_{13}^* N_{23} - N_{14}^* N_{24}) \sim \frac{g}{\cos\theta_W} |N_{13}^* N_{23}| \sim \mathcal{O}(0.1). \tag{9.47}$$

This coupling leads to $\sigma(pp \to Z \to \chi_1\chi_2 + j) \simeq 10^{-3}$ pb. Nevertheless, due to the decay $\chi_2 \to \chi_1 + \nu\nu$ with BR of the order of 10^{-2}, the total cross section is reduced to 10^{-5}, which is still very small compared to the SM backgrounds. In this regard, one may conclude that in MSSM there is little chance to probe the LSP from mono-jet, -γ and -Z signal at the LHC.

III

SUSY CP and Flavour

CHAPTER 10

CP Violation in Supersymmetric Models

The understanding of the nature of CP violation is another open question in particle physics that will be addressed by the LHCb experiment. The amount of CP violation present in the SM is not sufficient to explain the matter-antimatter asymmetry observed in our Universe. SUSY, however, predicts new sources of CP violation. Therefore, proving any deviation from the SM expectations for CP violating processes at LHCb would be a clear hint for SUSY.

10.1 CP VIOLATION IN THE SM

The SM Lagrangian consists of three parts. The first one, which is known as the gauge sector, is the part of the Lagrangian that contains the interactions of fermions with the gauge bosons and also the interactions of the gauge bosons among themselves. The second part is the Yukawa sector, which contains the interactions of fermions with the scalar Higgs doublet. The third part is the Higgs sector, which contains the scalar potential of the Higgs field. It is remarkable that the gauge sector is CP conserving and hence there is no complex phase in this sector. Also the Higgs sector of one Higgs doublet models can be made real by field redefinitions. Therefore, there is no CP violating phase in this sector either. Finally, the Yukawa couplings are generally complex and they are indeed the source of CP violation in the SM.

As is known, the fermion masses are induced by the Yukawa interactions after EWSB. For the up and down types of quarks, one finds

$$\mathcal{L}_{\text{mass}} = \frac{v}{\sqrt{2}} \left(Y_{ij}^u \bar{u}_L^i u_R^j + Y_{ij}^d \bar{d}_L^i d_R^j \right) + \text{h.c.} \tag{10.1}$$

Therefore, the mass matrices $M_{u,d}$ are defined in terms of the Yukawa matrices as $M_{u,d} = \frac{v}{\sqrt{2}} Y_{u,d}$. These mass matrices are generic 3×3 structures. The physical (or mass) eigenstates, where the mass matrices are diagonal, are obtained by the following bi-unitary rotations

$$u_L = V_L^u u_L' , \qquad u_R = V_R^u u_R' , \tag{10.2}$$
$$d_L = V_L^d d_L' , \qquad d_R = V_R^u d_R' , \tag{10.3}$$

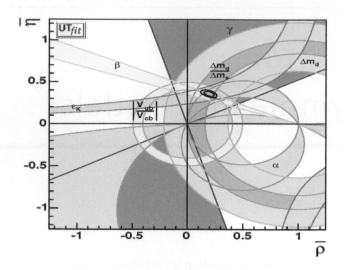

Figure 10.1 Experimental constraints on the $\rho - \eta$ plane as obtained by the UTfit collaboration [17].

where $V_{L,R}^{u,d}$ satisfy the following diagonalisation conditions

$$V_L^{u\dagger} M_u V_R^u = M_u^{\text{diag}}, \tag{10.4}$$

$$V_L^{d\dagger} M_d V_R^d = M_d^{\text{diag}}. \tag{10.5}$$

Under the change of basis from $u, d \to u', d'$, the EW charged current takes the form

$$\mathcal{L}_W = \frac{g}{\sqrt{2}} \bar{u}'^i_L \gamma^\mu V_{\text{CKM}}^{ij} d'^j_L W^{+\mu} + \text{h.c.} \tag{10.6}$$

where the V_{CKM} is the CKM mixing matrix, defined as

$$V_{\text{CKM}} = V_L^{u\dagger} V_L^d. \tag{10.7}$$

Note that the mismatch between the rotations of the up and down types of quark fields is the source of the mixing matrix V_{CKM}, which is generally non-diagonal. Therefore, flavour changes in the charged current are naturally implemented in the SM. Similarly, it is straightforward to show that there is no such mismatch in the neutral current ($\bar{u}uZ$ and $\bar{d}dZ$), hence it remains flavour diagonal.

The CKM matrix is a unitary matrix, since $V_{\text{CKM}} V_{\text{CKM}}^\dagger = (V_u^\dagger V_d)(V_d^\dagger V_u) = 1$. Thus, for three generations of quarks, the CKM matrix generally depends on 9 parameters: three real angles and six phases. However, five of these phases can be absorbed into the redefinition of quark phases. Therefore, in the SM with three families, CP violation is accommodated by a single complex phase in the CKM mixing matrix. Also, one can conclude that, with two quark generations, no CP violation can be generated. Indeed, a 2×2 unitary matrix is a real matrix, where all phases are rotated away by some field redefinitions.

The standard parameterisation of the the CKM matrix is given by [174]

$$V_{\text{CKM}} = \begin{pmatrix} c_{12}c_{13} & s_{12}c_{13} & s_{12}e^{-i\delta} \\ -s_{12}c_{23} - c_{12}s_{23}s_{13}e^{i\delta} & c_{12}c_{23} - s_{12}s_{23}s_{13}e^{i\delta} & s_{23}c_{13} \\ s_{12}s_{23} - c_{12}23s_{13}e^{i\delta} & -c_{12}s_{23} - s_{12}c_{23}s_{13}e^{i\delta} & c_{23}c_{13} \end{pmatrix}, \tag{10.8}$$

where $c_{ij} = \cos\theta_{ij}$, $s_{ij} = \sin\theta_{ij}$ and δ is the CP violating phase. Another useful parameterisation of the CKM matrix was introduced by Wolfenstein [175] and is based on experimental results and unitarity to express V_{CKM} as a series expansion in $\lambda \equiv \sin\theta_C \simeq 0.22$:

$$V_{\text{CKM}} = \begin{pmatrix} 1 - \frac{1}{2}\lambda^2 & \lambda & A\lambda^3(\rho - i\eta) \\ -\lambda & 1 - \frac{1}{2}\lambda^2 & A\lambda^2 \\ A\lambda^3(1 - \rho - i\eta) & -A\lambda^2 & 1 \end{pmatrix} + \mathcal{O}(\lambda^4). \tag{10.9}$$

The parameter A is determined by measuring V_{cb} from the inclusive leptonic b decays and the exclusive decay $B^0 \to D^{*+}l^-\bar{\nu}_l$. The (ρ, η) plane can be constrained from the combination of many experimental observables as in Fig. 10.1. Note that $\bar{\rho} = \rho(1 - \lambda^2/2)$ and $\bar{\eta} = \eta(1 - \lambda^2/2)$ are usually used instead of ρ and η to keep unitarity exact. The best determination of the Wolfenstein parameters is [17]

$$\rho = 0.135^{+0.031}_{-0.016}, \quad \eta = 0.349^{+0.015}_{-0.017}, \quad A = 0.814^{+0.021}_{-0.022}, \quad \lambda = 0.2257^{+0.0009}_{-0.0010}. \tag{10.10}$$

The uncertainty is still not negligible, in particular, the theoretical results suffer from significant uncertainties due to the hadronic matrix elements involved in the semileptonic decays, as we will show in the next chapter.

The unitarity of the CKM matrix leads to the following relations:

$$V_{ud}V_{us}^* + V_{cd}V_{cs}^* + V_{td}V_{ts}^* = 0, \tag{10.11}$$
$$V_{us}V_{ub}^* + V_{cs}V_{cb}^* + V_{ts}V_{tb}^* = 0, \tag{10.12}$$
$$V_{ud}V_{ub}^* + V_{cd}V_{cb}^* + V_{td}V_{tb}^* = 0. \tag{10.13}$$

Each of these relations can be represented graphically by a triangle. The last relation is commonly used to define what is called a 'unitarity triangle'. One can display this triangle in terms of the following three angles:

$$\alpha = \arg\left(-\frac{V_{td}V_{tb}^*}{V_{ud}V_{ub}^*}\right), \quad \beta = \arg\left(-\frac{V_{cd}V_{cb}^*}{V_{td}V_{tb}^*}\right), \quad \gamma = \arg\left(-\frac{V_{ud}V_{ub}^*}{V_{cd}V_{cb}^*}\right), \tag{10.14}$$

which satisfy the constraint $\alpha + \beta + \gamma = \pi$, as shown in Fig. 10.2. Multiplying Eq. (10.13) by $V_{us}^*V_{cs}$ and taking the imaginary part, one finds that

$$\text{Im}\left[V_{us}^*V_{cs}V_{ud}V_{cd}^*\right] = -\text{Im}\left[V_{us}^*V_{cs}V_{ub}V_{cb}^*\right]. \tag{10.15}$$

Also, from other orthogonal relations, one can show that the quartets $V_{ij}V_{kl}V_{il}^*V_{kj}^*$ have equal imaginary parts up to a sign. Therefore, one defines the Jarlskog invariant [176, 177]

$$J = \text{Im}\left[V_{us}V_{cb}V_{ub}^*V_{cs}^*\right] \tag{10.16}$$

as an invariant measure for the amount of CP violation in the SM. Using the Wolfenstein parameterisation for V_{CKM}, one can show that $J \simeq A^2\lambda^6\eta \lesssim 10^{-4}$. It is remarkable that the smallness of J is due to the smallness of the off-diagonal elements of V_{CKM}, in particular $|V_{ub}|$, and not because of a small value of the CP violating phase. In addition, the area of

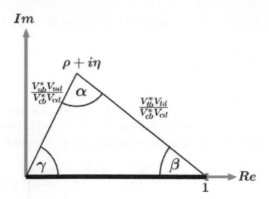

Figure 10.2 The unitarity triangle of the CKM matrix.

the unitarity triangle is given by $|J|/2$. Thus, the area of this triangle measures the amount of CP violation.

Finally, one should also recall (as previously intimated) that the SM has another source of CP violation. This arises from the term $\frac{\theta}{16\pi^2} F_{\mu\nu} \widetilde{F}^{\mu\nu}$ in the QCD Lagrangian. For non-zero θ, a large contribution to the EDM of the neutron is obtained. The current experimental limits on the EDMs imply that $\theta < 10^{-10}$. Such a small value induces a fine-tuning problem, known as the strong CP problem. The most elegant solution to overcome the strong CP problem is the PQ mechanism [178], where the smallness of the θ parameter is explained dynamically through a spontaneous breaking of a $U(1)$ symmetry, called PQ (as previously mentioned) and denoted by $U(1)_{\mathrm{PQ}}$.

10.2 CP VIOLATION IN THE MSSM

In SUSY theories there are several other possible sources for CP violation arising from the complexity of the soft SUSY breaking terms, namely, the phases associated with the gaugino masses, the Hermitian squark and slepton mass matrices, the trilinear couplings and the phase associated with bilinear coupling. In addition to all this, there is another CP phase of SUSY origin preserving the μ parameter. In general, these phases can be classified into the following two categories [179]:

- (*i*) *Flavour-independent phases* such as the phases of the μ parameter, B-parameter, gaugino masses M_a and the phases of diagonal elements of the trilinear couplings.

- (*ii*) *Flavour-dependent phases* such as the phases of the off-diagonal elements of A_{ij}^f, $i \neq j$, and phases in the Hermitian squark and slepton mass matrices $(m_{ij}^f)^2$.

In general, these phases are not all independent and one can eliminate some of them by field redefinitions. For instance, in the case of the cMSSM, with universal soft SUSY breaking terms, the SUSY phases arise from the parameters $M_{1/2}$, A and B in addition to μ. So let us assume that

$$M_{1/2} = |M_{1/2}|e^{i\alpha_M}, \qquad A = |A|e^{i\alpha_A}, \qquad B = |B|e^{i\alpha_B}, \qquad \mu = |\mu|e^{i\alpha_\mu}. \qquad (10.17)$$

However, one can make $M_{1/2}$ real by using an R transformation with R charge $Q_R = 1$ for lepton and quark superfields and $Q_R = 0$ for vector and Higgs superfields. Thus, under such an R transformation, the vector superfield transforms as

$$V(x, \theta, \bar{\theta}) \rightarrow V(x, e^{i\alpha_M/2}\theta, e^{-i\alpha_M/2}\bar{\theta}). \tag{10.18}$$

Under this transformation the gaugino $\lambda(x)$ transforms as $\lambda(x)e^{-i\alpha_M/2}$, which makes $M_{1/2}$ real. Note that in the case of non-universal gaugino masses with three different complex phases, one phase only can be rotated by the R transformation and two phases remain physical instead.

Also, under this R transformation, the Higgs and quark superfieds transform as

$$H_i(x, \theta) \quad \rightarrow \quad H_i(x, e^{i\alpha_M/2}\theta), \tag{10.19}$$
$$Q(x, \theta) \quad \rightarrow \quad e^{i\alpha_M/2}Q(x, e^{i\alpha_M/2}\theta). \tag{10.20}$$

Since the Higgs one is a chiral superfield, it has the form

$$H_i(x, \theta) = H_i(x) + \sqrt{2}\theta\tilde{H}_i + \theta^2 F_i(x), \tag{10.21}$$

where $\tilde{H}_i(x)$ are the Higgsino fields, which transform under R as $\tilde{H}_i(x)e^{i\alpha_M/2}$. Therefore, the μ term in the superpotential leads to

$$\mu H_u H_d|_{\theta\theta} = |\mu|e^{i\alpha_\mu}\tilde{H}_u\tilde{H}_d \rightarrow |\mu|e^{i(\alpha_\mu + \alpha_M)}\tilde{H}_u\tilde{H}_d. \tag{10.22}$$

Also, the bilinear coupling transforms as

$$B\mu H_u H_d \rightarrow |B\mu|e^{i(\alpha_B + \alpha_\mu)}H_u H_d. \tag{10.23}$$

Thus, by redefining the superfields H_u and H_d such that $H_i \rightarrow e^{-i(\alpha_B + \alpha_\mu)/2}H_i$, one finds that $B\mu H_u H_d$ is real, which ensures that the Higgs VEVs are real. Moreover, due to this redefinition, the phase of the μ term is changed to $-\phi_B = \alpha_M - \alpha_B$. Similarly, one can show that, under R transformation and superfield redefinitions, the phase of the trilinear coupling is changed to $\phi_A = \alpha_A - \alpha_M$. Therefore, only two phases are physical.

It is clear that, in the case of non-universal soft SUSY breaking terms, the number of CP violating phases increases. These new phases have significant implications and can modify the SM predictions in CP violating phenomena, like fermionic EDMs.

10.3 ELECTRIC DIPOLE MOMENTS AND SUSY CP PROBLEM

The most stringent constraints on models of new physics with additional sources of CP violation come from continued efforts to measure the EDMs of the neutron, electron and mercury atoms. The current experimental limits on EDMs are given by [180–182]

$$d_n \quad < \quad 2.9 \times 10^{-26} \text{ e cm}, \tag{10.24}$$
$$d_e \quad < \quad 8.7 \times 10^{-29} \text{ e cm}, \tag{10.25}$$
$$d_{Hg} \quad < \quad 3.1 \times 10^{-29} \text{ e cm}. \tag{10.26}$$

With the expected improvements in experimental precision, EDMs are likely to be one of the most important tests for physics beyond the SM for some time to come and will always remain a difficult hurdle for CP violating SUSY theories.

The EDM $\mathbf{D}$ of a pair of opposite charges of magnitude q is defined as

$$\mathbf{D} = q\,\mathbf{d}, \tag{10.27}$$

where **d** is the displacement vector pointing from the negative charge to the positive charge. For a classical charge distribution, the EDM is given by

$$\mathbf{D} = \int \mathbf{r}\rho(\mathbf{r})d^3r. \tag{10.28}$$

In the case of elementary particles, the EDM must be proportional to the spin **S**, since elementary particles have no intrinsic vectorial degree of freedom except their spin. Thus, one can write $\mathbf{D} = d\,\mathbf{S}$, where the proportional constant d is an intrinsic property of the particle that represents the size of the EDM. Since **D** and **S** have the same transformation under discrete symmetries, under parity, P, transformation one has $\mathbf{D} \to \mathbf{D}$ while the electric field **E** transforms as $\mathbf{E} \to -\mathbf{E}$. Also, under time reversal, T, transformation one has $\mathbf{D} \to -\mathbf{D}$ and $\mathbf{E} \to \mathbf{E}$. Therefore, the interaction **D.E** violates both P and T. In this case, if d is different from zero, then both P and T are violated and, hence, according to the CPT theorem, the CP symmetry is also violated.

In a QFT, the P and T violating effective interaction between a spin-1/2 particle and the EM field is given by

$$\mathcal{L}_{\text{eff}} = iF_3(q^2)\bar{\psi}(p_2)\sigma_{\mu\nu}\gamma_5\psi(p_1)F^{\mu\nu}, \tag{10.29}$$

where $q^2 = (p_2 - p_1)^2$. The magnitude of the EDM of ψ, d_ψ, corresponds to the on-shell limit of the form factor F_3, *i.e.*, $d_\psi = F_3(0)$. The renormalisability condition rules out the possibility of having this type of interaction at tree level. However, it can be generated at loop level.

The electron EDM is labelled d_e. It can be measured experimentally from the EDM of the thallium atom. The electron EDM calculation involves little uncertainty, therefore it allows us to extract reliable bounds on the CP-violating SUSY phases.

The neutron EDM has contributions from a number of CP-violating operators involving quarks, gluons and photons. The most important ones include the electric and chromo-electric dipole operators and the Weinberg three-gluon operator:

$$\mathcal{L} = -\frac{i}{2}d_q^E\,\bar{q}\sigma_{\mu\nu}\gamma_5 qF^{\mu\nu} - \frac{i}{2}d_q^C\,\bar{q}\sigma_{\mu\nu}\gamma_5 T^a qG^{a\mu\nu} - \frac{1}{6}d^G f_{abc}G_{a\mu\rho}G_{b\nu}^{\rho}G_{c\lambda\sigma}\epsilon^{\mu\nu\lambda\sigma}, \tag{10.30}$$

where $G_{\mu\nu}^a$ is the gluon field strength and T^a and f_{abc} are the $SU(3)$ generators and group structure coefficients, respectively. Given these operators, it is, however, a non-trivial task to evaluate the neutron EDM, since assumptions about the neutron's internal structure are necessary. Usually, two models, namely the quark chiral model and the quark parton model, are considered to relate the quark EDMs to the neutron EDM. Neither of these models is sufficiently reliable by itself, however, the power of the combined analysis should provide an insight into implications of the bound on the neutron EDM. It is important to note that, in any case, the neutron EDM calculations involve uncertain hadronic parameters, such as the quark masses, and thus these calculations have a status of estimates at best. However, the major conclusions of this section are independent of the specifics of the neutron model.

The chiral quark model is a non-relativistic model that relates the neutron EDM to the EDMs of valence quarks with the help of the $SU(6)$ coefficients:

$$d_n = \frac{4}{3}d_d - \frac{1}{3}d_u. \tag{10.31}$$

The quark EDMs can be estimated, via NDA, as [183]

$$d_q = \eta^E d_q^E + \eta^C \frac{e}{4\pi}d_q^C + \eta^G \frac{e\Lambda}{4\pi}d^G, \tag{10.32}$$

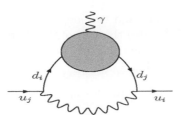

Figure 10.3 Possible two-loop SM contributions to the EDMs.

where the QCD correction factors are given by $\eta^E = 1.53$, $\eta^C \simeq \eta^G \simeq 3.4$, while $\Lambda \simeq 1.19$ GeV is the chiral symmetry breaking scale. The parameters $\eta^{C,G}$ involve considerable uncertainties stemming from the fact that the strong coupling constant at low energies is unknown. Another weak side of the model is that it neglects the sea quark contributions, which play an important role in the nucleon spin structure.

The quark parton model is based on the isospin symmetry and known contributions of different quarks to the spin of the proton. The quantities Δ_q, defined as $\langle n | \frac{1}{2} \bar{q} \gamma_\mu \gamma_5 q | n \rangle = \Delta_q S_\mu$, where S_μ is the neutron spin, are related by the isospin symmetry to the quantities $(\Delta_q)_p$, which are measured in deep inelastic scattering (and other) experiments, i.e., $\Delta_u = (\Delta_d)_p$, $\Delta_d = (\Delta_u)_p$ and $\Delta_s = (\Delta_s)_p$. The main assumption of the model is that the quark contributions to the neutron EDM are weighted by the same factors Δ_i, i.e.,

$$d_n = \eta^E (\Delta_d d_d^E + \Delta_u d_u^E + \Delta_s d_s^E) \,. \tag{10.33}$$

In our numerical analysis we use the standard values for these quantities: $\Delta_d = 0.746$, $\Delta_u = -0.508$ and $\Delta_s = -0.226$.

The major difference from the chiral quark model is a large strange quark contribution. In particular, due to the large strange and charm quark masses, the strange quark contribution dominates in most regions of the parameter space. This leads to considerable numerical differences between the predictions of the two models. It has been shown that the mercury EDM is primarily sensitive to the chromo-electric dipole moments of the quarks [184]:

$$d_{Hg} = -e \left(d_d^C - d_u^C - 0.012 d_s^C \right) \times 3.2 \times 10^{-2} \,. \tag{10.34}$$

Although the coefficient for d_s^C is suppressed with respect to those for the chromo-electric EDM of down and up quarks, this contribution is still important since d_s^C is enhanced by a heavier quark mass and large mixing in the second generation.

10.3.1 SM contributions to EDMs

As mentioned above, in the SM, CP violation comes from the CKM mixing matrix. Since $W^\pm$'s couple only to left-handed quarks and leptons, the one-loop contributions to d_e or d_q must have an external mass insertion. However, the CKM phase at the two $W^\pm$ vertices is cancelled. Hence, within the SM, a CP violating EDM cannot be generated at the one-loop level. At the two-loop level, it is possible to generate a non-self conjugate diagram, depicted in Fig. 10.3, that may contribute to the EDM of the electron and neutron. However, in 1978 it was shown that the sum of all two-loop contributions vanishes identically [185].

This cancellation was first shown by Donoghue for the electron and independently by Shabalin for quarks. However, there is no symmetry to explain this cancellation. In fact, the vanishing of the EDM in the SM at one and two loops is only due to the unitarity of the CKM matrix and to the purely left-handed chirality of the charged current. In this respect, one expects a non-vanishing three-loop contribution. Donoghue estimated this contribution to the electron EDM [186] and found that

$$d_e^{SM} < 10^{-50} \text{ e cm.} \tag{10.35}$$

Also, Shabalin showed that the corresponding neutron EDM is given by [185].

$$d_n^{SM} \leq 10^{-34} \text{ e cm.} \tag{10.36}$$

10.3.2 SUSY contribution to EDMs

In SUSY models, the new CP violating phases may generate EDMs for quarks and electrons at the one-loop level. The electron EDM arises due to CP-violating one-loop diagrams involving chargino and neutralino exchange, as shown in Fig. 10.4. Therefore, the SUSY electron EDM is given by

$$d_e = d_e^{\chi^+} + d_e^{\chi^0} \tag{10.37}$$

while the SUSY contributions to the EDMs of the individual quarks result from the one-loop gluino, chargino and/or neutralino exchange diagrams, also shown in Fig. 10.4. Thus,

$$d_q^{E,C} = d_q^{\tilde{g}\,(E,C)} + d_q^{\chi^+\,(E,C)} + d_q^{\chi^0\,(E,C)}. \tag{10.38}$$

In addition, one finds contributions from the two-loop gluino-quark-squark diagrams, shown in Fig. 10.5, which generate d^G.

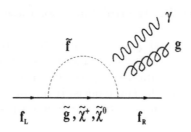

Figure 10.4 Leading SUSY contributions to the EDMs. The photon and gluon lines are to be attached to the loop in all possible ways.

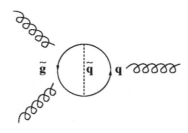

Figure 10.5 Two-loop diagram of the gluino-quark-squark contribution to EDMs.

The explicit computation of the SUSY contributions to the EDMs of electron and up and down quarks, using the full one-loop results, is straightforward. Here, we provide detailed calculations for the electron EDM and results for quarks can be derived similarly. As mentioned, the effective Lagrangian that defines the EDM of an electron is given by

$$\mathcal{L} = -\frac{i}{2}d_e\bar{\psi}\sigma_{\mu\nu}\gamma_5\psi F^{\mu\nu}, \tag{10.39}$$

which can be written for $\psi(x) = u(p)e^{-ip.x}$ and $A_\mu(x) = \varepsilon_\mu(q)e^{-iq.x}$ as

$$\mathcal{L} = -\frac{i}{2}d_e\bar{u}(p+q)\sigma_{\mu\nu}\varepsilon^\mu(q)q^\nu\gamma_5u(p). \tag{10.40}$$

Using Gordon decomposition, one can write this Lagrangian as

$$\mathcal{L} = -2im_ed_e\bar{u}(p+q)(\varepsilon.\gamma)\gamma_5u(p) + 2id_e\bar{u}(p+q)(\varepsilon.p)\gamma_5u(p). \tag{10.41}$$

As usual, the term proportional to the $\varepsilon.\gamma$ structure contributes to the wave function renormalisation. Thus, in our calculation of an invariant amplitude, we need only to concentrate on the $\varepsilon.p$ term:

$$\mathcal{L} = 2id_e\bar{u}(p+q)(\varepsilon.p)\gamma_5u(p). \tag{10.42}$$

The Feynman diagram of chargino contribution to the electron EDM is shown in Fig. 10.4. From the electron-chargino-sneutrino interaction in Chapter 5, one finds that the $e_L - e_R$ transition amplitude is given by

$$\mathcal{M} = \sum_{i=1}^{2}\int\frac{d^4k}{(2\pi)^4}\bar{u}(p+q)\frac{e}{\sin\theta_W}\left[(V^{i1})^*P_R + \frac{m_e}{\sqrt{2}M_W\cos\beta}U^{i2}P_L\right]$$
$$(\Gamma_\nu^{1I})\frac{1}{(\not{p} - \not{k} + \not{q}) - m_{\chi_i^+}}(e\varepsilon\gamma)\frac{1}{(\not{p} - \not{k}) - m_{\chi_i^+}}\left(\frac{e}{\sin\theta_W}\right)$$
$$\left[\left(V^{i1}P_L + \frac{m_e}{\sqrt{2}M_W\cos\beta}(U^{2i})^*P_R\right)(\Gamma_\nu^{1I})^*\right]\frac{1}{k^2 - m_{\tilde{\nu}^2}}u(p), \tag{10.43}$$

which leads to

$$\mathcal{M} = \frac{1}{8\pi^2}\frac{e^3}{\sin^2\theta_W}\frac{m_e}{\sqrt{2}M_W\cos\beta}(\varepsilon.p)\bar{u}(p+q)\gamma_5u(p)$$
$$\sum_{i=1}^{2}\frac{m_{\tilde{\chi}^+}}{m_{\tilde{\nu}}^2}\text{Im}\left[U^{21^*}V^{i1^*}\right]A\left(\frac{m_{\tilde{\chi}^+}^2}{m_{\tilde{\nu}}^2}\right), \tag{10.44}$$

where the loop function $A(x)$ is given by

$$A(r) = \frac{1}{2(1-r)^2}\left[3 - r + \frac{2\ln r}{1-r}\right]. \tag{10.45}$$

Thus one finds that the electron EDM is given by

$$d_e^{\tilde{\chi}^+}/e = \frac{\alpha_{em}}{4\pi\sin^2\theta_W}\frac{m_e}{\sqrt{2}M_W\cos\beta}\sum_{i=1}^{2}\frac{m_{\tilde{\chi}^+}}{m_{\tilde{\nu}}^2}\text{Im}\left[U^{2i^*}V^{i1^*}\right]A\left(\frac{m_{\tilde{\chi}^+}^2}{m_{\tilde{\nu}}^2}\right). \tag{10.46}$$

Similarly, one can show that the neutralino contribution to the electron (and also to the up and down quark) EDM is given by [187]

$$d_{\tilde{f}}^{\tilde{\chi}^0}/e = \frac{\alpha_{em}}{4\pi\sin^2\theta_W}\sum_{k=1}^{2}\sum_{i=1}^{4}\text{Im}(\eta_{fik})\frac{m_{\tilde{\chi}_i^0}}{M_{\tilde{f}_k}^2}Q_{\tilde{f}}\,B\left(\frac{m_{\tilde{\chi}_i^0}^2}{M_{\tilde{f}_k}^2}\right), \tag{10.47}$$

where η_{fik} is the neutralino vertex, which is given by

$$
\eta_{fik} = \left[-\sqrt{2}\{\tan\theta_W(Q_f - I_{3_f})N_{1i} + I_{3_f}N_{2i}\}D^*_{f1k} - \kappa_f N_{bi}D^*_{f2k} \right]
$$
$$
\times \left[\sqrt{2}\tan\theta_W Q_f N_{1i}D_{f2k} - \kappa_f N_{bi}D_{f1k} \right],
\tag{10.48}
$$

where I_3 is the third component of the isospin, $b = 3$ (4) for $I_3 = -1/2$ (1/2) and N is the unitary matrix diagonalising the neutralino mass matrix: $N^T M_{\chi^0} N = \text{diag}(m_{\chi^0_1}, m_{\chi^0_2}, m_{\chi^0_3}, m_{\chi^0_4})$. Here, we use our convention from Chapter 5, where the mass matrix eigenvalues are positive and ordered as $m_{\chi^0_1} < m_{\chi^0_2} < ...$. The loop function $B(r)$ is defined by

$$
B(r) = \frac{1}{2(r-1)^2}\left(1 + r + \frac{2r\ln r}{1-r}\right).
\tag{10.49}
$$

In a similar way, one can derive the following expression for the one-loop gluino contributions to the the EDM of the up and down quarks [187]:

$$
d_q^{\tilde{g}\,(E)}/e = \frac{-2\alpha_s}{3\pi}\sum_{k=1}^{2}\text{Im}\left[R^{\tilde{q}}_{2k}R^{\tilde{q}*}_{1k}\right]\frac{M_{\tilde{g}}}{M^2_{\tilde{q}_k}}Q_{\tilde{q}}\,B\left(\frac{M^2_{\tilde{g}}}{M^2_{\tilde{q}_k}}\right).
\tag{10.50}
$$

Here, $R^{\tilde{q}}$ is defined by $R^{\tilde{q}}M^2_{\tilde{q}}R^{\tilde{q}\dagger} = \text{diag}\left(M^2_{\tilde{q}_1}, M^2_{\tilde{q}_2}\right)$. Finally, the chargino contributions to the EDM of the up and down quarks are given by

$$
d_u^{\tilde{\chi}^+(E)}/e = \frac{-\alpha_{em}}{4\pi\sin^2\theta_W}\sum_{i,k=1}^{2}\text{Im}(\Gamma_{uik})\frac{m_{\tilde{\chi}^+_i}}{M^2_{\tilde{d}_k}}\left[Q_{\tilde{d}}B\left(\frac{m^2_{\tilde{\chi}^+_i}}{M^2_{\tilde{d}_k}}\right) + (Q_u - Q_{\tilde{d}})A\left(\frac{m^2_{\tilde{\chi}^+_i}}{M^2_{\tilde{d}_k}}\right)\right],
\tag{10.51}
$$

$$
d_d^{\tilde{\chi}^+(E)}/e = \frac{-\alpha_{em}}{4\pi\sin^2\theta_W}\sum_{i,k=1}^{2}\text{Im}(\Gamma_{dik})\frac{m_{\tilde{\chi}^+_i}}{M^2_{\tilde{u}_k}}\left[Q_{\tilde{u}}B\left(\frac{m^2_{\tilde{\chi}^+_i}}{M^2_{\tilde{u}_k}}\right) + (Q_d - Q_{\tilde{u}})A\left(\frac{m^2_{\tilde{\chi}^+_i}}{M^2_{\tilde{u}_k}}\right)\right],
\tag{10.52}
$$

where the chargino vertex Γ_{fik} is defined as

$$
\Gamma_{uik} = \kappa_u V^*_{i2}R^{\tilde{d}}_{1k}(U^*_{i1}R^{\tilde{d}*}_{1k} - \kappa_d U^*_{i2}R^{\tilde{d}*}_{2k}),
\tag{10.53}
$$
$$
\Gamma_{dik} = \kappa_d U^*_{i2}R^{\tilde{u}}_{1k}(V^*_{i1}R^{\tilde{u}*}_{1k} - \kappa_u V^*_{i2}R^{\tilde{u}*}_{2k}).
\tag{10.54}
$$

The quantities κ_f are the Yukawa couplings

$$
\kappa_u = \frac{m_u}{\sqrt{2}m_W\sin\beta}, \quad \kappa_{d,e} = \frac{m_{d,e}}{\sqrt{2}m_W\cos\beta}.
\tag{10.55}
$$

Also, the SUSY contributions to the quark chromo-electric dipole moment are given by [188]

$$
d_q^{\tilde{g}(C)} = \frac{g_s\alpha_s}{4\pi}\sum_{k=1}^{2}\text{Im}\left[R^{\tilde{q}}_{2k}R^{\tilde{q}*}_{1k}\right]\frac{M_{\tilde{g}}}{M^2_{\tilde{q}_k}}C\left(\frac{M^2_{\tilde{g}}}{M^2_{\tilde{q}_k}}\right),
\tag{10.56}
$$

$$
d_q^{\tilde{\chi}^+(C)} = \frac{-g^2 g_s}{16\pi^2}\sum_{k=1}^{2}\sum_{i=1}^{2}\text{Im}(\Gamma_{qik})\frac{m_{\tilde{\chi}^+_i}}{M^2_{\tilde{q}_k}}B\left(\frac{m^2_{\tilde{\chi}^+_i}}{M^2_{\tilde{q}_k}}\right),
\tag{10.57}
$$

$$
d_q^{\tilde{\chi}^0(C)} = \frac{g_s g^2}{16\pi^2}\sum_{k=1}^{2}\sum_{i=1}^{4}\text{Im}(\eta_{qik})\frac{m_{\tilde{\chi}^0_i}}{M^2_{\tilde{q}_k}}B\left(\frac{m^2_{\tilde{\chi}^0_i}}{M^2_{\tilde{q}_k}}\right),
\tag{10.58}
$$

where $C(r)$ is defined by

$$C(r) = \frac{1}{6(r-1)^2}\left(10r - 26 + \frac{2r\ln r}{1-r} - \frac{18\ln r}{1-r}\right).$$ (10.59)

Finally, the contribution to the Weinberg operator from the two-loop gluino-top-stop and gluino-bottom-sbottom diagrams reads [188]

$$d^G = -3\alpha_s m_t\left(\frac{g_s}{4\pi}\right)^3 \mathrm{Im}(\Gamma_t^{12})\frac{z_1 - z_2}{(M_3)^3}\,H(z_1, z_2, z_t) + (t \to b),$$ (10.60)

where $z_i = \left(\frac{M_{\tilde{t}_i}}{M_3}\right)^2$ and $z_t = \left(\frac{m_t}{M_3}\right)^2$. The two-loop function $H(z_1, z_2, z_t)$ is given by

$$H(z_1, z_2, z_t) = \frac{1}{2}\int_0^1 dx \int_0^1 du \int_0^1 dy\, x(1-x)u\frac{N_1 N_2}{D^4},$$ (10.61)

where

$$N_1 = u(1-x) + z_t x(1-x)(1-u) - 2ux[z_1 y + z_2(1-y)],$$ (10.62)

$$N_2 = (1-x)^2(1-u)^2 + u^2 - \frac{1}{9}x^2(1-u)^2,$$ (10.63)

$$D = u(1-x) + z_t x(1-x)(1-u) + ux[z_1 y + z_2(1-y)].$$ (10.64)

In addition to the Weinberg two-loop diagram, there is another (so-called Barr-Zee) two-loop contribution, which originates from CP-odd Higgs exchange. Its numerical effect is, however, negligible.

10.4 METHODS FOR EDM SUPPRESSION

Generic soft-SUSY breaking terms give EDMs that are many orders of magnitude larger than experimental bounds, the so-called SUSY CP problem. We now discuss possible ways to avoid overproduction of EDMs in SUSY models. There are four known scenarios allowing suppression of the EDM contributions [179]: small SUSY CP phases, heavy sfermions, EDM cancellations and flavour-off-diagonal CP violation.

Small CP phases
For a SUSY spectrum of order TeV (just above the current LHC bounds), the flavour-independent SUSY CP phases ϕ_μ and gaugino phases $\phi_{M_{1,3}}$ must be small in order to satisfy the experimental EDM bounds.

In Fig. 10.6 we illustrate the EDMs' behaviour as a function of ϕ_μ in the minimal SUGRA-type model, where we have set $m_0 = 500$ GeV, $m_{1/2} = 1.5$ TeV, $A_0 = 1$ TeV and $\tan\beta = 10$ (so that $m_h \sim 125$ GeV). As can be seen with such a heavy spectrum, the EDMs still impose the following stringent bound on the CP phase ϕ_μ:

$$\phi_\mu \leq 10^{-3} - 10^{-2}.$$ (10.65)

For $\tan\beta > 10$, this bound becomes even stricter. The phase ϕ_A is less constrained than ϕ_μ and with this heavy spectrum becomes essentially unconstrained. There are two reasons for that: firstly, ϕ_A is reduced by the RGE running from the GUT scale down to the EW scale (so it is at most of order $\mathcal{O}(0.1)$ at EW scale) and, secondly, the phase which gives

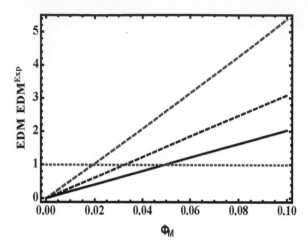

Figure 10.6 EDMs as a function of ϕ_μ. Curves from up to down are for electron, neutron and mercury EDMs, respectively. The experimental limit is given by the horizontal line. Here $\tan\beta = 10$, $m_0 = 500$ GeV, $m_{1/2} = 1.5$ TeV and $A_0 = 1$ TeV.

the dominant contribution to the EDMs is more sensitive to ϕ_μ (and ϕ_{M_i} for the general soft SUSY breaking scenario) due to the fact that $|A| < \mu\tan\beta$.

Generally, it is quite difficult to explain why the soft CP phases have to be small. In principle, small CP phases could appear if CP were an approximate symmetry of Nature. However, experimental results show that CP violation in $B - \bar{B}$ mixing is large and thus the approximate CP hypothesis cannot be motivated.

Heavy SUSY scalars
This possibility is based on the decoupling of heavy SUSY particles. Even if one allows $\mathcal{O}(1)$ CP violating phases, their effects will be negligible if the SUSY spectrum is sufficiently heavy. Generally, SUSY fermions are required to be lighter than the SUSY scalars by, for example, cosmological considerations. So, the decoupling scenario can be implemented with heavy sfermions only. Here, the SUSY contributions to the EDMs are suppressed even with maximal SUSY phases because the squarks in the loop are very heavy and the mixing angles are small.

In Fig. 10.7 we display the EDMs as functions of the universal scalar mass parameter m_0 for the mSUGRA model with maximal CP phases $\phi_\mu = \phi_A = \pi/2$ and $m_{1/2} = A = 200$ GeV. We observe that all EDM constraints require m_0 to be larger than 20 TeV. In this case, we encounter a serious fine-tuning problem. Recall that one of the primary motivations for SUSY was a solution to the naturalness problem. Certainly, this motivation will be entirely lost if a SUSY model reintroduces the same problem in a different sector, $e.g.$, a large hierarchy between the scalar masses and the EW scale.

EDM cancellations
The cancellation scenario is based on the fact that large cancellations among different contributions to the EDMs are possible in certain regions of the parameter space which allow for $\mathcal{O}(1)$ flavour-independent CP phases. However, it was shown that this possibility is practically ruled out if, in addition to the electron and neutron EDM constraints, the mer-

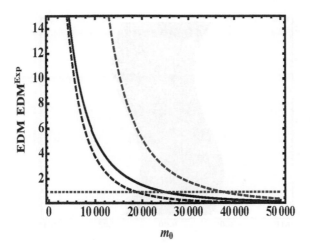

Figure 10.7 EDMs as a function of the universal mass parameter m_0. Curves from up to down are for electron, mercury and neutron EDMs, respectively. The experimental limit is given by the horizontal line. Here $\tan\beta = 10$, $m_{1/2} = A = 200$ GeV and $\phi_\mu = \phi_A = \pi/2$.

cury constraint is also imposed [179]. Also, the parameters allowing the EDM cancellations strongly depend on the neutron model. For example, in the parton model, it is more difficult to achieve these cancellations due to the large strange quark contribution. Therefore, one cannot restrict the parameter space in a model-independent way and caution is needed when dealing with the parameters allowed by the cancellations.

In the cMSSM, the EDM constraints can be satisfied simultaneously for the electron, neutron and mercury along a very narrow band in the (ϕ_A, ϕ_μ) plane (Fig. 10.8). However, in this case the electron constraint requires the μ phase to be $\mathcal{O}(10^{-2})$. It is important to note that the fact that the phase of the A-terms is unrestricted should not be attributed to the cancellations, but rather to the suppressed contribution of such terms.

If the gluino phase is turned on, simultaneous electron, neutron and mercury EDM cancellations are not possible. The gluino phase affects the neutron EDM cancellation while leaving the electron EDM band almost intact. An introduction of the bino phase ϕ_1 qualitatively has the same 'off-setting' effect on the electron EDM band as the gluino phase does on that of the neutron EDM. Moreover, any small values of $\phi_1 \sim 0.1$ lead to a very large electron EDM above the experimental limit by orders of magnitude. Note that these gaugino phases have no significant effect on the neutron and mercury cancellation bands since the neutralino contribution in both cases is small. When both the gluino and bino phases are present (and fixed), simultaneous electron, neutron and mercury EDMs cancellations do not appear to be possible along a band, as shown in Fig. 10.9 for neutron and mercury EDMs only, since the electron EDM is already above its experimental limit. Thus, we can conclude that non-zero gaugino phases do not allow EDM cancellations along a band.

Flavour-off-diagonal CP violation

Non-observation of EDMs may imply that CP-violation has a flavour-off-diagonal character just as in the SM. In the SM, CP violation appears in flavour-changing processes (apart from $\bar{\theta}_{\text{QCD}}$), which is one of the reasons why the predicted EDMs are small. A similar situation

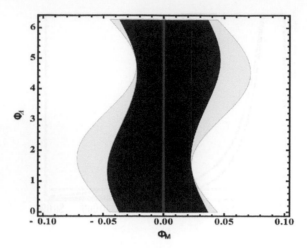

Figure 10.8 Phases allowed by simultaneous electron (red), neutron (blue) and mercury (yellow) EDM cancellations in mSUGRA. Here $\tan\beta = 10$, $m_0 = 500$ GeV and $m_{1/2} = A_0 = 1.5$ TeV.

may occur in SUSY models. The origin of CP-violation in this case would be closely related to the origin of the flavour structures rather than the origin of SUSY breaking. While models with flavour-off-diagonal CP violation avoid the EDM problem, they have testable effects in K and B physics.

In a sense, this scenario is similar to the one with small SUSY CP phases, since both assume absence (or suppression) of the flavour-independent phases. However, the latter typically implies that all SUSY CP phases are small (as could have been motivated by approximate CP conservation), whereas the former allows for $\mathcal{O}(1)$ flavour-off-diagonal phases. This class of models requires Hermitian Yukawa matrices and A-terms, which force the flavour-diagonal phases to vanish (up to small RGE corrections) in any basis. To see this, first note that the trilinear coupings

$$\hat{A}^\alpha_{ij} = A^\alpha_{ij} Y^\alpha_{ij} \qquad (10.66)$$

are also Hermitian . The quark Yukawa matrices are diagonalised by the unitary transformation

$$q \to V^q q \quad , \quad Y^q \to (V^q)^T Y^q (V^q)^* = \text{diag}(h^q_1, h^q_2, h^q_3), \qquad (10.67)$$

such that the CKM matrix is $V_{\text{CKM}} = (V^u)^\dagger V^d$. If we transform the squark fields in the same manner, which defines the super-CKM basis, we will have

$$\hat{A}^q \to (V^q)^T \hat{A}^q (V^q)^*. \qquad (10.68)$$

As a result, the trilinear couplings remain Hermitian and the flavour-diagonal CP phases inducing the EDMs vanish. Note that this argument would not work if, in the original basis, the trilinear couplings but not the Yukawas were Hermitian since the diagonalisation would require a bi-unitary transformation which would generally introduce the diagonal phases.

The hermiticity is generally spoiled by the RGE running from the high energy scale down to the EW scale. In the models under consideration, we have the following setting at

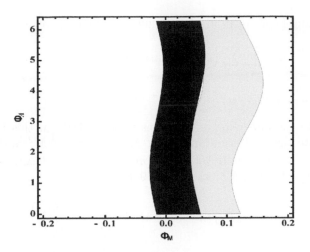

Figure 10.9 Bands allowed by neutron (blue) and mercury (yellow) EDM cancellations in the mSUGRA-type model with non-zero gluino and bino phases. Here $\tan\beta = 10$, $m_0 = 500$, $m_{1/2} = A_0 = 1.5$ TeV and $\phi_1 = \phi_3 = \pi/10$. The corresponding electron EDM is already above its experimental limit.

high energies:

$$Y^\alpha = Y^{\alpha\dagger}\,, A^\alpha = A^{\alpha\dagger}, \tag{10.69}$$

$$\mathrm{Arg}(M_k) = \mathrm{Arg}(B) = \mathrm{Arg}(\mu) = 0. \tag{10.70}$$

Generally, the off-diagonal elements of the A-terms can have $\mathcal{O}(1)$ phases without violating the EDM constraints. Due to RGE effects, large phases in the soft trilinear couplings involving the third generation generate small phases in the flavour-diagonal mass insertions for the light generations and thus induce the EDMs. However, the amount of generated "non-hermiticity" depends on $[Y^u, Y^d]$, $[\hat{A}^u, \hat{A}^d]$, etc. and thus is suppressed by the off-diagonal elements of the Yukawa matrices. In particular, the size of the flavour-diagonal phases is quite sensitive to Y_{13}. For $Y_{13} \leq \mathcal{O}(10^{-3})$, the RGE effects are unimportant and the induced EDMs are below the experimental limits.

Since CP violation is related to the flavour structures, the flavour-independent quantities such as the μ term, gaugino masses, etc. are real. This is naturally implemented in left-right symmetric models and models with a horizontal flavour symmetry. In left-right models, the hermiticity of the Yukawas and A-terms as well as the reality of the μ term is in fact forced by the left-right symmetry.

10.5 MASS INSERTION APPROXIMATION AND EDM CONSTRAINTS

As mentioned in Chapter 5, the MIA allows us to parameterise the SUSY effects in a model independent way. In this framework, one chooses a basis (the so-called super-CKM basis) where the couplings of fermions and sfermions to neutral gauginos are flavour diagonal. In this basis, the interacting Lagrangian involving gluinos and charginos is given by

$$
\begin{aligned}
\mathcal{L} \;=\;\; & g_s\,\bar{q}\,\tilde{g}^a T^a\,\tilde{q} - \sum_k \sum_{a,b} \Big(g V_{k1}\,(V_{\mathrm{CKM}})^*_{ba}\,\bar{d}^a_L(\tilde{\chi}^+)^*\tilde{u}^b_L \\
& - U^*_{k2}\,(Y^{\mathrm{diag}}_d.(V_{\mathrm{CKM}})^+_{ab}\bar{d}^a_R(\tilde{\chi}^+)^*\tilde{u}^b_L - V^*_{k2}(V_{\mathrm{CKM}}.Y^{\mathrm{diag}}_u)_{ab}\bar{d}^a_L(\tilde{\chi}^+)^*\tilde{u}^b_R \Big). \quad (10.71)
\end{aligned}
$$

Then one can compute the one-loop contributions to the EDMs in this basis. However, it is interesting to note that a Hermitian matrix $M^2 = M_0^2 + \Delta M^2$ is diagonalised by $U M^2 U^\dagger = \mathrm{diag}\{m_1^2, m_2^2, ..., m_n^2\}$ with $M_0^2 = \mathrm{diag}\{m_1^{0^2}, ..., m_n^{0^2}\}$ and ΔM^2 completely off-diagonal. Thus, at first order in ΔM^2, we have [189, 190]

$$
U^*_{ki} f(m_k^2) U_{kj} \simeq \delta_{ij} f(m_i^{0^2}) + \Delta M^2_{ij} \frac{f(m_i^{0^2}) - f(m_j^{0^2})}{m_i^{0^2} - m_j^{0^2}}. \tag{10.72}
$$

Furthermore, for small off-diagonal entries ΔM^2 and approximately degenerate squarks, one can replace the finite difference by the derivative of the function. In this case, one can show that the gluino contributions to the quark EDMs lead to [18]

$$
d^E_{d,u} \;=\; -\frac{2}{3}\frac{\alpha_s}{\pi}\,Q_{d,u}\,\frac{m_{\tilde{g}}}{m_{\tilde{d}}^2}\,\mathrm{Im}(\delta^{d,u}_{11})_{LR} M_1(x), \tag{10.73}
$$

$$
d^E_s \;=\; -\frac{2}{3}\frac{\alpha_s}{\pi}\,Q_s\,\frac{m_{\tilde{g}}}{m_{\tilde{d}}^2}\,\mathrm{Im}(\delta^d_{22})_{LR} M_1(x), \tag{10.74}
$$

$$
d^C_s \;=\; \frac{g_s \alpha_s}{4\pi}\,\frac{m_{\tilde{g}}}{m_{\tilde{d}}^2}\,\mathrm{Im}(\delta^d_{22})_{LR} M_2(x), \tag{10.75}
$$

where $x = m_{\tilde{g}}^2 / m_{\tilde{q}}^2$ and the functions $M_{1,2}(x)$ are given by [18]

$$
M_1(x) \;=\; \frac{1 + 4x - 5x^2 + 4x\ln x + 2x^2 \ln x}{2(1-x)^4}, \tag{10.76}
$$

$$
M_2(x) \;=\; -x^2 \frac{5 - 4x - x^2 + 2\ln x + 4x\ln x}{2(1-x)^4}. \tag{10.77}
$$

The experimental limits on EDMs impose stringent constraints on the imaginary parts of the involved mass insertions. In Tab. 10.1, we present the bounds for the chiral neutron model. In addition, in Tab. 10.2, we present the bounds on the mass insertions from the gluino contributions to the mercury EDM. As can be seen from these results, the constraints imposed on the mass insertions $|\mathrm{Im}(\delta^d_{11})_{LR}|$ and $|\mathrm{Im}(\delta^u_{11})_{LR}|$ by the mercury EDM limit are about an order of magnitude stricter than those imposed by the neutron EDM.

x	$\|\mathrm{Im}(\delta^d_{11})_{LR}\|$	$\|\mathrm{Im}(\delta^u_{11})_{LR}\|$	$\|\mathrm{Im}(\delta^l_{11})_{LR}\|$
0.1	1.9×10^{-7}	3.9×10^{-7}	3.5×10^{-9}
0.3	3.1×10^{-7}	6.1×10^{-7}	5.6×10^{-9}
1	6.8×10^{-7}	1.3×10^{-6}	1.2×10^{-8}

Table 10.1 Bounds on the imaginary parts of the mass insertions. The chiral quark model for the neutron is assumed. Here $x = m^2_{\tilde{g}}/m^2_{\tilde{q}} = m^2_{\tilde{\chi}}/m^2_{\tilde{l}}$ with $m_{\tilde{q}} = 1000$ GeV and $m_{\tilde{l}} = 500$ GeV. For different squark/slepton masses the bounds are to be multiplied by $m_{\tilde{q}}/1000$ GeV or $m_{\tilde{l}}/500$ GeV.

x	$\|\mathrm{Im}(\delta^d_{11})_{LR}\|$	$\|\mathrm{Im}(\delta^u_{11})_{LR}\|$	$\|\mathrm{Im}(\delta^d_{22})_{LR}\|$
0.1	5.2×10^{-8}	5.2×10^{-8}	4.4×10^{-6}
0.3	7.2×10^{-8}	7.2×10^{-8}	6.0×10^{-6}
1	1.3×10^{-7}	1.3×10^{-7}	1.1×10^{-5}

Table 10.2 Bounds on the imaginary parts of the mass insertions imposed by the mercury EDM. For the squark masses different from 1000 GeV, the bounds are to be multiplied by $m_{\tilde{q}}/1000$ GeV.

CHAPTER 11

SUSY CP Violation in K Mesons

In this chapter, we address the possible SUSY dynamics onsetting CP violation in the kaon system , where it was first observed. This discovery resulted in the Nobel Prize in Physics in 1980 for its discoverers, Cronin and Fitch. As experimental data are consistent with SM predictions, this amounts to impose constraints on the MSSM parameter space.

11.1 FLAVOUR VIOLATION IN THE MSSM

FCNCs provide an important test for any BSM physics. In the SM the baryon (B) and lepton (L) numbers are automatically conserved, therefore tree level FCNCs are absent. As discussed in Chapter 4, the problem of tree level B and L violation in SUSY theories is solved by introducing an additional discrete symmetry, R-parity (denoted by R_P). However, low energy SUSY still predicts new particles carrying flavour numbers with mass of order TeV that may lead to potentially large FCNC rates. Since the masses of the sparticles are mainly determined by the soft SUSY breaking parameters, the FCNC limits can put strong constraints on these parameters. Hence, these constraints may give some information on the fundamental theory at high energy, which determines the structure of the soft breaking parameters.

The quark mass matrices in the SM can be diagonalised by the bi-unitary rotations:

$$U_L^\dagger M_u U_R = M_u^{\text{diag}}, \tag{11.1}$$

$$V_L^\dagger M_d V_R = M_d^{\text{diag}}. \tag{11.2}$$

Since u_L and d_L sit in a doublet of the $SU(2)$ gauge group, their relative rotation ('difference' between U_L and V_L) gives rise to FCNC in $W^\pm$ vertices with fermions (*i.e.*, $W^\pm u_i d_j$, $i \neq j$, with $i, j = 1, 2, 3$ family indices). Analogously, in SUSY, we may have FCNC at vertices with f and $\tilde{f}$. The presence of off-diagonal terms in the sfermion mass matrix, at the EW scale, can be due to the following reasons.

1. The initial conditions, *i.e.*, that SUSY breaking terms are not universal.

2. The RGEs, which are an important source even in the case of starting with a universal mass condition for sfermions in the SUSY soft breaking sector.

The renormalisation of the squark mass matrices is different from the renormalisation of the quark masses. For instance, consider the mass matrix squared of the sfermion $\tilde{f}_L$ (forming a supermultiplet with the fermion f_L). At the scale of SUGRA breaking this matrix consists of the SUSY conserving contribution $M_f M_f^\dagger$ (where M_f denotes the f quark mass matrix) and the SUSY breaking condition $\tilde{m}_{ij}^2 \tilde{f}_i \tilde{f}_j^*$. To avoid FCNC in a vertex like $\tilde{\gamma} f_i \tilde{f}_j$, $i \neq j$, the mass matrix of the sfermion $M_{\tilde{f}}$ should be proportional to the mass matrix of the fermion M_f. However, even if one assumes $\tilde{m}_{ij}^2 = \tilde{m}^2 \delta_{ij}$, i.e.,

$$M_{\tilde{f}}^2 = M_f M_f^\dagger + \tilde{m}^2 \mathbf{1}, \tag{11.3}$$

so that M_f and $M_{\tilde{f}}$ are simultaneously diagonalised, one finds that the term $h_{f'} Q H F^c$ (for example, $h_d Q_L H_1 D^c$) of the superpotential generates a contribution which is proportional to $h_{f'} h_{f'}^\dagger$ and hence to $M_{f'} M_{f'}^\dagger$. Thus, the resulting $\tilde{f}_L$ mass matrix squared at the EW scale is

$$M_{\tilde{f}}^2 = M_f M_f^\dagger + \tilde{m}^2 \mathbf{1} + a \, M_{f'} M_{f'}^\dagger \tag{11.4}$$

For example, if $f = d$ and $f' = u$, one finds

$$M_{\tilde{d}_L}^2 = M_d M_d^\dagger + \tilde{m}^2 \mathbf{1} + a \, M_u M_u^\dagger, \tag{11.5}$$

where M_d and M_u denote the down- and up-quark mass matrices, respectively. Now, by rotating the $\tilde{q}$ fields together with q ones, one obtains off-diagonal mass terms due the presence of the last term in Eq. (11.5).

In this regard, one can conclude that the following types of FCNC and CP violating processes may take place in the MSSM.

1. CP violating processes without FCNCs, such as EDM of electron and neutron, which has been discussed in the previous chapter.

2. Processes with FCNCs but without CP violation, such as $K - \bar{K}$ or $B - \bar{B}$ mixing.

3. Processes with simultaneous FCNCs and CP violation such as ε and ε'/ε in K decays and CP asymmetries in B decays.

In this chapter we will focus on possible effects of SUSY on K decays and mixing. SUSY impact on B decays and mixing will be considered in the next chapter.

11.2 SUSY CONTRIBUTIONS TO $\Delta S = 2$

11.2.1 Basic formalism

In the K^0 and $\bar{K}^0$ system, the flavour eigenstates are given by $K^0 = (\bar{s}d)$ and $\bar{K}^0 = (s\bar{d})$. These states are not definite CP eigenstates but $CP|K^0\rangle = -|\bar{K}^0\rangle$. A definite CP eigenstate is constructed as

$$K_{1,2} = \frac{1}{\sqrt{2}} \left(K^0 \mp \bar{K}^0 \right) \tag{11.6}$$

with $CP|K_{1,2}\rangle = \pm|K_{1,2}\rangle$. Therefore, we have $K_1 \to \pi\pi$ and $K_2 \to \pi\pi\pi$. The time evolution of the $K^0 - \bar{K}^0$ system is given by [191, 192]

$$i\frac{d}{dt} \begin{pmatrix} K^0 \\ \bar{K}^0 \end{pmatrix} = \begin{pmatrix} M_{11} - i\Gamma_{11} & M_{12} - i\Gamma_{12} \\ M_{21} - i\Gamma_{21} & M_{22} - i\Gamma_{22} \end{pmatrix} \begin{pmatrix} K^0 \\ \bar{K}^0 \end{pmatrix}, \tag{11.7}$$

where $M = M_{ij}$ and $\Gamma = \Gamma_{ij}$ are Hermitian matrices. However, the Hamiltonian can be non-Hermitian since the probability is not conserved due to the possile decay of kaons into states out of this system (for more details, see [191]). The CPT invariance implies that $M_{11} = M_{22} = M$, $M_{12} = M_{21}^*$, $\Gamma_{11} = \Gamma_{22} = \Gamma$, and $\Gamma_{12} = \Gamma_{21}^*$. The physical eigenstates of the Hamiltonian, obtained by diagonalising the mixing matrix, are

$$K_S = pK^0 + q\bar{k}^0 \tag{11.8}$$

and

$$K_L = pK^0 - q\bar{k}^0, \tag{11.9}$$

where the indices L and S refer to long and short mass eigenstates, respectively, and

$$p = (1 + \varepsilon)/\sqrt{2(1 + |\varepsilon|^2)}, \qquad q = (1 - \varepsilon)/\sqrt{2(1 + |\varepsilon|^2)}, \tag{11.10}$$

where ε is the CP violating parameter in the $K^0 - \bar{K}^0$ system, since $K_{S,L}$ can now be written as

$$K_S = \frac{1}{\sqrt{1 + |\varepsilon|^2}} (K_1 + \varepsilon K_2), \tag{11.11}$$

$$K_L = \frac{1}{\sqrt{1 + |\varepsilon|^2}} (K_2 + \varepsilon K_1). \tag{11.12}$$

The corresponding eigenvalues of the Hamltonian are given by

$$\begin{aligned}
\lambda &= M - \frac{i}{2}\Gamma \pm \sqrt{(M_{12}^* - \frac{i}{2}\Gamma_{12}^*)(M_{12} - \frac{i}{2}\Gamma_{12})} \\
&= M - \frac{i}{2}\Gamma \pm \frac{1}{2}\left(\Delta M - \frac{i}{2}\Delta\Gamma\right),
\end{aligned} \tag{11.13}$$

with

$$(\Delta M)^2 - \frac{1}{4}(\Delta\Gamma)^2 = 4|M_{12}|^2 + |\Gamma_{12}|^2 \tag{11.14}$$

and

$$\Delta M \Delta\Gamma = 4\mathrm{Re}\,[M_{12}\Gamma_{12}]. \tag{11.15}$$

From the associate eigenvectors, one finds

$$\frac{1 - \varepsilon}{1 + \varepsilon} = \frac{M_{12}^* - \frac{i}{2}\Gamma_{12}^*}{M_{12} - \frac{i}{2}\Gamma_{12}}. \tag{11.16}$$

Therefore ε can be defined as

$$\varepsilon \simeq \frac{i\,\mathrm{Im}M_{12} + \frac{1}{2}\mathrm{Im}\Gamma_{12}}{\Delta M - \frac{i}{2}\Gamma_{12}}. \tag{11.17}$$

If $\Gamma_{12} \ll \Delta M$, then ε takes the form

$$|\varepsilon| \simeq \frac{\mathrm{Im}M_{12}}{\Delta M}. \tag{11.18}$$

Thus, if M_{12} and Γ_{12} are real, then ε will vanish and $K_{S,L}$ correspond to $K_{1,2}$. The mass difference, which measures the strength of $K^0 - \bar{K}^0$ mixing, is described by

$$\Delta M_K = M_{K_L} - M_{K_S}, \tag{11.19}$$

whose present experimental value is [174]

$$\Delta M_K = (3.490 \pm 0.006) \times 10^{-15} \text{ GeV}. \tag{11.20}$$

In addition, the experimental value of ε is given by [174]

$$|\varepsilon| = (2.28 \pm 0.02) \times 10^{-3}. \tag{11.21}$$

The measure of K_L decay into two pions through its small CP even component was the first observation of CP violation.

11.2.2 SM contribution

As explained in the previous section, ΔM_K and ε can be calculated from

$$\Delta M_K = 2|\langle K^0|H_{\text{eff}}^{\Delta S=2}|\bar{K}^0\rangle|\,, \tag{11.22}$$

$$|\varepsilon| = \frac{1}{\sqrt{2}\Delta M_K}\text{Im}\langle K^0|H_{\text{eff}}^{\Delta S=2}|\bar{K}^0\rangle\,, \tag{11.23}$$

where $H_{\text{eff}}^{\Delta S=2}$ is the effective Hamiltonian for the $\Delta S = 2$ transition. It can be expressed via the OPE as

$$H_{\text{eff}}^{\Delta S=2} = \sum_i C_i(\mu)Q_i, \tag{11.24}$$

where the $C_i(\mu)$s are the Wilson coefficients and Q_i are the relevant local operators. In the SM, the $K^0 - \bar{K}^0$ transition is generated through the $W^\pm$-box diagram with up-quarks exchanges as shown in Fig. 11.1. Therefore, the effective Hamiltonian is given in terms of the operator Q_1 which is defined as

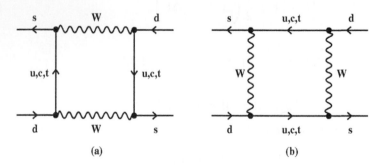

Figure 11.1 SM contribution to $K^0 - \bar{K}^0$ mixing.

$$Q_1 = \bar{d}_L^\alpha \gamma_\mu s_L^\alpha \, \bar{d}_L^\beta \gamma_\mu s_L^\beta\,. \tag{11.25}$$

The corresponding Wilson coefficient is given by [193]

$$C_1^{\text{SM}}(M_W) = \frac{G_F^2 M_W^2}{4\pi^2} \left[\eta_1 \lambda_c^2 S_0(x_c) + \eta_2 \lambda_t^2 S_0(x_t) + 2\eta_3 \lambda_c \lambda_t S_0(x_c, x_t)\right], \tag{11.26}$$

where $\lambda_i = V_{is}^* V_{id}$, η_i are the QCD correction factors with the NLO values $\eta_1 = 1.38 \pm 0.20$, $\eta_2 = 0.57 \pm 0.01$ and $\eta_3 = 0.47 \pm 0.04$, with $S_0(x_i, x_j)$ the loop functions given by [194]

$$S_0(x_c) = x_c\,, \qquad S_0(x_t) = 2.46 \left(\frac{m_t}{170\text{GeV}}\right)^{1.52}\,, \tag{11.27}$$

$$S_0(x_c, x_t) = x_c \left(\log\frac{x_t}{x_c} - \frac{3x_t}{4(1-x_t)} - \frac{3x_t^2 \log x_t}{4(1-x_t)^2}\right), \tag{11.28}$$

with $x_i = m_i^2/M_W^2$. The matrix element of the operator Q_1 is parameterised by a term $\hat{B}_1$ as

$$\langle \bar{K}^0 | Q_1 | K^0 \rangle = \frac{1}{3} f_K^2 m_K \hat{B}_1, \tag{11.29}$$

where f_K is the decay width, $f_K \simeq 160$ MeV. Therefore, one finds that $\mathcal{M}_{12}^{\text{SM}}$ is given by

$$\mathcal{M}_{12}^{\text{SM}} = \frac{G_F^2 M_W^2}{12\pi^2} f_K^2 m_K \hat{B}_K \mathcal{F}^*, \tag{11.30}$$

where

$$\mathcal{F}^* = \eta_1 \lambda_c^2 S_0(x_c) + \eta_2 \lambda_t^2 S_0(x_t) + 2\eta_3 \lambda_c \lambda_t S_0(x_c, x_t). \tag{11.31}$$

Taking $m_c = 1.4$ GeV, $m_t = 170$ GeV as the running quark masses at M_Z, one finds that the SM predictions for ε and ΔM_K are given by:

$$\varepsilon^{\text{SM}} \simeq 1.7 \times 10^{-3}, \tag{11.32}$$

$$\Delta M_K^{\text{SM}} \simeq 2.26 \times 10^{-15} \text{ GeV}. \tag{11.33}$$

These predictions lie in the ballpark of the measured values. However, a precise prediction cannot be made due to the hadronic and CKM uncertainties. In fact, the main uncertainty in this calculation arises from the matrix elements of Q_i, whereas the Wilson coefficients can be reliably calculated at high energies and evolved down to low energies via the RGE running.

11.2.3 Gluino contribution

In the MSSM, the $K^0 - \bar{K}^0$ transition can be generated through the box diagrams with gluino and chargino exchange, in addition to the SM $W^\pm$ diagram. We usually neglect the small effect due to charged Higgs and neutralino exchange. Thus, the off-diagonal entry in the kaon mass matrix, $\mathcal{M}_{12} = \langle K^0 | H_{\text{eff}}^{\Delta S=2} | \bar{K}^0 \rangle$, can be written as

$$\mathcal{M}_{12} = \mathcal{M}_{12}^{\text{SM}} + \mathcal{M}_{12}^{\tilde{g}} + \mathcal{M}_{12}^{\chi^+}. \tag{11.34}$$

The gluino contribution to the $\Delta S = 2$ effective Hamiltonian is given by the box diagrams in Fig. 11.2.

In this case, one finds that the effective Hamiltonian $H_{\text{eff}}^{\Delta S=2}$ is given in terms of the following local operators [18]:

$$Q_1 = \bar{d}_L^\alpha \gamma_\mu s_L^\alpha \, \bar{d}_L^\beta \gamma_\mu s_L^\beta, \tag{11.35}$$

$$Q_2 = \bar{d}_L^\alpha s_L^\alpha \, \bar{d}_R^\beta s_L^\beta, \tag{11.36}$$

$$Q_3 = \bar{d}_R^\alpha s_L^\beta \, \bar{d}_R^\beta s_L^\alpha, \tag{11.37}$$

$$Q_4 = \bar{d}_R^\alpha s_L^\alpha \, \bar{d}_L^\beta s_R^\beta, \tag{11.38}$$

$$Q_5 = \bar{d}_R^\alpha s_L^\beta \, \bar{d}_L^\beta s_R^\alpha. \tag{11.39}$$

In addition, the operators $\tilde{Q}_{1,2,3}$ are obtained from $Q_{1,2,3}$ by the exchange $L \leftrightarrow R$. Here α and β are the colour indices and $q_{R,L} = 1/2 \, (1 \pm \gamma_5) q$. The corresponding Wilson coefficients,

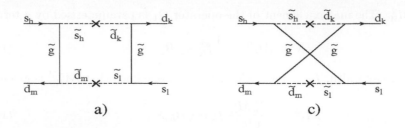

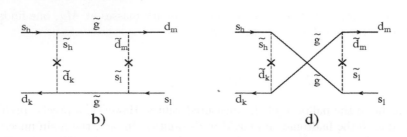

Figure 11.2 Gluino contribution to $K^0 - \bar{K}^0$ mixing, from [18].

taken at the SUSY scale M_{SUSY}, in the MIA approach, are given by [18]

$$C_1(M_{\text{SUSY}}) = \frac{-\alpha_s^2}{216m_{\tilde{q}}^2} \left[(\delta_{12}^d)_{LL}^2 \left(24x f_6(x) + 66\tilde{f}_6(x) \right) \right], \tag{11.40}$$

$$C_2(M_{\text{SUSY}}) = \frac{-\alpha_s^2}{216m_{\tilde{q}}^2} \left[(\delta_{12}^d)_{RL}^2 \left(204x f_6(x) \right) \right], \tag{11.41}$$

$$C_3(M_{\text{SUSY}}) = \frac{-\alpha_s^2}{216m_{\tilde{q}}^2} \left[(\delta_{12}^d)_{RL}^2 \left(-36x f_6(x) \right) \right], \tag{11.42}$$

$$C_4(M_{\text{SUSY}}) = \frac{-\alpha_s^2}{216m_{\tilde{q}}^2} \left[(\delta_{12}^d)_{LL}(\delta_{12}^d)_{RR} \left(504x f_6(x) - 72\tilde{f}_6(x) \right) \right], \tag{11.43}$$

$$C_5(M_{\text{SUSY}}) = \frac{-\alpha_s^2}{216m_{\tilde{q}}^2} \left[(\delta_{12}^d)_{LL}(\delta_{12}^d)_{RR} \left(24x f_6(x) + 120\tilde{f}_6(x) \right) \right], \tag{11.44}$$

where $x = m_{\tilde{g}}^2 / m_{\tilde{q}}^2$, $m_{\tilde{q}}$ is the average squark mass, $m_{\tilde{g}}$ is the gluino mass and the functions $f_6(x)$, $\tilde{f}_6(x)$ are given by [18]

$$f_6(x) = \frac{6(1 + 3x)\ln x + x^3 - 9x^2 - 9x + 17}{6(x-1)^5}, \tag{11.45}$$

$$\tilde{f}_6(x) = \frac{6(1 + 3x)\ln x + x^3 - 9x^2 - 9x + 17}{6(x-1)^5}. \tag{11.46}$$

x	$\sqrt{\left\|\mathrm{Re}\left(\delta_{12}^d\right)_{LL}^2\right\|}$	$\sqrt{\left\|\mathrm{Re}\left(\delta_{12}^d\right)_{LR}^2\right\|}$	$\sqrt{\left\|\mathrm{Re}\left(\delta_{12}^d\right)_{LL}\left(\delta_{12}^d\right)_{RR}\right\|}$
1	8.3×10^{-3}	1.2×10^{-3}	5.8×10^{-4}
2	2.3×10^{-2}	1.4×10^{-3}	6.7×10^{-4}
3	2.6×10^{-2}	1.6×10^{-3}	7.6×10^{-4}
4	1.9×10^{-2}	1.7×10^{-3}	8.3×10^{-4}

Table 11.1 Limits on $\mathrm{Re}\left(\delta_{12}^d\right)_{AB}\left(\delta_{12}^d\right)_{CD}$, with $A, B, C, D = (L, R)$, for an average squark mass $m_{\tilde{q}} = 1$ TeV and for different values of $x = m_{\tilde{g}}^2/m_{\tilde{q}}^2$. Since the LHC constraints imply that $m_{\tilde{g}} \gtrsim 1$ TeV, x must be larger than one.

The matrix elements of the operators Q_i between the K meson states can be written as [18]

$$\langle \bar{K}^0|Q_1(\mu)|K^0\rangle = \frac{1}{3}m_K f_K^2 B_1(\mu), \tag{11.47}$$

$$\langle \bar{K}^0|Q_2(\mu)|K^0\rangle = -\frac{5}{24}\left(\frac{m_K}{m_s(\mu)+m_d(\mu)}\right)^2 m_K f_K^2 B_2(\mu), \tag{11.48}$$

$$\langle \bar{K}^0|Q_3(\mu)|K^0\rangle = \frac{1}{24}\left(\frac{m_K}{m_s(\mu)+m_d(\mu)}\right)^2 m_K f_K^2 B_3(\mu), \tag{11.49}$$

$$\langle \bar{K}^0|Q_4(\mu)|K^0\rangle = \frac{1}{4}\left(\frac{m_K}{m_s(\mu)+m_d(\mu)}\right)^2 m_K f_K^2 B_4(\mu), \tag{11.50}$$

$$\langle \bar{K}^0|Q_5(\mu)|K^0\rangle = \frac{1}{12}\left(\frac{m_K}{m_s(\mu)+m_d(\mu)}\right)^2 m_K f_K^2 B_5(\mu). \tag{11.51}$$

Here the $Q_i(\mu)$'s refer to the operators renormalised at the scale μ and for $\mu = 2$ GeV we have [18]:

$$B_1(\mu) = 0.60, \quad B_2(\mu) = 0.66, \quad B_3(\mu) = 1.05, \quad B_4(\mu) = 1.03, \quad B_5(\mu) = 0.73. \tag{11.52}$$

Using these values, the gluino contribution to the $K - \bar{K}$ system can be calculated via Eq. (11.23). From the measured ΔM_K value, one obtains the bounds on the related mass insertions parameters, as given in Tab. 11.1.

Now we can estimate the gluino contribution to ε. The constraints imposed on the relevant mass insertions from the experimental limits of ε are reported in Tab. 11.2, for an average squark mass $m_{\tilde{q}} = 1$ TeV with different values of $x = m_{\tilde{g}}^2/m_{\tilde{q}}^2$. The bounds corresponding to different values of $m_{\tilde{q}}$ can be obtained by multiplying the ones in the table by $m_{\tilde{q}}(\text{GeV})/1000$.

11.2.4 Chargino contribution

Now we consider the chargino contribution to the $K - \bar{K}$ mixing in the MIA and derive the corresponding bounds on the mass insertion parameters [19]. The dominant chargino contribution to $H_{\mathrm{eff}}^{\Delta S=2}$ comes from the 'super-box' diagram in Fig. 11.3.

As in the gluino-dominated scenario, we perform calculations in the super-CKM basis, i.e., the basis in which the gluino-quark-squark vertices are flavour-diagonal. In this basis, the chargino-left quark-left squark vertices involve the usual CKM matrix:

$$\Delta \mathcal{L} = -g \sum_k \sum_{a,b} V_{k1} K_{ba}^* \, d_L^{a\dagger} i\sigma_2 (\tilde{\chi}_{kL}^+)^* \tilde{u}_L^b, \tag{11.53}$$

x	$\sqrt{\left|\mathrm{Im}\left(\delta_{12}^d\right)_{LL}^2\right|}$	$\sqrt{\left|\mathrm{Im}\left(\delta_{12}^d\right)_{LR}^2\right|}$	$\sqrt{\left|\mathrm{Im}\left(\delta_{12}^d\right)_{LL}\left(\delta_{12}^d\right)_{RR}\right|}$
1	6.6×10^{-4}	1×10^{-4}	4.6×10^{-5}
2	1.8×10^{-4}	1.2×10^{-4}	5.4×10^{-5}
3	2.1×10^{-3}	1.3×10^{-4}	6.1×10^{-5}
4	1.5×10^{-3}	1.4×10^{-4}	6.7×10^{-5}

Table 11.2 Limits on $\mathrm{Im}\left(\delta_{12}^d\right)_{AB}\left(\delta_{12}^d\right)_{CD}$, with $A, B, C, D = (L, R)$, for an average squark mass $m_{\tilde{q}} = 1$ TeV and for different values of $x = m_{\tilde{g}}^2 / m_{\tilde{q}}^2$.

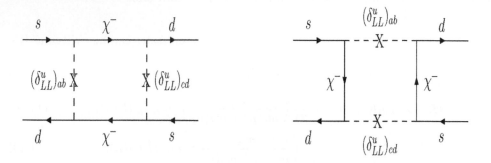

Figure 11.3 Leading chargino-up-squark contribution to $K - \bar{K}$ mixing, from [19].

where K is the CKM matrix, a, b are the flavour indices, $k = 1, 2$ labels the chargino mass eigenstates while V, U are the chargino mixing matrices. Only the gaugino components of the charginos lead to significant contributions to the $K - \bar{K}$ mixing, since the Higgsino couplings are suppressed by the quark masses (except for the stop coupling) and are not important even at large $\tan \beta$. The stop loop contribution is suppressed by the CKM mixing at the vertices: each vertex involving the stop is suppressed by λ^2 or λ^3 with λ being the Cabibbo mixing, whereas we will be working through $\mathcal{O}(\lambda)$ order. The super-box involving Higgsino interactions with the stops depends on the left-right mass insertions and, as it will be made clear later, does not lead to useful constraints on the SUSY flavour structures.

Due to the gaugino dominance, chargino-squark loops will generate a significant contribution to only one operator [19]

$$Q_1 = \bar{d}_L^\alpha \gamma^\mu s_L^\alpha \, \bar{d}_L^\beta \gamma_\mu s_L^\beta, \tag{11.54}$$

similar to the SM case (α, β are the colour indices). The corresponding Wilson coefficient is calculated to be [19]

$$C_1(M_W)^{\tilde{\chi}^+} = \frac{g^4}{768\pi^2 m^2} \left(\sum_{a,b} K_{a2}^* (\delta_{LL}^u)_{ab} K_{b1} \right)^2 \sum_{i,j} |V_{i1}|^2 |V_{j1}|^2 \frac{x_i h(x_i) - x_j h(x_j)}{x_i - x_j}, \tag{11.55}$$

where $x_i \equiv m_{\tilde{\chi}_i^+}^2 / m^2$ and

$$h(x) = \frac{2 + 5x - x^2}{(1-x)^3} + \frac{6x \ln x}{(1-x)^4}. \tag{11.56}$$

It is interesting to note that the 'flavour-conserving" mass insertions $(\delta_{LL}^u)_{aa}$ contribute to

M_2 \ m	300	500	700	900
150	0.04	0.06	0.08	0.09
250	0.07	0.08	0.09	0.11
350	0.09	0.10	0.11	0.12
450	0.12	0.12	0.13	0.14

Table 11.3 Bounds on $\sqrt{\left|\mathrm{Re}\left[(\delta_{LL}^u)_{21}\right]^2\right|}$ from ΔM_K (assuming a zero CKM phase). To obtain the corresponding bounds on $\delta \equiv (\delta_{LL}^u)_{11} - (\delta_{LL}^u)_{22}$, these entries are to be multiplied by 4.6. These bounds are largely insensitive to $\tan\beta$ in the range 3–40 and to μ in the range 200–500 GeV.

$C_1(M_W)$, unlike for the gluino case. Such mass insertions arise from non-degeneracy of the squark masses and are proportional to the difference of the average squark mass squared and the diagonal matrix elements of the squark mass matrix. If the diagonal elements are equal, the 'flavour-conserving' mass insertions drop out of the sum due to the GIM cancellations [195]. The flavour structure appearing in Eq. (11.55) can be expanded in powers of λ:

$$\sum_{a,b} K_{a2}^*(\delta_{LL}^u)_{ab}K_{b1} = (\delta_{LL}^u)_{21} + \lambda\left[(\delta_{LL}^u)_{11} - (\delta_{LL}^u)_{22}\right] + \mathcal{O}(\lambda^2). \qquad (11.57)$$

Assuming the presence of only one type of mass insertion at a time in Eq. (11.57) at each order in λ, one can derive constraints on $(\delta_{LL}^u)_{21}$ and $\delta \equiv (\delta_{LL}^u)_{11} - (\delta_{LL}^u)_{22}$ imposed by ΔM_K and ε. A much weaker constraint on $(\delta_{LL}^u)_{31}$ can also be obtained if we are to keep $\mathcal{O}(\lambda^2)$ terms in Eq. (11.57). In our analysis, we assume a zero CKM phase which corresponds to a conservative bound on the mass insertion. The Wolfenstein parameters are set to $A = 0.847$ and $\rho = 0.4$. The other relevant constants are $M_K = 0.498$ GeV and $f_K = 0.16$ GeV.

The resulting bounds on $(\delta_{LL}^u)_{21}$ and δ as functions of M_2 and the average squark mass m are presented in Tabs. 11.3 and 11.4 [19]. We find that these bounds are largely insensitive to $\tan\beta$ in the range 3–40 and to μ in the range 200–500 GeV. This can be understood since these parameters do not significantly affect the gaugino components of the charginos and their couplings. Note that δ is real due to the Hermiticity of the squark mass matrix and therefore does not contribute to ε. The presented bounds on the real part of $(\delta_{LL}^u)_{21}$ are a bit stronger than those derived from the gluino contribution to the $D - \bar{D}$ mixing [18], whereas the imaginary part of $(\delta_{LL}^u)_{21}$ is not constrained by any other FCNC processes.

M_2 \ m	300	500	700	900
150	5.3×10^{-3}	7.2×10^{-3}	9.1×10^{-3}	1.1×10^{-2}
250	7.8×10^{-3}	9.2×10^{-3}	1.1×10^{-2}	1.3×10^{-2}
350	1.1×10^{-2}	1.2×10^{-2}	1.3×10^{-2}	1.5×10^{-2}
450	1.5×10^{-2}	1.5×10^{-2}	1.6×10^{-2}	1.7×10^{-2}

Table 11.4 Bounds on $\sqrt{\left|\mathrm{Im}\left[(\delta_{LL}^u)_{21}\right]^2\right|}$ from ε. These bounds are largely insensitive to $\tan\beta$ in the range 3–40 and to μ in the range 200–500 GeV.

11.3 SUSY CONTRIBUTIONS TO $\Delta S = 1$

11.3.1 Basic formalism

CP violation can also be induced in direct K decays. In fact, the difference between the decay rate $\Gamma(K^0 \to fX)$ and the CP conjugate $\Gamma(\bar{K}^0 \to \bar{f}X)$ would be a clear indication for CP violation. The parameter ε'/ε that measures the direct CP violation in kaon ($\Delta S = 1$) decay is defined as [191]

$$\varepsilon'/\varepsilon = \frac{A(K_L \to (\pi\pi)_{I=2})}{A(K_L \to (\pi\pi)_{I=0})} - \frac{A(K_S \to (\pi\pi)_{I=2})}{A(K_S \to (\pi\pi)_{I=0})}. \tag{11.58}$$

In terms of the amplitudes of two isospin channels in $K^0 \to \pi\pi$, ε' can be approximated as [192]

$$\varepsilon'/\varepsilon \simeq \frac{i}{\sqrt{2}|\varepsilon|} e^{i(\delta_2 - \delta_0)} \mathrm{Im} \frac{A_2}{A_0}, \tag{11.59}$$

where $A_{0,2}$ are the amplitudes for the $\Delta I = 1/2, 3/2$ transitions, defined as

$$\langle (\pi\pi)_I | \mathcal{H} | K^0 \rangle = A_I e^{i\delta_I}, \qquad \langle (\pi\pi)_I | \mathcal{H} | \bar{K}^0 \rangle = \bar{A}_I e^{i\delta_I}. \tag{11.60}$$

The phases δ_I are the final state interaction (strong interaction) phase, where $\delta_2 - \delta_0 \sim -41°$. The average experimental results of $\mathrm{Re}(\varepsilon'/\varepsilon)$ is given by [196–199]

$$\mathrm{Re}(\varepsilon'/\varepsilon)_{\mathrm{exp}} = (1.66 \pm 0.16) \times 10^{-3}, \tag{11.61}$$

which provides firm evidence for the existence of direct CP violation. The effective Hamiltonian for the $\Delta S = 1$ transition is given by [200, 201]

$$H_{\mathrm{eff}}^{\Delta S=1} = \sum_{i=1}^{10} C_i(\mu) Q_i, \tag{11.62}$$

where the C_is are the Wilson coefficients and $Q_{1,2}$ refer to the current-current operators, Q_{3-6} to QCD penguin operators and Q_{7-10} to EW penguin operators and they are given as follows [201]:

$$Q_1 = (\bar{s}_\alpha u_\beta)_{V-A}(\bar{u}_\beta d_\alpha)_{V-A}, \qquad Q_2 = (\bar{s}u)_{V-A}(\bar{u}d)_{V-A}, \tag{11.63}$$

$$Q_3 = (\bar{s}d)_{V-A} \sum_q (\bar{q}q)_{V-A}, \qquad Q_4 = (\bar{s}_\alpha d_\beta)_{V-A} \sum_q (\bar{q}_\beta q_\alpha)_{V-A}, \tag{11.64}$$

$$Q_5 = (\bar{s}d)_{V-A} \sum_q (\bar{q}q)_{V+A}, \qquad Q_6 = (\bar{s}_\alpha d_\beta)_{V-A} \sum_q (\bar{q}_\beta q_\alpha)_{V+A}, \tag{11.65}$$

$$Q_7 = \frac{3}{2}(\bar{s}d)_{V-A} \sum_q e_q (\bar{q}q)_{V+A}, \qquad Q_8 = \frac{3}{2}(\bar{s}_\alpha d_\beta)_{V-A} \sum_q e_q (\bar{q}_\beta q_\alpha)_{V+A}, \tag{11.66}$$

$$Q_9 = \frac{3}{2}(\bar{s}d)_{V-A} \sum_q e_q (\bar{q}q)_{V-A}, \qquad Q_{10} = \frac{3}{2}(\bar{s}_\alpha d_\beta)_{V-A} \sum_q e_q (\bar{q}_\beta q_\alpha)_{V-A}, \tag{11.67}$$

where α, β are colour indices, e_q are quark charges and $(\bar{f}f)_{V\pm A} \equiv \bar{f}\gamma_\mu(1\pm\gamma_5)f$. In addition, the operators $\tilde{Q}_i$ are obtained from Q_i by the exchange $L \leftrightarrow R$.

11.3.2 SM contribution

The SM contribution to ε'/ε is dominated by the operators Q_6 and Q_8. They originate from the gluon and EW penguin diagrams and their matrix elements are enhanced by $(m_K/m_s)^2$ [202]:

$$\langle(\pi\pi)_{I=0}|Q_6|K^0\rangle = -4\sqrt{\frac{3}{2}}\left[\frac{m_K}{m_s(\mu)+m_d(\mu)}\right]^2 m_K^2(f_K - f_\pi) B_6^{(1/2)}, \quad (11.68)$$

$$\langle(\pi\pi)_{I=2}|Q_6|K^0\rangle = 0, \quad (11.69)$$

$$\langle(\pi\pi)_{I=0}|Q_8|K^0\rangle \simeq 2\sqrt{\frac{3}{2}}\left[\frac{m_K}{m_s(\mu)+m_d(\mu)}\right]^2 m_K^2 f_K\ B_8^{(1/2)}, \quad (11.70)$$

$$\langle(\pi\pi)_{I=2}|Q_8|K^0\rangle \simeq \sqrt{3}\left[\frac{m_K}{m_s(\mu)+m_d(\mu)}\right]^2 m_K^2 f_\pi\ B_8^{(3/2)}, \quad (11.71)$$

where $B_{6,8}^{(1/2)}$ and $B_8^{(3/2)}$ are the 'so-called' bag parameters wherein the matrix elements for the tilde operators come with an opposite sign. In addition, the contributions of the operators Q_6 and Q_8 are enhanced by QCD corrections. Although the Wilson coefficient of Q_8 is suppressed by α/α_s compared to that of Q_6, its contribution to ε' is enhanced by $1/\omega \sim 22$ and is significant. The SM contribution to ε'/ε can be expressed as [202, 203]

$$\varepsilon'/\varepsilon = \mathrm{Im}\lambda_t\ F_{\varepsilon'} \quad (11.72)$$

with

$$F_{\varepsilon'} = P_0 + P_X X_0(x_t) + P_Y Y_0(x_t) + P_Z Z_0(x_t) + P_E E_0(x_t). \quad (11.73)$$

The gauge independent loop functions in the SM are, to a very good approximation, given by

$$X_0(x_t) = 1.57\left(\frac{m_t}{170\mathrm{GeV}}\right)^{1.15}, \quad Y_0(x_t) = 1.02\left(\frac{m_t}{170\mathrm{GeV}}\right)^{1.56}, \quad (11.74)$$

$$Z_0(x_t) = 0.71\left(\frac{m_t}{170\mathrm{GeV}}\right)^{1.86}, \quad E_0(x_t) = 0.26\left(\frac{m_t}{170\mathrm{GeV}}\right)^{-1.02}. \quad (11.75)$$

The coefficients P_i are functions of the non-perturbative parameters $B_6^{(1/2)}$ and $B_8^{(3/2)}$:

$$P_i = r_i^{(0)} + \left(\frac{137\mathrm{MeV}}{m_s(m_c)+m_d(m_c)}\right)^2 \left\{B_6^{(1/2)} r_i^{(6)} + B_8^{(3/2)} r_i^{(8)}\right\}. \quad (11.76)$$

We will use the central values $B_6^{(1/2)} \sim 1.0$ and $B_8^{(3/2)} \sim 0.8$. The numerical factors $r_i^{(x)}$ are given in [203]. Here we use the following values:

$$r_i^0 = -3.122,\ 0.556,\ 0.404,\ 0.412,\ 0.204, \quad (11.77)$$

$$r_i^6 = 10.905,\ 0.019,\ 0.080,\ -0.015,\ -1.276, \quad (11.78)$$

and

$$r_i^8 = 1.423,\ 0,\ 0,\ -9.363,\ 0.409, \quad (11.79)$$

for $i = 0, X, Y, Z$ and E, respectively. The uncertainty in the SM prediction is increased by the fact that there is a large cancellation between the QCD and EW penguin contributions. The estimate of the SM prediction to $\mathrm{Re}(\varepsilon'/\varepsilon)$ leads to [203]

$$\mathrm{Re}(\varepsilon'/\varepsilon)_{\mathrm{SM}} \approx 7.5 \times 10^{-4}. \quad (11.80)$$

This result is particularly sensitive to the bag parameters. Given the theoretical and experimental uncertainties, the SM prediction is consistent with the measured value.

11.3.3 Gluino contribution

As in the case of ε, the SUSY contribution to ε'/ε is dominated by the gluino-mediated diagrams. In this scenario, the LR mass insertions can have large imaginary parts and the chromo-magnetic operator O_{8g} gives the dominant contribution to ε'/ε. The relevant Wilson coefficient is given by [18]

$$C_{8g} = \frac{\alpha_s \pi}{m_{\tilde{q}}^2} \left[(\delta_{12}^d)_{LL} \left(-\frac{1}{3} M_3(x) - 3 M_4(x) \right) + (\delta_{12}^d)_{LR} \frac{m_{\tilde{g}}}{m_s} \left(-\frac{1}{3} M_1(x) - 3 M_2(x) \right) \right],$$
(11.81)

with $x = m_{\tilde{g}}^2 / m_{\tilde{q}}^2$. The Wilson coefficients $\tilde{C}_{8g}$ can be obtained from C_{8g} by exchanging $L \leftrightarrow R$. The box-diagram loop functions $M_i(x)$ are given by

$$M_1(x) = \frac{1 + 4x - 5x^2 + 4x\ln(x) + 2x^2\ln(x)}{2(1-x)^4},$$
(11.82)

$$M_2(x) = -x^2 \frac{5 - 4x - x^2 + 2x\ln(x) + 4x\ln(x)}{2(1-x)^4},$$
(11.83)

$$M_3(x) = \frac{-1 + 9x + 9x^2 - 17x^3 + 18x^2\ln(x) + 6x^3\ln(x)}{12(x-1)^5},$$
(11.84)

$$M_4(x) = \frac{-1 - 9x + 9x^2 + x^3 - 6x\ln(x) - 6x^2\ln(x)}{6(x-1)^5},$$
(11.85)

where $x = m_{\tilde{g}}^2 / m_{\tilde{q}}^2$. Therefore, the gluino contribution to ε'/ε can be written as

$$\left(\frac{\varepsilon'}{\varepsilon} \right)^g \simeq \frac{11\sqrt{3}}{64\pi^2 |\varepsilon| \mathrm{Re} A_0} \frac{m_s}{m_s + m_d} \frac{F_k^2}{F_\pi^3} m_K^2 \, m_\pi^2 \, \mathrm{Im} \left[C_g - \tilde{C}_g \right].$$
(11.86)

In Tab. 11.5, the bounds imposed on the imaginary part of the relevant mass insertions due to experimental limits of ε'/ε for an average squark mass $m_{\tilde{q}} = 500$ GeV and different values of $x = m_{\tilde{g}}^2 / m_{\tilde{q}}^2$ are presented. As usual, the bounds for different values of $m_{\tilde{q}}$ can be obtained by multiplying the ones in the table times $(m_{\tilde{q}}/(500 \text{ GeV}))^2$.

x	$\left|\mathrm{Im}(\delta_{12}^d)_{LL}\right|$	$\left|\mathrm{Im}(\delta_{12}^d)_{LR}\right|$
0.3	1.0×10^{-1}	1.1×10^{-5}
1.0	4.8×10^{-1}	2.0×10^{-5}
4.0	2.6×10^{-1}	6.3×10^{-5}

Table 11.5 Limits from $\varepsilon'/\varepsilon < 2.7 \times 10^{-3}$ on $\mathrm{Im}\left(\delta_{12}^d\right)$, for an average squark mass $m_{\tilde{q}} = 500$ GeV and for different values of $x = m_{\tilde{g}}^2 / m_{\tilde{q}}^2$. For different values of $m_{\tilde{q}}$, the limits can be obtained by multiplying the ones in the table by $(m_{\tilde{q}}(\text{GeV})/500)^2$.

Clearly, ε'/ε is much more sensitive to new flavour structures in the A-terms than ΔM_K and ε are. Let us roughly estimate how large $(\delta_{12}^d)_{LR}$ can be expected in models with non-universal A-terms. Here, $(\delta_{12}^d)_{LR}$ is proportional to a linear combination of the down-type quark masses: $(\delta_{12}^d)_{LR} \sim (a_1 m_d + a_2 m_s + a_3 m_b)/m_{\tilde{q}}$. For a_2 of order Cabibbo mixing, we get a number of order 10^{-4}. If there is a considerable mixing with the third generation, this number will even increase. We see that, in principle, ε'/ε can be generated entirely by supersymmetric contributions.

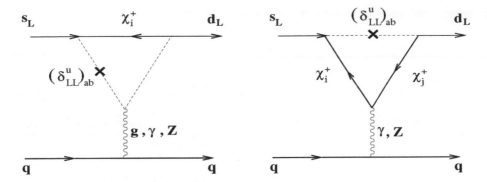

Figure 11.4 Leading chargino-up-squark contributions to ε'/ε, from [19].

11.3.4 Chargino contribution

The chargino contributions to ε'/ε, using the same approximations, is found to be given by [19]

$$\left(\frac{\varepsilon'}{\varepsilon}\right)^{\chi^{\pm}} = \mathrm{Im}\left(\sum_{a,b} K^*_{a2}(\delta^u_{ab})_{LL} K_{b1}\right) F_{\varepsilon'}(x_{q\chi}), \tag{11.87}$$

where

$$F_{\varepsilon'} = (P_X + P_Y + P_Z)\, F_Z + \frac{1}{4} P_Z\, F_\gamma + P_E\, F_g. \tag{11.88}$$

Here we have omitted the box diagram contributions which are negligible. The parameters P_i include the relevant matrix elements and NLO QCD corrections, they are given by $P_X = 0.58$, $P_Y = 0.48$, $P_Z = -7.67$, and $P_E = -0.82$. The quantities F_i are functions of supersymmetric parameters resulting from the gluon, photon as well as Z penguin diagrams (Fig. 11.4) and are calculated in the MIA [19]:

$$F_g = 2\frac{M_W^2}{m^2}\sum_i |V_{i1}|^2\, f_g(x_i), \tag{11.89}$$

$$F_\gamma = 2\frac{M_W^2}{m^2}\sum_i |V_{i1}|^2\, f_\gamma(x_i), \tag{11.90}$$

$$\begin{aligned}
F_Z = {} & \frac{1}{8} - 2\sum_i |V_{i1}|^2 f_Z^{(1)}\left(\frac{1}{x_i}, \frac{1}{x_i}\right) \\
& + 2\sum_{i,j} V_{j1}^* V_{i1}\left[U_{i1}U_{j1}^*\, f_Z^{(2)}(x_j, x_i) - V_{j1}V_{i1}^*\, f_Z^{(1)}(x_j, x_i)\right],
\end{aligned} \tag{11.91}$$

where $x_i \equiv m_{\tilde{\chi}_i^+}^2/m^2$ and the loop functions are given by [19]

$$f_g(x) = \frac{1 - 6x + 18x^2 - 10x^3 - 3x^4 + 12x^3 \ln x}{18(x-1)^5}, \tag{11.92}$$

$$f_\gamma(x) = \frac{22 - 60x + 45x^2 - 4x^3 - 3x^4 + 3(3 - 9x^2 + 4x^3)\ln x}{27(x-1)^5}, \tag{11.93}$$

$$f_Z^{(1)}(x,y) = \frac{(y-1)\left[(x-1)(x^2 - x^2 y + xy^2 - y^2) + x^2(y-1)\ln x\right] - (x-1)^2 y^2 \ln y}{16(x-1)^2(y-1)^2(y-x)}, \tag{11.94}$$

$$f_Z^{(2)}(x,y) = \sqrt{xy}\,\frac{(y-1)\left[(x-1)(x-y) + x(y-1)\ln x\right] - (x-1)^2 y \ln y}{8(x-1)^2(y-1)^2(y-x)}. \tag{11.95}$$

One finds that the dominant contribution typically comes from the Z penguin diagram, especially if the SUSY particles are heavy. This can be seen as follows. Due to gauge invariance, the $g\bar{s}_L d_L$ and $\gamma\bar{s}_L d_L$ vertices are proportional to the second power of the momentum transfer, *i.e.*, $(q_\mu q_\nu - g_{\mu\nu}q^2)/m^2$. This momentum dependence is cancelled by the gluon (photon) propagator which leads to a suppression factor $1/m^2$ in the final result. In contrast, the $Z\bar{s}_L d_L$ vertex exists at $q^2 = 0$ due to the weak current non-conservation and is momentum-independent to leading order. It is given by a dimensionless function of the ratios of the SUSY particles' masses. The Z propagator then leads to the suppression factor $1/M_Z^2$, which is much milder than $1/\tilde{m}_q^2$ appearing in the gluon and photon contributions.

In Tab. 11.6, we present the constraints on the imaginary parts of $(\delta_{LL}^u)_{21}$ due to the experimental bound on ε'/ε. It is clear that a considerably large mass insertion is required in order to saturate the experimental measurement of ε'/ε by chargino contribution. However, for universal GUT scale soft terms, the bounds on $\mathrm{Im}(\delta_{LL}^u)_{21}$ are slightly stronger than those on $\mathrm{Im}(\delta_{LL}^d)_{21}$ derived from the gluino contribution to ε'.

$M_2 \,\backslash\, m$	300	500	700	900
150	0.11	0.11	0.13	0.16
250	0.17	--	0.87	0.64
350	0.12	0.29	0.74	--
450	0.12	0.23	0.42	0.79

Table 11.6 Bounds on $\left|\mathrm{Im}(\delta_{LL}^u)_{21}\right|$ from ε'. For some parameter values the mass insertions are unconstrained due to the cancellations of different contributions to ε'. These bounds are largely insensitive to $\tan\beta$ in the range 3–40. (Notice that μ is set to 200 GeV.)

These results show that to have a chargino-induced ε' would require a relatively large LL mass insertion (of $\mathcal{O}(10^{-1})$), which typically violates the constraints from ΔM_K and ε. Yet, it is possible to saturate ε and ε' with the chargino contributions in corners of the parameter space [19]. In any case, non-universality among the soft scalar masses is necessary to get large values of ε'/ε and ε.

CHAPTER 12

SUSY CP Violation in B Mesons

In this chapter we analyse the severe constraints imposed on the squark masses and mixings, in terms of the mass insertions, by the experimental bounds of FCNC processes with $\Delta B = 2$ and $\Delta B = 1$ transitions.

12.1 SUSY CONTRIBUTIONS TO $\Delta B = 2$

There are two neutral $B^0 - \bar{B}^0$ meson systems: $B_q^0 - \bar{B}_q^0$, with $q = d, s$. In these systems, the flavour eigenstates are given by $B_q = (\bar{b}q)$ and $\bar{B}_q = (b\bar{q})$. Like the $K^0 - \bar{K}^0$ system, the $B_q^0 - \bar{B}_q^0$ oscillations are described by the Schrödinger equation:

$$i\frac{d}{dt}\left[\begin{array}{c} |B_q(t)\rangle \\ |\bar{B}_q(t)\rangle \end{array}\right] = \left(M^q - \frac{i}{2}\Gamma^q\right)\left[\begin{array}{c} |B_q(t)\rangle \\ |\bar{B}_q(t)\rangle \end{array}\right]. \tag{12.1}$$

It is customary to denote the corresponding mass eigenstates by $B_H^q = pB_q + q\bar{B}_q$ and $B_L^q = pB_q - q\bar{B}_q$, where the indices H and L refer to heavy and light mass eigenstates, respectively, and

$$\frac{q}{p} = \left[\frac{\mathcal{M}_{12}^* - \frac{i}{2}\Gamma_{12}^*}{\mathcal{M}_{12} - \frac{i}{2}\Gamma_{12}}\right]^{1/2}. \tag{12.2}$$

Thus, the mass and width differences between B_L^q and B_H^q are given by [204]

$$\begin{align}
\Delta M_{B_q} &= M_{B_H}^q - M_{B_L}^q = 2|\mathcal{M}_{12}^q|, \tag{12.3} \\
\Delta \Gamma_q &= \Gamma_L^q - \Gamma_H^q = 2|\Gamma_{12}^q|\cos\phi_q, \tag{12.4}
\end{align}$$

where $\phi_q = \arg\{-\mathcal{M}_{12}^q/\Gamma_{12}^q\}$. Therefore, ΔM_{B_q} can be calculated via

$$\Delta M_{B_q} = 2|\langle B_q^0|H_{\text{eff}}^{\Delta B=2}|\bar{B}_q^0\rangle|, \tag{12.5}$$

where $H_{\text{eff}}^{\Delta B=2}$ is the effective Hamiltonian responsible for the $\Delta B = 2$ transitions. The SM expression for $H_{\text{eff}}^{\Delta B=2}$ is

$$H_{\text{eff}}^{\Delta B=2} = C_1^{\text{SM}}Q_1 + \text{h.c.}, \tag{12.6}$$

where the four quark operator Q_1 is given by

$$Q_1 = \bar{q}_L\gamma_\mu b_L \ \bar{q}_L\gamma^\mu b_L \tag{12.7}$$

and the Wilson coefficient C_1^{SM} is defined as

$$C_1^{\text{SM}} = \frac{G_F^2}{4\pi^2} M_W^2 (V_{tq}V_{tb}^*)^2 S_0(x_t). \tag{12.8}$$

The loop function $S_0(x_t)$ of the $\Delta B_q = 2$ box diagram with $W^\pm$ exchange is as given in Eq. (11.28). The SM contribution is known at NLO accuracy in QCD. It is given by [201,204]

$$\mathcal{M}_{12}^{\text{SM}}(B_q) = \frac{G_F^2}{12\pi^2} \eta_B \hat{B}_{B_q} f_{B_q}^2 M_{B_q} M_W^2 (V_{tq}V_{tb}^*)^2 S_0(x_t), \tag{12.9}$$

where f_{B_q} is the B_q meson decay constant, $\hat{B}_{B_q}$ is the renormalisation group invariant B parameters and $\eta_B = 0.8393 \pm 0.0034$.

The dominant SUSY contribution to the effective Hamiltonian of $\Delta B = 2$ transitions can be generated through box diagrams mediated by gluino and chargino exchanges. Thus, one can write the off-diagonal entry in the B_q-meson mass matrix, $\mathcal{M}_{12}(B_q) = \langle B_q^0 | H_{\text{eff}}^{\Delta B=2} | \bar{B}_q^0 \rangle$, as follows [205]:

$$M_{12}(B_q) = M_{12}^{\text{SM}}(B_q) + M_{12}^{\tilde{g}}(B_q) + M_{12}^{\tilde{\chi}^+}(B_q), \tag{12.10}$$

where $M_{12}^{\text{SM}}(B_q)$, $M_{12}^{\tilde{g}}(B_q)$ and $M_{12}^{\tilde{\chi}^+}(B_q)$ indicate the SM, gluino and chargino contributions, respectively.

In a model independent way, the effect of SUSY can simply be described by a dimensionless parameter r_q^2 and a phase $2\theta_q$ defined as follows:

$$r_q^2 e^{2i\theta_q} = \frac{\mathcal{M}_{12}(B_q)}{M_{12}^{\text{SM}}(B_q)} = 1 + \frac{M_{12}^{\text{SUSY}}(B_q)}{M_{12}^{\text{SM}}(B_q)}. \tag{12.11}$$

Therefore,

$$\Delta M_{B_q} = 2|M_{12}^{\text{SM}}(B_q)|r_q^2 = \Delta M_{B_q}^{SM} r_q^2. \tag{12.12}$$

In this respect, r_q^2 is bounded by

$$r_q^2 \lesssim \Delta M_{B_q}^{exp} / \Delta M_{B_q}^{SM}, \tag{12.13}$$

which leads to stringent constraints on the SUSY contributions.

12.1.1 Gluino contribution

Similar to the gluino contribution to $\Delta S = 2$, the induced effective Hamiltonian for $\Delta B = 2$ processes, generated by $\Delta B_q = 2$ box diagrams, can be expressed as

$$H_{\text{eff}}^{\Delta B_q=2} = \sum_{i=1}^{5} C_i(\mu) Q_i(\mu) + \sum_{i=1}^{3} \tilde{C}_i(\mu) \tilde{Q}_i(\mu) + \text{h.c.}, \tag{12.14}$$

where $C_i(\mu)$, $\tilde{C}_i(\mu)$ and $Q_i(\mu)$, $\tilde{Q}_i(\mu)$ are the Wilson coefficients and operators, respectively, renormalised at the scale μ, with

$$\begin{aligned}
Q_1 &= \bar{q}_L^\alpha \gamma_\mu b_L^\alpha \; \bar{q}_L^\beta \gamma_\mu b_L^\beta, & Q_2 &= \bar{q}_R^\alpha b_L^\alpha \; \bar{q}_R^\beta b_L^\beta, & Q_3 &= \bar{q}_R^\alpha b_L^\beta \; \bar{q}_R^\beta b_L^\alpha, \\
Q_4 &= \bar{q}_R^\alpha b_L^\alpha \; \bar{q}_L^\beta b_R^\beta, & Q_5 &= \bar{q}_R^\alpha b_L^\beta \; \bar{q}_L^\beta b_R^\alpha.
\end{aligned} \tag{12.15}$$

In addition, the operators $\tilde{Q}_{1,2,3}$ are obtained from $Q_{1,2,3}$ by exchanging $L \leftrightarrow R$. The corresponding Wilson coefficients can be obtained from $C_i(M_{\text{SUSY}})$ given in Eq. (11.44), with exchanging $(\delta_{12}^d)_{AB}$ to $(\delta_{13}^d)_{AB}$ in the case of $q = d$ (i.e., $B_d^0 - \bar{B}_d^0$) and $(\delta_{12}^d)_{AB}$ to $(\delta_{23}^d)_{AB}$ in the case of $q = s$ (i.e., $B_s^0 - \bar{B}_s^0$). In this case, the off-diagonal matrix elements of the operators Q_i are given by

$$\langle B_d | Q_1 | \bar{B}_d \rangle = \frac{1}{3} m_{B_d} f_{B_d}^2 B_1(\mu), \tag{12.16}$$

$$\langle B_d | Q_2 | \bar{B}_d \rangle = -\frac{5}{24} \left(\frac{m_{B_d}}{m_b(\mu) + m_d(\mu)} \right)^2 m_{B_d} f_{B_d}^2 B_2(\mu), \tag{12.17}$$

$$\langle B_d | Q_3 | \bar{B}_d \rangle = \frac{1}{24} \left(\frac{m_{B_d}}{m_b(\mu) + m_d(\mu)} \right)^2 m_{B_d} f_{B_d}^2 B_3(\mu), \tag{12.18}$$

$$\langle B_d | Q_4 | \bar{B}_d \rangle = \frac{1}{4} \left(\frac{m_{B_d}}{m_b(\mu) + m_d(\mu)} \right)^2 m_{B_d} f_{B_d}^2 B_4(\mu), \tag{12.19}$$

$$\langle B_d | Q_5 | \bar{B}_d \rangle = \frac{1}{12} \left(\frac{m_{B_d}}{m_b(\mu) + m_d(\mu)} \right)^2 m_{B_d} f_{B_d}^2 B_4(\mu). \tag{12.20}$$

As in $K^0 - \bar{K}^0$ mixing, the Wilson coefficients at the scale $\mu = m_b$ are given in of terms the Wilson coefficients at the SUSY scale as follows [206]:

$$C_r(\mu) = \sum_i \sum_s \left(b_i^{(r,s)} + \eta\, c_i^{(r,s)} \right) \eta^{a_i} C_s(M_{\text{SUSY}}), \tag{12.21}$$

where $\eta = \alpha_s(M_{\text{SUSY}})/\alpha_s(m_t)$ and the parameters a_i, $b^{r,s}$, and $c^{r,s}$ are given by [206]

$$a_i = (0.286, -0.692, 0.787, -1.143, 0.143)$$

$$
\begin{aligned}
b_i^{(11)} &= (0.865, 0, 0, 0, 0), & c_i^{(11)} &= (-0.017, 0, 0, 0, 0), \\
b_i^{(22)} &= (0, 1.879, 0.012, 0, 0), & c_i^{(22)} &= (0, -0.18, -0.003, 0, 0), \\
b_i^{(23)} &= (0, -0.493, 0.18, 0, 0), & c_i^{(23)} &= (0, -0.014, 0.008, 0, 0), \\
b_i^{(32)} &= (0, -0.044, 0.035, 0, 0), & c_i^{(32)} &= (0, 0.005, -0.012, 0, 0), \\
b_i^{(33)} &= (0, 0.011, 0.54, 0, 0), & c_i^{(33)} &= (0, 0.000, 0.028, 0, 0), \\
b_i^{(44)} &= (0, 0, 0, 2.87, 0), & c_i^{(44)} &= (0, 0, 0, -0.48, 0.005), \\
b_i^{(45)} &= (0, 0, 0, 0.961, -0.22), & c_i^{(45)} &= (0, 0, 0, -0.25, -0.006), \\
b_i^{(54)} &= (0, 0, 0, 0.09, 0), & c_i^{(54)} &= (0, 0, 0, -0.013, -0.016), \\
b_i^{(55)} &= (0, 0, 0, 0.029, 0.863), & c_i^{(55)} &= (0, 0, 0, -0.007, 0.019).
\end{aligned}
\tag{12.22}
$$

The latest estimations for the the B parameters lead to the following values:

$$
\begin{aligned}
B_1(m_b) &= 0.87(4), & B_2(m_b) &= 0.82(3), & B_3(m_b) &= 1.02(6), \\
B_4(m_b) &= 1.16(3), & B_5(m_b) &= 1.91(4).
\end{aligned}
\tag{12.23}
$$

Using these expressions, one can compute the SM and SUSY results for ΔM_{B_d}. We find that $(\Delta M_{B_d})^{\text{SM}}$ is given by

$$(\Delta M_{B_d})^{\text{SM}} \approx 0.5 \ (\text{ps})^{-1}. \tag{12.24}$$

Note that the present experimental value of ΔM_{B_d} is given by

$$(\Delta M_{B_d})^{\text{Exp}} = 0.484 \pm 0.010 \ (\text{ps})^{-1}. \tag{12.25}$$

Therefore, the small difference between the SM result and the experimental value of ΔM_{B_d} implies stringent constraints on the relevant mass insertions in the down-squark sector, mediated by gluino exchange [205]. The upper bounds on the relevant combinations of mass insertions $(\delta^d_{AB})_{13}$ (with $A, B = (L, R)$) from the gluino corrections to ΔM_{B_d} are presented in Tab. 12.1.

x	$\sqrt{\left|\mathrm{Re}\left[(\delta^d_{LL})^2_{31}\right]\right|}$	$\sqrt{\left|\mathrm{Re}\left[(\delta^d_{RL})^2_{31}\right]\right|}$	$\sqrt{\left|\mathrm{Re}\left[(\delta^d_{LL})_{31}(\delta^d_{RR})_{31}\right]\right|}$	$\sqrt{\left|\mathrm{Re}\left[(\delta^d_{LR})_{31}(\delta^d_{RL})_{31}\right]\right|}$
1	0.29	6.4×10^{-2}	2.4×10^{-2}	4.9×10^{-2}
1.5	0.45	6.9×10^{-2}	2.7×10^{-3}	6.1×10^{-2}
2	0.8	7.4×10^{-2}	2.9×10^{-2}	7.2×10^{-2}
2.5	2.2	7.8×10^{-2}	3.1×10^{-2}	8.3×10^{-2}

Table 12.1 Upper bounds on real parts of combinations of mass insertions $(\delta^d_{AB})_{31}$, with $(A, B) = L, R$, from gluino contributions to ΔM_{B_d} (assuming zero SM contribution), evaluated at $m_{\tilde{q}} = 1$ TeV. As usual, $x = (m_{\tilde{g}}/m_{\tilde{q}})^2$.

Now we turn to $B^0_s - \bar{B}^0_s$ mixing. It turns out that the SM contribution to ΔM_{B_s} can be estimated more accurately from the ratio $\Delta M^{\mathrm{SM}}_{B_s}/\Delta M^{\mathrm{SM}}_{B_d}$, in which all short-distance effects cancel [204]:

$$\frac{\Delta M^{\mathrm{SM}}_{B_s}}{\Delta M^{\mathrm{SM}}_{B_d}} = \frac{M_{B_s}}{M_{B_d}} \frac{B_{B_s} f^2_{B_s}}{B_{B_d} f^2_{B_d}} \frac{|V_{ts}|^2}{|V_{td}|^2}. \tag{12.26}$$

The remaining ratio of hadronic parameters has been calculated on the lattice yielding

$$\frac{B_{B_s}(m_b) f^2_{B_s}}{B_{B_d}(m_b) f^2_{B_d}} = (1.15 \pm 0.06^{+0.07}_{-0.00})^2, \tag{12.27}$$

where the asymmetric error is due to the effect of chiral logarithms in the quenched approximation. From the fact that $\Delta M^{\mathrm{SM}}_{B_d} \simeq \Delta M^{\mathrm{Exp}}_{B_d}$ and $|V_{ts}|^2/|V_{td}|^2$ can be determined from a process which is not constrained by new physics, one finds $\Delta M^{SM}_{B_s} \simeq 15$ ps^{-1} for a quark mixing angle $\gamma \simeq 67°$. The most recent results of ΔM_{B_s} reported by CDF and D0, are given by:

$$\Delta M_{B_s} = 17.77 \pm 0.10(\mathrm{stat.}) \pm 0.07(\mathrm{syst.}) \qquad \text{(CDF)}, \tag{12.28}$$

$$\Delta M_{B_s} = 18.53 \pm 0.93(\mathrm{stat.}) \pm 0.30(\mathrm{syst.}) \qquad \text{(D0)}. \tag{12.29}$$

12.1.2 Chargino contribution

The leading diagrams of chargino contributions to the effective Hamiltonian $H^{\Delta B_q=2}_{\mathrm{eff}}$ are given by box diagrams similar to those of $K^0 - \bar{K}^0$ mixing, upon replacing the s quark by the b quark and the d quark by the q quark. One can show that the dominant chargino exchange gives significant contributions to the operators Q_1 and Q_3 in Eq. (12.15). Recall that, in $K - \bar{K}$ mixing, the relevant chargino exchange affects only the operator Q_1 as in the SM.

In the framework of the MIA, the interacting Lagrangian involving charginos is given by

$$\mathcal{L}_{q\tilde{q}\tilde{\chi}^+} = -g \sum_k \sum_{a,b} \left(V_{k1} K^*_{ba} \bar{d}^a_L (\tilde{\chi}^+)^* \tilde{u}^b_L - U^*_{k2} (Y^{\mathrm{diag}}_d . K^+)_{ab} \bar{d}^a_R (\tilde{\chi}^+)^* \tilde{u}^b_L \right.$$

$$\left. -V^*_{k2} (K . Y^{\mathrm{diag}}_u)_{ab} \bar{d}^a_L (\tilde{\chi}^+)^* \tilde{u}^b_R \right), \tag{12.30}$$

where $Y_{u,d}^{\text{diag}}$ are the diagonal Yukawa matrices and K is the usual CKM matrix. The indices a, b and k label flavour and chargino mass eigenstates, respectively, and V, U are the chargino mixing matrices. As one can see from Eq. (12.61), the Higgsino couplings are suppressed by the Yukawas of the light quarks, and therefore they are negligible, except for the stop coupling to the bottom quark, which is directly enhanced by the top Yukawa coupling Y_t. The other vertex involving the d quark and stop squark could also be enhanced by Y_t, but one should pay the price of a λ^3 suppression, with λ being the Cabibbo mixing. Since in our analysis we will work in the approximation of retaining only terms proportional to order λ, we will neglect the effect of this vertex. Moreover, we will also set to zero the Higgsino contributions proportional to the Yukawa couplings of light quarks with the exception of the bottom Yukawa Y_b, since its effect could be enhanced by large $\tan\beta$. In this respect, it is clear that the chargino contribution to the Wilson coefficients C_4 and C_5 is negligible. Furthermore, due to the colour structure of the chargino box diagrams, there is no contribution to C_2 or $\tilde{C}_2$. However, they are induced at low energy by QCD corrections through the mixing with C_3.

Now we calculate the relevant Wilson coefficients $C_{1,3}^\chi(M_{\text{SUSY}})$ at the SUSY scale M_{SUSY}, by using the MIA. At the first order in the MIA, one finds that the Wilson coefficients of $B_d^0 - \bar{B}_d^0$ are given by

$$
\begin{aligned}
C_1^\chi(M_{\text{SUSY}}) &= \frac{g^4}{768\pi^2 m_{\tilde{q}}^2} \sum_{i,j} \Big\{ |V_{i1}|^2 |V_{j1}|^2 \left((\delta_{LL}^u)_{31}^2 + 2\lambda(\delta_{LL}^u)_{31}(\delta_{LL}^u)_{32} \right) \\
&\quad - 2Y_t |V_{i1}|^2 V_{j1} V_{j2}^* \left((\delta_{LL}^u)_{31}(\delta_{RL}^u)_{31} + \lambda(\delta_{LL}^u)_{32}(\delta_{RL}^u)_{31} + \lambda(\delta_{LL}^u)_{31}(\delta_{RL}^u)_{32} \right) \\
&\quad + Y_t^2 V_{i1} V_{i2}^* V_{j1} V_{j2}^* \left((\delta_{RL}^u)_{31}^2 + 2\lambda(\delta_{RL}^u)_{31}(\delta_{RL}^u)_{32} \right) \Big\} L_2(x_i, x_j), \quad (12.31)
\end{aligned}
$$

$$
C_3^\chi(M_{\text{SUSY}}) = \frac{g^4 Y_b^2}{192\pi^2 m_{\tilde{q}}^2} \sum_{i,j} U_{i2} U_{j2} V_{j1} V_{i1} \left((\delta_{LL}^u)_{31}^2 + 2\lambda(\delta_{LL}^u)_{31}(\delta_{LL}^u)_{32} \right) L_0(x_i, x_j),
$$

$$(12.32)$$

where $x_i = m_{\tilde{\chi}_i^+}^2 / m_{\tilde{q}}^2$, and the functions $L_0(x, y)$ and $L_2(x, y)$ are given by

$$
L_0(x, y) = \sqrt{xy} \left(\frac{x\, h_0(x) - y\, h_0(y)}{x - y} \right), \quad (12.33)
$$

$$
h_0(x) = \frac{-11 + 7x - 2x^2}{(1 - x)^3} - \frac{6 \ln x}{(1 - x)^4}, \quad (12.34)
$$

$$
L_2(x, y) = \frac{x\, h_2(x) - y\, h_2(y)}{x - y}, \quad (12.35)
$$

$$
h_2(x) = \frac{2 + 5x - x^2}{(1 - x)^3} + \frac{6x \ln x}{(1 - x)^4}. \quad (12.36)
$$

As in the gluino case, the corresponding results for the $\tilde{C}_1$ and $\tilde{C}_1$ coefficients are simply obtained by interchanging $L \leftrightarrow R$ in the mass insertions appearing in the expressions for $C_{1,3}$.

The Wilson coefficients $C_{1,3}^\chi(M_{\text{SUSY}})$ of $B_s^0 - \bar{B}_s^0$ can be obtained from the above expression by the exchange $(\delta_{AB}^u)_{31} \leftrightarrow (\delta_{AB}^u)_{32}$. Now we analyse the upper bounds imposed on the mass insertions $(\delta_{AB}^u)_{ij}$ from the experimental limits of $\Delta M_{B_{d,s}}$. These constraints will depend on the relevant MSSM low energy parameters, in particular, by $m_{\tilde{q}}$, M_2, μ and $\tan\beta$. Note that with respect to the gluino-mediated FCNC processes, which are parameterised by $m_{\tilde{q}}$ and $m_{\tilde{g}}$, the chargino-mediated ones contain two free parameters more.

M_2	$\sqrt{\left|\text{Re}\left[(\delta^u_{LL})_{31}\right]^2\right|}$	$\sqrt{\left|\text{Re}\left[(\delta^u_{RL})_{31}\right]^2\right|}$	$\sqrt{\left|\text{Re}\left[(\delta^u_{LL})_{31}(\delta^u_{LL})_{32}\right]\right|}$
200	0.35	7.48	0.53
400	0.46	6.46	0.69
600	0.59	6.57	0.88
800	0.74	7.04	1.11

Table 12.2 Upper bounds on mass insertions from ΔM_{B_d} versus M_2, for $M_{\tilde{q}} = 1$ TeV, $\mu = 500$ GeV, and $\tan\beta = 10$.

$m_{\chi^+_1}$	$\sqrt{\left|\text{Re}\left[(\delta^u_{LL})_{31}(\delta^u_{RL})_{31}\right]\right|}$	$\sqrt{\left|\text{Re}\left[(\delta^u_{LL})_{31}(\delta^u_{RL})_{32}\right]\right|}$	$\sqrt{\left|\text{Re}\left[(\delta^u_{RL})_{31}(\delta^u_{RL})_{32}\right]\right|}$
200	1.22	2.59	11.22
400	1.30	2.76	9.68
600	1.49	3.17	9.86
800	1.74	3.70	10.55

Table 12.3 Upper bounds on mass insertions from ΔM_{B_d} versus M_2, for $M_{\tilde{q}} = 1$ TeV, $\mu = 500$ GeV, and $\tan\beta = 10$.

In Tabs. (12.2) and (12.3), we present our results for the upper bounds on the mass insertions coming from ΔM_{B_d}, for four values of M_2 and fixed values of $m_{\tilde{q}} = 1$ TeV, $\mu = 500$ GeV and $\tan\beta = 10$. It is important to note that the bounds presented in these tables are less stringent than those reported in [20], since the SUSY spectrum is now much heavier. Also, for larger values of μ and M_2, these bounds become clearly less severe due to decoupling. Notice that they are also quite insensitive to $\tan\beta$, since no mass insertion receives leading contributions from bottom Yukawa couplings. It is also worth mentioning that the bounds on the mass insertion $(\delta^u_{LL})_{32}(\delta^u_{RL})_{31}$ are identical to the bounds on $(\delta^u_{LL})_{31}(\delta^u_{RL})_{32}$. Therefore, here we just present the bounds on one of them.

12.2 SUSY CONTRIBUTIONS TO $\Delta B = 1$

We start our analysis by considering the supersymmetric effect in the non-leptonic $\Delta B = 1$ processes. Such an effect could be a probe for any testable SUSY implications in CP-violating experiments. The most general effective Hamiltonian $H^{\Delta B=1}_{\text{eff}}$ for these processes can be expressed via the OPE as [207]

$$H^{\Delta B=1}_{\text{eff}} = \left\{\frac{G_F}{\sqrt{2}}\sum_{p=u,c}\lambda_p\left(C_1 Q^p_1 + C_2 Q^p_2 + \sum_{i=3}^{10} C_i Q_i + C_{7\gamma}Q_{7\gamma} + C_{8g}Q_{8g}\right)\right\}$$
$$+ \left\{Q_i \to \tilde{Q}_i\,,\; C_i \to \tilde{C}_i\right\}, \tag{12.37}$$

where $\lambda_p = V_{pb}V^*_{ps}$, with V_{pb} the unitary CKM matrix elements satisfying the unitarity triangle relation $\lambda_t + \lambda_u + \lambda_c = 0$, and $C_i \equiv C_i(\mu_b)$ are the Wilson coefficients at the low energy scale $\mu_b \simeq \mathcal{O}(m_b)$. The basis $Q_i \equiv Q_i(\mu_b)$ is given by the relevant local operators renormalised at the same scale μ_b, namely,

$$Q_2^p = (\bar{p}b)_{V-A}\ (\bar{s}p)_{V-A}, \qquad Q_1^p = (\bar{p}_\alpha b_\beta)_{V-A}\ (\bar{s}_\beta p_\alpha)_{V-A}, \tag{12.38}$$

$$Q_3 = (\bar{s}b)_{V-A} \sum_q (\bar{q}q)_{V-A}, \qquad Q_4 = (\bar{s}_\alpha b_\beta)_{V-A} \sum_q (\bar{q}_\beta q_\alpha)_{V-A}, \tag{12.39}$$

$$Q_5 = (\bar{s}b)_{V-A} \sum_q (\bar{q}q)_{V+A}, \qquad Q_6 = (\bar{s}_\alpha b_\beta)_{V-A} \sum_q (\bar{q}_\beta q_\alpha)_{V+A}, \tag{12.40}$$

$$Q_7 = (\bar{s}b)_{V-A} \sum_q \frac{3}{2}e_q(\bar{q}q)_{V+A}, \qquad Q_8 = (\bar{s}_\alpha b_\beta)_{V-A} \sum_q \frac{3}{2}e_q(\bar{q}_\beta q_\alpha)_{V+A}, \tag{12.41}$$

$$Q_9 = (\bar{s}b)_{V-A} \sum_q \frac{3}{2}e_q(\bar{q}q)_{V-A}, \qquad Q_{10} = (\bar{s}_\alpha b_\beta)_{V-A} \sum_q \frac{3}{2}e_q(\bar{q}_\beta q_\alpha)_{V-A}, \tag{12.42}$$

$$Q_{7\gamma} = \frac{e}{8\pi^2}m_b\bar{s}\sigma^{\mu\nu}(1+\gamma_5)F_{\mu\nu}b, \qquad Q_{8g} = \frac{g_s}{8\pi^2}m_b\bar{s}_\alpha\sigma^{\mu\nu}(1+\gamma_5)G^A_{\mu\nu}t^A_{\alpha\beta}b_\beta. \tag{12.43}$$

Here α and β stand for colour indices and $t^A_{\alpha\beta}$ are the $SU(3)_c$ colour matrices, $\sigma^{\mu\nu} = \frac{1}{2}i[\gamma^\mu, \gamma^\nu]$. Moreover, e_q are quark electric charges in unity of e, $(\bar{q}q)_{V\pm A} \equiv \bar{q}\gamma_\mu(1 \pm \gamma_5)q$, and q runs over the u, d, s, c and b quark labels. In the SM, only the first part of the right-hand side of Eq. (12.37) (inside the first curly brackets) containing the operators Q_i will contribute, where $Q_{1,2}^p$ refer to the current-current operators, Q_{3-6} to the QCD penguin operators and Q_{7-10} to the EW penguin operators while $Q_{7\gamma}$ and Q_{8g} are the magnetic and the chromo-magnetic dipole operators, respectively. In addition, the operators $\tilde{Q}_i \equiv \tilde{Q}_i(\mu_b)$ are obtained from Q_i by the chirality exchange $(\bar{q}_1 q_2)_{V\pm A} \to (\bar{q}_1 q_2)_{V\mp A}$. Notice that in the SM the coefficients $\tilde{C}_i$ identically vanish due to the $V - A$ structure of charged weak currents, while in the MSSM they can receive contributions from both chargino and gluino exchanges.

Due to the asymptotic freedom of QCD, the calculation of hadronic weak decay amplitudes can be factorised by the product of *long* and *short* distance contributions. The first ones, which will be analysed in the next section, are related to the evaluation of hadronic matrix elements of Q_i and contain the main uncertainty of our predictions. The second ones are instead contained in the Wilson coefficients C_i's and they can be evaluated in perturbation theory with high precision. For instance, all the relevant contributions of particle spectra above the $W^\pm$ mass (m_W) scale, including SUSY particle exchanges, will enter in $C_i(\mu_W)$ at the $\mu_W \simeq \mathcal{O}(m_W)$ scale.

The low energy coefficients $C_i(\mu_b)$ can be extrapolated from the high energy ones $C_i(\mu_W)$ by solving the RGEs for QCD and QED in the SM. The solution is generally expressed as follows:

$$C_i(\mu) = \sum_j \hat{U}_{ij}(\mu, \mu_W)\, C_j(\mu_W), \tag{12.44}$$

where $\hat{U}_{ij}(\mu, \mu_W)$ is the evolution matrix, which takes into account the re-summation of the terms proportional to large logs $(\alpha_s(\mu_W)\log(\mu_W/\mu_b))^n$ (leading), $(\alpha_s^2(\mu_W)\log(\mu_W/\mu_b))^n$ (next-to-leading), *etc.*, in QCD. In our analysis we include the NLO corrections in QCD and QED for the Wilson coefficients $C_{i=1-10}$, while for $C_{7\gamma}(\mu)$ and $C_{8g}(\mu)$ we include only the LO ones. The reason for retaining only the LO accuracy in $C_{7\gamma}(\mu)$ and $C_{8g}(\mu)$ is that the matrix elements of the dipole operators enter the decay amplitudes only at the NLO. The expressions for the evolution matrix $\hat{U}_{ij}(\mu, \mu_W)$ at NLO in QCD and QED can be found in [207].

Next we discuss the SUSY contributions to the effective Hamiltonian in Eq. (12.37). The modifications caused by SUSY appear only in the boundary conditions of the Wilson coefficients at the μ_W scale and they can be computed through the appropriate matching with one-loop Feynman diagrams where Higgs, neutralino, gluino and chargino states are exchanged. Only chargino and gluino contributions can provide a potential source of new CP violating phases in the MSSM. In principle, the neutralino exchange diagrams involve the same mass insertions as the gluino ones, but they are strongly suppressed compared to the latter. For these reasons we neglect neutralinos in our analysis. The charged Higgs contributions cannot generate any new source of CP violation in addition to the SM ones, or any sizable effect to operators beyond the SM basis Q_i. However, when charged Higgs contributions are taken into account together with chargino or gluino exchanges, their effect is relevant. In particular, as we will show in the next sections, due to destructive interferences with chargino and gluino amplitudes, $b \to s\gamma$ constraints can be relaxed, allowing sizable contributions to the CP asymmetries.

The results for the Wilson coefficients at the μ_W scale can be expressed as follows [20]

$$C_i(\mu_W) = C_i^W + C_i^H + \lambda_t^{-1} \left\{ C_i^\chi + C_i^{\tilde{g}} \right\}, \tag{12.45}$$

where C_i^W, C_i^H, C_i^χ and $C_i^{\tilde{g}}$ correspond to the $W^\pm$, charged Higgs, chargino and gluino exchanges, respectively. In our analysis we will impose the boundary conditions for C_i^χ and $C_i^{\tilde{g}}$ at the scale $\mu_W = M_W$, although they should apply to the energy scale at which SUSY particles are integrated out, namely M_{SUSY}. However, these threshold corrections, originating from the mismatch of energy scales, are numerically not significant since the running of α_s from M_{SUSY} to M_W is not very steep.

Finally, the EW contributions to the Wilson coefficients are given by [203, 208, 209]

$$C_1^W = \frac{14\alpha_s}{16\pi}, \tag{12.46}$$

$$C_2^W = 1 - \frac{11}{6}\frac{\alpha_s}{4\pi}, \tag{12.47}$$

$$C_3^{(W,H,\chi)} = \frac{\alpha}{6\pi}\frac{1}{\sin^2\theta_w}\left(B_d^{(W,H,\chi)} + \frac{1}{2}B_u^{(W,H,\chi)} + C^{(W,H,\chi)}\right) - \frac{\alpha_s}{24\pi}E^{(W,H,\chi)}, \tag{12.48}$$

$$C_4^{(W,H,\chi)} = \frac{\alpha_s}{8\pi}E^{(W,H,\chi)}, \tag{12.49}$$

$$C_5^{(W,H,\chi)} = -\frac{\alpha_s}{24\pi}E^{(W,H,\chi)}, \tag{12.50}$$

$$C_6^{(W,H,\chi)} = \frac{\alpha_s}{8\pi}E^{(W,H,\chi)}, \tag{12.51}$$

$$C_7^{(W,H,\chi)} = \frac{\alpha}{6\pi}\left(4C^{(W,H,\chi)} + D^{(W,H,\chi)}\right), \tag{12.52}$$

$$C_8^{(W,H,\chi)} = 0, \tag{12.53}$$

$$C_9^{(W,H,\chi)} = \frac{\alpha}{6\pi}\left(4C^{(W,H,\chi)} + D^{(W,H,\chi)} + \frac{1}{\sin^2\theta_w}\left(-B_d^{(W,H,\chi)} + B_u^{(W,H,\chi)} - 4C^{(W,H,\chi)}\right)\right), \tag{12.54}$$

$$C_{10}^{(W,H,\chi)} = 0, \tag{12.55}$$

$$C_{7\gamma}^{(W,H,\chi)} = M_\gamma^{(W,H,\chi)}, \tag{12.56}$$

$$C_{8g}^{(W,H,\chi)} = M_g^{(W,H,\chi)}, \tag{12.57}$$

where α_s and α are evaluated at M_W scale. The functions appearing above include the contributions from photon-penguins (D), Z-penguins (C), gluon-penguins (E), boxes with external down quarks (B_d) and up quarks (B_u), the magnetic- (M_γ) plus the chromomagnetic-penguins (M_g). The corresponding SM results are $D^W \equiv \mathrm{D}(x_t)$, $C^W \equiv \mathrm{C}(x_t)$, $E^W \equiv \mathrm{E}(x_t)$, and $B^W \equiv \mathrm{B}(x_t)$ with $x_t = m_t^2/M_W^2$, and analogously for the charged Higgs boson, $D^H \equiv \mathrm{D_H}(x_H)$, $C^H \equiv \mathrm{C_H}(x_H)$, $E^H \equiv \mathrm{E_H}(x_H)$, and $B^H \equiv \mathrm{B_H}(x_H)$ with $x_H = m_t^2/m_H^2$, where the loop functions $\mathrm{B, C, D, E}$ and $\mathrm{B_H, C_H, D_H, E_H}$ are provided in Appendix C. Regarding the SM and charged Higgs contributions to all magnetic-penguins, we have

$$M_\gamma^W = -x_t \left(F_1(x_t) + \frac{3}{2} F_2(x_t) \right), \qquad M_g^W = -\frac{3}{2} x_t F_1(x_t), \tag{12.58}$$

$$M_\gamma^H = -\frac{x_H}{2} \left(\left(\frac{2}{3} F_1(x_H) + F_2(x_H) \right) \cot^2 \beta + \frac{2}{3} F_3(x_H) + F_4(x_H) \right), \tag{12.59}$$

$$M_g^H = -\frac{x_H}{2} \left(F_1(x_H) \cot^2 \beta + F_3(x_H) \right), \tag{12.60}$$

where the functions $F_i(x)$ are reported in Appendix C. Finally, the gluino and chargino exact contributions to the expressions appearing in Eq. (12.57) can be found in [208, 209], while here we will provide only the corresponding results in the MIA.

Regarding the SUSY contributions to the opposite chirality operators $\tilde{Q}_i$, we stress that, while in the SM the Wilson coefficients $\tilde{C}_i$'s identically vanish, in SUSY models chargino and gluino exchanges could sizably affect these coefficients. However, in the case of charginos, these effects are quite small, being proportional to the Yukawa couplings of the light quarks [210], so that we will not include them in our analysis. Moreover, we have also neglected the small contributions to $C_{1,2}^\chi$ coming from box diagrams, where both charginos and gluinos are exchanged [209, 210].

12.2.1 Chargino contribution

As mentioned above, in the super-CKM basis, the interacting Lagrangian involving charginos is given by

$$\mathcal{L}_{q\tilde{q}\tilde{\chi}^+} = -g \sum_k \sum_{a,b} \left(V_{k1} K_{ba}^* \, \bar{d}_L^a \, (\tilde{\chi}_k^+)^* \, \tilde{u}_L^b - U_{k2}^* \, (Y_d^{\mathrm{diag}}.K^+)_{ab} \, \bar{d}_R^a \, (\tilde{\chi}_k^+)^* \, \tilde{u}_L^b \right.$$
$$\left. - V_{k2} \, (K.Y_u^{\mathrm{diag}})_{ab} \, \bar{d}_L^a \, (\tilde{\chi}_k^+)^* \, \tilde{u}_R^b \right), \tag{12.61}$$

where $q_{R,L} = \frac{1}{2}(1 \pm \gamma_5)q$ and contraction of colour and Dirac indices is understood. Here $Y_{u,d}^{\mathrm{diag}}$ are the diagonal Yukawa matrices and K stands for the CKM matrix. Again, the indices a, b and k label flavour and chargino mass eigenstates, respectively, while V, U are the chargino mixing matrices. As one can see from this equation, the Higgsino couplings are suppressed by Yukawas of the light quarks and therefore they are negligible, except for the stop-bottom interaction, which is directly enhanced by the top Yukawa (Y_t).

At the first order in the MIA, the penguin and box diagrams which contribute to the $\Delta B = 1$ effective Hamiltonian are given in Figs. 12.1 and 12.2, respectively. Evaluating the diagrams in Figs. 12.1 and 12.2 by retaining only terms proportional to bottom- and top-quark Yukawa couplings and performing the matching, the chargino contributions to

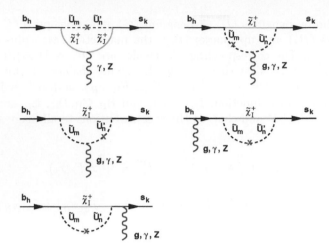

Figure 12.1 Penguin diagrams for $\Delta B = 1$ transitions with chargino (χ_I^+) exchanges at the first order in mass insertion. Here $\tilde{U}, \tilde{U}' = \{\tilde{u}, \tilde{c}, \tilde{t}\}$, with indices $h, k, m, n = \{L, R\}$ and $I, J = \{1, 2\}$. The cross symbol in the squark propagator indicates the mass insertion. The corresponding diagrams at zero order in mass insertion are simply obtained by removing the mass insertion in the propagators of up-type squarks $(\tilde{U})$.

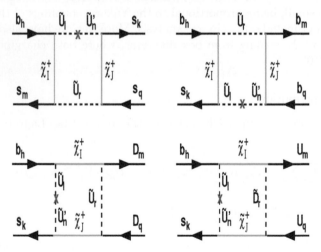

Figure 12.2 Box diagrams for $\Delta B = 1$ transitions with chargino exchanges at the first order in mass insertion, where $\tilde{U}, \tilde{U}' = \{\tilde{t}, \tilde{c}, \tilde{u}\}$, $U = \{c, u\}$, and $D = \{b, s, d\}$, where $h, k, l, n, r, m, q = \{L, R\}$.

the Wilson coefficients in Eq. (12.57) can be determined from the following relations [210]

$$
\begin{aligned}
F^\chi &= \Big[\sum_{a,b} K_{a2}^\star K_{b3} (\delta_{LL}^u)_{ba} \Big] R_F^{LL} + \Big[\sum_a K_{a2}^\star K_{33} (\delta_{RL}^u)_{3a} \Big] Y_t\, R_F^{RL} \\
&+ \Big[\sum_a K_{32}^\star K_{a3} (\delta_{LR}^u)_{a3} \Big] Y_t\, R_F^{LR} + \Big[K_{32}^\star K_{33} \big((\delta_{RR}^u)_{33} R_F^{RR} + R_F^0 \big) \Big] Y_t^2,
\end{aligned}
$$

$$(12.62)$$

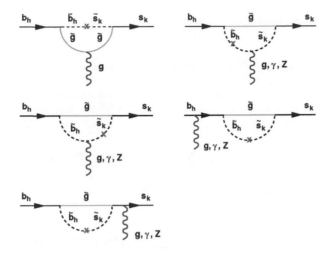

Figure 12.3 Penguin diagrams for $\Delta B = 1$ transitions with gluino exchanges at the first order in mass insertion, where $h, k = \{L, R\}$.

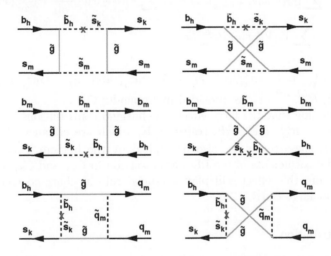

Figure 12.4 Box diagrams for $\Delta B = 1$ transitions with gluino exchanges at the first order in mass insertion, where $q = \{b, c, s, d, u\}$ and $h, k, m = \{L, R\}$.

where the symbol $F \equiv \{D, C, E, B_d, B_u, M_\gamma, M_g\}$ and the detailed expressions for R_F^{LL}, R_F^{LR}, R_F^{RL}, R_F^{RR} and R_F^0 can be found in Appendix C. Notice that the last term in Eq. (12.62), proportional to R_F^0, is independent of mass insertions. This is due to the fact that, for chargino exchanges, the super-GIM mechanism is only partially effective, when only squarks but not quarks are taken to be degenerate.

Here, we will just concentrate on the dominant contributions, which turn out to be due to the chromo-magnetic (M_g) penguin and Z-penguin (C) diagrams [210]. From the above expressions, it is clear that LR and RR contributions are suppressed by order λ^2 or λ^3, where $\lambda = \sin \theta_c \simeq 0.22$, with θ_c the Cabibbo angle. In our analysis we adopt the approximation of retaining only terms proportional to order $\lambda = \sin \theta_c$. In this case, Eq. (12.62) simplifies as follows [210]:

$$F^X = \xi_{LL} R_F^{LL} + Y_t \, \xi_{RL} R_F^{RL}, \tag{12.63}$$

where $\xi_{LL} = (\delta_{LL}^u)_{32} + \lambda \, (\delta_{LL}^u)_{31}$ and $\xi_{RL} = (\delta_{RL}^u)_{32} + \lambda \, (\delta_{RL}^u)_{31}$.

The functions R_F^{LL} and R_F^{RL} depend on the SUSY parameters through the chargino masses (m_{χ_i}), squark masses $(\tilde{m})$ and the entries of the chargino mass matrix. For instance, for Z and magnetic (chromo-magnetic) dipole penguins $R_C^{LL,RL}$ and $R_{M_{\gamma(g)}}^{LL,RL}$, respectively, we have

$$R_C^{LL} = \sum_{i=1,2} |V_{i1}|^2 P_C^{(0)}(\bar{x}_i) + \sum_{i,j=1,2} \Big[U_{i1} V_{i1} U_{j1}^\star V_{j1}^\star P_C^{(2)}(x_i, x_j) \tag{12.64}$$

$$+ |V_{i1}|^2 |V_{j1}|^2 \Big(\frac{1}{8} - P_C^{(1)}(x_i, x_j) \Big) \Big], \tag{12.65}$$

$$R_C^{RL} = -\frac{1}{2} \sum_{i=1,2} V_{i2}^\star V_{i1} P_C^{(0)}(\bar{x}_i) - \sum_{i,j=1,2} V_{j2}^\star V_{i1} \Big(U_{i1} U_{j1}^\star P_C^{(2)}(x_i, x_j)$$

$$+ V_{i1}^\star V_{j1} P_C^{(1)}(x_i, x_j) \Big), \tag{12.66}$$

$$R_{M_{\gamma,g}}^{LL} = \sum_i |V_{i1}|^2 \, x_{Wi} \, P_{M_{\gamma,g}}^{LL}(x_i) - Y_b \sum_i V_{i1} U_{i2} \, x_{Wi} \, \frac{m_{\chi_i}}{m_b} P_{M_{\gamma,g}}^{LR}(x_i), \tag{12.67}$$

$$R_{M_{\gamma,g}}^{RL} = -\sum_i V_{i1} V_{i2}^\star \, x_{Wi} \, P_{M_{\gamma,g}}^{LL}(x_i), \tag{12.68}$$

where $x_{Wi} = M_W^2/m_{\chi_i}^2$, $x_i = m_{\chi_i}^2/\tilde{m}^2$, $\bar{x}_i = \tilde{m}^2/m_{\chi_i}^2$, and $x_{ij} = m_{\chi_i}^2/m_{\chi_j}^2$. The loop functions $P_C^{(1,2)}(x, y)$, $P_{M_{\gamma,g}}^{LL(LR)}(x)$ are provided in Appendix C.

It is worth mentioning that the large effects of chargino contributions to $C_{7\gamma}$ and C_{8g} come from the terms in $R_{M_\gamma}^{LL}$ and $R_{M_g}^{LL}$, respectively, which are enhanced by m_{χ_i}/m_b in Eq. (12.68). However, these terms are also multiplied by the bottom Yukawa Y_b, which leads to enhancing the coefficients of the LL mass insertion in $C_{7\gamma}$ and C_{8g} at large $\tan\beta$. As we will see later on, this effect will play a crucial role in chargino contributions to $B \to \phi(\eta')K$ decays at large $\tan\beta$.

12.2.2 Gluino contribution

Now let us turn to the gluino contributions in the $b \to s$ transition. In the super-CKM basis, the quark-squark-gluino interaction is given by:

$$\mathcal{L}_{dd\tilde{g}} = \sqrt{2} \, g_s \, T_{\alpha\beta}^A \Big[(\bar{d}^\alpha \, P_L \, \tilde{g}^A) \, \tilde{d}_R^\beta - (\bar{d}^\alpha \, P_R \, \tilde{g}^A) \, \tilde{d}_L^\beta + \text{h.c.} \Big], \tag{12.69}$$

where $\tilde{g}^A$ are the gluino Majorana fields, $\tilde{d}_{R,L}^\beta$ are the squark fields, T^A are the $SU(3)_C$ generators and α, β are colour indices. The dominant gluino contributions are due to the QCD penguin diagrams, the magnetic and chromo-magnetic dipole operators. At the first order in MIA, the penguin and box diagrams are shown in Figs. 12.3 and 12.4, respectively. Performing the matching, the gluino contributions to the corresponding Wilson coefficients at the SUSY scale are given by [18]

$$C_3^{\tilde{g}} = -\frac{\alpha_s^2}{2\sqrt{2}G_F m_{\tilde{q}}^2}(\delta_{LL}^d)_{23}\left[-\frac{1}{9}B_1(x) - \frac{5}{9}B_2(x) - \frac{1}{18}P_1(x) - \frac{1}{2}P_2(x)\right], \tag{12.70}$$

$$C_4^{\tilde{g}} = -\frac{\alpha_s^2}{2\sqrt{2}G_F m_{\tilde{q}}^2}(\delta_{LL}^d)_{23}\left[-\frac{7}{3}B_1(x) + \frac{1}{3}B_2(x) + \frac{1}{6}P_1(x) + \frac{3}{2}P_2(x)\right], \tag{12.71}$$

$$C_5^{\tilde{g}} = -\frac{\alpha_s^2}{2\sqrt{2}G_F m_{\tilde{q}}^2}(\delta_{LL}^d)_{23}\left[\frac{10}{9}B_1(x) + \frac{1}{18}B_2(x) - \frac{1}{18}P_1(x) - \frac{1}{2}P_2(x)\right], \tag{12.72}$$

$$C_6^{\tilde{g}} = -\frac{\alpha_s^2}{2\sqrt{2}G_F m_{\tilde{q}}^2}(\delta_{LL}^d)_{23}\left[-\frac{2}{3}B_1(x) + \frac{7}{6}B_2(x) + \frac{1}{6}P_1(x) + \frac{3}{2}P_2(x)\right], \tag{12.73}$$

$$C_{7\gamma}^{\tilde{g}} = \frac{8\alpha_s\pi}{9\sqrt{2}G_F m_{\tilde{q}}^2}\left[(\delta_{LL}^d)_{23}M_3(x) + (\delta_{LR}^d)_{23}\frac{m_{\tilde{g}}}{m_b}M_1(x)\right], \tag{12.74}$$

$$C_{8g}^{\tilde{g}} = \frac{\alpha_s\pi}{\sqrt{2}G_F m_{\tilde{q}}^2}\left[(\delta_{LL}^d)_{23}\left(\frac{1}{3}M_3(x) + 3M_4(x)\right) + (\delta_{LR}^d)_{23}\frac{m_{\tilde{g}}}{m_b}\left(\frac{1}{3}M_1(x) + 3M_3(x)\right)\right], \tag{12.75}$$

where $\tilde{C}_{i,8g}$ are obtained from $C_{i,8g}$ by exchanging $L \leftrightarrow R$ in $(\delta_{AB}^d)_{23}$. The functions appearing in these expressions can be found in Appendix C, with $x = m_{\tilde{g}}^2/m_{\tilde{q}}^2$. As for charginos in Eq. (12.68), the term proportional to $(\delta_{LR}^d)_{23}$ in $C_{7\gamma,8g}^{\tilde{g}}$ in Eq. (12.75) has the large enhancement factor $m_{\tilde{g}}/m_b$ in front. Moreover, contrary to the chargino case, this term is not suppressed by the bottom Yukawa coupling. This enhancement factor will be responsible for the dominant gluino effects in $B \to \phi(\eta')K$ decays.

12.3 SUSY CONTRIBUTIONS TO CP ASYMMETRY OF B MESONS

Here we analyse the supersymmetric contributions to the time dependent CP asymmetries in $B \to \phi K_S$ and $B \to \eta'K_S$ decays in the MIA framework, in gluino and chargino dominated scenarios.

New physics could in principle affect the B meson decay by means of a new source of CP-violating phase in the corresponding amplitude. In general, this phase is different from the corresponding SM one. If so, then deviations on CP asymmetries from the SM expectations can be sizable, depending on the relative magnitude of SM and NP amplitudes. For instance, in the SM the $B \to \phi K_S$ decay amplitude is generated at one loop and therefore it is very sensitive to NP contributions. In this respect, SUSY models with non-minimal flavour structure and new CP-violating phases in the squark mass matrices can easily generate large deviations in the $B \to \phi K_S$ asymmetry.

The time dependent CP asymmetry for $B \to \phi K_S$ can be described by

$$a_{\phi K_S}(t) = \frac{\Gamma(\overline{B}^0(t) \to \phi K_S) - \Gamma(B(t) \to \phi K_S)}{\Gamma(\overline{B}^0(t) \to \phi K_S) + \Gamma(B(t) \to \phi K_S)}$$
$$= C_{\phi K_S}\cos\Delta M_{B_d}t + S_{\phi K_S}\sin\Delta M_{B_d}t, \tag{12.76}$$

where $C_{\phi K_S}$ and $S_{\phi K_S}$ represent the direct and the CP mixing asymmetry, respectively, and they are given by

$$C_{\phi K_S} = \frac{|\overline{\rho}(\phi K_S)|^2 - 1}{|\overline{\rho}(\phi K_S)|^2 + 1}, \quad S_{\phi K_S} = \frac{2\mathrm{Im}\left[\frac{q}{p}\overline{\rho}(\phi K_S)\right]}{|\overline{\rho}(\phi K_S)|^2 + 1}. \tag{12.77}$$

The parameter $\bar{\rho}(\phi K_S)$ is defined by

$$\bar{\rho}(\phi K_S) = \frac{\overline{A}(\phi K_S)}{A(\phi K_S)}, \tag{12.78}$$

where $\overline{A}(\phi K_S)$ and $A(\phi K_S)$ are the decay amplitudes of $\overline{B}^0$ and B^0 mesons, which can be written in terms of the matrix element of the $\Delta B = 1$ transition as

$$\overline{A}(\phi K_S) = \langle \phi K_S | H_{eff}^{\Delta B=1} | \overline{B}^0 \rangle, \quad A(\phi K_S) = \langle \phi K_S | \left(H_{eff}^{\Delta B=1} \right)^\dagger | B^0 \rangle. \tag{12.79}$$

In order to simplify our analysis, it is useful to parameterise the SUSY effects by introducing the ratio of SM and SUSY amplitudes as follows:

$$\left(\frac{A^{\text{SUSY}}}{A^{\text{SM}}} \right)_{\phi K_S} \equiv R_\phi \, e^{i\theta_\phi} \, e^{i\delta_\phi}, \tag{12.80}$$

and analogously for the $\eta' K_S$ decay mode

$$\left(\frac{A^{\text{SUSY}}}{A^{\text{SM}}} \right)_{\eta' K_S} \equiv R_{\eta'} \, e^{i\theta_{\eta'}} \, e^{i\delta_{\eta'}}, \tag{12.81}$$

where R_i stands for the corresponding absolute values of $|\frac{A^{\text{SUSY}}}{A^{\text{SM}}}|$, the angles $\theta_{\phi, \eta'}$ are the corresponding SUSY CP-violating phases while $\delta_{\phi, \eta'} = \delta_{\phi, \eta'}^{SM} - \delta_{\phi, \eta'}^{SUSY}$ parameterise the strong (CP conserving) ones. In this case, the mixing CP asymmetry $S_{\phi K_S}$ in Eq. (12.76) takes the form

$$S_{\phi K_S} = \frac{\sin 2\beta + 2R_\phi \cos \delta_\phi \sin(\theta_\phi + 2\beta) + R_\phi^2 \sin(2\theta_\phi + 2\beta)}{1 + 2R_\phi \cos \delta_\phi \cos \theta_\phi + R_\phi^2}, \tag{12.82}$$

and, analogously for $B \to \eta' K_S$, one also has

$$S_{\eta' K_S} = \frac{\sin 2\beta + 2R_{\eta'} \cos \delta_{\eta'} \sin(\theta_{\eta'} + 2\beta) + R_{\eta'}^2 \sin(2\theta_{\eta'} + 2\beta)}{1 + 2R_{\eta'} \cos \delta_{\eta'} \cos \theta_{\eta'} + R_{\eta'}^2}. \tag{12.83}$$

Assuming that the SUSY contribution to the amplitude is smaller than the SM one, i.e., $R_{\phi(\eta')} \ll 1$, one can simplify the above expressions as:

$$S_{\phi(\eta')K_S} = \sin 2\beta + 2 \cos 2\beta \sin \theta_{\phi(\eta')} \cos \delta_{\phi(\eta')} R_{\phi(\eta')} + \mathcal{O}(R_{\phi,(\eta')}^2). \tag{12.84}$$

It is now clear that, in order to render $S_{\phi K_S}$ smaller than $\sin 2\beta$, the relative sign of $\sin \theta_\phi$ and $\cos \delta_\phi$ has to be negative. If one assumes that $\sin \theta_\phi \cos \delta_\phi \simeq -1$, then $R_\phi \geq 0.1$ is required in order to get $S_{\phi K_S}$ within 1σ of the experimental range.

12.3.1   CP asymmetry in $B \to \phi K_S$

The world average measurement of the CP asymmetry in $B \to \phi K_S$ is given by

$$S_{\phi K_S} = 0.34 \pm 0.20, \tag{12.85}$$

which is about a 2σ deviation from the SM expectation. Several studies have been performed to explain such deviations in SUSY models [20, 210–221]. It was emphasised that

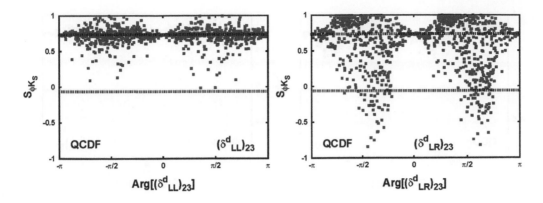

Figure 12.5 $S_{\Phi K_S}$ as a function of $\arg[(\delta^d_{LL})_{23}]$ and $\arg[(\delta^d_{LR})_{23}]$ with the gluino contribution of one mass insertion $(\delta^d_{LL})_{23}$ and $(\delta^d_{LR})_{23}$. The region inside the two horizontal lines corresponds to the allowed experimental region at 2σ level (from [20]).

the amplitude of $B \to \phi K$ can be parameterised in terms of the Wilson coefficients as follows [20]:

$$A(B \to \phi K) = -i\frac{G_F}{\sqrt{2}}m_B^2 F_+^{B \to K} f_\phi \sum_{i=1..10,7\gamma,8g} H_i(\phi)(\mathbf{C}_i + \tilde{\mathbf{C}}_i), \qquad (12.86)$$

where $F_+^{B \to K}$ is the transition form factor, which is given by $F_+^{B \to K} = 0.35 \pm 0.05$, and f_ϕ is the decay constant of the ϕ meson, which is given by $f_\phi = 0.233$ GeV. The numerical values of $H_i(\phi)$ can be calculated using naive [222] or QCDF [223, 224] for the hadronic matrix elements of the B meson. In Appendix D, we give an example of $H_i(\phi)$ that corresponds to QCDF and fixes all relevant parameters with their central values.

In this regard, one can compute SUSY contributions to the $B \to \phi K_S$ process and check their impact on the CP asymmetry $S_{\Phi K_S}$ and the BR($B \to \phi K_S$). The numerical results for the gluino contributions to the CP asymmetry $S_{\Phi K_S}$ are presented in Fig. 12.5. In these plots, regions inside the horizontal lines indicate the allowed 2σ experimental range. Only one mass insertion per time is taken active, in particular, this means that we scanned over $|(\delta^d_{LL})_{23}| < 1$ and $|(\delta^d_{LR})_{23}| < 1$. Then, $S_{\Phi K_S}$ is plotted versus θ_ϕ, which in the case of one dominant mass insertion should be identified here as $\theta_\phi = \arg[(\delta^d_{AB})_{ij}]$. We have scanned over the relevant SUSY parameter space, in this case the average squark mass $\tilde{m}$ and gluino mass $m_{\tilde{g}}$, assuming SM central values [174]. In addition, we imposed that the BR of $b \to s\gamma$ and the $B - \bar{B}$ mixing are satisfied at 95% CL [225], namely $2 \times 10^{-4} \leq \text{BR}(b \to s\gamma) < 4.5 \times 10^{-4}$.

As can be seen from this figure, only the gluino contributions proportional to $(\delta^d_{LR})_{23}$ have chances to drive $S_{\Phi K_S}$ towards the region of large and negative values, while the pure effect of $(\delta^d_{LL})_{23}$ just approaches the negative values region. This result can be easily understood by noticing that the dominant SUSY source of the $B \to \phi K_S$ decay amplitude is provided by the chromo-magnetic operator Q_{8g}. As mentioned above, the gluino contribution to C_{8g}, which is proportional to $(\delta^d_{LR})_{23}$, can be very large with respect to the SM ones, being enhanced by terms of order $m_{\tilde{g}}/m_b$. In addition, large gluino effects in C_{8g} may still escape $b \to s\gamma$ constraints [226].

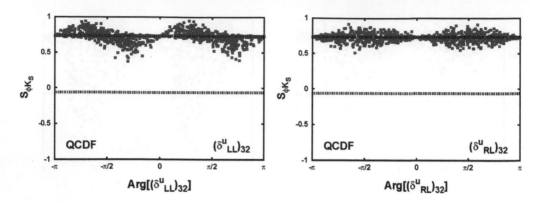

Figure 12.6 $S_{\Phi K_S}$ as a function of $\arg[(\delta^u_{LL})_{32}]$ and $\arg[(\delta^u_{RL})_{32}]$ with the chargino contribution of one mass insertion $(\delta^u_{LL})_{32}$ and $(\delta^u_{RL})_{32}$ (from [20]).

Chargino effects on the CP asymmetry $S_{\phi K_S}$ are summarised in Fig. 12.6. In this figure, $S_{\phi K_S}$ is plotted versus the argument of the relevant chargino mass insertions, namely, $(\delta^u_{LL})_{32}$ and $(\delta^u_{RL})_{32}$. As in the gluino dominated scenario, we have scanned over the relevant SUSY parameter space, in particular, the average squark mass $\tilde{m}$, the weak gaugino mass M_2, the μ term and the light right stop mass $\tilde{m}_{\tilde{t}_R}$. Further, $\tan\beta = 40$ has been assumed. We also scanned over the real and imaginary parts of the mass insertions $(\delta^u_{LL})_{32}$ and $(\delta^u_{RL})_{32}$, by considering the constraints on BR($b \to s\gamma$) and $B - \bar{B}$ mixing at 95% CL. The $b \to s\gamma$ constraints impose stringent bounds on $(\delta^u_{LL})_{32}$, especially at large $\tan\beta$ [210].

The reason why extensive regions of negative values for $S_{\phi K_S}$ are excluded here is only due to the $b \to s\gamma$ constraints. As shown in [210], the inclusion of $(\delta^u_{LL})_{32}$ mass insertion can generate large and negative values of $S_{\phi K_S}$ via the chargino contributions to chromomagnetic operator Q_{8g} which are enhanced by terms of order $m_{\chi^\pm}/m_b$. However, contrary to the gluino scenario, the ratio $|C_{8g}/C_{7\gamma}|$ is not enhanced by colour factors and large contributions to C_{8g} lead unavoidably to the breaking of $b \to s\gamma$ constraints. Moreover, the contribution of $(\delta^u_{RL})_{32}$ is independent of $\tan\beta$ and so large effects in R_ϕ that could drive $S_{\phi K_S}$ towards the region of negative values cannot be achieved.

12.3.2 CP asymmetry in $B \to \eta' K_S$

The measurements of the CP asymmetry in $B \to \eta' K_S$ show another discrepancy with respect to SM predictions. In particular, the world average of the CP asymmetry $S_{\eta' K_S}$ is given by

$$S_{\eta' K_S} = 0.41 \pm 0.11, \tag{12.87}$$

which is about a 2.5σ deviation from the SM expectations. From these results we see that, similarly to what happens in the $B \to \phi K_S$ decay, large deviations from the SM are possible. As in the case of $B \to \phi K$, the amplitude of $B \to \eta' K_S$ can be parameterised in terms of the Wilson coefficients as

$$A(B \to \eta' K) = -i\frac{G_F}{\sqrt{2}}m_B^2 F_+^{B\to K} f_{\eta'}^s \sum_{i=1..10,7\gamma,8g} H_i(\eta')(\mathbf{C}_i - \tilde{\mathbf{C}}_i), \tag{12.88}$$

where the decay constant $f_{\eta'}^s$ is given by $f_{\eta'}^s = 174.2$. The numerical values of $H_i(\eta')$ that correspond to QCDF conditions can be found in Appendix D. Note that the sign difference between $\mathbf{C}_i$ and $\tilde{\mathbf{C}}_i$ appearing in Eq. (12.88) is due to the fact that, contrary to the $B \to \phi K$ transition, initial and final states have opposite parity here. Thus, due to the invariance of strong interactions under parity transformations, only $V \times A$ or $A \times V$ structures of four-fermion operators will contribute to the hadronic matrix elements, so that $\langle \eta' K | Q_i | B \rangle = -\langle \eta' K | \tilde{Q}_i | B \rangle$.

Since SUSY contributes to both CP asymmetries $S_{\phi K_S}$ and $S_{\eta' K_S}$ with the same CP-violating source, it is possible that the SUSY effects driving $S_{\phi K_S}$ towards negative values could also sizably decrease $S_{\eta' K_S}$. The main reason for that is because the leading SUSY contributions to the amplitudes of $B \to \phi K_S$ and $B \to \eta' K_S$ enter through the Wilson coefficient C_{8g} and the operator Q_{8g} has comparable matrix elements in both processes. However, since NP corrections enter through the quantity $R_{\eta'}$, the role of the SM contribution will be crucial. Indeed, while the $B \to \phi K_S$ amplitude is generated purely at one-loop in the SM, the $B \to \eta' K_S$ amplitude receives tree-level contributions from the SM by means of a non-vanishing matrix element of Q_2. Therefore, the increase of SUSY contributions to C_{8g} is now compensated in $R_{\eta'}$ by the large SM amplitude contribution.

We show the results for gluinos in Fig. 12.7, where the same analysis of $B \to \phi K_S$ is extended. The same conventions as in the figures for $B \to \phi K_S$ have been adopted here. As we can see from these results, there is a depletion of the gluino contribution in $S_{\eta' K_S}$ precisely for the reasons explained above. Regions of negative values of $S_{\eta' K_S}$ are more disfavoured with respect to $S_{\phi K_S}$, but a minimum of $S_{\eta' K_S} \simeq 0$ can easily be achieved.

Finally, in Fig. 12.8 we present our results for chargino contributions. Here we see that charginos can produce at most a deviation from the SM predictions of about $\pm 20\%$ and the most conspicuous effect is achieved by $(\delta_{LL}^u)_{32}$. These results again show the relevant role played by the chromo-magnetic operator.

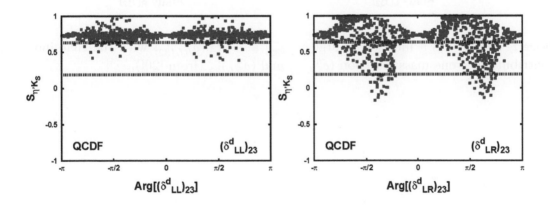

Figure 12.7 $S_{\eta' K_S}$ as a function of $\arg[(\delta_{LL}^d)_{23}]$ and $\arg[(\delta_{LR}^d)_{23}]$ with the gluino contribution of one mass insertion $(\delta_{LL}^d)_{23}$ and $(\delta_{LR}^d)_{23}$. The region inside the two horizontal lines corresponds to the allowed experimental region at 2σ level (from [20]).

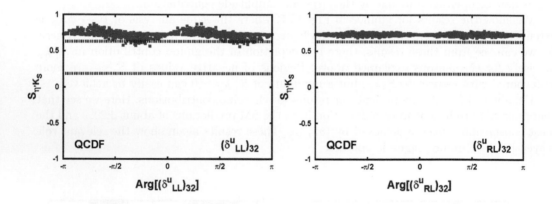

Figure 12.8 $S_{\Phi K_S}$ as a function of $\arg[(\delta^u_{LL})_{32}]$ and $\arg[(\delta^u_{RL})_{32}]$ with the chargino contribution of one mass insertion $(\delta^u_{LL})_{32}$ and $(\delta^u_{RL})_{32}$ (from [20]).

IV

MSSM Extensions

CHAPTER 13

Neutrino Masses in Supersymmetric Models and Seesaw Mechanisms

Neutrinos are massless in the MSSM. However, from neutrino oscillations, we have the following experimental values for the mixing angles [227]:

$$\sin^2\theta_{12} = 0.307^{+0.018}_{-0.016}, \tag{13.1}$$

$$\sin^2\theta_{23} = 0.38^{+0.024}_{-0.021}, \tag{13.2}$$

$$\sin^2\theta_{13} = 0.0244 \pm 0.0025. \tag{13.3}$$

From neutrino oscillations we can measure the neutrino squared mass differences only. The latest measurements lead to

$$\Delta m^2_{21} = 7.54^{+0.26}_{-0.22} \times 10^{-5} \text{eV}^2, \tag{13.4}$$

$$\Delta m^2_{32} = 2.43^{+0.1}_{-0.06} \times 10^{-3} \text{eV}^2. \tag{13.5}$$

In this chapter we review several possible modifications of the MSSM in order to overcome this problem. The first two sections are based on the review in [228].

13.1 NEUTRINO MASSES

The most straightforward way to generate neutrino masses is to introduce a new MSSM singlet superfield $\hat{N}$, whose fermionic member accounts for a RH neutrino ν_R. This allows the formulation of the Yukawa term $Y_\nu L N^c H_u$ in the superpotential. After EWSB, a Dirac mass term

$$m_D \bar{\nu}_L \nu_R \tag{13.6}$$

is formed, where $m_D = Y_\nu v_u$. Now it is worth mentioning that, in order to get the tiny neutrino masses given by the experimental results, we should impose the condition that the neutrino Yukawa coupling must be small, $Y_\nu \sim \mathcal{O}(10^{-11})$, which is far from the smallest known Yukawa coupling $Y_e \sim \mathcal{O}(10^{-6})$. This is the reason why physicists does not favour this scenario, though.

If neutrinos are Majorana particles, their mass at low energy is described by a unique dimension-5 operator [229]

$$m_\nu = \frac{f}{\Lambda}(HL)(HL). \tag{13.7}$$

Using only renormalisable interactions, there are exactly three tree-level models leading to this operator [230]. The first one is the exchange of a heavy fermionic singlet, called the RH neutrino. This is the celebrated seesaw mechanism [231–235], nowadays called seesaw Type I. The second possibility is the exchange of a scalar $SU(2)_L$ triplet [236,237]. This is commonly known as seesaw Type II. And lastly, one could also add one (or more) fermionic triplets to the field content of the SM [238]. This is known as seesaw Type III. The seesaw mechanisms provide a rationale for the observed smallness of neutrino masses, by the introduction of the inverse of some large scale Λ. In seesaw Type I, for example, Λ is equal to the mass(es) of the RH neutrinos. Since these are $SU(2)_L$ singlets, their masses can take any value, and with neutrino masses as indicated by the results from oscillation experiments, $m_\nu \sim \sqrt{\Delta m_A^2} \sim 0.05$ eV, where Δm_A^2 is the atmospheric neutrino mass splitting, and couplings of order $\mathcal{O}(1)$, the scale of the seesaw is estimated to be very large, roughly $m_R \sim 10^{15}$ GeV. This value is close to, but slightly lower than the GUT scale. In addition, there exist seesaw models with large couplings at the EW scale, such as the linear [239] and inverse [240] seesaw models.

13.2   SEESAW MECHANISMS

In this section, we discuss different implementations of the seesaw mechanism. As mentioned above, the aim of the seesaw mechanism is to explain the neutrino masses and mixing angles. This is done by linking the tiny masses to other parameters which are of the naturally expected order. The general idea can be summarised by writing down the most general mass matrix combining LH neutrino (L), RH neutrino (R) and additional singlet fields carrying lepton number (S) [228]:

$$\begin{pmatrix} m_{LL} & m_{LR} & m_{LS} \\ m_{LR}^T & m_{RR} & m_{RS} \\ m_{LS}^T & m_{RS}^T & m_{SS} \end{pmatrix}. \tag{13.8}$$

Looking at specific limits of this matrix, we can recover the different seesaw realisations: $m_{LL} = m_{LS} = m_{RS} = 0$ leads to Type I. Type III is obtained in the same limit, as Type II, but with m_{RR} stemming from $SU(2)_L$ triplets. Then $m_{LL} = m_{RR} = m_{LS} = 0$ is the characteristic matrix for inverse seesaw, while $m_{LL} = m_{RR} = m_{SS} = 0$ is the standard parameterisation of the linear seesaw. What most of these different seesaw models have in common is the way the tiny neutrino masses are recovered, just by suppressing them with very high scales for the new fields. This is strictly true for the Type I/II/III models. The linear and inverse seesaw versions work slightly differently: the heaviness of the new fields is reduced at the price of introducing a relatively small dimensionful parameter, usually connected to an explicit violation of the lepton number.

In the following subsection we discuss models which can explain the origin of the distinct neutrino mass matrices.

13.2.1   Type I/II/III

The simplest seesaw models describe neutrino masses with an effective operator arising after integrating out heavy superfields. In order to maintain gauge unification, it would be convenient to add complete multiplets of a GUT group. For example, in the case of $SU(5)$, the seesaw Type II can arise in models with $\mathbf{15}$ and $\overline{\mathbf{15}}$, while for Type III $\mathbf{24}$ has to be considered. In Tab. 13.1 we introduce different models with minimal addition of superfields that account for seesaw mechanisms.

Type I					
					R-parity
SF	Spin 0	Spin 1/2	Generations	$U(1)_Y \times SU(2)_L \times SU(3)_C$	of fermion
N^c	$\tilde{\nu}^c$	ν^c	n_1	$(0, \mathbf{1}, \mathbf{1})$	$+$
Type II					
					R-parity
SF	Spin 0	Spin 1/2	Generations	$U(1)_Y \times SU(2)_L \times SU(3)_C$	of fermion
$\hat{T}$	$\tilde{T}$	T	n_{15}	$(1, \mathbf{3}, \mathbf{1})$	$-$
$\hat{\bar{T}}$	$\tilde{\bar{T}}$	$\bar{T}$	n_{15}	$(-1, \mathbf{3}, \mathbf{1})$	$-$
$\hat{S}$	$\tilde{S}$	S	n_{15}	$(-\frac{2}{3}, \mathbf{1}, \mathbf{6})$	$-$
$\hat{\bar{S}}$	$\tilde{\bar{S}}^*$	$\bar{S}^*$	n_{15}	$(\frac{2}{3}, \mathbf{1}, \bar{\mathbf{6}})$	$-$
$\hat{Z}$	$\tilde{Z}$	Z	n_{15}	$(\frac{1}{6}, \mathbf{2}, \mathbf{3})$	$-$
$\hat{\bar{Z}}$	$\tilde{\bar{Z}}$	$\bar{Z}$	n_{15}	$(-\frac{1}{6}, \mathbf{2}, \bar{\mathbf{3}})$	$-$
Type III					
					R-parity
SF	Spin 0	Spin 1/2	Generations	$U(1)_Y \times SU(2)_L \times SU(3)_C$	of fermion
$\hat{W}_M$	$\tilde{W}_M$	W_M	n_{24}	$(0, \mathbf{3}, \mathbf{1})$	$+$
$\hat{G}_M$	$\tilde{G}_M$	G_M	n_{24}	$(0, \mathbf{1}, \mathbf{8})$	$+$
$\hat{B}_M$	$\tilde{B}_M$	B_M	n_{24}	$(0, \mathbf{1}, \mathbf{1})$	$+$
$\hat{X}_M$	$\tilde{X}_M$	X_M	n_{24}	$(\frac{5}{6}, \mathbf{2}, \bar{\mathbf{3}})$	$+$
$\hat{\bar{X}}_M$	$\tilde{\bar{X}}_M$	$\bar{X}_M$	n_{24}	$(-\frac{5}{6}, \mathbf{2}, \mathbf{3})$	$+$

Table 13.1 New chiral superfields appearing in the effective Type I/II/III seesaw models. While $n_{15} = 1$ is sufficient to explain neutrino data, n_1 and n_{24} must be at least 2.

The combined superpotential of all three types can be written as

$$W = W_{\text{MSSM}} + W_{\text{I}} + W_{\text{II}} + W_{\text{III}}, \tag{13.9}$$

where

$$W_{\text{I}} = Y_\nu N^c L H_u + \frac{1}{2} M_{\nu^c} N^c N^c, \tag{13.10}$$

$$W_{\text{II}} = \frac{1}{\sqrt{2}} Y_T L \hat{T} L + \frac{1}{\sqrt{2}} Y_S D^c \hat{S} D^c + Y_{\hat{Z}} D^c \hat{Z} L + \frac{1}{\sqrt{2}} \lambda_1 H_d \hat{T} H_d$$

$$+ \frac{1}{\sqrt{2}} \lambda_2 H_u \hat{\bar{T}} H_u + M_T \hat{T} \hat{\bar{T}} + M_{\hat{Z}} \hat{Z} \hat{\bar{Z}} + M_S \hat{S} \hat{\bar{S}}, \tag{13.11}$$

$$W_{\text{III}} = \sqrt{\frac{3}{10}} Y_B\, H_u\, \hat{B}_M\, L + Y_{\hat{W}}\, H_u\, \hat{W}_M\, L + Y_X\, H_u\, \hat{X}_M\, D^c + M_X\, \hat{X}_M\, \hat{X}_M$$

$$+ \frac{1}{2} M_{\hat{W}}\, \hat{W}_M\, \hat{W}_M + \frac{1}{2} M_G\, \hat{G}_M\, \hat{G}_M + \frac{1}{2} M_B\, \hat{B}_M\, \hat{B}_M. \qquad (13.12)$$

The soft-breaking terms can be split into three categories: terms stemming from the superpotential couplings when replacing the fermions with their scalar superpartners ($L_{SB,W}$), the scalar soft-breaking masses for each chiral superfield ($L_{SB,\phi}$) and the soft-breaking masses for the gauginos ($L_{SB,\lambda}$). Since the gauge sector is not modified, $L_{SB,\lambda}$ reads as in the MSSM. The soft-breaking terms stemming from the superpotential are

$$\mathcal{L}_{\text{soft}} = \mathcal{L}_{\text{soft}}^{\text{MSSM}} + \mathcal{L}_{\text{soft}}^{\text{I}} + \mathcal{L}_{\text{soft}}^{\text{II}} + \mathcal{L}_{\text{soft}}^{\text{III}},$$

where

$$\mathcal{L}_{\text{soft}}^{\text{I}} = T_\nu\, \tilde{\nu}^c\, \tilde{L}\, H_u + \frac{1}{2} B_{\nu^c}\, \tilde{\nu}^c\, \tilde{\nu}^c + \text{h.c.}, \qquad (13.13)$$

$$\mathcal{L}_{\text{soft}}^{\text{II}} = \frac{1}{\sqrt{2}} T_T\, \tilde{L}\, \tilde{T}\, \tilde{L} + \frac{1}{\sqrt{2}} T_S\, \tilde{d}^c\, \tilde{S}\, \tilde{d}^c + T_Z\, \tilde{d}^c\, \tilde{Z}\, \tilde{L} + \frac{1}{\sqrt{2}} T_1\, H_d\, \tilde{T}\, H_d$$

$$+ \frac{1}{\sqrt{2}} T_2\, H_u\, \tilde{\bar{T}}\, H_u + B_T\, \tilde{T}\, \tilde{\bar{T}} + B_Z\, \tilde{Z}\, \tilde{\bar{Z}} + B_S\, \tilde{S}\, \tilde{\bar{S}} + \text{h.c.}, \qquad (13.14)$$

$$\mathcal{L}_{\text{soft}}^{\text{III}} = \sqrt{\frac{3}{10}} T_B\, H_u\, \tilde{B}_M\, \tilde{L} + T_W\, H_u\, \tilde{W}_M\, \tilde{L} + T_X\, H_u\, \tilde{X}_M\, \tilde{d}^c + B_X\, \tilde{X}_M\, \tilde{X}_M,$$

$$+ \frac{1}{2} B_W\, \tilde{W}_M\, \tilde{W}_M + \frac{1}{2} B_G\, \tilde{G}_M\, \tilde{G}_M + \frac{1}{2} B_B\, \tilde{B}_M\, \tilde{B}_M + \text{h.c.}, \qquad (13.15)$$

while the soft-breaking scalar masses read as

$$\mathcal{L}_{\text{soft},\phi} = \mathcal{L}_{\text{soft},\phi}^{\text{MSSM}} + \mathcal{L}_{\text{soft},\phi}^{\text{I}} + \mathcal{L}_{\text{soft},\phi}^{\text{II}} + L_{\text{soft},\phi}^{\text{III}},$$

where

$$\mathcal{L}_{\text{soft},\phi}^{\text{I}} = -(\tilde{\nu}^c)^\dagger m_{\nu^c}^2 \tilde{\nu}^c \qquad (13.16)$$

$$\mathcal{L}_{\text{soft},\phi}^{\text{II}} = -m_S^2 \tilde{S}^* \tilde{S} - m_{\tilde{S}}^2 \tilde{\bar{S}}^* \tilde{\bar{S}} - m_T^2 \tilde{T}^* \tilde{T} - m_{\tilde{T}}^2 \tilde{\bar{T}}^* \tilde{\bar{T}} - m_{\tilde{Z}}^2 \tilde{Z}^* \tilde{Z} - m_{\tilde{\bar{Z}}}^2 \tilde{\bar{Z}}^* \tilde{\bar{Z}} \qquad (13.17)$$

$$L_{\text{soft},\phi}^{\text{III}} = -\tilde{B}_M^\dagger m_B^2 \tilde{B}_M - \tilde{W}_M^\dagger m_{\tilde{W}}^2 \tilde{W}_M - \tilde{G}_M^\dagger m_G^2 \tilde{G}_M - \tilde{X}_M^\dagger m_X^2 \tilde{X}_M - \tilde{\bar{X}}_M^\dagger m_{\tilde{X}}^2 \tilde{\bar{X}}_M. \qquad (13.18)$$

Since the new interactions in Eqs. (13.11) to (13.12) are the result of $SU(5)$-invariant terms, it is natural to assume a unification of the different couplings at the GUT scale:

$$M_T = M_{\hat{Z}} = M_S \equiv M_{15}, \qquad Y_S = Y_T = Y_{\hat{Z}} \equiv Y_{15}, \qquad (13.19)$$

$$Y_B = Y_{\hat{W}} = Y_X \equiv Y_{24}, \qquad M_X = M_{\hat{W}} = M_G = M_B \equiv M_{24}. \qquad (13.20)$$

In the same way, the bi- and tri-linear soft-breaking terms unify and are connected to the superpotential parameters by

$$B_{\nu^c} \equiv B_0 M_{\nu^c}, \qquad T_\nu \equiv A_0 Y_\nu, \qquad (13.21)$$

$$B_{15} \equiv B_0 M_{15}, \qquad T_{15} \equiv A_0 Y_{15}, \qquad (13.22)$$

$$B_{24} \equiv B_0 M_{24}, \qquad T_{24} \equiv A_0 Y_{24}. \qquad (13.23)$$

In the case of cMSSM-like boundary conditions, this leads to the following free parameters

$$B_0, \ M_{\nu^c}, \ Y_\nu, \ M_{15}, \lambda_1, \lambda_2, \ Y_{15}, \ M_{24}, \ Y_{24} \qquad (13.24)$$

in addition to the well-known MSSM parameters

$$m_0, \ m_{1/2}, \ A_0, \ \tan\beta, \ \mathrm{sign}(\mu). \qquad (13.25)$$

In principle, this B_0 is not the same as the B for the Higgs, though in a minimal case they may be defined to be equal at the GUT scale. Furthermore, $T_i = A_0 Y_i$ holds at the GUT scale.

Finally, the effective neutrino mass matrices appearing in Type I/II/III at the SUSY scale are

$$m_\nu^{\mathrm{I}} = -\frac{v_u^2}{2} Y_\nu^T M_R^{-1} Y_\nu, \qquad (13.26)$$

$$m_\nu^{\mathrm{II}} = \frac{v_u^2}{2} \frac{\lambda_2}{M_T} Y_T, \qquad (13.27)$$

$$m_\nu^{\mathrm{III}} = -\frac{v_u^2}{2} \left(\frac{3}{10} Y_B^T M_B^{-1} Y_B + \frac{1}{2} Y_W^T M_{\hat{W}}^{-1} Y_W \right). \qquad (13.28)$$

13.2.2 Inverse and linear seesaw

The inverse and linear seesaw realisations are obtained in models that provide three generations of a further gauge singlet carrying a lepton number in addition to three generations of the well-known RH neutrino superfields, here $\hat{\nu}^c$ (see Tab. 13.2).

SF	Spin 0	Spin 1/2	Generations	$U(1)_Y \times SU(2)_L \times SU(3)_C$	lepton number	R-parity of fermion
$\hat{\nu}^c$	$\tilde{\nu}^c$	ν^c	n_{ν^c}	$(0, \mathbf{1}, \mathbf{1})$	$+1$	$+$
$\hat{N}_S$	$\tilde{N}_S$	N_S	n_{N_S}	$(0, \mathbf{1}, \mathbf{1})$	-1	$+$

Table 13.2 New chiral superfields appearing in models with inverse and linear seesaw.

The only additional terms in the superpotential which are allowed by conservation of gauge quantum numbers are

$$W - W_{\mathrm{MSSM}} = Y_\nu \, \hat{\nu}^c \, \hat{L} \, \hat{H}_u + M_R \, \hat{\nu}^c \, \hat{N}_S + \begin{cases} \frac{1}{2} \mu_N \, \hat{N}_S \, \hat{N}_S & \text{inverse seesaw,} \\ Y_{LN} \, \hat{N}_S \, \hat{L} \, \hat{H}_u & \text{linear seesaw.} \end{cases} \qquad (13.29)$$

It is important to note that the last term in each model breaks the lepton number explicitly, but this is expected for different reasons.

The soft-breaking terms read

$$\mathcal{L}_{\text{soft}} = \mathcal{L}_{\text{soft}}^{\text{MSSM}} + T_\nu\, \tilde{\nu}^c\, \tilde{L}\, H_u + B_R\, \tilde{\nu}^c\, \tilde{N}_S + \left\{ \begin{array}{ll} \frac{1}{2} B_N\, \tilde{N}_S\, \tilde{N}_S & \text{inverse seesaw} \\ T_{LN}\, \tilde{N}_S\, \tilde{L} H_u & \text{linear seesaw} \end{array} \right\} + \text{h.c.},$$

(13.30)

$$\mathcal{L}_{\text{soft},\phi} = \mathcal{L}_{\text{soft},\phi}^{\text{MSSM}} - (\tilde{\nu}^c)^\dagger m_{\nu^c}^2 \tilde{\nu}^c - \tilde{N}_S m_N^2 \tilde{N}_S^*,$$

(13.31)

while $\mathcal{L}_{\text{soft},\lambda}$ is again the same as for the MSSM. It is necessary to split the sneutrinos and the singlets into their scalar and pseudoscalar components:

$$\tilde{\nu}_L = \frac{1}{\sqrt{2}} \left(\sigma_L + i\phi_L \right), \ \tilde{\nu}^c = \frac{1}{\sqrt{2}} \left(\sigma_R + i\phi_R \right), \ \tilde{N}_S = \frac{1}{\sqrt{2}} \left(\sigma_S + i\phi_S \right).$$

(13.32)

In comparison to the MSSM, additional mixings between fields take place: the LH and RH scalar components mix with the scalar component of the singlet fields. The same holds for the pseudoscalar components. Furthermore, the neutrinos mix with the fermionic singlet fields to build up 9 Majorana fermions. All three appearing 9×9 mass matrices can be diagonalised by unitary matrices. We define the basis for the mass matrices as follows.

- Scalar sneutrinos: $(\sigma_L, \sigma_R, \sigma_S)^T$.

- Pseudoscalar sneutrinos: $(\phi_L, \phi_R, \phi_S)^T$.

- Neutrinos: $(\nu_L, \nu^c, N_S)^T$.

Thus, the neutrino mass matrix for the inverse and linear seesaw is given, respectively, by

$$\begin{pmatrix} 0 & \frac{v_u}{\sqrt{2}} Y_\nu & 0 \\ \frac{v_u}{\sqrt{2}} Y_\nu^T & 0 & M_R \\ 0 & M_R^T & \mu_N \end{pmatrix}, \qquad \begin{pmatrix} 0 & \frac{v_u}{\sqrt{2}} Y_\nu & \frac{v_u}{\sqrt{2}} Y_{LN} \\ \frac{v_u}{\sqrt{2}} Y_\nu^T & 0 & M_R \\ \frac{v_u}{\sqrt{2}} Y_{LN}^T & M_R^T & 0 \end{pmatrix}.$$

(13.33)

Note that the presence of v_u in all terms of the first column and row is just a coincidence caused by the given minimal particle content. For more general models, different VEVs can appear.

If cMSSM-like boundary conditions are assumed, the following new free parameters arise:

$$M_R, \ Y_{LN}, \ \mu_N, \ B_0,$$

(13.34)

in addition to those given in Eq. (13.25). Calculating the eigenvalues of the above mass matrices, it can be seen that the light neutrino masses are linear functions of Y_{LN} in the linear seesaw models, while the neutrino masses are linearly proportional to μ_N, as in the inverse seesaw models. The neutrino masses in the two models read as

$$m_\nu^{\text{LS}} \simeq \frac{v_u^2}{2} \left[Y_\nu (Y_{LN} M_R^{-1})^T + (Y_{LN} M_R^{-1}) Y_\nu^T \right],$$

(13.35)

$$m_\nu^{\text{IS}} \simeq \frac{v_u^2}{2} Y_\nu^T (M_R^T)^{-1} \mu_N M^{-1} Y_\nu^T.$$

(13.36)

Hence it has been proposed that both models, and any combination of the two, be specified by extending Y_ν to a $3 \times (n_{\nu^c} + n_{N_S})$ Yukawa matrix, incorporating Y_{LN} as Y_ν^{ij} with i running from $n_{\nu^c} + 1$ to $n_{\nu^c} + n_{N_S}$. Implementation of each model consists of zeros being specified in the appropriate entries in the relevant matrices (e.g., specifying that the elements of Y_ν corresponding to Y_{LN} are zero recovers the inverse seesaw model, while specifying that μ_N is zero recovers the linear seesaw model).

13.3   *R*-PARITY VIOLATING SCENARIO

For completeness, we also mention another way of generating neutrino masses in the MSSM without introducing any singlet, but by violating R-parity. As mentioned in Eq. (4.11) in Chapter 4, the MSSM gauge group can allow for terms that violate lepton number or baryon number or both. Thus, if we care about the lepton and baryon numbers, we would apply the discrete symmetry $(-1)^{3B+L+2S}$ to prevent such terms. This discrete symmetry is called the R-parity. Now, if we seek an extension or a modification in the MSSM to allow for a mass term for the neutrino (which violates, in its own right, the lepton number), we may allow for terms that violate R-parity explicitly in the superpotential.

The violation of R-parity can be done via extra bilinear or trilinear terms in the superpotential. Here we will briefly review the neutrino masses in the BRPV scenario, because it is the simplest way to include such effects into the MSSM. In this model, which is called BRVP-MSSM for short, we extend the superpotential of the MSSM by adding the term

$$\varepsilon_i L_i H_u, \qquad i = 1, 2, 3. \tag{13.37}$$

For simplicity, we will assume from now on that $\varepsilon_1 = \varepsilon_2 = 0$. That is, the full superpotential is given by

$$W = Y_u Q_L U_L^c H_u + Y_d Q_L D_L^c H_d + Y_e L_L E_L^c H_d - \mu H_d H_u + \varepsilon_3 L_{L_3} H_u. \tag{13.38}$$

The relevant soft supersymmetry breaking terms are given by

$$V_{\text{soft}} = m_{H_d}^2 |H_d|^2 + M_{L_3}^2 |\tilde{L}_3|^2 - \left[B\mu H_d H_u - B_2 \varepsilon_3, \tilde{L}_3 H_u + \text{h.c.} \right] + \cdots, \tag{13.39}$$

where $m_{H_d}^2$ and $M_{L_3}^2$ are the soft masses corresponding to H_d and $\tilde{L}_3$, respectively, while B and B_2 are the bilinear soft mass parameters associated to the last two terms in the superpotential in Eq. (13.38).

From a first look at the superpotential Eq. (13.38), one may claim that the BRPV term, the last term, can be rotated away by a suitable choice of basis [241]. Indeed, consider the following rotation of the superfields

$$H_d' = \frac{\mu H_d - \varepsilon_3 L_3}{\sqrt{\mu^2 + \varepsilon_3^2}}, \qquad L_3' = \frac{\varepsilon_3 H_d + \mu L_3}{\sqrt{\mu^2 + \varepsilon_3^2}}. \tag{13.40}$$

Then the superpotential of the third generation, written by means of the new basis, is given by

$$W = Y_t Q_3 U_3 H_u + Y_b \frac{\mu}{\mu'} Q_3 D_3 H_d' + Y_\tau L_3' R_3 H_d' - \mu' H_d' H_u + Y_b \frac{\varepsilon_3}{\mu'} Q_3 D_3 L_3', \tag{13.41}$$

where $\mu'^2 = \mu^2 + \epsilon_3^2$. This succeeds in rotating the BRPV term away in the new basis. But, unfortunately, RPV is reintroduced from the trilinear RPV term $Y_b \frac{\varepsilon_3}{\mu'} Q_3 D_3 L_3'$. This means that the BRPV term is physical and describes a true interaction. The soft terms in the rotated basis are given by

$$
\begin{aligned}
V_{\text{soft}} &= \frac{m_{H_d}^2 \mu^2 + M_{L_3}^2 \varepsilon_3^2}{\mu'^2} |H_d'|^2 + \frac{m_{H_d}^2 \varepsilon_3^2 + M_{L_3}^2 \mu^2}{\mu'^2} |\tilde{L}_3'|^2 - \frac{B\mu^2 + B_2 \varepsilon_3^2}{\mu'} H_d' H_u \\
&+ \frac{\varepsilon_3 \mu}{\mu'^2} (m_{H_d}^2 - M_{L_3}^2) \tilde{L}_3' H_d' + \frac{\varepsilon_3 \mu}{\mu'^2} (B_2 - B) \tilde{L}_3' H_u + \text{h.c.} + \cdots.
\end{aligned} \tag{13.42}
$$

As usual, the VEVs are derived from equating the tadpole equations to zero. Thus, the VEVs are given from the solutions of the following equations:

$$0 = \frac{\mu'^4 + m_{H_d}^2 \mu^2 + M_{L_3}^2 \varepsilon_3^2}{\mu'^2} v_d' - \frac{B\mu^2 + B_2 \varepsilon_3^2}{\mu'} v_u + \frac{(m_{H_d}^2 - M_{L_3}^2)\varepsilon_3 \mu}{\mu'^2} v_3'$$
$$+ \frac{g_2^2 + g_1^2}{8} v_d'(v_d'^2 - v_u^2 + v_3'^2), \tag{13.43}$$

$$0 = \mu'^2 v_u + m_{H_u}^2 v_u - \frac{B\mu^2 + B_2 \varepsilon_3^2}{\mu'} v_d'$$
$$+ (B_2 - B)\frac{\varepsilon_3 \mu}{\mu'} v_3' - \frac{g_2^2 + g_1^2}{8} v_u(v_d'^2 - v_u^2 + v_3'^2), \tag{13.44}$$

$$0 = (m_{H_d}^2 - M_{L_3}^2)\frac{\varepsilon_3 \mu}{\mu'^2} v_d' + (B_2 - B)\frac{\varepsilon_3 \mu}{\mu'} v_u + \frac{m_{H_d}^2 \varepsilon_3^2 + M_{L_3}^2 \mu^2}{\mu'^2} v_3'$$
$$+ \frac{g_2^2 + g_1^2}{8} v_3'(v_d'^2 - v_u^2 + v_3'^2), \tag{13.45}$$

where

$$\langle H_d' \rangle = v_d^{(\prime)}/\sqrt{2}, \qquad \langle H_u \rangle = v_u, \qquad \langle L_3' \rangle = v_3'/\sqrt{2}, \tag{13.46}$$

which means that

$$v_d' = \frac{(\mu v_d - \epsilon_3 v_3)}{\mu'}, \tag{13.47}$$

$$v_3' = \frac{(\epsilon_3 v_d + \mu v_3)}{\mu'}. \tag{13.48}$$

From Eq. (13.45), it is clear that if $\Delta m^2 \equiv m_{H_1}^2 - M_{L_3}^2 = 0$ and $\Delta B \equiv B_2 - B = 0$, then we have $v_3' = 0$. Therefore, if we assume the universality condition, which contains the assumptions $m_{H_1}^2 = M_{L_3}^2 = m_0^2$ and $B_2 = B = B_0$, at the GUT scale, then $v_3' = 0$ at the GUT scale. However, the running of the RGE of the soft mass parameters as well as the bilinear soft breaking parameters from the GUT scale to the EW scale cannot preserve those aforementioned differences as zero. If small Δm^2 and ΔB are assumed, the value of v_3' is also small and can be approximated to be [242]

$$v_3' \approx -\frac{\varepsilon_3 \mu}{\mu'^2 m_{\tilde{\nu}_\tau^0}^2} \left(v_d' \Delta m^2 + \mu' v_2 \Delta B \right), \tag{13.49}$$

where

$$m_{\tilde{\nu}_\tau^0}^2 \equiv \frac{m_{H_d}^2 \varepsilon_3^2 + M_{L_3}^2 \mu^2}{\mu'^2} + \tfrac{1}{8}(g_2^2 + g_1^2)(v_d'^2 - v_u^2). \tag{13.50}$$

It is interesting to note that, if we set $\varepsilon_3 = 0$ in the last equation, it reduces to the τ sneutrino mass in view of the MSSM.

The solution of the RGEs for the soft mass parameters $m_{H_d}^2$, $m_{L_3}^2$, B, and B_2 can be approximated to be

$$m_{H_1}^2 - M_{L_3}^2 \approx -\frac{3Y_b^2}{8\pi^2}\left(m_{H_d}^2 + M_Q^2 + M_D^2 + A_D^2 \right) \ln \frac{M_{\text{GUT}}}{M_Z}, \tag{13.51}$$

$$B_2 - B \approx \frac{3Y_b^2}{8\pi^2} A_D \ln \frac{M_{\text{GUT}}}{M_Z}. \tag{13.52}$$

This illustrates that the sneutrino VEV v_3' is radiatively generated and, by using Eq. (13.49), it can be estimated easily that the maximum value of v_3' is a few GeV or so.

In fact, the neutrino mass is quite interesting in this model. The τ neutrino is coupled with the usual MSSM neutralinos via the BRPV mixing term $\varepsilon_3 \hat{L}_3 \hat{H}_u$. That is, the neutralino mass matrix in this model is a 5×5^1 matrix instead of the usual MSSM 4×4 matrix. As a result, the lightest eigenstate would be identified as the τ neutrino. In the original (un-rotated) basis, the neutralino mass matrix is given, for the basis $(-i\lambda', -i\lambda^3, \tilde{H}_d^1,, \tilde{H}_u^2, \nu_\tau)$, by

$$
\mathbf{M}_{\chi^0} = \begin{pmatrix}
M_1 & 0 & -\frac{1}{2}g'v_d & \frac{1}{2}g'v_u & -\frac{1}{2}g'v_3 \\
0 & M_2 & \frac{1}{2}gv_d & -\frac{1}{2}gv_u & \frac{1}{2}gv_3 \\
-\frac{1}{2}g'v_d & \frac{1}{2}gv_d & 0 & -\mu & 0 \\
\frac{1}{2}g'v_u & -\frac{1}{2}gv_u & -\mu & 0 & \epsilon_3 \\
-\frac{1}{2}g'v_3 & \frac{1}{2}gv_3 & 0 & \epsilon_3 & 0
\end{pmatrix},
\tag{13.53}
$$

where M_2 and M_1 are the $SU(2)$ and $U(1)$ gaugino masses, respectively. It is remarkable that the neutralino mass matrix in the rotated basis can be obtained easily from the mass matrix Eq. (13.53) by making the substitution $(v_d, v_3, \epsilon_3, \mu) \to (v_d', v_3', 0, \mu')$. In this basis the τ neutrino mass can be approximated by

$$
m_{\nu_\tau} \approx -\frac{(g_2^2 M_2 + g_1^2 M_1)\mu'^2 v_3'^2}{4M_1 M_2 \mu'^2 - 2(g_2^2 M_2 + g_1^2 M_1)v_d' v_u \mu'}.
\tag{13.54}
$$

Now, substituting from Eqs. (13.49), (13.51) and (13.52) into Eq. (13.54), we get

$$
m_{\nu_\tau} \approx \frac{\left[\mu' v_u A_D - v_d'\left(m_{H_d}^2 + M_Q^2 + M_D^2 + A_D^2\right)\right]^2}{\left[2v_d' v_u - 4M_1 M_2 \mu'/(g_2^2 M_2 + g_1^2 M_1)\right]\mu' m_{\tilde{\nu}_\tau^0}^2} \left(\frac{\epsilon_3 \mu}{\mu'^2}\right)^2 \left(\frac{3Y_b^2}{8\pi^2}\ln\frac{M_{\text{GUT}}}{M_Z}\right)^2.
\tag{13.55}
$$

This means that

$$
m_{\nu_\tau} \approx \mathcal{O}\left(\frac{M_Z^2}{M_{\text{SUSY}}}\left(\frac{\epsilon_3}{M_{\text{SUSY}}}\right)^2 Y_b^4\right).
\tag{13.56}
$$

Therefore, if $M_{\text{SUSY}} \sim$ TeV, $\epsilon_3 \sim M_Z$ and $Y_b \sim 10^{-2}$, then a τ neutrino mass will be of order $\sim \mathcal{O}(1\,\text{eV})$. The interesting remark here is that m_{ν_τ} can be naturally small, even though the BRPV parameter is large. A last note to be mentioned here is that, in contrast to the MSSM case, where the lightest neutralino is stable and hence a DM candidate, the lightest neutralino in BRPV-MSSM is no longer stable because it can decay into a τ neutrino and hence it stops being a DM candidate.

13.4 SUPERSYMMETRIC SEESAW MODEL

As advocated above, the seesaw mechanism is an elegant way to generate very light neutrino masses by introducing heavy SM singlets (RH Majorana neutrinos). However, the SM Higgs hierarchy problem becomes more severe in the seesaw extension of the SM. The very massive RH neutrinos that couple to the Higgs field would generate radiative corrections to the Higgs mass of the order of RH neutrino masses, which are much higher than its tree level value. As explained in Chapter 4, in SUSY models, this problem is solved by the cancellation between this correction and the contributions of the supersymmetric bosonic partner RH neutrinos (called RH sneutrinos).

---

[1]It is remarkable that if the assumption of $\varepsilon_1 = \varepsilon_2 = 0$ is dropped, the neutralino mass matrix would be a 7×7 instead.

In a SS model we have three families of RH neutrino superfields, $N_i^c, i = 1, 2, 3$, so that the superpotential for the lepton sector is given by

$$W = (Y_\ell)_{ij} E_i^c L_j H_d + (Y_\nu)_{ij} N_i^c L_j H_u + \frac{1}{2}(M_R)_{ij} N_i^c N_j^c, \qquad (13.57)$$

where i, j run over generations. M_R is the heavy RH neutrino mass matrix. By integrating out the heavy RH neutrino, the following effective superpotential is obtained:

$$W_{\text{eff}} = W_{\text{MSSM}} + \frac{1}{2}(Y_\nu L H_u)^T M_R^{-1}(Y_\nu L H_u). \qquad (13.58)$$

After EWSB, one finds the following neutrino mass

$$m_\nu = m_D^T M_R^{-1} m_D, \qquad (13.59)$$

where m_D is the Dirac mass matrix, given by $m_D = Y_\nu \langle H_u \rangle = v \sin \beta Y_\nu$. It is now clear that light neutrinos can be naturally explained if the Majorana mass is much larger than the Dirac mass m_D. One can easily get

$$m_\nu = 0.1 \text{eV} \left(\frac{m_D}{100 \text{ GeV}} \right)^2 \left(\frac{M_R}{10^{14} \text{ GeV}} \right)^{-1}. \qquad (13.60)$$

The symmetric light neutrino mass matrix m_ν can be diagonalised by a unitary matrix U, such that $U^T m_\nu U = \text{diag}(m_{\nu_1}, m_{\nu_2}, m_{\nu_3})$. In general, the MNS mixing matrix, U_{MNS}, is given by $U_{\text{MNS}} = U U_\ell^+$, where U_ℓ is the charged lepton mixing matrix. In the basis of a diagonal charged lepton, $U = U_{\text{MNS}}$. A popular parameterisation for U is

$$U = V \text{ diag}(e^{i\phi_1}, e^{i\phi_2}, 1), \qquad (13.61)$$

where $\phi_{1,2}$ are the CP-violating Majorana phases and V can be written in the standard parameterisation of the CKM, with three mixing angles θ_{12}, θ_{13} and θ_{23} that are measured in the neutrino oscillation experiments and one CP-violating Dirac phase. It is also worth mentioning that one of the interesting parameterisation for the Dirac neutrino mass matrix is given by [243]:

$$m_D = U_{\text{MNS}} \sqrt{m_\nu^{\text{diag}}} R \sqrt{M_R}, \qquad (13.62)$$

where m_ν^{diag} is the physical light neutrino mass matrix and U_{MNS} is the lepton mixing matrix. The matrix R is an arbitrary orthogonal matrix which can in general be parameterised in terms of three complex angles.

The Lagrangian of the soft SUSY breaking terms is given by

$$
\begin{aligned}
-\mathcal{L}_{\text{soft}} = & -\mathcal{L}_{\text{soft}}^{\text{MSSM}} + (m_\nu^2)_{ij} \tilde{\nu}_i^* \tilde{\nu}_j + \left[(A_\nu)_{ij} \tilde{\nu}_i^* \tilde{L}_j H_u + \text{h.c.} \right] \\
& + \left[\frac{1}{2}(B_\nu)_{ij} M_{ij} \tilde{\nu}_i \tilde{\nu}_j + \text{h.c.} \right].
\end{aligned} \qquad (13.63)
$$

It is clear that lepton flavour conservation can be violated by the off-diagonal elements of the slepton matrices. Therefore, the sizes of these elements are strongly constrained from experiments. In the cMSSM, $m_{\tilde{E}}^2$ and $m_{\tilde{L}}^2$ are proportional to the unity matrix and A_ℓ is proportional to Y_ℓ. Therefore, the lepton flavour number is essentially conserved. In the SS model, this is no longer true. Indeed, the neutrino Yukawa coupling and soft terms induce off-diagonal elements of $m_{\tilde{L}}^2$ through the radiative corrections. The RGE of $m_{\tilde{L}}^2$ is then modified to be

$$
\begin{aligned}
\frac{dm_{\tilde{L}}^2}{dt} &= (\frac{dm_{\tilde{L}}^2}{dt})_{\text{MSSM}} + \frac{1}{16\pi^2} \{(m_{\tilde{L}}^2 Y_\nu^+ Y_\nu) + (Y_\nu^+ Y_\nu m_{\tilde{L}}^2) + 2(Y_\nu^+ Y_\nu)m_H^2 \\
&\quad + (Y_\nu^+ m_{\tilde{\nu}}^2 Y_\nu) + 2(A_\nu^+ A_\nu)\},
\end{aligned}
\tag{13.64}
$$

where $(dm_{\tilde{L}}^2/dt)_{\text{MSSM}}$ is given in Appendix B. Thus, off-diagonal elements of $m_{\tilde{L}}^2$ are induced by the renormalisation effects if non-vanishing off-diagonal elements of Y_ν and A_ν exist. In contrast, in the limit of decoupling heavy RH (s)neutrinos, the 3×3 light sneutrino mass matrix of the MSSM is obtained.

13.5 LEPTON FLAVOUR VIOLATION

The lepton flavour symmetry is an accidental symmetry in the SM and it may be violated in BSM scenarios. There is no fundamental reason for lepton flavour to be conserved. In fact, several SM extensions (GUT, Technicolour, Compositeness other than SUSY) exploit the LFV possibility. Therefore, a signal of LFV in a charged lepton sector would be a clear signature of BSM physics. The present bounds for LFV processes are as follows [244–247]:

$$
\begin{aligned}
\text{BR}(\mu \to e\gamma) &= 5.7 \times 10^{-13}, & (13.65) \\
\text{BR}(\tau \to \mu\gamma) &= 4.4 \times 10^{-8}, & (13.66) \\
\text{BR}(\tau \to e\gamma) &= 3.3 \times 10^{-8}, & (13.67) \\
\text{BR}(\mu \to 3e) &= 1.0 \times 10^{-12}, & (13.68) \\
\text{BR}(\tau \to 3e) &= 2.7 \times 10^{-8}, & (13.69) \\
\text{BR}(\tau \to 3\mu) &= 2.1 \times 10^{-8}. & (13.70)
\end{aligned}
$$

In the SM, with non-zero neutrino masses, the $\text{BR}(\mu \to e\gamma)$ is given by

$$
\text{BR}(\mu \to e\gamma) \simeq \frac{\alpha}{4\pi} \left(\frac{m_\nu}{M_W}\right)^4 \simeq 10^{-43} \left(\frac{m_\nu}{1\,\text{TeV}}\right)^4,
\tag{13.71}
$$

which confirms the conclusion that LFV in the charged sector is forbidden in the SM. In the MSSM, the amplitude for the decay $\ell_i^- \to \ell_j^- \gamma$ can be written as [190, 248, 249]

$$
\mathcal{M}_{ij} = i\, e\, m_{\ell_i}\, \bar{u}_j(p')\, \varepsilon_\alpha^* \sigma^{\alpha\beta} q_\beta \left(A_{ij}^L + A_{ij}^R\right) u_i(p),
\tag{13.72}
$$

where $\sigma_{\alpha\beta} = \frac{i}{2}[\gamma_\alpha, \gamma_\beta]$, q is the photon momentum and ε is the polarisation vector of the photon. In the limit of a massless final lepton, $m_{\ell_j} = 0$, the decay rate is given by [248]

$$
\Gamma(\ell_i^- \to \ell_j^- \gamma) = \frac{e^2}{16\pi} m_{\ell_i}^5 \left(|A_{ij}^L|^2 + |A_{ij}^R|^2\right).
\tag{13.73}
$$

The coefficients $A^{L,R}$ are determined by calculating the photon penguin diagram shown in Fig. 13.1, with charginos/sneutrinos or neutralinos/charged sleptons in the loop [248]:

$$
\begin{aligned}
A_{ij}^L &= \frac{1}{32\pi^2} \left\{ \sum_{k=1}^{6}\sum_{a=1}^{2} \frac{1}{m_{\tilde{\ell}_k}^2} \left[N_{jka}^L N_{ika}^{L*} F_1^N \left(\frac{m_{\chi_a^0}^2}{m_{\tilde{\ell}_k}^2}\right) + \frac{m_{\chi_a}^0}{m_{\ell_i}} N_{jka}^L N_{ika}^{R*} F_2^N \left(\frac{m_{\chi_a^0}^2}{m_{\tilde{\ell}_k}^2}\right) \right] \right. \\
&\quad \left. - \sum_{k=1}^{3}\sum_{a=1}^{2} \frac{1}{m_{\tilde{\nu}_k}^2} \left[C_{jka}^L C_{ika}^{L*} F_1^C \left(\frac{m_{\chi_a^-}^2}{m_{\tilde{\nu}_k}^2}\right) + \frac{m_{\chi_a}^-}{m_{\ell_i}} C_{jka}^L C_{ika}^{R*} F_2^C \left(\frac{m_{\chi_a^-}^2}{m_{\tilde{\nu}_k}^2}\right) \right] \right\}, \\
A_{ij}^R &= A_{ij}^L \Big|_{L\leftrightarrow R},
\end{aligned}
\tag{13.74}
$$

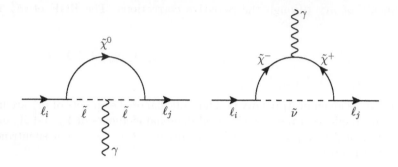

Figure 13.1 Feynman diagrams of $\ell_i \to \ell_j \gamma$ in the SS model.

where the interaction couplings for the neutralino-lepton-slepton, $N^{L,R}$, and for the chargino-lepton-sneutrino, $C^{L,R}$, are given in Chapter 5. The loop functions $F_{1,2}^{N,C}$ are given by [248]:

$$F_1^N = \frac{1 - 6x + 3x^2 + 2x^3 - 6x^2 \ln x}{6(1-x)^4}, \tag{13.75}$$

$$F_2^N = \frac{1 - x^2 + 2x \ln x}{(1-x)^3}, \tag{13.76}$$

$$F_1^C = \frac{2 + 3x - 6x^2 + x^3 + 6x \ln x}{6(1-x)^4}, \tag{13.77}$$

$$F_2^C = \frac{-3 + 4x - x^2 - 2 \ln x}{(1-x)^3}. \tag{13.78}$$

Finally, the BRs for $\ell_i \to \ell_j \gamma$ are given by

$$\mathrm{BR}(\ell_i \to \ell_j \gamma) = \mathrm{BR}(\ell_i \to \ell_j \nu_i \bar{\nu}_j) \frac{48\pi^3 \alpha}{G_{\mathrm{F}}^2} \left(|A_{ij}^L|^2 + |A_{ij}^R|^2 \right). \tag{13.79}$$

In the MSSM, with universal soft SUSY breaking terms, the LFV limits are easily satisfied, especially with the new LHC bounds on the SUSY spectrum. With non-universal soft SUSY breaking terms, in particular non-universal trilinear couplings A_{ij}^ℓ and $m_{\tilde{\ell}}^2 = \mathrm{diag}(m_1^2, m_2^2, m_3^2)$, one finds that $\mu \to e\gamma$ imposes stringent constraints on the SUSY parameter space [250]. As discussed in Chapter 10 in terms of the MIA, it implies that $|(\delta_{12}^\ell)_{LL}| \lesssim 10^{-4}$ and $|(\delta_{12}^\ell)_{LR}| \lesssim 10^{-7}$. With the current LHC limits on the SUSY spectrum, these bounds are not so severe and can be satisfied by a vast number of points in the parameter space.

In the SS model, it has been shown that neutrino flavour mixing radiatively induces slepton flavour mixing from the GUT to RH neutrino mass scale (M_R) [190, 249, 251, 252]. As emphasised in the previous section, in this class of models, $m_{\tilde{\ell}}^2$ at the EW scale can be written as

$$m_{\tilde{\ell}}^2 = m_0^2 \delta_{ij} + (\Delta m^{\tilde{\ell}})_{ij}^2, \tag{13.80}$$

where the slepton off-diagonal mass corrections, with universal soft terms, are given by

$$(\Delta m_{LL}^{\tilde{\ell}})_{ij}^2 = -\frac{1}{8\pi^2}(3m_0^2 + A_0^2)(Y_\nu^+ Y_\nu)_{ij}\ln\left(\frac{M_{\text{GUT}}}{M_R}\right), \tag{13.81}$$

$$(\Delta m_{LR}^{\tilde{\ell}})_{ij}^2 = -\frac{3}{8\pi^2}A_0 Y_\ell(Y_\nu^+ Y_\nu)_{ij}\ln\left(\frac{M_{\text{GUT}}}{M_R}\right). \tag{13.82}$$

It is clear that, depending on the Y_ν texture, the constraints imposed by $\mu \to e\gamma$ could be very stringent [251]. In general, the neutrino Yukawa matrix Y_ν can be parameterised in terms of M_R and an arbitrary orthogonal matrix, as shown in Eq. (13.62). Therefore, one can scan over M_{R_i} and the three free angles of the matrix R to consider all possible types of Y_ν structure. As shown in Fig. 13.2, in this class of models, significant enhancements for $\mu \to e\gamma$ can occur.

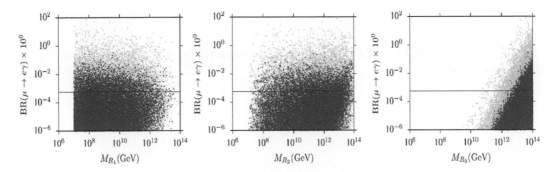

Figure 13.2 $\text{BR}(\mu \to e\gamma)$ as a function of M_{R_i} in the SS model.

However, with current LHC limits for the SUSY spectrum, one can easily show that the current LFV constraints are satisfied in a wide range of the parameter space of the SS model, especially if universal soft terms are assumed as indicated in Fig. 13.3.

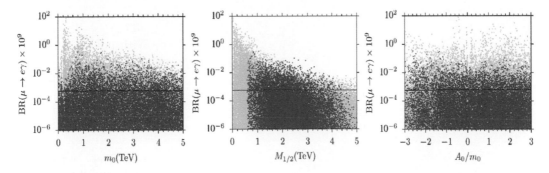

Figure 13.3 $\text{BR}(\mu \to e\gamma)$ as a function of the universal soft SUSY breaking terms $m_0, M_{1/2}$ and A_0 in the SS model.

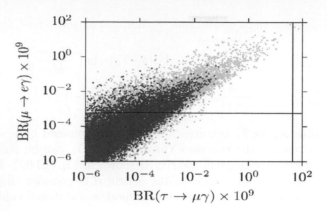

Figure 13.4 The correlation between $BR(\mu \to e\gamma)$ and $BR(\tau \to \mu\gamma)$ in the SS model.

Finally, we display in Fig. 13.4 the correlation between $BR(\mu \to e\gamma)$ and $BR(\tau \to \mu\gamma)$ in this class of models.

CHAPTER 14

Minimal Gauge Extensions of the MSSM

With this chapter we start the treatment of non-minimal SUSY models, viewed here from the perspective of an enlarged gauge structure. In fact, we will concentrate on the case of an additional $U(1)$ group above and beyond the SM gauge structure. We will eventually construct the BLSSM (with two possible seesaw realisations) as a phenomenological benchmark for this category of BSM scenarios. Finally, the ESSM will also be introduced. Much of the content of this chapter is based on [21, 22, 24, 228, 253].

14.1 SEESAW MODELS WITH $U(1)_R \times U(1)_{B-L}$ GAUGE SECTOR

Models based on a $SO(10)$ GUT theory can lead to a gauge sector containing the product group $U(1)_R \times U(1)_{B-L}$, through the breaking pattern

$$SO(10) \to SU(3)_C \times SU(2)_L \times SU(2)_R \times U(1)_{B-L}$$
$$\to SU(3)_C \times SU(2)_L \times U(1)_R \times U(1)_{B-L}. \tag{14.1}$$

The $U(1)_R \times U(1)_{B-L}$ factors will be subsequently broken down into the hypercharge $U(1)_Y$ of the SM. However, it is possible that this final breaking scale is just around the TeV scale without spoiling gauge unification [254]. This can therefore lead to an interesting phenomenology and can have an important impact on the Higgs sector [255, 256]. The first version of these models included a linear seesaw mechanism, but it has been shown that the inverse seesaw can be also included [257]. Further, the minimal (Type I) seesaw can also be included if the Higgs fields responsible for the $U(1)_R \times U(1)_{B-L}$ breaking carry twice the traditional $U(1)$ charges.

Notice that, in general, these models contain not only gauge couplings per each Abelian gauge group, but also so-called 'off-diagonal couplings', as discussed in [228]. The minimal particle content for such a model, extending the MSSM and leading to the spontaneous breaking of $U(1)_R \times U(1)_{B-L}$ and to neutrino masses, is given in Tab. 14.1. This particle content consists of 3 generations of 16-plets of $SO(10)$, 2 additional Higgs fields and 3 generations of a singlet field. The vector superfields are given in Tab. 14.2.

SF	Spin 0	Spin $\frac{1}{2}$	Generations	$U(1)_{B-L} \times SU(2) \times$ $U(1)_R \times SU(3)$
Matter fields (fermionic components have positive R-parity)				
$\hat{Q}$	$\tilde{Q} = \begin{pmatrix} \tilde{u}_L \\ \tilde{d}_L \end{pmatrix}$	$Q = \begin{pmatrix} u_L \\ d_L \end{pmatrix}$	3	$(\frac{1}{6}, \mathbf{2}, 0, \mathbf{3})$
$\hat{L}$	$\tilde{L} = \begin{pmatrix} \tilde{\nu}_L \\ \tilde{e}_L \end{pmatrix}$	$L = \begin{pmatrix} \nu_L \\ e_L \end{pmatrix}$	3	$(-\frac{1}{2}, \mathbf{2}, 0, \mathbf{1})$
$\hat{u}^c$	$\tilde{u}^c$	u^c	3	$(-\frac{1}{6}, \mathbf{1}, -\frac{1}{2}, \mathbf{\bar{3}})$
$\hat{d}^c$	$\tilde{d}^c$	d^c	3	$(-\frac{1}{6}, \mathbf{1}, \frac{1}{2}, \mathbf{\bar{3}})$
$\hat{\nu}^c$	$\tilde{\nu}^c$	ν^c	3	$(\frac{1}{2}, \mathbf{1}, -\frac{1}{2}, \mathbf{1})$
$\hat{e}^c$	$\tilde{e}^c$	e^c	3	$(\frac{1}{2}, \mathbf{1}, \frac{1}{2}, \mathbf{1})$
$\hat{N}_S$	$\tilde{N}_S$	N_S	n_{N_S}	$(0, \mathbf{1}, 0, \mathbf{1})$
Higgs fields (scalar components have positive R-parity)				
$\hat{H}_d$	$H_d = \begin{pmatrix} H_d^0 \\ H_d^- \end{pmatrix}$	$\tilde{H}_d = \begin{pmatrix} \tilde{H}_d^0 \\ \tilde{H}_d^- \end{pmatrix}$	1	$(0, \mathbf{2}, -\frac{1}{2}, \mathbf{1})$
$\hat{H}_u$	$H_u = \begin{pmatrix} H_u^+ \\ H_u^0 \end{pmatrix}$	$\tilde{H}_u = \begin{pmatrix} \tilde{H}_u^+ \\ \tilde{H}_u^0 \end{pmatrix}$	1	$(0, \mathbf{2}, \frac{1}{2}, \mathbf{1})$
For minimal seesaw				
$\hat{\delta}_R$	δ_R^0	$\tilde{\delta}_R^0$	1	$(-1, \mathbf{1}, 1, \mathbf{1})$
$\hat{\bar{\delta}}_R$	$\bar{\delta}_R^0$	$\tilde{\bar{\delta}}_R^0$	1	$(1, \mathbf{1}, -1, \mathbf{1})$
For linear and inverse seesaw				
$\hat{\xi}_R$	ξ_R^0	$\tilde{\xi}_R^0$	1	$(-\frac{1}{2}, \mathbf{1}, \frac{1}{2}, \mathbf{1})$
$\hat{\bar{\xi}}_R$	$\bar{\xi}_R^0$	$\tilde{\bar{\xi}}_R^0$	1	$(\frac{1}{2}, \mathbf{1}, -\frac{1}{2}, \mathbf{1})$
Fields integrated out (scalar components have positive R-parity)				
$\hat{\xi}_L$	ξ_L^0	$\tilde{\xi}_L^0$	1	$(\frac{1}{2}, \mathbf{2}, 0, \mathbf{1})$
$\hat{\bar{\xi}}_L$	$\bar{\xi}_L^0$	$\tilde{\bar{\xi}}_L^0$	1	$(-\frac{1}{2}, \mathbf{2}, 0, \mathbf{1})$

Table 14.1 Chiral superfields appearing in models with $U(1)_R \times U(1)_{B-L}$ gauge sector, which incorporate minimal, linear and inverse seesaw mechanisms.

SF	Spin $\frac{1}{2}$	Spin 1	$SU(N)$	Coupling	Name
$\hat{B}'$	$\lambda_{\tilde{B}'}$	B'	$U(1)$	g_{BL}	$B-L$
$\hat{W}_L$	λ_L	W_L	$SU(2)$	g_L	left
$\hat{B}_R$	$\lambda_{\tilde{B}_R}$	B_R	$U(1)$	g_R	right
$\hat{g}$	$\lambda_{\tilde{g}}$	g	$SU(3)$	g_s	colour

Table 14.2 Vector superfields appearing in models with $U(1)_R \times U(1)_{B-L}$ gauge sector.

14.1.1 Superpotential

We assume for the following discussion that the superpotential can contain the following terms:

$$W - W_{\text{MSSM}} = Y_\nu \, \hat{\nu}^c \, \hat{L} \, \hat{H}_u - \mu_\delta \, \hat{\bar{\delta}}_R \, \hat{\delta}_R + Y_M \hat{\nu}^c \hat{\delta}_R \hat{\nu}^c \tag{14.2}$$

for the realisation of the minimal seesaw and

$$W - W_{\text{MSSM}} = Y_\nu\, \hat{\nu}^c\, \hat{L}\, \hat{H}_u\, - \mu_\xi\, \hat{\bar{\xi}}_R\, \hat{\xi}_R + Y_{N\nu^c} \hat{N}_S \hat{\nu}^c \hat{\xi}_R$$

$$+ \left\{ \begin{array}{ll} \frac{1}{2}\mu_N \hat{N}_S \hat{N}_S & \text{inverse seesaw} \\ Y_{LS}\, \hat{L}\, \hat{\xi}_L\, \hat{N}_S + Y_{LR}\hat{\xi}_L\hat{\xi}_R\hat{H}_d + \mu_L\hat{\xi}_L\hat{\bar{\xi}}_L & \text{linear seesaw} \end{array} \right. \tag{14.3}$$

for linear and inverse seesaw, respectively. Notice that, again, the $\hat{N}_S$ superfield carries a lepton number and that therefore the term μ_N provides its explicit violation. Since $\mu_L \gg M_{\text{SUSY}}$, the fields $\hat{\xi}_L$ and $\hat{\bar{\xi}}_L$ are integrated out and create an effective operator $\frac{Y_{LR}Y_{LS}}{\mu_L}\hat{L}\,\hat{N}_S\hat{H}_d\hat{\xi}_R$.

14.1.2 Soft-breaking terms

The soft-breaking terms in the matter sector are

$$\mathcal{L}_{\text{SB},W} - \mathcal{L}_{\text{SB},W,\text{MSSM}} = T_\nu\, \tilde{\nu}^c\, \tilde{L}\, H_u\, - B_\delta\, \bar{\delta}_R\, \delta_R, + T_M\tilde{\nu}^c\delta^0_R\tilde{\nu}^c + \text{h.c.},$$

$$\mathcal{L}_{\text{SB},\phi} - \mathcal{L}_{\text{SB},\phi,\text{MSSM}} = -m^2_\delta|\delta_R|^2 - m^2_{\bar{\delta}}|\bar{\delta}_R|^2 - (\tilde{\nu}^c)^\dagger m^2_{\nu^c}\tilde{\nu}^c, \tag{14.4}$$

respectively, given by

$$\mathcal{L}_{\text{SB},W} - \mathcal{L}_{\text{SB},W,\text{MSSM}} = T_\nu\, \tilde{\nu}^c\, \tilde{L}\, H_u\, - B_\xi\, \bar{\xi}_R\, \xi_R, + T_{N\nu^c}\tilde{N}_S\tilde{\nu}^c\xi_R$$

$$+ \left\{ \begin{array}{ll} \frac{1}{2}B_N \tilde{N}_S \tilde{N}_S & \text{inverse seesaw} \\ T_{LS}\, \tilde{L}\, \xi_L\, \tilde{N}_S + T_{LR}\xi_L\xi_R H_d + B_L\xi_L\bar{\xi}_L & \text{linear seesaw} \end{array} \right\} + \text{h.c.}, \tag{14.5}$$

$$\mathcal{L}_{\text{SB},\phi} - \mathcal{L}_{\text{SB},\phi,\text{MSSM}} = -\tilde{N}^\dagger_S m^2_N \tilde{S} - m^2_\xi|\xi_R|^2 - m^2_{\bar{\xi}}|\bar{\xi}_R|^2 - (\tilde{\nu}^c)^\dagger m^2_{\nu^c}\tilde{\nu}^c$$

$$- m^2_{\xi_L}|\xi_L|^2 - m^2_{\bar{\xi}_L}|\bar{\xi}_L|^2, \tag{14.6}$$

while the soft-breaking gaugino sector reads

$$\mathcal{L}_{\text{SB},\lambda} = \frac{1}{2}\left(-\lambda^2_{\tilde{B}'}M_{BL} - 2\lambda_{\tilde{B}'}\lambda_{\tilde{B}_R}M_{RB} - M_2\lambda^2_{\tilde{W},i} - M_3\lambda^2_{\tilde{g},\alpha} - \lambda^2_{\tilde{B}_R}M_R + \text{h.c.} \right). \tag{14.7}$$

The term $\lambda_{\tilde{B}'}\lambda_{\tilde{B}_R}M_{RB}$ is due to the presence of two Abelian gauge groups [228].

14.1.3 Symmetry breaking

Since it is assumed that in these models the scale of spontaneous symmetry breaking to the SM gauge group is near the TeV scale, it is possible to restrict ourselves to a direct one-step breaking pattern, i.e., $SU(2)_L \times U(1)_R \times U(1)_{B-L} \to U(1)_{\text{EM}}$. This breaking pattern takes place when the Higgs fields in the left and right sectors receive VEVs. We can parameterise the scalar fields as follows:

$$H^0_d = \frac{1}{\sqrt{2}}\left(v_d + \sigma_d + i\phi_d \right), \quad H^0_u = \frac{1}{\sqrt{2}}\left(v_u + \sigma_u + i\phi_u \right), \tag{14.8}$$

$$X_R = \frac{1}{\sqrt{2}}\left(v_{X_R} + \sigma_{X_R} + i\phi_{X_R} \right), \quad \bar{X}_R = \frac{1}{\sqrt{2}}\left(v_{\bar{X}_R} + \sigma_{\bar{X}_R} + i\phi_{\bar{X}_R} \right), \tag{14.9}$$

with $X = \xi, \delta$. It is useful to define the quantities $v_R = v^2_{X_R} + v^2_{\bar{X}_R}$ and $\tan\beta_R = \frac{v_{\bar{X}_R}}{v_{X_R}}$, in analogy to $v^2 = v^2_d + v^2_u$ and to $\tan\beta = \frac{v_u}{v_d}$.

14.1.4 Particle mixing

Additional mixing effects take place in the gauge and Higgs sectors due to the additional gauge fields considered, besides the neutrino and sneutrino ones. The 3 neutral gauge bosons B', B_R and W^3 mix to form 3 mass eigenstates: the massless photon, the well-known Z boson and a new Z' boson. This mixing can be parameterised by a unitary 3×3 matrix which diagonalises the mass matrix of the gauge bosons, such as

$$(\gamma, Z, Z')^T = U^{\gamma Z Z'}(B', B_R, W^3)^T. \tag{14.10}$$

Similarly, this model contains 7 neutralinos which are an admixture of the 3 neutral gauginos, of the 2 neutral components of the Higgsino doublets and 2 additional fermions coming from the right sector. The mass matrix, written in the basis $\left(\lambda_{\tilde{B}'}, \tilde{W}^0, \tilde{H}_d^0, \tilde{H}_u^0, \lambda_{\tilde{B}_R}, \tilde{X}_R, \tilde{\tilde{X}}_R\right)$, with $X = \xi, \delta$, can be diagonalised by a unitary matrix, here denoted with Z^N. In the Higgs sector we choose the mixing basis and rotation matrices to be, respectively,

- Scalar Higgs fields: $\left(\sigma_d, \sigma_u, \sigma_{X_R}, \sigma_{\bar{X}_R}\right)^T$ and Z^H,

- Pseudoscalar Higgs fields: $\left(\phi_d, \phi_u, \phi_X, \phi_{\bar{X}_R}\right)^T$ and Z^A.

The neutrino and sneutrino sectors are similar to the case for inverse and linear seesaw discussed in the previous chapter: the scalar fields are decomposed into their CP-even and CP-odd components according to Eq. (13.32) in Chapter 13. The mass matrices are defined in the same basis. If all terms of Eq. (14.3) are present, the resulting masses of the light neutrinos are a result of a mixed linear and inverse seesaw. The mass matrix for inverse seesaw is analogue to the left matrix given in Eq. (13.33) in Chapter 13 with M_R replaced by $\frac{1}{\sqrt{2}} Y_{N\nu^c} v_{\xi_R} \equiv \tilde{Y}$, while the mass matrix for linear seesaw is given by the right matrix in Eq. (13.33) and the replacement $Y_{LS} \to \frac{Y_{LS} Y_{LR}}{\sqrt{2}\mu_L} v_\eta$. For the minimal realisation with the superpotential given in Eq. (14.2), the neutrino mass matrix is given by

$$\begin{pmatrix} 0 & \frac{v_u}{\sqrt{2}} Y_\nu \\ \frac{v_u}{\sqrt{2}} Y_\nu^T & \frac{2v_{\delta_R}}{\sqrt{2}} Y_M \end{pmatrix}. \tag{14.11}$$

14.1.5 Free parameters

If cMSSM-like boundary conditions are assumed, new free parameters arise in addition to those given in Eq. (13.25) in Chapter 13 for inverse seesaw,

$$Y_{N\nu^c}, \ \mu_N, \ B_0, \ \tan \beta_R, \ \mathrm{sign}(\mu_\xi) \ M_{Z'}, \tag{14.12}$$

and for linear seesaw,

$$Y_{LS}, \ Y_{LR}, \ \mu_L. \tag{14.13}$$

Here we have assumed that the parameters μ_ξ and B_ξ are fixed by the tadpole equations. The relationships of the soft trilinear terms to the Yukawa couplings are as before and $B_N = B_0 \mu_N$.

14.2 SEESAW MODELS WITH $U(1)_Y \times U(1)_{B-L}$ GAUGE SECTOR

The final category of models considered here includes an additional $B - L$ gauge group tensored to the SM gauge groups, i.e., $SU(3)_C \times SU(2)_L \times U(1)_Y \times U(1)_{B-L}$. The corresponding vector superfields are given in Tab. 14.3. The minimal version of these models [21, 228, 258, 259] extends the MSSM particle content with three generations of RH superfields. Two additional scalars, singlets with respect to SM gauge interactions but carrying $B - L$ charge , are added to break $U(1)_{B-L}$, as well as allowing for a Majorana mass term for the RH neutrino superfields. Furthermore, two new lepton fields per generation can be included to specifically implement the inverse seesaw mechanism [260, 261], as well as the linear seesaw realisation if a further two doublet fields ($\hat{\rho}$ and $\hat{\bar{\rho}}$) are considered, to be integrated out. All particles and their quantum numbers are given in Tab. 14.4. This table also contains the charge assignment under a Z_2 symmetry, which is just present in the case of the inverse and linear seesaw models[1].

SF	Spin $\frac{1}{2}$	Spin 1	$SU(N)$	Coupling	Name
$\hat{B}$	$\lambda_{\tilde{B}}$	B	$U(1)$	g_1	hypercharge
$\hat{W}$	$\lambda_{\tilde{W}}$	W^-	$SU(2)$	g_2	left
$\hat{g}$	$\lambda_{\tilde{g}}$	g	$SU(3)$	g_3	colour
$\hat{B}'$	$\lambda_{\tilde{B}'}$	B'	$U(1)$	g_B	$B - L$

Table 14.3 Vector superfields appearing in models with $U(1)_Y \times U(1)_{B-L}$ gauge sector.

The additional terms for the superpotential in comparison to the MSSM read as

$$W - W_{\text{MSSM}} = Y_\nu \hat{\nu}^c \hat{l} \hat{H}_u + \begin{cases} Y_{\eta\nu^c} \hat{\nu}^c \hat{\eta} \hat{\nu}^c - \mu_\eta \hat{\eta} \hat{\bar{\eta}} & \text{minimal seesaw,} \\ Y_{IS} \hat{\nu}^c \hat{\eta} \hat{N}_S + \mu_N \hat{N}_S \hat{N}_S & \text{inverse seesaw,} \\ Y_{IS} \hat{\nu}^c \hat{\eta} \hat{N}_S + Y_{LS} \hat{L} \hat{\rho} \hat{N}_S + Y_{LR} \hat{\rho} \hat{\eta} \hat{H}_d \\ \quad + \overline{Y}_{LR} \hat{\bar{\rho}} \hat{\bar{\eta}} \hat{H}_u + \mu_\rho \hat{\rho} \hat{\bar{\rho}} & \text{linear seesaw.} \end{cases}$$
$$(14.14)$$

The extra parity is not required in the minimal case, thus the seesaw term $Y_{\eta\nu^c} \hat{\nu}^c \hat{\eta} \hat{\nu}^c$ is allowed, whereas the latter is forbidden in the other seesaw realisations precisely due to the Z_2 symmetry. In the inverse seesaw model, the $\mu_N \hat{N}_S \hat{N}_S$ term plays an important role, even though, since $\hat{N}_S$ carries $B - L$ charge , this term violates $B - L$ charge conservation similarly to the terms in Eq. (13.29) and Eq. (14.3). This term is assumed to come from higher-order effects. (A similar mass term for $\hat{N}'_S$ is possible, however, it is not relevant since the N'_S does not take part in the mixing with the neutrinos.) A possible bilinear term $\hat{N}_S \hat{N}'_S$ is forbidden by the Z_2 symmetry given in the last column of Tab. 14.4. This discrete symmetry also forbids terms like $\hat{N}_S \hat{N}_S \eta$ or $\hat{N}'_S \hat{N}'_S \bar{\eta}$ as well as the μ_η-term that is necessary to obtain a pure minimal seesaw scenario. In the case of a linear seesaw, the heavy fields $\hat{\rho}$ and $\hat{\bar{\rho}}$ are present and get integrated at $\mu_\rho \gg M_{\text{SUSY}}$ similarly to the linear seesaw case in $U(1)_R \times U(1)_{B-L}$. This creates an effective operator of the form $\frac{Y_{LR} Y_{LS}}{\mu_\rho} \hat{L} \hat{N}_S \hat{H}_d \hat{\eta}$. Notice that it is assumed that in the linear seesaw the term μ_N is not generated at higher loop-level.

___

[1] Notice that, in comparison to [260], the charge assignments of the new particles in the inverse seesaw model, as well as the Z_2 symmetry, have been redefined for consistency with the similar minimal model of [258].

SF	Spin 0	Spin $\frac{1}{2}$	Generations	$U(1)_Y \times$ SU(2)$\times$ SU(3) $\times U(1)_{B-L}$	Z_2 inverse SS
			Matter fields (fermionic components have positive R-parity)		
$\hat{Q}$	$\tilde{Q} = \begin{pmatrix} \tilde{u}_L \\ \tilde{d}_L \end{pmatrix}$	$Q = \begin{pmatrix} u_L \\ d_L \end{pmatrix}$	3	$(\frac{1}{6}, \mathbf{2}, \mathbf{3}, \frac{1}{6})$	$+$
$\hat{L}$	$\tilde{L} = \begin{pmatrix} \tilde{\nu}_L \\ \tilde{e}_L \end{pmatrix}$	$L = \begin{pmatrix} \nu_L \\ e_L \end{pmatrix}$	3	$(-\frac{1}{2}, \mathbf{2}, \mathbf{1}, -\frac{1}{2})$	$+$
$\hat{d}^c$	$\tilde{d}^c$	d^c	3	$(\frac{1}{3}, \mathbf{1}, \overline{\mathbf{3}}, -\frac{1}{6})$	$+$
$\hat{u}^c$	$\tilde{u}^c$	u^c	3	$(-\frac{2}{3}, \mathbf{1}, \overline{\mathbf{3}}, -\frac{1}{6})$	$+$
$\hat{e}^c$	$\tilde{e}^c$	e^c	3	$(1, \mathbf{1}, \mathbf{1}, \frac{1}{2})$	$+$
$\hat{\nu}^c$	$\tilde{\nu}^c$	ν^c	3	$(0, \mathbf{1}, \mathbf{1}, \frac{1}{2})$	$+$
			Higgs fields (scalar components have positive R-parity)		
$\hat{\eta}$	η	$\tilde{\eta}$	1	$(0, \mathbf{1}, \mathbf{1}, -1)$	$-$
$\hat{\bar{\eta}}$	$\bar{\eta}$	$\tilde{\bar{\eta}}$	1	$(0, \mathbf{1}, \mathbf{1}, +1)$	$-$
$\hat{H}_d$	H_d	$\tilde{H}_d$	1	$(-\frac{1}{2}, \mathbf{2}, \mathbf{1}, 0)$	$+$
$\hat{H}_u$	H_u	$\tilde{H}_u$	1	$(\frac{1}{2}, \mathbf{2}, \mathbf{1}, 0)$	$+$
			Additional field for inverse seesaw (fermionic components have positive R-parity)		
$\hat{N}_S$	$\tilde{N}_S$	N_S	n_{N_S}	$(0, \mathbf{1}, \mathbf{1}, -\frac{1}{2})$	$-$
$\hat{N}'_S$	$\tilde{N}'_S$	N'_S	n_{N_S}	$(0, \mathbf{1}, \mathbf{1}, \frac{1}{2})$	$+$
			Fields integrated out (for linear seesaw)		
$\hat{\rho}$	ρ	$\tilde{\rho}$	1	$(\frac{1}{2}, \mathbf{2}, \mathbf{1}, 1)$	$-$
$\hat{\bar{\rho}}$	$\bar{\rho}$	$\tilde{\bar{\rho}}$	1	$(-\frac{1}{2}, \mathbf{2}, \mathbf{1}, -1)$	$-$

Table 14.4 Chiral superfields appearing in models with $U(1)_Y \times U(1)_{B-L}$ gauge sector. The minimal particle content is needed for Type I seesaw, the additional fields can be used to incorporate inverse or linear seesaw. The $SU(2)_L$ doublets are named as in Tab. 14.1. The Z_2 in the last column is not present in the minimal model but just in the model with inverse and linear seesaw. The number of generations of $\hat{N}'_S$ must match those of $\hat{N}_S$ for anomaly cancellation.

The additional soft-breaking terms are written as follows:

$$L_{\text{SB},w} - L_{\text{SB},w,\text{MSSM}} = T_\nu \tilde{\nu}^c \tilde{l} H_u$$

$$+ \left\{ \begin{array}{ll} T_{\eta\nu^c} \tilde{\nu}^c \eta \tilde{\nu}^c - B_\eta \eta \bar{\eta} & \text{minimal seesaw} \\ T_{IS} \tilde{\nu}^c \eta \tilde{N}_S + B_N \tilde{N}_S \tilde{N}_S & \text{inverse seesaw} \\ T_{IS} \tilde{\nu}^c \eta \tilde{N}_S + T_{LS} \tilde{L} \rho \tilde{N}_S + T_{LR} \rho \eta H_d & \\ \quad + \overline{T}_{LR} \bar{\rho} \bar{\eta} H_u + B_\rho \rho \bar{\rho} & \text{linear seesaw} \end{array} \right\} + \text{h.c.} \qquad (14.15)$$

$$L_{\text{SB},\phi} - L_{\text{SB},\phi,\text{MSSM}} = -m_\eta^2 |\eta|^2 - m_{\bar{\eta}}^2 |\bar{\eta}|^2$$

$$- (\tilde{N}_S^* m_N^2 \tilde{N}_S + \tilde{N}_{S'}^* m_{N'}^2 \tilde{N}'_S + m_\rho^2 |\rho|^2 + m_{\bar{\rho}}^2 |\bar{\rho}|^2) \qquad (14.16)$$

$$L_{\text{SB},\lambda} = \frac{1}{2} \left(-\lambda_{\tilde{B}}^2 M_1 - \lambda_{\tilde{B}} \lambda_{\tilde{B}'} M_{BB'} - M_2 \lambda_{\tilde{W},i}^2 - M_3 \lambda_{\tilde{g},\alpha}^2 - \lambda_{\tilde{B}'}^2 M_{BL} + \text{h.c.} \right). \qquad (14.17)$$

The soft-breaking terms m_ρ^2 and $m_{\bar{\rho}}^2$ are just present in linear seesaw while m_N^2 and $m_{N'}^2$ exist for both linear and inverse seesaw. To break $SU(2)_L \times U(1)_Y \times U(1)_{B-L}$ to $U(1)_{\text{EM}}$, the neutral MSSM Higgs fields and the new SM scalar singlets acquire VEVs:

$$H_d^0 = \frac{1}{\sqrt{2}} \left(v_d + \sigma_d + i\phi_d \right), \; H_u^0 = \frac{1}{\sqrt{2}} \left(v_u + \sigma_u + i\phi_u \right), \tag{14.18}$$

$$\eta = \frac{1}{\sqrt{2}} \left(\sigma_\eta + v_\eta + i\phi_\eta \right), \; \bar\eta = \frac{1}{\sqrt{2}} \left(\sigma_{\bar\eta} + v_{\bar\eta} + i\phi_{\bar\eta} \right). \tag{14.19}$$

We also define here

$$v^2 = v_d^2 + v_u^2, \qquad \tan\beta = \frac{v_u}{v_d} \tag{14.20}$$

as well as

$$v'^2 = v_\eta^2 + v_{\bar\eta}^2, \qquad \tan\beta' = \frac{v_{\bar\eta}}{v_\eta}. \tag{14.21}$$

The LH and RH sneutrinos are decomposed into their scalar and pseudoscalar components according to Eq. (13.32) of Chapter 13. Similarly, in the inverse seesaw model, the scalar component of $\hat N_S$ reads

$$\tilde N_S = \frac{1}{\sqrt{2}} \left(\sigma_S + i\phi_S \right). \tag{14.22}$$

The additional mixing effects which take place in this model are similar to the case of $U(1)_R \times U(1)_{B-L}$ discussed in Sect. 14.1. In the gauge sector, three neutral gauge bosons appear, which mix to give rise to the massless photon, the Z boson and a Z' boson:

$$(\gamma, Z, Z')^T = U^{\gamma Z} (B, W^3, B')^T. \tag{14.23}$$

Notice that, when the kinetic mixing is neglected, the B' field decouples and the mass matrix of the gauge bosons becomes block diagonal, where the upper 2×2 block reads as in the SM. In the matter sector we choose the basis and the mixing matrices, which diagonalise the mass matrices, respectively, as follows:

- Neutralinos: $\left(\lambda_{\tilde B}, \tilde W^0, \tilde H_d^0, \tilde H_u^0, \lambda_{\tilde B'}, \tilde\eta, \tilde{\bar\eta} \right)^T$ and Z^N

- Scalar Higgs fields: $(\sigma_d, \sigma_u, \sigma_\eta, \sigma_{\bar\eta})^T$ and Z^H

- Pseudoscalar Higgs fields: $(\phi_d, \phi_u, \phi_\eta, \phi_{\bar\eta})^T$ and Z^A

- Scalar sneutrinos: $(\sigma_L, \sigma_R, \sigma_S)^T$ and $Z^{\sigma v}$

- Pseudoscalar sneutrinos: $(\phi_L, \phi_R, \phi_S)^T$ and $Z^{\phi v}$

- Neutrinos: $(\nu_L, \nu^c, N_S)^T$ and U^V

where the σ_S, ϕ_S and N_S are only present in the inverse seesaw case. The neutrino mass matrices for the minimal, inverse and linear seesaw cases can be written as

$$\begin{pmatrix} 0 & \frac{v_u}{\sqrt{2}} Y_\nu \\ \frac{v_u}{\sqrt{2}} Y_\nu^T & \frac{2v_\eta}{\sqrt{2}} Y_{\eta\nu^c} \end{pmatrix} \text{ or } \begin{pmatrix} 0 & \frac{v_u}{\sqrt{2}} Y_\nu & 0 \\ \frac{v_u}{\sqrt{2}} Y_\nu^T & 0 & \frac{v_\eta}{\sqrt{2}} Y_{IS} \\ 0 & \frac{v_\eta}{\sqrt{2}} Y_{IS}^T & \mu_N \end{pmatrix} \text{ or } \begin{pmatrix} 0 & \frac{v_u}{\sqrt{2}} Y_\nu & \frac{\tilde Y}{2} v_d v_\eta \\ \frac{v_u}{\sqrt{2}} Y_\nu^T & 0 & \frac{v_\eta}{\sqrt{2}} Y_{IS} \\ \frac{\tilde Y^T}{2} v_d v_\eta & \frac{v_\eta}{\sqrt{2}} Y_{IS}^T & 0 \end{pmatrix}$$
$$\tag{14.24}$$

respectively. The $\tilde Y$ is the running value of the effective $\frac{Y_{LR} Y_{LS}}{\mu_\rho}$ caused by the heavy superfields ρ and $\bar\rho$. In the minimal case, the seesaw of Type I is recovered. The light neutrino mass matrix can be approximated in this case by

$$m_\nu \simeq -\frac{v_u^2}{2\sqrt{2} v_\eta} Y_\nu Y_{\eta\nu^c}^{-1} Y_\nu. \tag{14.25}$$

In the inverse seesaw case, the light neutrino masses can instead be written as

$$m_\nu \simeq -\frac{v_u^2}{2} Y_\nu^T (Y_{IS}^T)^{-1} \mu_N Y_{IS} Y_\nu. \qquad (14.26)$$

14.2.1 Free parameters

If the minimum conditions for the vacuum are solved with respect to μ, B_μ, μ_η and B_η, the following parameters can be treated as free in addition to those given in Eq. (13.25) of Chapter. 13:

$$Y_{\eta\nu^c}, \ Y_{IS}, \ \mu_N, \ B_0, \ \tan\beta', \ \text{sign}(\mu_\eta), \ M_{Z'}, \qquad (14.27)$$

where we have again used $B_N = B_0 \mu_N$.

14.3 THE BLSSM

The BLSSM (at times, also acronymed as $(B-L)$ SSM in the literature) is representative of the class of supersymmetric seesaw scenarios with the $U(1)_Y \times U(1)_{B-L}$ gauge sector discussed in Sect. 14.2. We will discuss two of its possible realisations, Type I and inverse seesaw, in the two upcoming subsections.

14.3.1 The BLSSM with Type I seesaw

The Type I BLSSM (BLSSM-I for short) was first introduced in [258, 259] and later worked out in detail (i.e., spectra of masses and couplings were derived) in [21, 262]. While to a large extent the Higgs sector of the BLSSM is similar to the same sector of the MSSM, an important aspect is the $U(1)$ gauge kinetic mixing, which leads to significant changes in the spectrum involving the Z' boson and its supersymmetric counterparts. We will illustrate this for the case of Z' and –ino states in the upcoming subsections, whereas we will treat the Higgs sector in Chapter 15. As usual, unless otherwise indicated, we restrict ourselves to tree-level expressions, as this is sufficient for discussing the main differences with respect to the MSSM.

14.3.1.1 Particle content and superpotential

Here we follow the comprehensive review of the BLSSM-I spectrum given in [21]. The model consists of three generations of matter particles including RH neutrinos. Moreover, below the GUT scale the usual MSSM Higgs doublets are present as well as two fields η and $\bar{\eta}$ responsible for the breaking of the $U(1)_{B-L}$. Furthermore, η is responsible for generating a Majorana mass term for the RH neutrinos and thus we interpret the $B-L$ charge of this field as its lepton number, likewise for $\bar{\eta}$, and call these fields bileptons since they carry twice the lepton number of (anti-)neutrinos.

14.3.1.2 Gauge kinetic mixing

It is well known that in models with several $U(1)$ gauge groups, the following kinetic mixing terms

$$-\chi_{ab} \hat{F}^{a,\mu\nu} \hat{F}^b_{\mu\nu}, \quad a \neq b \qquad (14.28)$$

between the field strength tensors are allowed by gauge and Lorentz invariance [263], as $\hat{F}^{a,\mu\nu}$ and $\hat{F}^{b,\mu\nu}$ are gauge invariant quantities by themselves, see, e.g., [264]. Even if these terms are absent at tree level at a particular scale, they might be generated by RGE effects [265,266]. This can be seen most easily by inspecting the matrix of the anomalous dimension, which at one loop is given by

$$\gamma_{ab} = \frac{1}{16\pi^2} \mathrm{Tr} Q_a Q_b, \tag{14.29}$$

where the indices a and b run over all $U(1)$ groups and the trace runs over all fields charged under the corresponding $U(1)$ group.

For our model we obtain

$$\gamma = \frac{1}{16\pi^2} N \begin{pmatrix} 11 & 4 \\ 4 & 6 \end{pmatrix} N \tag{14.30}$$

and we see that there are sizable off-diagonal elements. Here, N contains the GUT normalisation of the two Abelian gauge groups. We will take as in [259] $\sqrt{\frac{3}{5}}$ for $U(1)_Y$ and $\sqrt{\frac{3}{2}}$ for $U(1)_{B-L}$, i.e., $N = \mathrm{diag}(\sqrt{\frac{3}{5}}, \sqrt{\frac{3}{2}})$. Hence, we finally obtain

$$\gamma = \frac{1}{16\pi^2} \begin{pmatrix} \frac{33}{5} & 6\sqrt{\frac{2}{5}} \\ 6\sqrt{\frac{2}{5}} & 9 \end{pmatrix}. \tag{14.31}$$

Therefore, even if at the GUT scale the $U(1)$ kinetic mixing terms are zero, they are induced via RGE evolution to lower scales. In practice, it turns out that it is easier to work with non-canonical covariant derivatives instead of off-diagonal field-strength tensors such as in Eq. (14.28). However, both approaches are equivalent [267]. Hence, in the following, we consider covariant derivatives of the form

$$D_\mu = \partial_\mu - iQ_\phi^T GA, \tag{14.32}$$

where Q_ϕ is a vector containing the charges of the field ϕ with respect to the two Abelian gauge groups, G is the gauge coupling matrix

$$G = \begin{pmatrix} g_{YY} & g_{YB} \\ g_{BY} & g_{BB} \end{pmatrix}, \tag{14.33}$$

and A contains the gauge bosons $A = (A_\mu^Y, A_\mu^B)^T$.

As long as the two Abelian gauge groups are unbroken, we still have the freedom to perform a change of basis: $A = (A_\mu^Y, A_\mu^B) \to A' = ((A_\mu^Y)', (A_\mu^B)') = RA$ where R is an orthogonal matrix. It is possible to absorb this rotation of the gauge fields completely in the definition of the gauge couplings without the necessity of changing the charges, which can easily be seen using Eq. (14.32)

$$Q_\phi^T GA = Q_\phi^T G(R^T R)A = Q_\phi^T (GR^T)A' = Q_\phi^T \tilde{G}A'. \tag{14.34}$$

This freedom can be used to choose a basis such that EW precision data can be accommodated in an easy way. A convenient choice is the basis where $g_{BY} = 0$, as in this basis only the Higgs doublets contribute to the entries in the gauge boson mass matrix of the $U(1)_Y \times SU(2)_L$ sector and the impact of η and $\bar\eta$ is only in the off-diagonal elements as discussed in Sect. 14.3.1.4. Therefore, we choose the following basis at the EW scale [268]:

$$g'_{YY} = \frac{g_{YY}g_{BB} - g_{YB}g_{BY}}{\sqrt{g_{BB}^2 + g_{BY}^2}} = g_1, \tag{14.35}$$

$$g'_{BB} = \sqrt{g_{BB}^2 + g_{BY}^2} = g_{BL}, \tag{14.36}$$

$$g'_{YB} = \frac{g_{YB}g_{BB} + g_{BY}g_{YY}}{\sqrt{g_{BB}^2 + g_{BY}^2}} = \tilde{g}, \tag{14.37}$$

$$g'_{BY} = 0. \tag{14.38}$$

This also leads to our condition for finding the GUT scale in the numerical analysis:

$$g_2 \equiv \frac{g_{YY}g_{BB} - g_{YB}g_{BY}}{\sqrt{g_{BB}^2 + g_{BY}^2}}. \tag{14.39}$$

This is equivalent to a rotation of the general 2×2 gauge coupling matrix at each energy scale to the triangle form and using $g_1 = g_2$ as the GUT condition. Neglecting threshold corrections, this leads in the case of kinetic mixing to exactly the same GUT scale, as in the MSSM [269].

Immediate interesting consequences of the gauge kinetic mixing arise in various sectors of the model, as discussed in the subsequent subsections: *(i)* it induces mixing at tree level between the H_u, H_d and η, $\bar{\eta}$; *(ii)* additional D-terms contribute to the mass matrices of the squarks and sleptons; *(iii)* off-diagonal soft-SUSY breaking terms for the gauginos are induced via RGE evolution [267,270] with important consequences for the neutralino sector as discussed in Sect. 14.3.1.5, even if at some fixed scale $M_{ab} = 0$ for $a \neq b$.

14.3.1.3 Tadpole equations

We find for the four minimisation conditions at tree level

$$t_d = v_d \left(m_{H_d}^2 + |\mu|^2 + \frac{1}{8} \left(g_1^2 + g_2^2 + \tilde{g}^2 \right) \left(v_d^2 - v_u^2 \right) + \frac{1}{4} \tilde{g} g_{BL} \left(v_\eta^2 - v_{\bar{\eta}}^2 \right) \right) - v_u B_\mu = 0, \tag{14.40}$$

$$t_u = v_u \left(m_{H_u}^2 + |\mu|^2 + \frac{1}{8} \left(g_1^2 + g_2^2 + \tilde{g}^2 \right) \left(v_u^2 - v_d^2 \right) + \frac{1}{4} \tilde{g} g_{BL} \left(v_{\bar{\eta}}^2 - v_\eta^2 \right) \right) - v_d B_\mu = 0, \tag{14.41}$$

$$t_\eta = v_\eta \left(m_\eta^2 + |\mu'|^2 + \frac{1}{4} \tilde{g} g_{BL} \left(v_d^2 - v_u^2 \right) + \frac{1}{2} g_{BL}^2 \left(v_\eta^2 - v_{\bar{\eta}}^2 \right) \right) - v_{\bar{\eta}} B_{\mu'} = 0, \tag{14.42}$$

$$t_{\bar{\eta}} = v_{\bar{\eta}} \left(m_{\bar{\eta}}^2 + |\mu'|^2 + \frac{1}{4} \tilde{g} g_{BL} \left(v_u^2 - v_d^2 \right) + \frac{1}{2} g_{BL}^2 \left(v_{\bar{\eta}}^2 - v_\eta^2 \right) \right) - v_\eta B_{\mu'} = 0. \tag{14.43}$$

We solve them with respect to μ, B_μ, μ' and $B_{\mu'}$, as these parameters do not enter any of the RGEs of the other parameters. We obtain

$$|\mu|^2 = \frac{1}{8}\left(\left(2\tilde{g}g_{BL}v'^2\cos(2\beta') - 4m_{H_d}^2 + 4m_{H_u}^2\right)\sec(2\beta) - 4\left(m_{H_d}^2 + m_{H_u}^2\right),\right. \tag{14.44}$$

$$\left. - \left(g_1^2 + \tilde{g}^2 + g_2^2\right)v^2\right) \tag{14.45}$$

$$B_\mu = -\frac{1}{8}\left(-2\tilde{g}g_{BL}v'^2\cos(2\beta') + 4m_{H_d}^2 - 4m_{H_u}^2 + \left(g_1^2 + \tilde{g}^2 + g_2^2\right)v^2\cos(2\beta)\right)\tan(2\beta), \tag{14.46}$$

$$|\mu'|^2 = \frac{1}{4}\left(-2\left(g_{BL}^2 v'^2 + m_\eta^2 + m_{\bar{\eta}}^2\right) + \left(2m_\eta^2 - 2m_{\bar{\eta}}^2 + \tilde{g}g_{BL}v^2\cos(2\beta)\right)\sec(2\beta'),\right) \tag{14.47}$$

$$B_{\mu'} = \frac{1}{4}\left(-2g_{BL}^2 v'^2\cos(2\beta') + 2m_\eta^2 - 2m_{\bar{\eta}}^2 + \tilde{g}g_{BL}v^2\cos(2\beta)\right)\tan(2\beta'). \tag{14.48}$$

With $M_Z' \simeq g_{BL}v'$, as we will show in Sect. 14.3.1.4, we find an approximate relation between M_Z' and μ':

$$M_{Z'}^2 \simeq -2|\mu'|^2 + \frac{4(m_{\bar{\eta}}^2 - m_\eta^2\tan^2\beta') - v^2\tilde{g}g_{BL}\cos\beta(1 + \tan\beta')}{2(\tan^2\beta' - 1)}. \tag{14.49}$$

A closer inspection of the system shows that either $m_{\bar{\eta}}^2$ or m_η^2 has to become negative to break $U(1)_{B-L}$. For both parameters, gauge couplings enter the RGEs, increasing their values when evolving from the GUT scale to the EW scale. The Yukawa couplings Y_ν and Y_x as well as the trilinear couplings T_ν and T_x lead to a decrease, but at the one-loop level they only affect the RGE for m_η^2. However, neutrino data require $|Y_{\nu,ij}|$ to be very small in this model and thus they can be neglected for these considerations. Therefore, $m_{\bar{\eta}}$ will always be positive whereas m_η^2 can become negative for sufficiently large Y_x and T_x.

We can roughly estimate the contribution of these couplings to the running value of m_η^2 by a one-step integration assuming mSUGRA-like GUT conditions (see Sect. 14.3.1.7) to

$$\Delta m_\eta^2 \simeq -\frac{1}{4\pi^2}\text{Tr}(Y_x Y_x^\dagger)(3m_0^2 + A_0^2)\log\left(\frac{M_{\text{GUT}}}{M_{\text{SUSY}}}\right), \tag{14.50}$$

with $T_x \simeq A_0 Y_x$. Therefore, we expect that large values of m_0 and A_0 will be preferred, implying heavy sfermions. Moreover, $\tan\beta'$ has to be small and of $\mathcal{O}(1)$ in order to get a small denominator in the second term of Eq. (14.49). One last comment concerning the effect of gauge kinetic mixing: g_{YB} is always negative below the GUT scale, if it is zero at the GUT scale, as can be seen by the following argument. For vanishing off-diagonal gauge couplings, the β-functions (see [21]) will always be positive, i.e., g_{BY} and g_{YB} are driven negative. Using Eq. (14.37), one can see that this also drives $\tilde{g}$ negative. Therefore, the second term will give a positive contribution. From this point of view, one might expect that small m_0 for given $\tan\beta'$ would be sufficient to get the same size of $|\mu'|$. However, as can be seen in Fig. 14.1, where we plot the tree-level value of μ' in the $(m_0, \tan\beta')$-plane for the cases with and without kinetic mixing, the opposite effect takes place. The reason is the contribution of the kinetic mixing to the evaluation of m_η^2 and $m_{\bar{\eta}}^2$. One can also see in this figure that the upper limit of $\tan\beta'$ for a given value of m_0 decreases with increasing $M_{Z'}$, as expected. Even if one might get the impression from this figure that the effects of kinetic mixing are in general small as they slightly shift the region where breaking of $U(1)_{B-L}$ can occur, one can show that it can have a significant impact on the masses.

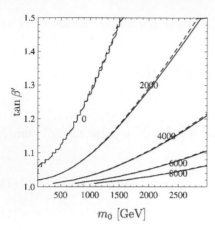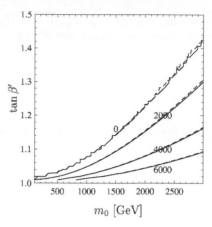

Figure 14.1 Contour plots of μ' at tree-level in the $(m_0, \tan \beta')$-plane for $M_{Z'} = 2000$ GeV (left) and $M_{Z'} = 4000$ GeV (right). The other parameters are $M_{1/2} = 0.5$ TeV, $\tan(\beta) = 10$, $A_0 = 1.5$ TeV, $Y_{x,ii} = 0.42$. The full lines correspond to the case including gauge kinetic mixing, the dashed lines are without kinetic mixing, taken from [21].

14.3.1.4 Gauge boson mixing

Due to the presence of the kinetic mixing terms, the B' boson mixes at tree level with the B and W^3 bosons. Requiring the conditions of Eqs. (14.35)–(14.38) means that the corresponding mass matrix reads, in the basis (B, W^3, B'),

$$
\begin{pmatrix}
\frac{1}{4} g_1^2 v^2 & -\frac{1}{4} g_1 g_2 v^2 & \frac{1}{4} g_1 \tilde{g} v^2 \\
-\frac{1}{4} g_1 g_2 v^2 & \frac{1}{4} g_2^2 v^2 & -\frac{1}{4} \tilde{g} g_2 v^2 \\
\frac{1}{4} g_1 \tilde{g} v^2 & -\frac{1}{4} \tilde{g} g_2 v^2 & \left(g_{BL}^2 v'^2 + \frac{1}{4} \tilde{g}^2 v^2 \right)
\end{pmatrix} .
\tag{14.51}
$$

In the limit $\tilde{g} \to 0$ both sectors decouple and the upper 2×2 block is just the standard mass matrix of the neutral gauge bosons in EWSB. This mass matrix can be diagonalised by a unitary mixing matrix to get the physical mass eigenstates γ, Z and Z'. The rotation matrix can be expressed by two mixing angles θ_W and θ'_W as

$$
\begin{pmatrix}
B \\
W \\
B'
\end{pmatrix}
=
\begin{pmatrix}
\cos \theta_W & \cos \theta'_W \sin \theta_W & -\sin \theta_W \sin \theta'_W \\
\sin \theta_W & -\cos \theta_W \cos \theta'_W & \cos \theta_W \sin \theta'_W \\
0 & \sin \theta'_W & \cos \theta'_W
\end{pmatrix}
\begin{pmatrix}
\gamma \\
Z \\
Z'
\end{pmatrix} .
\tag{14.52}
$$

The third angle is zero due to the special form of this matrix. Here, θ'_W can be approximated by [271, 272]

$$
\tan 2\theta'_W \simeq \frac{2\tilde{g} \sqrt{g_1^2 + g_2^2}}{\tilde{g}^2 + 16 \left(\frac{x}{v} \right)^2 g_{BL}^2 - g_2^2 - g_1^2} .
\tag{14.53}
$$

The exact eigenvalues of Eq. (14.51) are given by

$$M_\gamma = 0, \tag{14.54}$$

$$M_{Z,Z'} = \frac{1}{8}\Big((g_1^2 + g_2^2 + \tilde{g}^2)v^2 + 4g_{BL}^2 v'^2 \mp$$
$$\sqrt{(g_1^2 + g_2^2 + \tilde{g}^2)^2 v^4 - 8(g_1^2 + g_2^2)g_{BL}^2 v^2 v'^2 + 16g_{BL}^2 v'^4}\,\Big). \tag{14.55}$$

Expanding these formulae in powers of v^2/v'^2, we find up to first order:

$$M_Z = \frac{1}{4}\left(g_1^2 + g_2^2\right)v^2\,, \qquad M_{Z'} = g_{BL}^2 v'^2 + \frac{1}{4}\tilde{g}^2 v^2. \tag{14.56}$$

All parameters in Eqs. (14.40)–(14.43) as well as in the following mass matrices are understood as running parameters at a given renormalisation scale Q. Note that the VEVs v_d and v_u are obtained from the running mass $M_Z(Q)$ of the Z boson, which is related to the pole mass M_Z through

$$M_Z^2(Q) = \frac{g_1^2 + g_2^2}{4}(v_u^2 + v_d^2) = M_Z^2 + \mathrm{Re}\{\Pi_{ZZ}^T(M_Z^2)\}. \tag{14.57}$$

Here, Π_{ZZ}^T is the transverse self-energy of the Z boson. (See [107] for more details.)

The mass of additional vector bosons as well as their mixing with the SM Z boson, which imply, for example, a deviation of the fermion couplings to the Z boson compared to SM expectations, are severely constrained by precision measurements from the LEP experiments [273–275]. The bounds are on both the mass of the Z' and the mixing with the standard Z boson, where the latter is constrained by $|\sin(\theta_{W'}) < 0.0002|$. Using Eq. (14.53) together with Eq. (14.56) as well as the values of the running gauge couplings, a limit on the Z' mass of about 1.2 TeV is obtained. Taking in addition the bounds obtained from U, T and S parameters into account [276], one gets $\frac{M_{Z'}}{Q_e^{B-L}g_{BL}} > 7.1$ TeV, which, for $g_{BL} \simeq 0.52$, implies $M_{Z'} \gtrsim 1.8$ TeV. Therefore, even accounting for most recent bounds obtained by the ATLAS and CMS [277], values of $M_{Z'} \geq 2$ TeV are satisfactory for phenomenological purposes.

14.3.1.5 Neutralinos

In the neutralino sector we find that the gauge kinetic effects lead to a mixing between the usual MSSM neutralinos with the additional states, similar to the mixing in the CP-even Higgs sector. In other words, were these to be neglected, both sectors would decouple. The mass matrix reads, in the basis $\left(\lambda_{\tilde{B}}, \tilde{W}^0, \tilde{H}_d^0, \tilde{H}_u^0, \lambda_{\tilde{B}'}, \tilde{\eta}, \tilde{\bar{\eta}}\right)$, as follows:

$$m_{\tilde{\chi}^0} = \begin{pmatrix} M_1 & 0 & -\frac{1}{2}g_1 v_d & \frac{1}{2}g_1 v_u & \frac{1}{2}M_{BB'} & 0 & 0 \\ 0 & M_2 & \frac{1}{2}g_2 v_d & -\frac{1}{2}g_2 v_u & 0 & 0 & 0 \\ -\frac{1}{2}g_1 v_d & \frac{1}{2}g_2 v_d & 0 & -\mu & -\frac{1}{2}\tilde{g}v_d & 0 & 0 \\ \frac{1}{2}g_1 v_u & -\frac{1}{2}g_2 v_u & -\mu & 0 & \frac{1}{2}\tilde{g}v_u & 0 & 0 \\ \frac{1}{2}M_{BB'} & 0 & -\frac{1}{2}\tilde{g}v_d & \frac{1}{2}\tilde{g}v_u & M_B & -g_{BL}v_\eta & g_{BL}v_{\bar{\eta}} \\ 0 & 0 & 0 & 0 & -g_{BL}v_\eta & 0 & -\mu' \\ 0 & 0 & 0 & 0 & g_{BL}v_{\bar{\eta}} & -\mu' & 0 \end{pmatrix}. \tag{14.58}$$

It is well known that for real parameters such a matrix can be diagonalised by an orthogonal mixing matrix N such that $N^* M_T^{\tilde{\chi}^0} N^\dagger$ is diagonal. For complex parameters one has to diagonalise $M_T^{\tilde{\chi}^0}(M_T^{\tilde{\chi}^0})^\dagger$. We obtain, via a straightforward generalisation of the formulae given in [107], at the one-loop level,

$$
\begin{aligned}
M_{1L}^{\tilde{\chi}^0}(p_i^2) = {} & M_T^{\tilde{\chi}^0} - \frac{1}{2}\Big[\Sigma_S^0(p_i^2) + \Sigma_S^{0,T}(p_i^2) + \left(\Sigma_L^{0,T}(p_i^2) + \Sigma_R^0(p_i^2)\right)M_T^{\tilde{\chi}^0} \\
& + M_T^{\tilde{\chi}^0}\left(\Sigma_R^{0,T}(p_i^2) + \Sigma_L^0(p_i^2)\right)\Big],
\end{aligned}
\tag{14.59}
$$

where we have denoted the wave-function corrections by Σ_R^0, Σ_L^0 and the direct one-loop contribution to the mass by Σ_S^0.

In this model, for the chosen boundary conditions, the LSP, and therefore the DM candidate, is always either the lightest neutralino or the lightest sneutrino. The reason is that m_0 must be very heavy in order to solve the tadpole equations, therefore, all sfermions are heavier than the lightest neutralino, with the possible exception of the sneutrinos. As described in Sect. 14.3.1.6 below, the splitting of the CP-even and CP-odd components of the sneutrinos can be very large, pushing the mass of the lighter eigenstate down even to the point of being lighter than the lightest neutralino. However, here, we take only points with neutralino LSP as our benchmark scenarios. A neutralino LSP is in general a mixture of all seven gauge eigenstates. However, normally the character is dominated by only one or two constituents. In that context, we can distinguish the following extreme cases:

1. $M_1 \ll M_2, \mu, M_B, \mu'$: Bino-like LSP ,

2. $M_2 \ll M_1, \mu, M_B, \mu'$: Wino-like LSP ,

3. $\mu \ll M_1, M_2, M_B, \mu'$: Higgsino-like LSP ,

4. $M_B \ll M_1, M_2, \mu, \mu'$: BLino-like LSP ,

5. $\mu' \ll M_1, M_2, \mu, M_B$: Bileptino-like LSP .

As mentioned in [21], although the gauge kinetic effects do lead to sizable effects in the spectrum, they are not large enough to lead to a large mixing between the usual MSSM-like states and the new ones. Therefore, we find that the LSP is either mainly an MSSM-like state or mainly an admixture between the BLino and bileptinos.

14.3.1.6 Charginos and sfermions

For completeness we also give a short summary of the other sectors of the model. The chargino mass matrix at tree level is exactly the same as for the MSSM:

$$
M_T^{\tilde{\chi}^+} = \begin{pmatrix} M_2 & \frac{1}{\sqrt{2}}g_2 v_u \\ \frac{1}{\sqrt{2}}g_2 v_d & \mu \end{pmatrix}.
\tag{14.60}
$$

This mass matrix is diagonalised by a bi-unitary transformation such that $U^* M_T^{\tilde{\chi}^+} V^\dagger$ is diagonal. The matrices U and V are obtained by diagonalising $M_T^{\tilde{\chi}^+}$.

14.3.1.7 Boundary conditions at the GUT scale

We will consider in the following an mSUGRA inspired scenario. This means that we assume a GUT unification of all soft-breaking scalar masses as well as a unification of all gaugino mass parameters

$$m_0^2 = m_{H_d}^2 = m_{H_u}^2 = m_\eta^2 = m_{\bar\eta}^2, \tag{14.61}$$

$$m_0^2 \delta_{ij} = m_D^2 \delta_{ij} = m_U^2 \delta_{ij} = m_Q^2 \delta_{ij} = m_E^2 \delta_{ij} = m_L^2 \delta_{ij} = m_\nu^2 \delta_{ij}, \tag{14.62}$$

$$M_{1/2} = M_1 = M_2 = M_3 = M_{\tilde B'}. \tag{14.63}$$

Also, for the trilinear soft-breaking coupling, the ordinary mSUGRA conditions are assumed to be

$$T_i = A_0 Y_i, \qquad i = e, d, u, x, \nu. \tag{14.64}$$

We do not fix the parameters μ, B_μ, μ' and $B_{\mu'}$ at the GUT scale but determine them from the tadpole equations. The reason is that they do not enter the RGEs of the other parameters and thus can be treated independently. The corresponding formulae are given in Sect. 14.3.1.3.

In addition, we consider the mass of the Z' and $\tan\beta'$ as inputs and use the following set of free parameters

$$m_0, \ M_{1/2}, \ A_0, \ \tan\beta, \ \tan\beta', \ \mathrm{sign}(\mu), \ \mathrm{sign}(\mu'), \ M_{Z'}, \ Y_x \ \text{and} \ Y_\nu. \tag{14.65}$$

Y_ν is constrained by neutrino data and must therefore be very small in comparison to the other couplings; thus it can be neglected in the following. In addition, Y_x can always be taken diagonal and thus effectively we have 9 free parameters and two signs.

Furthermore, we assume that there are no off-diagonal gauge couplings or gaugino mass parameters present at the GUT scale:

$$g_{BY} = g_{YB} = 0, \tag{14.66}$$

$$M_{BB'} = 0. \tag{14.67}$$

However, such parameters become non-zero, upon RGE running, at the EW scale. Notably, $M_{BB'}$ becomes sizable (and negative) [21].

14.3.2 The BLSSM with Inverse Seesaw

In the Type I seesaw mechanism discussed in the previous section, the RH neutrinos acquire Majorana masses at the $B - L$ symmetry breaking scale, which can be related to the SUSY breaking scale, i.e., $\mathcal{O}(1 \text{ TeV})$ [258]. In contrast, in the IS case, these Majorana masses are not allowed by the $B - L$ gauge group and another pair of SM gauge singlet fermions with tiny masses ($\mathcal{O}(1 \text{ keV})$) must be introduced. One of these two singlets fermions couples to RH neutrinos and is involved in generating the light neutrino masses. The other singlet (which is usually called inert or sterile neutrino) is completely decoupled and interacts only through the $B - L$ gauge boson , therefore it may account for warm DM [278], also see [262, 279]. In both scenarios, this $B - L$ model induces several testable signals at the LHC involving the new predicted particles: a Z' (neutral gauge boson associated with the $U(1)_{B-L}$ group), an extra Higgs state (an additional singlet state introduced to break the gauge group $U(1)_{B-L}$ spontaneously) and three (Type I) or six (IS) heavy neutrinos, ν_h (that are required to cancel the associated anomaly and are necessary for the consistency of the model). This is the setup for the non-SUSY sector of the $B - L$ scenario, which is well established in the literature (see [280, 281] for a review of its main phenomenological manifestations).

14.3.2.1 Constructing the model

The particle content of this model includes the following superfields in addition to those of the MSSM: (i) two SM singlet chiral Higgs superfields $\eta, \bar{\eta}$, whose VEVs of their scalar components spontaneously break the $U(1)_{B-L}$, where $\bar{\eta}$ is required to cancel the $U(1)_{B-L}$ anomaly; (ii) three sets of SM singlet chiral superfields, $\nu_i, s_{1_i}, s_{2_i} (i = 1, 2, 3)$, to implement the IS mechanism (also in order to cancel the $B - L$ anomaly).

$$W = Y_u \, \hat{u} \, \hat{q} \, \hat{H}_u - Y_d \, \hat{d} \, \hat{q} \, \hat{H}_d - Y_e \, \hat{e} \, \hat{l} \, \hat{H}_d + Y_\nu \, \hat{\nu} \, \hat{l} \, \hat{H}_u + Y_s \, \hat{\nu} \, \hat{\chi}_1 \, \hat{s}_2 + \mu \, \hat{H}_u \, \hat{H}_d \, - \mu' \, \hat{\chi}_1 \, \hat{\chi}_2, \quad (14.68)$$

In order to prevent a possible large mass term $M s_1 s_2$, we assume that the superfields $\hat{\nu}$, $\chi_{1,2}$ and s_2 are even under matter parity, while s_1 is an odd particle. By assuming an mSUGRA inspired universality of parameters at the scale of a GUT, we obtain that the SUSY soft breaking Lagrangian is given by

$$-\mathcal{L}_{\text{soft}} = m_0^2 \Big[|\tilde{q}|^2 + |\tilde{u}|^2 + |\tilde{d}|^2 + |\tilde{l}|^2 + |\tilde{e}^c|^2 + |\tilde{\nu}^c|^2 + |\tilde{S}_1|^2 + |\tilde{S}_2|^2 + |H_d|^2 + |H_u|^2 + |\chi_1|^2$$

$$+ |\chi_2|^2 \Big] + \Big[Y_u^A \tilde{q} H_u \tilde{u}^c + Y_d^A \tilde{q} H_d \tilde{d}^c + Y_e^A \tilde{l} H_d \tilde{e}^c + Y_\nu^A \tilde{l} H_u \tilde{\nu}^c + Y_s^A \tilde{\nu}^c \chi_1 \tilde{S}_2 \Big]$$

$$+ [B(\mu H_1 H_2 + \mu' \chi_1 \chi_2) + h.c.] + \frac{1}{2} M_{1/2} \Big[\tilde{g}^a \tilde{g}^a + \tilde{W}^a \tilde{W}^a + \tilde{B} \tilde{B} + \tilde{B}' \tilde{B}' + h.c. \Big], \quad (14.69)$$

where the trilinear terms are defined as $(Y_f^A)_{ij} = (Y_f A)_{ij}$.

The $B - L$ symmetry is radiatively broken by the non-vanishing VEVs $\langle \chi_1 \rangle = v_1'$ and $\langle \chi_2 \rangle = v_2'$ [258]. After $B - L$ and EW symmetry breaking, the neutrino Yukawa interaction terms lead to the following expression:

$$\mathcal{L}_m^\nu = m_D \, \bar{\nu}_L \nu^c + M_R \, \bar{\nu}^c S_2 + \text{ h.c.}, \quad (14.70)$$

where $m_D = \frac{1}{\sqrt{2}} Y_\nu v$ and $M_R = \frac{1}{\sqrt{2}} Y_s v'$ with $v = \sqrt{v_1^2 + v_2^2} = 246$ GeV and $v' = \sqrt{v_1'^2 + v_2'^2} \simeq \mathcal{O}(1 \text{ TeV})$. Also, the ratio of these VEVs are defined as $\tan \beta = v_2 / v_1$ and $\tan \beta' = v_1' / v_2'$. In this framework, the light neutrino masses are related to a small mass term $\mu_s S_2^2$ in the Lagrangian, with $\mu_s \sim \mathcal{O}(1 \text{ KeV})$, which can be generated at the $B - L$ breaking scale through a non-renormalisable higher order term $\frac{\chi_1^4 S_2^2}{M^3}$, where M is the renormalisation scale, which could be of order $\mathcal{O}(10^3 \text{ GeV})$ if the coupling associated to this interaction (up to some power) is $\sim \mathcal{O}(0.1)$. Therefore, one finds that the neutrinos mix with the fermionic singlet fields to build up the following 9×9 mass matrix, in the basis (ν_L, ν^c, S_2):

$$\mathcal{M}_\nu = \begin{pmatrix} 0 & m_D & 0 \\ m_D^T & 0 & M_R \\ 0 & M_R^T & \mu_s \end{pmatrix}. \quad (14.71)$$

The diagonalisation of the mass matrix, Eq. (14.71), leads to the following light and heavy neutrino masses, respectively:

$$m_{\nu_l} = m_D M_R^{-1} \mu_s (M_R^T)^{-1} m_D^T, \quad (14.72)$$

$$m_{\nu_h} = m_{\nu_{H'}} = \sqrt{M_R^2 + m_D^2}. \quad (14.73)$$

Thus, one finds that the light neutrino masses can be of order eV, with a TeV scale M_R, if $\mu_s \ll M_R$, and an order one Yukawa coupling Y_ν. Such a large coupling is crucial for testing the BLSSM-IS and probing the heavy neutrinos at the LHC. As shown in [282], the mixings between light and heavy neutrinos are of order $\mathcal{O}(0.01)$. Therefore, the decay widths of these heavy neutrinos into SM fermions are sufficiently large. It is worth mentioning that the second SM singlet fermion, S_1, remains light with mass given by

$$m_{S_1} = \mu_s \simeq \mathcal{O}(1 \text{ keV}), \tag{14.74}$$

where S_1 is a sort of inert/sterile neutrino that has no mixing with the active ones.

14.3.2.2 Radiative $B - L$ symmetry breaking

The breaking of $B - L$ can spontaneously occur through the VEV of the scalar field $\chi_{1,2}$ or the RH sneutrino $\tilde{\nu}_3^c$ [258], depending on the initial values of the Yukawa couplings and soft terms involved in the RGEs of the parameters in the scalar potential $V(\chi_1, \chi_2)$, where

$$V(\chi_1, \chi_2) = \mu_1^2 |\chi_1|^2 + \mu_2^2 |\chi_2|^2 - \mu_3^2 (\chi_1 \chi_2 + h.c.) + \frac{1}{2} g_{BL}^2 \left(|\chi_2|^2 - |\chi_1|^2 \right)^2, \tag{14.75}$$

where $\mu_{1,2}^2 = m_{\chi_{1,2}}^2 + |\mu'|^2$, $\mu_3^2 = -B'\mu'$ and g_{BL} is the gauge coupling of $U(1)_{B-L}$. The stablity condition of $V(\chi_1, \chi_2)$ is given by

$$2\mu_3^2 < \mu_1^2 + \mu_2^2. \tag{14.76}$$

The minimisation of this potential, $\frac{\partial V}{\partial \chi_i} = 0$, $i = 1, 2$, implies that

$$\mu_1^2 = \mu_3^2 \cot \beta' + \frac{M_{Z'}^2}{4} \cos 2\beta', \tag{14.77}$$

$$\mu_2^2 = \mu_3^2 \tan \beta' - \frac{M_{Z'}^2}{4} \cos 2\beta', \tag{14.78}$$

where $M_{Z'} = g_{BL}^2 v'^2$ (no mixing between $U(1)$ and $U(1)_{B-L}$ is assumed here). From Eqs. (14.77)–(14.78), one gets

$$\sin 2\beta' = \frac{2\mu_3^2}{m_{A_0'}^2}, \tag{14.79}$$

where $m_{A_0'}^2 = \mu_1^2 + \mu_2^2$. Note that, again from (14.77)–(14.78), we also get

$$v'^2 = \frac{(\mu_1^2 - \mu_2^2) - (\mu_1^2 + \mu_2^2) \cos 2\beta'}{2g_{BL}^2 \cos 2\beta'}. \tag{14.80}$$

Now we complete our analysis of symmetry breaking. We have

$$V_{11}(v_1', v_2') = 2\mu_1^2 - 2g_{BL}^2 (v_2'^2 - 3v_1'^2), \tag{14.81}$$
$$V_{12}(v_1', v_2') = -2\mu_3^2 - 4g_{BL}^2 v_1' v_2', \tag{14.82}$$
$$V_{22}(v_1', v_2') = 2\mu_2^2 + 2g_{BL}^2 (3v_2'^2 - v_1'^2), \tag{14.83}$$

where $V_{ij} = \frac{\partial^2 V(\chi_1, \chi_2)}{\partial \chi_i \partial \chi_j}$. To show that the symmetry will be broken spontaneously, we must ensure that the point $(v_1', v_2') = (0,0)$ is not a local minimum of the potential V. Since $\left(V_{11} V_{22} - V_{12}^2 \right)(0,0) = (2\mu_1^2)(2\mu_2^2) - (2\mu_3^2)^2$ and $V_{11}(0,0) = 2\mu_1^2 > 0$, we should impose a condition to make $(0,0)$ a saddle point. This condition is

$$\mu_1^2 \, \mu_2^2 < \mu_3^4. \tag{14.84}$$

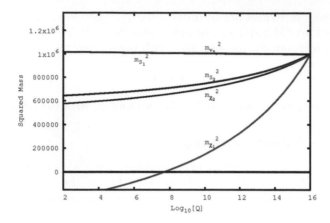

Figure 14.2 The evolution of the $B-L$ scalar masses from the GUT to the TeV scale for $m_0 = M_{1/2} = A_0 = 1$ TeV and $Y_{s_3} \sim \mathcal{O}(1)$, taken from [22].

It is worth noting that it is impossible to simultaneously fulfil both the conditions (14.76) and (14.84) for positive values of μ_1^2 and μ_2^2. However, one should note that the condition (14.84) is valid at the $B-L$ breaking scale, where the running of the RGEs, from the GUT scale down to the $B-L$ breaking scale, may induce a negative squared mass for χ_2. At this scale both conditions (14.76) and (14.84) are satisfied and the underlying symmetry is spontaneously broken with a stable potential. This can be seen as follows. The RGEs for $m_{\chi_{1,2}}^2$ are given by

$$16\pi^2 \frac{dm_{\chi_1}^2}{dt} = 12g_{BL}^2 M_{BL}^2 - 6Y_{s_3}^2 \left(m_{\chi_1}^2 + m_{\tilde{\nu}_3^c}^2 + m_{\tilde{S}_{2_3}}^2 + A_{s_3}^2\right), \qquad (14.85)$$

$$16\pi^2 \frac{dm_{\chi_2}^2}{dt} = 12g_{BL}^2 M_{BL}^2, \qquad (14.86)$$

where $t = \ln\left(\frac{M_X^2}{Q^2}\right)$ and M_{BL} is the gaugino mass of $\tilde{B}'$, which is given by $M_{1/2}$ at the GUT scale, M_X. In order to solve these equations, we should take into account all involved RGEs, given in [283]. Fig. 14.2 reports the result of the running. In plotting this figure, we set mSUGRA inspired conditions at the high scale, $e.g.$, $m_0 = M_{1/2} = A_0 = 100$ GeV, and an order-one ratio, $Y_{s_3} \simeq M_{R_3}/v'$. As can be seen from the plot, $m_{\chi_1}^2$ drops rapidly to a negative mass region whereas $m_{\chi_2}^2$ remains positive. Also in Fig. 14.2, we plot the scale evolution for the scalar mass $m_{\tilde{\nu}_3^c}$ as well as for $m_{\tilde{S}_{1_3}}$ and $m_{\tilde{S}_{2_3}}$. The figure illustrates that they remain positive at the TeV scale. Therefore, the $B-L$ breaking via a non-vanishing VEV for RH sneutrinos does not occur in the present framework.

Before closing this section, let us emphasise that in the BLSSM with Type I seesaw, the $B-L$ symmetry is spontaneously broken by the VEV of the scalar singlet χ_2 only if the Yukawa coupling Y_{ν_R} of the term $Y_{\nu_R}\nu_R^c\nu_R^c\chi_2$ is assumed to be degenerate, $i.e.$, $Y_{\nu_R} = Y_0 \operatorname{diag}\{1,1,1\}$. If one assumes a hierarchical texture, where, for instance, only $Y_{\nu_{R_3}}$ gives important contributions, one finds that the RH sneutrino may acquire a VEV before χ_2 and breaks both $B-L$ and R-parity. However, here, within the BLSSM-IS, the situation is different. One can show that, even with one non-vanishing Yukawa, χ_2 will acquire a VEV before the RH sneutrino such that $B - L$ is spontaneously broken while R-parity remains conserved [284].

14.3.2.3 The Z' gauge boson

The $U(1)_Y$ and $U(1)_{B-L}$ gauge kinetic mixing can be absorbed in the covariant derivative redefinition, where the gauge coupling matrix will be transformed as follows:

$$G = \begin{pmatrix} g_{YY} & g_{YB} \\ g_{BY} & g_{BB} \end{pmatrix} \implies \tilde{G} = \begin{pmatrix} g_1 & \tilde{g} \\ 0 & g_{BL} \end{pmatrix}, \tag{14.87}$$

where

$$g_1 = \frac{g_{YY}g_{BB} - g_{YB}g_{BY}}{\sqrt{g_{BB}^2 + g_{BL}^2}}, \tag{14.88}$$

$$g_{BL} = \sqrt{g_{BB}^2 + g_{BY}^2}, \tag{14.89}$$

$$\tilde{g} = \frac{g_{YB}g_{BB} + g_{BY}g_{YY}}{\sqrt{g_{BB}^2 + g_{BY}^2}}. \tag{14.90}$$

In this basis, one finds

$$M_Z^2 = \frac{1}{4}(g_1^2 + g_2^2)v^2, \quad M_{Z'}^2 = g_{BL}^2 v'^2 + \frac{1}{4}\tilde{g}^2 v^2. \tag{14.91}$$

Furthermore, the mixing angle between Z and Z' is given by

$$\tan 2\theta' = \frac{2\tilde{g}\sqrt{g_1^2 + g_2^2}}{\tilde{g}^2 + 16(\frac{v'}{v})^2 g_{BL}^2 - g_2^2 - g_1^2}. \tag{14.92}$$

14.3.2.4 Sneutrino masses

Now we turn to the sneutrino mass matrix. If we write $\tilde{\nu}_{L,R}$ and $\tilde{S}_2$ as $\tilde{\nu}_{L,R} = \frac{1}{\sqrt{2}}(\phi_{L,R} + i\sigma_{L,R})$ and $\tilde{S}_2 = \frac{1}{\sqrt{2}}(\phi_S + i\sigma_S)$, then we get the following mass matrix for the CP-odd sneutrinos:

$$m_{\tilde{\nu}i}^2 = \begin{pmatrix} m_{\sigma_L \sigma_L} & m_{\sigma_L \sigma_R}^T & \frac{1}{2}v_2 v_1' Re(Y_\nu^T Y_s^*) \\ m_{\sigma_L \sigma_R} & m_{\sigma_R \sigma_R} & m_{\sigma_R \sigma_S}^T \\ \frac{1}{2}v_2 v_1' Re(Y_s^T Y_\nu^*) & m_{\sigma_R \sigma_S} & m_{\sigma_S \sigma_S} \end{pmatrix}, \tag{14.93}$$

where $m_{\sigma_L \sigma_L}$, $m_{\sigma_L \sigma_R}$, $m_{\sigma_R \sigma_S}$ and $m_{\sigma_S \sigma_S}$ are given in [260] and are proportional to v^2, vA_0, v'^2, $v'\mu'$ and v'^2, respectively. The mass matrix for the CP-even sneutrino ($m_{\tilde{\nu}R}$) is obtained by changing $\sigma_{L,R,S} \to \phi_{L,R,S}$. The sneutrino mass eigenstates can be obtained by diagonalising the mass matrices as

$$U_{\tilde{\nu}i} m_{\tilde{\nu}i}^2 U_{\tilde{\nu}i}^\dagger = m_{\tilde{\nu}i}^{\text{dia}}, \quad U_{\tilde{\nu}R} m_{\tilde{\nu}R}^2 U_{\tilde{\nu}R}^\dagger = m_{\tilde{\nu}R}^{\text{dia}}. \tag{14.94}$$

However, the diagonalisation of these matrices is not an easy task and can only be performed numerically. It turns out that the mass of the lightest CP-odd sneutrino, $\tilde{\nu}_I$, is almost equal to the mass of the lightest CP-even sneutrino, $\tilde{\nu}_R$. Both $\tilde{\nu}_{I,R}$ are generated from the mixing between $\tilde{\nu}_R$ and $\tilde{S}_2$. The mass of these lightest sneutrinos can be of order $\mathcal{O}(100 \text{ GeV})$, as shown in Fig. 14.3.

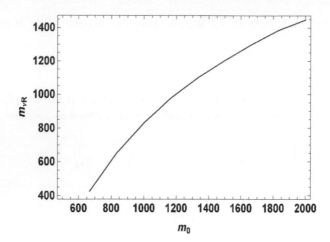

Figure 14.3 The lightest sneutrino mass as function of m_0 for $m_{1/2} = 1.5$ TeV and $A_0 = 2.5$ TeV, so that the SM-like Higgs boson mass is within experimental limits, taken from [22].

14.3.3 Search for a BLSSM Z'

A local, spontaneously broken $U(1)$ structure above and beyond the SM symmetry group calls for evidence of a Z' state. While Z' searches in the DY channel are always possible and a $B - L$ quantum structure can be accessible through sophisticated, post-discovery diagnostic procedures [285–287] (see [279, 288–294] for $B - L$ specific studies), it is through exotic decays, i.e., via (s)particles which do not exist in the MSSM, that the BLSSM nature of a Z' would readily be manifest. In fact, while the BLSSM clearly represents an appealing framework for non-minimal SUSY, both theoretically and experimentally, it is crucial to find a way of disentangling its experimental manifestations from those of other non-minimal SUSY realisations. The Z' and (s)neutrino sectors would naturally be the ideal hallmark manifestations of the BLSSM. However, if one investigates the Z' and (s)neutrino dynamics separately, there is little in the way of disentangling the BLSSM Z' from that of popular extended gauge models (with and without SUSY) or distinguishing the BLSSM (s)neutrinos from those of other SUSY scenarios (minimal or not).

It may be different, however, if Z' and (s)neutrino dynamics (of the BLSSM) are, somehow, tested together. We look separately at this case below for both visible and invisible decays.

14.3.3.1 Visible Z' decays

Now we will study the signatures of the extra neutral gauge boson Z' in the BLSSM-IS at the CERN machine[2]. The possibility of a Z' decay into a pair of heavy (inert) neutrinos would increase the total decay width of the Z'. Therefore, the BR of $Z' \to l^+ l^-$ ($l = e, \mu$), the prime Z' signal at the LHC, is suppressed with respect to the prediction of, e.g., the

---

[2]In fact, we assume here that all SUSY particles (including sneutrinos) are heavy enough so that the Z' cannot decay into these objects, thereby implying that our analysis can also be applied to the standard $B - L$ scenario. We will look at some SUSY effects later on in this chapter.

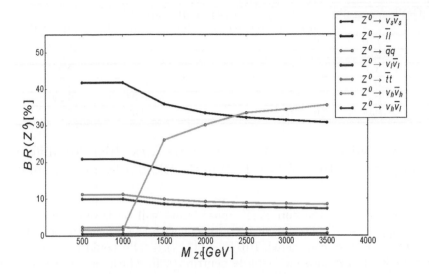

Figure 14.4 BRs of the Z' decays in the BLSSM-IS as a function of $M_{Z'}$ (note that fermion species are summed over), for $g_{BL} = 0.5$ and $\tilde{g} = 0.1$, taken from [22].

SSM, which is usually considered as a benchmark in experimental searches for a Z'. Fig. 14.4 shows the BRs of all Z' decays. Note that we have assumed that the sfermions are quite heavy, so that the Z' decay is dominated by SM particles and light inert/sterile neutrinos. According to this plot, the BRs of the non-SUSY Z' decays are given by [295]

$$\sum_l BR(Z' \to l\bar{l}) \sim 16.1\%,$$

$$\sum_{\nu_l} BR(Z' \to \nu_l \bar{\nu}_l) \sim 7.8\%,$$

$$\sum_q BR(Z' \to q\bar{q}) \sim 8.92\%,$$

$$\sum_{\nu_h} BR(Z' \to \nu_h \bar{\nu}_h) \sim 33.4\%,$$

$$\sum_{\nu_s} BR(Z' \to \nu_s \bar{\nu}_s) \sim 32.1\%,$$

$$\sum_{\nu_l, \nu_h} BR(Z' \to \nu_l \bar{\nu}_h) \sim 0.6\%, \tag{14.95}$$

where l, q and ν_h refer to the charged leptons, the six quarks and the six heavy neutrinos, respectively, whereas ν_s stands for the three inert neutrinos. In this example we have assumed $M_{Z'} = 2.5$ TeV, $g_{BL} = 0.5$, $\tilde{g} = 0.1$ and heavy neutrino masses are set at 200, 430 and 600 GeV, respectively.

It is worth noting that, in our model, the Z' cross sections (σ's) that were used to derive the ATLAS and CMS current mass limit could simply be rescaled by a factor of $(g_{BL}/g_Z)^2 \times (1 - BR(Z' \to \text{new decay channels}))$. If $g_{BL} = g_Z$ and $BR(Z' \to \text{new decay channels}) = 0$, this reproduces the SSM cross sections that were used by ATLAS and CMS. Considering

$M_{Z'}$ [GeV]	σ_{SSM} [fb]	σ_{B-L} [fb] (with IS)		
		$g_{BL} = g_Z = e/\sin\theta_W\cos\theta_W$	$g_{BL} = 0.5$	$g_{BL} = 0.8$
1000	170	6	41	105.7
1500	21.7	0.58	4.5	13.2
2000	3.4	0.087	0.72	2.3
2500	0.8	0.015	0.15	0.58
3000	0.21	0.003	0.04	0.19
3500	0.06	6×10^{-4}	0.009	0.06

Table 14.5 Representative $pp \rightarrow Z' \rightarrow ee$ rates ($\sigma \times$ BR) for different Z' masses/couplings at the LHC ($\sqrt{s} = 8$ TeV) in the BLSSM-IS, taken from [22].

the scaling of cross sections, the current Z' mass limits will be lowered by a factor of $\sigma_{B-L}(Z' \rightarrow ll)/\sigma_{\mathrm{SSM}}(Z' \rightarrow ll)$. This result is consistent with the conclusion of [296].

If $M_Z' = 1000$ GeV were considered, BR($Z' \rightarrow l^+l^-$) $\sim 14\%$ could be achieved (e.g., through onsetting Z' decays into inert/sterile neutrinos), in which case $\sigma \times$ BR $= 16$ fb when $g_{BL} = g_Z = 0.188$ and $\sigma \times$ BR $= 82$ fb when $g_{BL} = 0.5$, while in the SSM the BR($Z' \rightarrow l^+l^-$) $\sim 7.6\%$, giving $\sigma \times$ BR $= 340$ fb for both electron and muon channels. In this respect, the experimental limit $M_{Z'} \gtrsim 2.5$ TeV by ATLAS [297] (2.8 TeV [298] by CMS) will be lowered, because of a 0.241[0.035] rescaling of the cross section when, e.g., $g_{BL} = 0.5[0.188]$. This yields a new limit of 1.9[0.7] TeV (2.2[0.81] TeV). For reference, Tab. 14.3.3.1 gives $\sigma \times$ BR($Z' \rightarrow ee$) for the SSM and BLSSM-IS at different g_{BL} values.

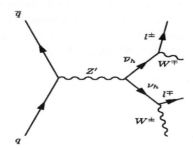

Figure 14.5 Feynman diagram for $q\bar{q} \rightarrow Z' \rightarrow \nu_h\bar{\nu}_h \rightarrow WWll$.

Detecting a Z' signal for such small Z' masses, of order 1 TeV, obtainable for $g_{BL} \approx g_Z$, would only be circumstantial evidence for the BLSSM-IS, however, as other Z' models may well feature a similar mass spectrum. A truly smoking-gun signature of the BLSSM-IS would be to produce a Z' and heavy neutrinos simultaneously. Indeed, it turns out that the dominant production mode for heavy neutrinos at the LHC would be through the DY mechanism itself, mediated by the Z'. The mixing between light and heavy neutrinos generates new couplings between the heavy neutrinos, the weak gauge bosons $Z, W^{\pm}$ and the associated leptons. These couplings are crucial for the decay of the heavy neutrinos. The main decay channel is through a $W^{\pm}$ gauge boson, which may decay leptonically or hadronically. We sketch this production and decay channel via the Feynman diagram given in Fig. 14.5. Unfortunately, the very fact that g_{BL} ought to be small to comply with current LHC data, which rule out Z' detection in the DY induced di-lepton channel for $\mathcal{O}(1$ TeV) masses, in turn means that such $Z' \rightarrow \nu_h\bar{\nu}_h$ decays also remains unaccessible. One needs a

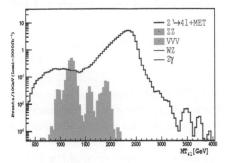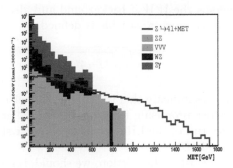

Figure 14.6 Number of events versus the transverse mass of the '4lepton+ E_T^{miss}' system (left) and the missing transverse energy (right). The expected SM backgrounds are included. The luminosity assumed here is 3000 fb^{-1}. Note that the bin width is 10 GeV, taken from [22].

stronger g_{BL} coupling to access the latter, which thus requires higher Z' masses. Hence, for the remainder of this study, we will adopt a BLSSM-IS benchmark where $M_{Z'} = 2.5$ TeV.

Once such a Z' state is produced and decays into $\nu_h \bar{\nu}_h \to WWll$, one has to further sample WW decays. In the case of a multi-lepton final state, one ends up with four leptons plus missing energy ($4l + 2\nu_l$), while in the case of a multi-hadronic final state, one ends up with four jets plus two leptons ($4j + 2l$). In addition, it is also possible to have a mixed final state ($2j + 3l + \nu_l$). (Notice that one or more neutrinos would appear in the detector as missing transverse energy/momentum, E_T^{miss}.) If two flavours of the heavy neutrinos are assumed to be degenerate in mass, one gets the same final states for the produced heavy neutrino pair with similar event rates. This will double the number of final state events but will make it difficult to distinguish between final state leptons. Therefore, throughout the current study, we consider non-degenerate heavy neutrino masses also including the interference between every two different flavours. (See [288, 292, 294, 299–301] for alternative phenomenological analyses in the case of the standard $B - L$ model.)

We thus focus on the possibilities of the LHC in accessing Z' decays into heavy neutrinos in the BLSSM-IS. In doing so, we have carried out full MC event generation using PYTHIA [302] to simulate the initial and final state radiation, fragmentation and hadronisation effects. For detector effects, we have used Delphes [303].

We consider the following benchmark: $M_{Z'} = 2.5$ TeV, $M_{\nu_4} = M_{\nu_5} = 250$ GeV, $M_{\nu_6} = M_{\nu_7} = 400$ GeV and $M_{\nu_8} = M_{\nu_9} = 630$ GeV. Of the three decay signatures of the WW pair discussed in [295], *i.e.*, $4j$, $2jl\nu$ and $2l2\nu$ ($l = e$), only the latter appeared promising, hence we only focus here on this case, by highlighting the main results. In doing so, we produce our results at $\sqrt{s} = 14$ TeV assuming a variable luminosity, ranging from the standard 300 fb^{-1} to the tenfold increase forseen at the Super-LHC [304]. The selections assumed in our analysis are as follows [295]: a transverse momentum, p_T, cut of 10 GeV and a pseudo-rapidity, η, cut of 2 were set on each electron while the separation between two electrons, R_{ll}, was enforced to be 0.2. We have assumed no restrictions on E_T^{miss}, generated for this signature by two neutrinos escaping detection.

The key advantage of this channel is that it is almost background free. The main SM noise comes from WWZ (three gauge boson) production with $\sigma(WWZ) \sim 200$ fb at 14 TeV [305, 306]. In Fig. 14.6 we show the invariant mass of the '4 lepton' system from the Z'

signal versus the WWZ background and also the transverse mass of the 4 lepton + E_T^{miss} system, where such a variable is defined as

$$M_T = \sqrt{(\sqrt{M^2(4l) + p_T^2(4l)} + |p_T^{\text{miss}}|)^2 - (\vec{p}_T(4l) + \vec{p}_T^{\,\text{miss}})^2}. \qquad (14.96)$$

These figures indicate that the decay channel 4 lepton + E_T^{miss} yields a quite clean signature and is rather promising for probing both Z' and ν_h afterwords by the end of the standard luminosity run of the LHC by using only few simple cuts to extract the Signal (S) from the Background (B).

Before closing this section, we should, however, comment on the influence of the SUSY spectrum on the scope of the $Z' \to \nu_h \bar{\nu}_h$ signal within the BLSSM. Clearly, once the assumption made so far (that all sparticles are heavier than the Z') is dismissed, the Z' boson can decay via SUSY objects. This will correspond to an increased value of its total width $\Gamma_{Z'}^{\text{tot}}$, which would then reflect onto the event rates of such a signal. In fact, the latter will scale as the inverse of $\Gamma_{Z'}^{\text{tot}}$. In order to quantify this (reduction) effect induced by a low mass spectrum within the BLSSM, we have revisited the benchmarks introduced in Tab. 14.3.3.1, excluding only the 1 TeV mass point (now ruled out by data, as discussed) and thus including the 2.5 TeV benchmark considered so far in our MC analysis. For each of these scenarios we have computed the ratio

$$R(Z') = \frac{\Gamma(Z' \to \text{all BLSSM decays})}{\Gamma(Z' \to \text{SM} - \text{like decays})}, \qquad (14.97)$$

the inverse of which corresponds to the rescaling factor to be applied to our event rates for the $Z' \to \nu_h \bar{\nu}_h$ signal in the presence of a low-lying SUSY spectrum in the BLSSM. We can see from Fig. 14.7 that $R(Z')$, obtained from the average total Z' width after scanning the entire BLSSM parameter space compatible with current experimental and theoretical constraints, is typically smaller than 3, so that the $Z' \to \nu_h \bar{\nu}_h$ rates will generally not be smaller than a factor 1/3 with respect to the values considered here. In fact, also recall that we have not allowed for decays into muons ($l = \mu^\pm$), which would contribute a factor of ≈ 2 towards the signal rates in the '4 lepton + E_T^{miss}' channel. The SM rates presented here would of course be unchanged. One should, however, include intrinsic SUSY backgrounds in the full analysis. We can anticipate that the latter would not spoil the feasibility of such a signal over a sizable portion of the full BLSSM parameter space.

14.3.3.2 Invisible Z' decays

The reviews in [307, 308] dealt with the case of invisible Z' decays, which are very peculiar to the BLSSM. In this respect, from a phenomenological point of view, an intriguing signal, both for experimental cleanliness and theoretical naturalness, would be the one involving totally invisible decays of a Z' into (s)neutrinos, thereby accessible, e.g., in mono-jet analyses[3]. Contrary to SUSY models which do not have a Z' in their spectra or where the invisible final state is induced by direct couplings of the lightest neutralino pair to (light and potentially highly off-shell) Z bosons, in the BLSSM one can afford resonant Z' production and decay into heavy (s)neutrinos, which can in turn decay, again on-shell, into an invisible final state. Under these circumstances, one would expect the typical distributions

---

[3]We do not consider here the case of mono-top and $W^\pm$ -ISR probes, which have also been used experimentally.

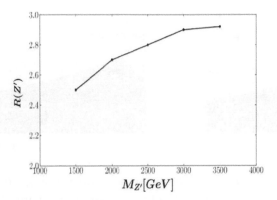

Figure 14.7 The ratio $R(Z')$ (as defined in the text) versus the Z' mass for the four heaviest benchmark points in Tab. 14.3.3.1, taken from [22].

of the visible probe (whether it be mono-jet, single-photon, or Z-ISR) to be substantially different from the case of other SUSY scenarios. This remains true even if the Z' decays into gauginos, either directly into the lightest neutralinos or else into heavier -ino states cascading down (invisibly) to the LSP and even (both light and heavy) neutrinos.

We thus look at the process (hereafter, j=jet),

$$q\bar{q} \to Z'(\to \tilde{\nu}_R \tilde{\nu}_R^* \to \tilde{\chi}_1^0 \tilde{\chi}_1^0 \nu\bar{\nu}) + j, \qquad (14.98)$$

which is generally dominated by sneutrino decays[4]. The SM backgrounds are the following: $Z(\to \nu\bar{\nu}) + j$ (irreducible) plus $W(\to l\nu) + j$, $W(\to \tau\nu) + j$, $t\bar{t}$ and $ZZ(\to 2\nu2\bar{\nu}) + j$ (all reducible). We closely follow here the selection of [172, 173]. Further, in order to increase the MC efficiency and thus obtain reasonable statistics, we have applied a parton level cut of $p_T(j_1) > 120$ GeV for both signals and backgrounds (here $p_T(j_1)$ is the highest jet transverse momentum). According to the estimation of the QCD background based on the full detector simulation of [309, 310], such a noise can be reduced to a negligible level by requiring a large $\not{E}_T$ cut. Thanks to the heavy Z' mediation, we can afford to set here $\not{E}_T > 500$ GeV [311]. The beneficial effect of the $\not{E}_T$ selection is evident from the left plot in Fig. 14.8. In contrast, a similar cut on $p_T(j_1)$ is not as selective (and is at any rate correlated), see the right plot in Fig. 14.8, yet it pays off to also enforce it.

14.3.4 DM in the BLSSM

In this section, we consider the scenario where the extra $B - L$ neutralinos (three extra neutral fermions: $U(1)_{B-L}$ gaugino $\tilde{Z}_{B-L}$ and two extra Higgsinos $\tilde{\chi}_{1,2}$) can be cold DM candidates. In particular, we examine the thermal relic abundance of these particles. As intimated in Chapter 9, assuming the lightest neutralino in the MSSM as a DM candidate implies severe constraints on the parameter space of this model. Indeed, in the case of universal soft-breaking terms, the MSSM is almost ruled out by combining collider, astrophysics

---

[4]Also the $Z' \to \nu\bar{\nu}$ and $\tilde{\chi}_1^0 \tilde{\chi}_1^0$ invisible decays are present, but they generically are subleading.

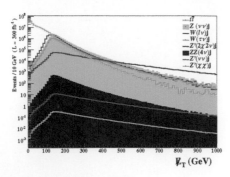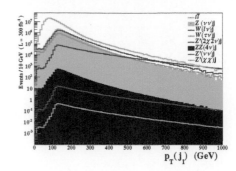

Figure 14.8 (Left panel) Number of events versus the missing transverse energy. (Right panel) Number of events versus the transverse momentum of the leading jet. Distributions are for the mono-jet case given after the jet selection only. The energy is 14 TeV whereas the integrated luminosity is 300 fb^{-1}. Here, $M_{Z'} = 2.5$ TeV and $g_{BL} = 0.4$. Plots taken from [22].

and rare decay constraints [12]. Therefore, it is important to explore very well mSUGRA extensions of the MSSM, such as the SUSY $B - L$ model, which provides new DM candidates that may account for the relic density with no conflict with other phenomenological constraints [312].

Here, we focus on the cases where the LSP is pure $\tilde{Z}_{B-L}$ or $\tilde{\chi}_{1(2)}$. In this case, the relevant Lagrangian is given by [312]

$$-\mathcal{L}_{\tilde{Z}_{B-L}} \simeq i\sqrt{2}g_{B-L}Y^f_{B-L}\overline{\tilde{Z}}_{B-L}P_R f\tilde{f}_L + i\sqrt{2}g_{B-L}Y^f_{B-L}\overline{\tilde{Z}}_{B-L}P_L f\tilde{f}_R + \text{ h.c.,} \quad (14.99)$$

$$-\mathcal{L}_{\tilde{\chi}_1} \simeq i\sqrt{2}g_{B-L}Y^{\chi_1}_{B-L}\overline{\tilde{\chi}}_1 Z\!\!\!\!/_{B-L}\gamma_5\tilde{\chi}_1 + i\sqrt{2}g_{B-L}Y^{\chi_1}_{B-L}\overline{\tilde{\chi}}_1\tilde{Z}_{B-L}\chi_1$$
$$+ (Y_N)_{ij}\tilde{\chi}_1 N^c{}_i\tilde{N}^c_j + \text{ h.c.,} \quad (14.100)$$

$$-\mathcal{L}_{\tilde{\chi}_2} \simeq i\sqrt{2}g_{B-L}Y^{\chi_2}_{B-L}\overline{\tilde{\chi}}_2 Z\!\!\!\!/_{B-L}\gamma_5\tilde{\chi}_2 + i\sqrt{2}g_{B-L}Y^{\chi_2}_{B-L}\overline{\tilde{\chi}}_2\tilde{Z}_{B-L}\chi_2 + \text{ h.c.,} \quad (14.101)$$

where f refers to all the SM fermions, including the RH neutrinos. $\tilde{f}_L$ and $\tilde{f}_R$ are the LH and RH sfermion mass eigenstates, respectively. We assume the first RH neutrino N_1's mass to be at the EW scale, therefore the annihilation channel of the LSP into $N_1 N_1$ is also considered [312].

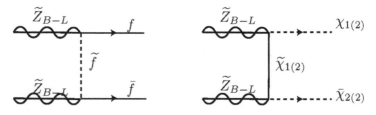

Figure 14.9 The dominant annihilation diagrams in the case of a $\tilde{Z}_{B-L}$-like LSP. Note that the u-channel is also taken into consideration for each diagram.

From Eq. (14.99), one finds that the dominant annihilation processes for the case when $\tilde{Z}_{B-L}$ is the LSP are given in Fig. 14.9.

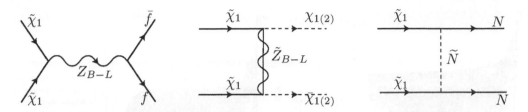

Figure 14.10 The dominant annihilation diagrams in the case of a $\tilde{\chi}_1$-like LSP. For the last two diagrams, the u-channel is also considered.

In the case of Higgsino DM, from Eq. (14.100), one finds that the dominant annihilation processes of $\tilde{\chi}_1$'s are given in Fig. 14.10.

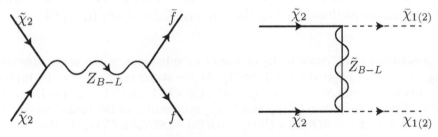

Figure 14.11 The dominant annihilation diagrams in the case of a $\tilde{\chi}_2$-like LSP. Note that u-channel topologies are also taken into consideration alongside the t-channel ones.

Finally, when $\tilde{\chi}_2$ states are considered as DM candidates, from Eq. (14.101), one finds that $\tilde{\chi}_2\tilde{\chi}_2$ annihilation is dominated by the diagrams in Fig. 14.11.

It is evident from the above that the extended particle spectrum of the BLSSM, in turn translating into a more varied nature of the LSP as well as a more numerous combination of DM annihilation diagrams, can play a significant role in dramatically changing the response of the model to the cosmological data, in comparison to the much constrained MSSM. This is well manifested in Fig. 14.12. From here, it is obvious how the BLSSM offers a variety of solutions to the relic abundance constraint, whether taken at 2σ from the central value measured by experiment or as an absolute upper limit, precluded to the MSSM. In the former, different DM incarnations (Bino-, BLino-, Bleptino-like and mixed neutralino alongside the sneutrino) can comply with experimental evidence over an $M_{\rm DM}$ interval which extends up to 2 TeV or so, while in the MSSM case, solutions can only be found for much lighter LSP masses and are limited to one nature (the usual Bino-like neutralino).

14.4 THE ESSM

As in the last chapter of this book we will be looking at connections between SUSY and string theories, it is instructive to introduce here a construct of the former that may serve the purpose of being circumstantial evidence of the latter, the so-called Exceptional Supersymmetric Standard Model (ESSM) of [24, 253, 313], specifically, of a ten dimensional heterotic superstring theory based on $E_8 \times E_8'$. Compactification of the extra dimensions

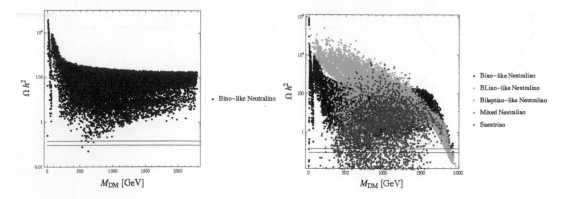

Figure 14.12 Distribution of the DM candidates over the MSSM (left) and BLSSM (right) parameter space compatible with all other experimental bounds against the relic abundance constraint at 2σ level (*i.e.*, $0.09 < \Omega h^2 < 0.14$). The colour scheme refers to the DM nature as illustrated in the legends. Taken from Ref. [23].

results in the breakdown of E_8 down to E_6 or one of its subgroups in the observable sector [314] (hence, it is often referred to as E$_6$SSM). At the string scale, E_6 can be broken directly to the rank-6 subgroup $SU(3)_C \times SU(2)_L \times U(1)_Y \times U(1)_\psi \times U(1)_\chi$ via the Hosotani mechanism [315]. Two anomaly-free $U(1)_\psi$ and $U(1)_\chi$ symmetries of the rank-6 model are defined by [285, 316–318]: $E_6 \rightarrow SO(10) \times U(1)_\psi$, $SO(10) \rightarrow SU(5) \times U(1)_\chi$. In this section we explore a particular E_6 inspired supersymmetric model with one extra $U(1)_N$ gauge symmetry defined by:

$$U(1)_N = \frac{1}{4}U(1)_\chi + \frac{\sqrt{15}}{4}U(1)_\psi \,, \tag{14.102}$$

under which RH neutrinos have no charge and thus may gain large Majorana masses in accordance with the seesaw mechanism (for a review see, *e.g.*, [319]). The extra $U(1)_N$ gauge symmetry survives to low energies and serves to forbid an elementary μ term as well as terms like S^n in the superpotential but allows the interaction $\lambda S H_d H_u$. After EWSB the scalar component of the singlet superfield acquires a non-zero VEV, $\langle S \rangle = s/\sqrt{2}$, breaking $U(1)_N$ and an effective $\mu = \lambda s/\sqrt{2}$ term is automatically generated. Clearly there are no domain wall problems in such a model since there is no discrete Z_3 symmetry and, instead of a global symmetry, there is a gauged $U(1)_N$. Anomalies are cancelled by complete 27 representatations of E_6 which survive to low energies, even though E_6 is broken at the GUT scale.

As discussed repeatedly, one of the most important issues in models with additional Abelian gauge symmetries is the cancellation of anomalies. In E_6 theories the anomalies are cancelled automatically. Therefore any model based on E_6 subgroups which contains complete representations should be anomaly-free. Thus, in order to ensure anomaly cancellation, the particle content of the ESSM should include the complete fundamental 27 representations of E_6. These multiplets decompose under the $SU(5) \times U(1)_N$ subgroup of E_6 [320] as follows:

$$27_i \rightarrow (10, 1)_i + (5^*, 2)_i + (5^*, -3)_i + (5, -2)_i + (1, 5)_i + (1, 0)_i \,. \tag{14.103}$$

The first and second quantities in the brackets are the $SU(5)$ representation and extra $U(1)_N$ charge, while i is a family index that runs from 1 to 3. An ordinary SM family which contains the doublets of LH quarks Q_i and leptons L_i, RH up- and down-quarks (u_i^c and d_i^c) as well as RH charged leptons, is assigned to $(10, 1)_i + (5^*, 2)_i$. Right-handed neutrinos N_i^c should be associated with the last term in Eq. (14.103), $(1, 0)_i$. The next-to-last term in Eq. (14.103), $(1, 5)_i$, represents SM-type singlet fields S_i which carry non-zero $U(1)_N$ charges and therefore survive down to the EW scale. The pair of $SU(2)$-doublets (H_{1i} and H_{2i}) that are contained in $(5^*, -3)_i$ and $(5, -2)_i$ have the quantum numbers of Higgs doublets. Other components of these $SU(5)$ multiplets form colour triplets of exotic quarks D_i and $\overline{D}_i$ with electric charges $-1/3$ and $+1/3$, respectively. The matter content and correctly normalised Abelian charge assignments are in Tab. 14.6.

	Q	u^c	d^c	L	e^c	N^c	S	H_2	H_1	D	$\overline{D}$	H'	$\overline{H'}$
$\sqrt{\frac{5}{3}}Q_i^Y$	$\frac{1}{6}$	$-\frac{2}{3}$	$\frac{1}{3}$	$-\frac{1}{2}$	1	0	0	$\frac{1}{2}$	$-\frac{1}{2}$	$-\frac{1}{3}$	$\frac{1}{3}$	$-\frac{1}{2}$	$\frac{1}{2}$
$\sqrt{40}Q_i^N$	1	1	2	2	1	0	5	-2	-3	-2	-3	2	-2

Table 14.6   The $U(1)_Y$ and $U(1)_N$ charges of matter fields in the ESSM, where Q_i^N and Q_i^Y are defined here with the correct E_6 normalisation factor required for the RG analysis.

The most general renormalisable superpotential which is allowed by the E_6 symmetry can be written in the following form:

$$\begin{aligned} W_{E_6} &= W_0 + W_1 + W_2\,, \\ W_0 &= \lambda_{ijk}S_i(H_{1j}H_{2k}) + \kappa_{ijk}S_i(D_j\overline{D}_k) + h_{ijk}^N N_i^c(H_{2j}L_k) + h_{ijk}^U u_i^c(H_{2j}Q_k) + \\ &\quad + h_{ijk}^D d_i^c(H_{1j}Q_k) + h_{ijk}^E e_i^c(H_{1j}L_k)\,, \\ W_1 &= g_{ijk}^Q D_i(Q_jQ_k) + g_{ijk}^q \overline{D}_i d_j^c u_k^c\,, \\ W_2 &= g_{ijk}^N N_i^c D_j d_k^c + g_{ijk}^E e_i^c D_j u_k^c + g_{ijk}^D (Q_iL_j)\overline{D}_k. \end{aligned}$$

$$(14.104)$$

Although $B-L$ is conserved automatically, some Yukawa interactions in Eq. (14.104) violate baryon number conservation, resulting in rapid proton decay. The baryon and lepton number violating operators can be suppressed by imposing an appropriate Z_2 symmetry which is usually called R-parity. But the straightforward generalisation of the definition of R-parity, assuming $B_D = 1/3$ and $B_{\overline{D}} = -1/3$, implies that W_1 and W_2 are forbidden by this symmetry and the lightest exotic quark is stable. Models with stable charged exotic particles are ruled out by different experiments [321–323].

To prevent rapid proton decay in E_6 supersymmetric models, a generalised definition of R-parity should be used. There are two ways to do that. If H_{1i}, H_{2i}, S_i, D_i, $\overline{D}_i$, and the quark superfields (Q_i, u_i^c, d_i^c) are even under a discrete Z_2^L symmetry while the lepton superfields (L_i, e_i^c, N_i^c) are odd, all terms in W_2 are forbidden (Model I). Then the remaining superpotential is invariant with respect to a $U(1)_B$ global symmetry if the exotic quarks $\overline{D}_i$ and D_i are diquark and anti-diquark, i.e., $B_D = -2/3$ and $B_{\overline{D}} = 2/3$. An alternative possibility is to assume that the exotic quarks D_i and $\overline{D}_i$ as well as lepton superfields are all odd under Z_2^B whereas the others remain even. Then we get Model II in which all Yukawa interactions in W_1 are forbidden by the discrete Z_2^B symmetry. Here, exotic quarks are leptoquarks. The two possible models are summarised as:

$$W_{\text{ESSM I}} = W_0 + W_1\,, \qquad\qquad W_{\text{ESSM II}} = W_0 + W_2\,. \qquad (14.105)$$

In addition to the complete 27_i representations, some components of the extra $27'$

and $\overline{27}'$ representations must survive to low energies in order to preserve gauge coupling unification. We assume that an additional $SU(2)$ doublet components H' of $(5^*, 2)$ from $27'$ and corresponding anti-doublet $\overline{H}'$ from $\overline{27}'$ survive to low energies. In either model the superpotential involves a lot of new Yukawa couplings in comparison to the SM. In general these new interactions induce non-diagonal flavour transitions. To suppress the flavour changing processes, one can postulate a Z_2^H symmetry under which all superfields except one pair of H_{1i} and H_{2i} (say $H_d \equiv H_{13}$ and $H_u \equiv H_{23}$) and one SM-type singlet field $S \equiv S_3$ are odd. Then only one Higgs doublet H_d interacts with the down-type quarks and charged leptons and only one Higgs doublet H_u couples to up-type quarks while the couplings of all other exotic particles to ordinary quarks and leptons are forbidden. This eliminates any problem related to non-diagonal flavour transitions. The $SU(2)$ doublets H_u and H_d play the role of Higgs fields generating the masses of quarks and leptons after EWSB. Thus it is natural to assume that only S, H_u and H_d acquire non-zero VEVs.

The Z_2^H symmetry reduces the structure of the Yukawa interactions in (14.105):

$$W_{\mathrm{ESSM\,I,II}} \longrightarrow \lambda_i S(H_{1i}H_{2i}) + \kappa_i S(D_i \overline{D}_i) + f_{\alpha\beta} S_\alpha (H_d H_{2\beta})$$
$$+ \tilde{f}_{\alpha\beta} S_\alpha (H_{1\beta} H_u) + W_{\mathrm{MSSM}}(\mu = 0), \qquad (14.106)$$

where $\alpha, \beta = 1, 2$ and $i = 1, 2, 3$. In Eq. (14.106) we choose the basis of $H_{1\alpha}$, $H_{2\alpha}$, D_i and $\overline{D}_i$ so that the Yukawa couplings of the singlet field S have flavour diagonal structure. Here we define $\lambda \equiv \lambda_3$ and $\kappa \equiv \kappa_3$. If λ or κ_i are large at the GUT scale M_X they affect the evolution of the soft scalar mass m_S^2 of the singlet field S rather strongly, resulting in negative values of m_S^2 at low energies that trigger the breakdown of the $U(1)_N$ symmetry. To guarantee that only H_u, H_d and S acquire a VEV, we impose a certain hierarchy between the couplings H_{1i} and H_{2i} to the SM-type singlet superfields S_i: $\lambda \gg \lambda_{1,2}, f_{\alpha\beta}$ and $\tilde{f}_{\alpha\beta}$. Although $\lambda_{1,2}, f_{\alpha\beta}$ and $\tilde{f}_{\alpha\beta}$ are expected to be considerably smaller than λ they must be large enough to generate sufficiently large masses for the exotic particles to avoid conflict with direct particle searches at present and former accelerators. Keeping only Yukawa interactions whose couplings are allowed to be of order unity gives approximately:

$$W_{\mathrm{ESSM\,I,II}} \approx \lambda S(H_d H_u) + \kappa_i S(D_i \overline{D}_i) + h_t (H_u Q) t^c + h_b (H_d Q) b^c + h_\tau (H_d L)\tau^c.$$
$$(14.107)$$

The Z_2^H symmetry discussed above forbids all terms in W_1 or W_2 that would allow the exotic quarks to decay. Therefore the discrete Z_2^H symmetry can only be approximate although Z_2^B or Z_2^L must be exact to prevent proton decay. In our model we allow only the third family $SU(2)$ doublets H_d and H_u to have Yukawa couplings to the ordinary quarks and leptons of order unity. This is a self-consistent assumption since the large Yukawa couplings of the third generation (in particular, the top-quark Yukawa coupling) provide a radiative mechanism for generating the Higgs VEVs [89, 324–326]. The Yukawa couplings of two other pairs of $SU(2)$ doublets H_{1i} and H_{2i} as well as H' and exotic quarks to the quarks and leptons of the third generation are supposed to be significantly smaller ($\lesssim 0.1$) so that none of the other exotic bosons gain VEVs. These couplings break the Z_2^H symmetry explicitly, resulting in FCNC. In order to suppress the contribution of new particles and interactions to $K^0 - \overline{K}^0$ oscillations and to the muon decay channel $\mu \to e^- e^+ e^-$ in accordance with experimental limits, it is necessary to assume that the Yukawa couplings of new exotic particles to the quarks and leptons of the first and second generations are less than or of order 10^{-4}.

Previously, the implications of SUSY models with an additional $U(1)_N$ gauge symmetry had been studied in the context of leptogenesis [327], EW baryogenesis [328] and neutrino physics [329]. Supersymmetric models with a $U(1)_N$ gauge symmetry under which RH neutrinos are neutral have been specifically considered in [330, 331] from the point of view of $Z - Z'$ mixing and the neutralino sector, in [320] where an RGE analysis was performed, and in [332] where a one-loop Higgs mass upper bound was presented. Here, we are mainly concerned with the collider implications of the ESSM, as the latter contains a rich phenomenology accessible to the LHC in the form of a Z' plus three families of exotic quarks and non-Higgs doublets as well as an enriched Higgs sector. We will tackle the first aspect here, while we will defer the illustration of Higgs phenomenology to the next chapter.

14.4.1 Z' and exotica phenomenology at the LHC

The presence of a relatively light Z' or of exotic multiplets of matter permits us to distinguish the ESSM from the MSSM. Collider experiments and precision EW tests imply that a narrow Z' is typically heavier than 2 TeV, while the mixing angle between Z and Z' is smaller than a few $\times 10^{-3}$. The analysis performed in [286, 333, 334] revealed that a Z' boson in E_6 inspired models can be discovered at the LHC if its mass is less than $4 - 4.5$ TeV. At the same time the determination of the Z' couplings should be possible up to $M_{Z'} \sim 2 - 2.5$ TeV using DY production [335].

Fig. 14.13 shows the differential distribution in invariant mass of the lepton pair l^+l^- (for one species of lepton $l = e, \mu$) in DY production at the LHC, with and without light exotic quarks with representative masses $\mu_{D_i} = 250$ GeV for all three generations and with $M_{Z'} = 1.2$ TeV[5]. This distribution is promptly measurable at the CERN collider with a high resolution and would enable one to not only confirm the existence of a Z' state but also to establish the possible presence of additional exotic matter, by simply fitting to the data the width of the Z' resonance [335]. In order to perform such an exercise, the Z' couplings to ordinary matter ought to have been previously established elsewhere, as a modification of the latter may well lead to effects similar to those induced by our exotic matter.

The exotic quarks can also be pair produced directly and decay with novel signatures. In the ESSM the exotic quarks and squarks receive their masses from the large VEV of the singlet S, according to the terms $\kappa_i S(D_i\overline{D}_i)$ in Eq. (14.107). Their couplings to the quarks and leptons of the first and second generation should be rather small, as discussed in [24, 253]. The exotic quarks can be relatively light in the ESSM since their masses are set by the Yukawa couplings κ_i and λ_i, which may be small. This happens, for example, when the Yukawa couplings of the exotic particles have a hierarchical structure which is similar to the one observed in the ordinary quark and lepton sectors. Since the exotic squarks also receive soft masses from SUSY breaking, they are expected to be much heavier and their production cross sections will be considerably smaller.

If exotic quarks of the nature described here do exist at low scales, they could possibly be accessed through direct pair hadroproduction at the LHC. Fig. 14.14 shows the production cross section of exotic quark pairs at the LHC as a function of the invariant mass of the final state. The lifetime and decay modes of these particles are determined by the operators that break the Z_2^H symmetry. If Z_2^H is only slightly broken, then exotic quarks may live for a long time. Then they will form bound states with ordinary quarks. It means that at the LHC it may be possible to study the spectroscopy of new composite scalar leptons or baryons. When Z_2^H is broken significantly, exotic quarks can also produce a remarkable signature. Since according to our initial assumptions the Z_2^H symmetry is mostly broken by

---

[5]A smaller $M_{Z'}$ value is typically allowed in the ESSM with respect to the case of the $U(1)$ model previously discussed, owing to a larger Z' width induced by decays into the additional matter states present in the spectrum, which relaxes current experimental limits.

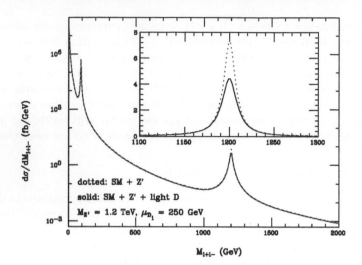

Figure 14.13 Differential cross section in the final state invariant mass, denoted by $M_{l^+l^-}$, at the LHC ($\sqrt{s} = 14$ TeV) for DY production ($l = e$ or μ only) in presence of a Z' with and without the (separate) contribution of exotic D-quarks with $\mu_{Di} = 250$ GeV for $M_{Z'} = 1.2$ TeV. Plot taken from [24].

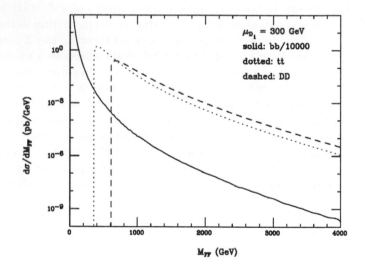

Figure 14.14 Cross section at the LHC ($\sqrt{s} = 14$ TeV) for pair production of exotic D-quarks as a function of the invariant mass of a $D\overline{D}$ pair. Similar cross sections of $t\bar{t}$ and $b\bar{b}$ production are also included for comparison. Plot taken from [24].

the operators involving quarks and leptons of the third generation, the exotic quarks decay either via $\overline{D} \to t + \tilde{b}$, $\overline{D} \to b + \tilde{t}_2$ if the exotic quarks $\overline{D}$ are diquarks, or via $D \to t + \tilde{\tau}$, $D \to \tau + \tilde{t}$, $D \to b + \tilde{\nu}_\tau$, $D \to \nu_\tau + \tilde{b}$ if they are are leptoquarks. Because in general, sfermions decay into corresponding fermions and neutralinos, one can expect that each diquark will decay further into t- and b-quarks while a leptoquark will produce a t-quark and a τ-lepton in the final state with rather high probability.

As each t-quark decays into a b-quark whereas an τ-lepton gives one charged lepton l in the final state with a probability of 35%, both these scenarios would generate an excess in the b-quark production cross section. In this respect SM data samples which should be altered by the presence of exotic D-quarks are those involving $t\bar{t}$ production and decay as well as direct $b\bar{b}$ production. For this reason, Fig. 14.14 shows the cross sections for these two genuine SM processes alongside those for the exotica. Detailed LHC analyses will be required to establish the feasibility of extracting the excess due to the light exotic particles predicted by our model. However, Fig. 14.14 should clearly make the point that-for the discussed parameter configuration -one is in a favourable position in this respect, as the decay BRs of the exotic objects are much larger than the expected four-body cross section involving heavy quarks and/or leptons, as discussed in [253]. Thus the presence of light exotic quarks in the particle spectrum could result in an appreciable enhancement of the cross section of either $pp \to t\bar{t}b\bar{b} + X$ and $pp \to b\bar{b}b\bar{b} + X$ if exotic quarks are diquarks or $pp \to t\bar{t}l^+l^- + X$ and $pp \to b\bar{b}l^+l^- + X$ if new quark states are leptoquarks.

CHAPTER 15

Higgs Extensions

In the present chapter, it is the Higgs sector which is enlarged, rather than the gauge sector discussed in the previous chapter, through the addition of a singlet Higgs state (and supersymmetric counterparts) alongside the two doublet Higgs states of the minimal version of SUSY. After an introduction to the main theoretical problems of the MSSM, we discuss various formulations of non-minimal SUSY scenarios of this kind and their most important phenomenological implications.

15.1 THE μ PROBLEM OF THE MSSM

Due to the non-observation of superpartners at the LHC, SUSY has to be broken at a scale $M_{\rm SUSY}$ not larger than $\mathcal{O}({\rm TeV})$, so that it still provides a natural solution to the hierarchy problem. Unfortunately, a phenomenologically acceptable realisation of EWSB in the MSSM requires the presence of the so-called μ term, a direct SUSY mass term for the Higgs(ino) fields, with values of the (theoretically arbitrary) parameter μ close to $M_{\rm SUSY}$, when its natural value would be either 0 or $M_{\rm Pl}$. Of course, there can exist explanations for an $\mathcal{O}(M_{\rm SUSY})$ value of the μ term: needless to say, they all occur in extended settings [336–341]. We shall detail in the forthcoming sections some possible solutions to the μ problem of the MSSM and highlight their key phenomenological features. Much of the material of this chapter reflects the content of [342] and [343–347] (see also [25]).

15.2 THE MSSM WITH AN EXTRA SINGLET

The simplest approach is to introduce an extra Higgs iso-singlet superfield, $\hat{S}$, within the minimal SUSY model. If one replaces the μ term of the MSSM with a term coupling this new superfield to the Higgs boson doublets, i.e.,

$$W_\lambda^{\rm MSSM} = Y_u \hat{Q} \hat{H}_u \hat{u}^C + Y_d \hat{H}_d \hat{Q} \hat{d}^C + Y_e \hat{H}_d \hat{L} \hat{e}^C + \lambda \hat{S} \hat{H}_u \hat{H}_d, \tag{15.1}$$

where λ is some dimensionless coupling, then an effective μ term will be generated if the real scalar component of $\hat{S}$ develops a VEV. Such an effective μ parameter is then given by

$$\mu \equiv \mu_{\rm eff} = \lambda < S > . \tag{15.2}$$

The constraints which arise when the resulting Higgs potential is minimised to find the vacuum state relate the VEVs of the three neutral scalars, H_u^0, H_d^0 and S, to their soft SUSY-breaking masses. Therefore, in the absence of fine tuning, one expects that these

VEVs should all be of order M_{SUSY}, so that the μ problem is solved. These three VEVs are usually then replaced by the phenomenological parameters μ_{eff}, M_Z and $\tan\beta$.

Additionally, one must introduce soft SUSY-breaking terms into the Lagrangian. These must be of the same form as for the MSSM except that the term involving the SUSY-breaking parameter B must be removed (since it was associated with the μ term) and instead a soft mass for the new singlet should be added, together with soft SUSY-breaking terms for the extra interactions. The soft SUSY-breaking terms associated with the Higgs sector are then

$$-\mathcal{L}_{\mathrm{soft}}^{\mathrm{MSSM}} \supset m_{H_u}^2 |H_u|^2 + m_{H_d}^2 |H_d|^2 + m_S^2 |S|^2 + (\lambda A_\lambda S H_u H_d + \mathrm{h.c.}), \qquad (15.3)$$

where A_λ is a dimensionful parameter of order M_{SUSY}. The new singlet superfield provides an additional scalar Higgs field, a pseudoscalar Higgs field and an accompanying Higgsino. The new Higgs fields will mix with the neutral Higgs fields from the usual Higgs doublets, so that the model will in total have five neutral Higgs bosons (three scalars and two pseudoscalars if CP is conserved). The extra Higgsino states will mix with the Higgsinos from the doublets and the gauginos to provide an extra neutralino state, for a total of five. The charged Higgs and chargino mass spectrum remain unchanged.

15.3 THE PQ SYMMETRY

Now, one should notice that the potential given in Eq. (15.1) contains an extra global $U(1)$ invariance, known as PQ symmetry [178]. The PQ charges, Q_i^{PQ} ($i = \hat{H}_u, \hat{H}_d, \hat{S}, \hat{Q}, \hat{U}^C, \hat{D}^C, \hat{L}, \hat{E}^C$), can be as assigned as follows:

$$\hat{H}_u \to 1, \ \hat{H}_d \to 1, \ \hat{S} \to -2, \ \hat{Q} \to -1, \ \hat{U}^C \to 0, \ \hat{D}^C \to 0, \ \hat{L} \to -1, \ \hat{E}^C \to 0, \qquad (15.4)$$

so that the superpotential becomes invariant upon the phase transformation

$$\hat{\Psi} \equiv (\hat{\psi}_i) \longrightarrow \hat{\Psi}' \equiv (\hat{\psi}_i') = (\exp(iQ_i^{\mathrm{PQ}})\hat{\psi}_i). \qquad (15.5)$$

Following EWSB, when the Higgs fields acquire VEVs, a spontaneous breaking of the $U(1)_{\mathrm{PQ}}$ symmetry also occurs, so that a so-called PQ axion is generated. This state, originally a pNGB emerging from the QCD triangle anomaly that breaks the $U(1)_{\mathrm{PQ}}$ invariance, thereby inducing a small mass for the axion via the mixing with the pion, is effectively one of the ensuing CP-odd Higgs bosons. Being very light, though, which occurs for $\lambda \sim \mathcal{O}(1)$, this state should have been detected by innumerable experiments, which is not the case. Hence, the scenario is ruled out. This is rather unfortunate, as a light axion coupling to gluons would automatically solve the strong CP problem. One may assume λ to be very small, so as to effectively decouple the axion, but this would be to no avail as there would be no justification for such a choice, just like there is no reason for the μ scale to be near the SUSY scale, our starting point, hence this is nothing but a replacement problem.

Alternatively, one could turn the PQ symmetry from global to local. The emerging Goldstone boson would then be absorbed by the gauge boson associated to the spontaneously broken $U(1)$ group, denoted by Z', in turn acquiring a mass and acting as the mediator of a new force. The existence of additional $U(1)$ gauge groups at TeV energies can be well motivated within GUTs and string models as we have discussed in the previous chapter.

In the remainder of this chapter, we will concentrate on the third possible way of rescuing the $U(1)_{\mathrm{PQ}}$ symmetry, which is essentially to break it explicitly. We will therefore look at two specific solutions proposed in the literature so far, by concentrating on their key phenomenological features. Both will feature a very distinctive signature in their Higgs sector: light pseudoscalars will continue to easily appear in the form of pNGBs.

15.4 THE NMSSM

As intimated, the most straightforward solution to the μ problem of the MSSM is to promote the μ parameter into a field whose VEV is determined, like the other scalar field VEVs, from the minimisation of the scalar potential along the new direction [88, 90, 348–357]. Naturally, it is expected to fall in the range of the other VEVs, i.e., of order $\mathcal{O}(M_{\text{SUSY}})$. Such a superfield has to be a singlet under the SM gauge group, hence, it has no gauge couplings but has a PQ charge, so one can naively introduce any term of the form $\sim \hat{S}^n$, with n integer, in the superpotential in order to break the PQ symmetry. However, since the superpotential is of dimension 3, any power with $n \neq 3$ will require a dimensionful coefficient naturally at the GUT or Planck scale, naively making the term either negligible (for $n > 3$) or unacceptably large (for $n < 3$). For this reason, it is usual to postulate some extra discrete symmetry, e.g., Z_3, in order to forbid terms with dimensionful coefficients. The superpotential of the model then becomes

$$W_\lambda^{\text{NMSSM}} = W_\lambda^{\text{MSSM}} + \frac{\kappa}{3}\hat{S}^3, \tag{15.6}$$

where κ is a dimensionless constant which measures the strength of the PQ symmetry breaking. Additionally, one must also introduce an extra soft SUSY-breaking term to accompany the new trilinear self-coupling. The complete soft SUSY-breaking Higgs sector then becomes

$$-\mathcal{L}_{\text{soft}}^{\text{NMSSM}} \supset -\mathcal{L}_{\text{soft}}^{\text{MSSM}} + (\frac{1}{3}\kappa A_\kappa S^3 + \text{h.c.}), \tag{15.7}$$

where, like A_λ, A_κ is a dimensionful coefficient of order M_{SUSY}. This model is known as the NMSSM and has generated much interest in the literature (see [358] for a recent review).

The Z_3 symmetry is then spontaneously broken at the EW scale when the Higgs fields acquire non-zero VEVs. It is well known, however, that the spontaneous breaking of such a discrete symmetry results in disastrous cosmological domain walls [359], unless this symmetry is explicitly broken by the non-renormalisable sector of the theory. Domain walls can be tolerated if there is a discrete-symmetry-violating contribution to the scalar potential larger than the scale $\mathcal{O}(1\text{ MeV})$ set by nucleosynthesis [359].

Historically, it was always assumed that the Z_3 symmetry could be broken by an appropriate type of unification with gravity at the Planck scale. Non-renormalisable operators will generally be introduced into the superpotential and Kähler potential such that they break Z_3 and lead to a preference for one particular vacuum, thereby solving the problem.

Heavy fields interacting with the standard light fields generate in the effective low-energy theory an infinite set of non-renormalisable operators of the light fields scaled by powers of the characteristic mass-scale of the heavy sector (M_{Pl}, M_{GUT},...). These terms appear either as D-terms in the Kähler potential or as F-terms in the superpotential. It is known, however, that gauge singlet superfields do not obey decoupling [336–341], so that, when SUSY is either spontaneously or softly broken, in addition to the suppressed non-renormalisable terms, they can in general give rise to a large (quadratically divergent) tadpole term in the potential proportional to the heavy scale, which is of the form [337, 360–367]

$$\mathcal{L}_{\text{soft}} \supset t_S S \sim \frac{1}{(16\pi^2)^n}M_{\text{Pl}}M_{\text{SUSY}}^2 S, \tag{15.8}$$

where n is the number of loops.

Technically, the tadpole is generated through higher-order loop diagrams in which the non-renormalisable interactions participate as vertices together with the renormalisable ones. A discrete global symmetry like the one discussed above would forbid this term but would lead to the appearance of the aforementioned domain walls upon its unavoidable spontaneous breakdown. The generated large tadpole reintroduces the hierarchy problem [337,360–365], since due to its presence the singlet VEV gets a value $\langle S \rangle^2 \sim M_{\text{SUSY}} M_{\text{Pl}}$. It appears that $N = 1$ supergravity, spontaneously broken by a set of hidden sector fields, is the natural setting to study the generation of the destabilising tadpoles. A thorough analysis carried out in [368] shows that the only harmful non-renormalisable interactions are either even superpotential terms or odd Kähler potential ones. In addition, operators with six or more powers of the cut-off in the denominator are harmless. Finally, a tadpole diagram is divergent only if it contains an odd number of 'dangerous' vertices.

Clearly, the tadpole given above breaks the Z_3 symmetry as desired, but if $n < 5$, t_S is several orders of magnitude larger than the soft-SUSY-breaking scale M_{SUSY}. This leads to an unacceptably large would-be μ term since $t_S S$ combines with the $\sim M_{\text{SUSY}}^2 S^* S$ soft mass term to produce a shift in the VEV of S to a value of order $< S > \sim \frac{t_S}{M_{\text{SUSY}}^2} \sim \frac{1}{(16\pi^2)^n} M_{\text{Pl}}$ and correspondingly $\mu_{\text{eff}} \sim \lambda S$. For example, if the tadpole were generated at the one-loop level, the effective μ term would be huge, of order $10^{16} - 10^{17}$ GeV, *i.e.*, close to the GUT scale, whereas μ should be of the order of the EW scale to realise a natural Higgs mechanism, as explained. Hence, it was argued in [368] that the NMSSM is either ruled out cosmologically or suffers from a naturalness problem related to the destabilisation of the gauge hierarchy. However, there are at least two simple escapes.

One obvious way out of this problem would be to gauge the $U(1)_{\text{PQ}}$ symmetry [369–371]. In this case, the Z_3 symmetry is embedded into the local $U(1)$ symmetry. The would-be PQ axion is then eaten by the longitudinal component of the extra gauge boson. However, from a low-energy perspective, the price one has to pay here is that the field content needs to be extended by adding new chiral quark and lepton states in order to ensure anomaly cancellation related to the gauged $U(1)_{\text{PQ}}$ symmetry (as we touched upon in a previous chapter).

The second approach is to find symmetry scenarios where all harmful destabilising tadpoles are absent. What is needed is a suitable symmetry that forbids the dangerous non-renormalisable terms and allows only for tadpoles of order $M_{\text{SUSY}}^3 (S + S^*)$. This symmetry should at the same time allow for a large enough Z_3-breaking term in the scalar potential in order to destroy the unwanted domain walls [372]. This can be achieved by imposing a discrete Z_2^R symmetry under which all superfields and the superpotential flip sign. To avoid destabilisation while curing the domain wall problem, this symmetry has to be extended to the non-renormalisable part of the superpotential and to the Kähler potential. As happens to all R-symmetries, the Z_R^2 symmetry is broken by the soft-SUSY-breaking terms, giving rise to harmless tadpoles of order $\frac{1}{(16\pi^2)^n} M_{\text{SUSY}}^3$ with $2 \leq n \leq 4$. Although these terms are phenomenologically irrelevant, they are entirely sufficient to break the global Z_3 symmetry and make the domain walls collapse.

15.5 THE nMSSM

Indeed, a potentially interesting alternative for breaking the PQ symmetry is based on the adoption of discrete R-symmetries, such that the destabilising tadpoles do appear but are naturally suppressed because they arise at loops higher than 5 [373]. In particular, these symmetries may lead to a superpotential whose renormalisable part has exactly the form in Eq. (15.1). A notable example is the so-called MNSSM [374]. Herein, the effective renormalisable superpotential reads as

$$W_\lambda^{\text{MNSSM}} = W_\lambda + t_F \hat{S}, \tag{15.9}$$

where t_F is a radiatively induced tadpole of EW scale strength. In addition, there will be a soft SUSY-breaking tadpole term $t_S S$ as given above. The key point in the construction of the renormalisable superpotential is that the simple form (15.9) can be enforced by discrete R-symmetries [373–376]. The latter govern the complete gravity-induced non-renormalisable superpotential and Kähler potential. Within the SUGRA framework of SUSY-breaking, it has then been possible to show [374] that the potentially dangerous tadpole t_S will appear at a loop level n higher than 5. From Eq. (15.12), the size of the tadpole parameter t_S can be estimated to be in the right range, i.e., $|S| < 1 - 10$ TeV^3 for $n = 6, 7$, such that the gauge hierarchy does not get destabilised. To be specific, the tadpole $t_S S$ together with the soft SUSY-breaking mass term $m_S^2 S^* S \sim M_{\text{SUSY}}^2 S^* S$ lead to a VEV for S, $< S > = \frac{1}{\sqrt{2}} v_S$, of order $M_{\text{SUSY}} \sim 1$ TeV. The latter gives rise to a μ parameter at the required EW scale. Thus, another natural explanation for the origin of the μ parameter can be obtained.

A useful discrete R-symmetry to remove the singlet S induced divergent tadpoles is Z_{5R} of a non-anomalous R-symmetry $U(1)_{R'}$ with charges

$$\hat{H}_u \to 0, \ \hat{H}_d \to 0, \ \hat{S} \to 2, \ \hat{Q} \to 1, \ \hat{u}^C \to 1, \ \hat{d}^C \to 1, \ \hat{L} \to 1, \ \hat{e}^C \to 1, \tag{15.10}$$

hence the combination $R' = 3R + PQ$ is imposed on the complete theory (including non-renormalizable operators), such that, ultimately, the tadpole terms generated at high loop order [372] are of the form

$$\delta W = t_F S, \quad \delta V = t_S (S + S^*), \quad \text{where} \ \ t_F \lesssim M_{\text{SUSY}}^2 \ \text{and} \ \ t_S \lesssim M_{\text{SUSY}}^3. \tag{15.11}$$

In the case where $t_F \sim M_{\text{SUSY}}^2$ and $t_S \sim M_{\text{SUSY}}^3$, the singlet cubic self interaction in the NMSSM superpotential is not even phenomenologically required and can be omitted [373]. The resulting model has been denoted as the nMSSM as, in the limit where SUSY is unbroken, the MSSM μ term is only traded for the dimensionless λ coupling. Once SUSY is softly broken, the generated tadpole terms t_F and t_S break both the Z_3 and the PQ symmetry and an adequately suppressed linear term is generated at six-loop level by combining the non-renormalisable Kähler potential terms $\lambda_1 S^2 H_1 H_2 / M_{\text{Pl}}^2 +$ h.c. and $\lambda_2 S (H_1 H_2)^3 / M_{\text{Pl}}^5 +$ h.c. with the renormalisable superpotential term $\lambda S H_1 H_2$:

$$V_{\text{tadpole}} \sim \frac{1}{(16\pi^2)^6} \lambda_1 \lambda_2 \lambda^4 M_{\text{SUSY}}^2 M_{\text{Pl}} (S + S^*). \tag{15.12}$$

This tadpole has the desired order of magnitude $\mathcal{O}(M_{\text{SUSY}})$ if $\lambda_1 \lambda_2 \lambda^4 \sim 10^{-3}$.

15.6 THE HIGGS SECTOR OF THE NMSSM, nMSSM AND ESSM

Before proceeding with the study of the Higgs sector, let us recall their relevant super-potential terms specific to the NMSSM, nMSSM and UMSSM, which we collect together for convenience in Tab. 15.1, alongside those of the MSSM, also including the underlying symmetries and particle content (here, as usual, H_i refers to a CP-even Higgs boson and A_j to a CP-odd one, $H^\pm$ being the charged Higgs states).

A convenient way of comparing the Higgs sector of the NMSSM, nMSSM and UMSSM with respect to the MSSM is by defining the F-, D- and soft-terms in the Lagrangian in a cumulative form, as nicely done in [25] (an approach that we borrow here then), i.e., as

Model	Symmetry	Superpotential	CP-even	CP-odd	Charged
MSSM	–	$\mu \hat{H}_u \hat{H}_d$	H_1, H_2	A_2	$H^\pm$
NMSSM	Z_3	$\lambda \hat{S} \hat{H}_u \hat{H}_d + \frac{\kappa}{3} \hat{S}^3$	H_1, H_2, H_3	A_1, A_2	$H^\pm$
nMSSM	Z_{5R}	$\lambda \hat{S} \hat{H}_u \hat{H}_d + t_F \hat{S}$	H_1, H_2, H_3	A_1, A_2	$H^\pm$
UMSSM	$U(1)'$	$\lambda \hat{S} \hat{H}_u \hat{H}_d$	H_1, H_2, H_3	A_2	$H^\pm$

Table 15.1 Higgs bosons of the MSSM and several of its extensions. We denote the single CP-odd state in the MSSM and UMSSM by A_2 for easier comparison with the other models.

follows:

$$V_F = |\lambda H_u H_d + t_F + \kappa S^2|^2 + |\lambda S|^2 \left(|H_d|^2 + |H_u|^2\right), \tag{15.13}$$

$$V_D = \frac{G^2}{8} \left(|H_d|^2 - |H_u|^2\right)^2 + \frac{g_2^2}{2} \left(|H_d|^2 |H_u|^2 - |H_u H_d|^2\right) \tag{15.14}$$

$$+ \frac{g_1'^2}{2} \left(Q_{H_d}|H_d|^2 + Q_{H_u}|H_u|^2 + Q_S|S|^2\right)^2,$$

$$V_{\text{soft}} = m_{H_d}^2 |H_d|^2 + m_{H_u}^2 |H_u|^2 + m_S^2 |S|^2 + (\lambda A_\lambda S H_u H_d + \frac{\kappa}{3} A_\kappa S^3 + t_S S + \text{h.c.}). \tag{15.15}$$

Here, the two Higgs doublets with hypercharge $Y = -1/2$ and $Y = +1/2$, respectively, are

$$H_d = \begin{pmatrix} H_d^0 \\ H^- \end{pmatrix}, \qquad H_u = \begin{pmatrix} H^+ \\ H_u^0 \end{pmatrix}, \tag{15.16}$$

and $H_u H_d = \epsilon_{ij} H_u^i H_d^j$. For a particular model, the model-dependent parameters g_1', κ, A_κ and $t_{F,S}$ in V_F, V_D and V are understood to be turned-off appropriately, according to Tab. 15.1:

$$\begin{aligned} \text{NMSSM} &: g_{1'} = 0, t_{F,S} = 0, \\ \text{nMSSM} &: g_{1'} = 0, \kappa = 0, A_\kappa = 0, \\ \text{UMSSM} &: t_{F,S} = 0, \kappa = 0, A_\kappa = 0. \end{aligned} \tag{15.17}$$

The couplings g_1, g_2 and $g_{1'}$ are for the $U(1)_Y, SU(2)_L$ and $U(1)'$ gauge symmetries, respectively, and the parameter G is defined as $G^2 = g_1^2 + g_2^2$. The NMSSM model-dependent parameters are κ and A_κ, while the free nMSSM parameters are t_F and t_S. The model dependence of the UMSSM is expressed by the D-term that has the $U(1)'$ charges of the Higgs fields, Q_{H_d}, Q_{H_u} and Q_S. In general, these charges are free parameters with the restriction[1] that $Q_{H_d} + Q_{H_u} + Q_S = 0$ to preserve gauge invariance. In any particular $U(1)'$ construction, the charges have specified values. The F-term and the soft terms contain the model dependence of the NMSSM and nMSSM. The soft terms A_κ of the NMSSM and t_S of the nMSSM are new to V_{soft}. The B-term of the MSSM is expressed in V_{soft} as $A_\lambda \lambda S H_u H_d$ after we identify

$$B\mu = A_\lambda \mu_{\text{eff}}. \tag{15.18}$$

---

[1]Additional restrictions on the charges of the ordinary and exotic particles come from the cancellation of anomalies.

The other terms in V_{soft} are the usual MSSM soft mass terms.

The minimum of the potential is found explicitly using the minimisation conditions found in, *e.g.*, [25]. The conditions found allow us to express the soft mass parameters in terms of the VEVs of the Higgs fields. At the minimum of the potential, the Higgs fields are expanded as

$$H_d^0 = \frac{1}{\sqrt{2}}(v_d + \phi_d + i\varphi_d), \quad H_u^0 = \frac{1}{\sqrt{2}}(v_u + \phi_u + i\varphi_u), \quad S = \frac{1}{\sqrt{2}}(s + \sigma + i\xi), \quad (15.19)$$

with $v^2 \equiv v_d^2 + v_u^2 = (246 \text{ GeV})^2$ and $\tan\beta \equiv v_u/v_d$. We write the Higgs mass-squared matrix in a compact form that includes all the extended models under consideration. The CP-even tree-level matrix elements in the H_d^0, H_u^0, S basis are:

$$
\begin{aligned}
(\mathcal{M}_+^0)_{11} &= \left[\frac{G^2}{4} + Q_{H_d}^2 g_{1'}^2\right] v_d^2 + \left(\frac{\lambda A_\lambda}{\sqrt{2}} + \frac{\lambda\kappa v_s}{2} + \frac{\lambda t_F}{v_s}\right)\frac{v_u v_s}{v_d}, \\
(\mathcal{M}_+^0)_{12} &= -\left[\frac{G^2}{4} - \lambda^2 - Q_{H_d}Q_{H_u}g_{1'}^2\right] v_d v_u - \left(\frac{\lambda A_\lambda}{\sqrt{2}} + \frac{\lambda\kappa v_s}{2} + \frac{\lambda t_F}{v_s}\right) v_s, \\
(\mathcal{M}_+^0)_{13} &= \left[\lambda^2 + Q_{H_d}Q_S g_{1'}^2\right] v_d v_s - \left(\frac{\lambda A_\lambda}{\sqrt{2}} + \lambda\kappa v_s\right) v_u, \\
(\mathcal{M}_+^0)_{22} &= \left[\frac{G^2}{4} + Q_{H_u}^2 g_{1'}^2\right] v_u^2 + \left(\frac{\lambda A_\lambda}{\sqrt{2}} + \frac{\lambda\kappa v_s}{2} + \frac{\lambda t_F}{v_s}\right)\frac{v_d v_s}{v_u}, \\
(\mathcal{M}_+^0)_{23} &= \left[\lambda^2 + Q_{H_u}Q_S g_{1'}^2\right] v_u v_s - \left(\frac{\lambda A_\lambda}{\sqrt{2}} + \lambda\kappa v_s\right) v_d, \\
(\mathcal{M}_+^0)_{33} &= \left[Q_S^2 g_{1'}^2 + 2\kappa^2\right] s^2 + \left(\frac{\lambda A_\lambda}{\sqrt{2}} - \frac{\sqrt{2}t_S}{v_d v_u}\right)\frac{v_d v_u}{v_s} + \frac{\kappa A_\kappa}{\sqrt{2}}v_s. \quad (15.20)
\end{aligned}
$$

The tree-level CP-odd matrix elements are:

$$
\begin{aligned}
(\mathcal{M}_-^0)_{11} &= \left(\frac{\lambda A_\lambda}{\sqrt{2}} + \frac{\lambda\kappa v_s}{2} + \frac{\lambda t_F}{v_s}\right)\frac{v_u v_s}{v_d}, \\
(\mathcal{M}_-^0)_{12} &= \left(\frac{\lambda A_\lambda}{\sqrt{2}} + \frac{\lambda\kappa v_s}{2} + \frac{\lambda t_F}{v_s}\right) v_s, \\
(\mathcal{M}_-^0)_{13} &= \left(\frac{\lambda A_\lambda}{\sqrt{2}} - \lambda\kappa v_s\right) v_u, \\
(\mathcal{M}_-^0)_{22} &= \left(\frac{\lambda A_\lambda}{\sqrt{2}} + \frac{\lambda\kappa v_s}{2} + \frac{\lambda t_F}{v_s}\right)\frac{v_d v_s}{v_u}, \\
(\mathcal{M}_-^0)_{23} &= \left(\frac{\lambda A_\lambda}{\sqrt{2}} - \lambda\kappa v_s\right) v_d, \\
(\mathcal{M}_-^0)_{33} &= \left(\frac{\lambda A_\lambda}{\sqrt{2}} + 2\lambda\kappa s - \frac{\sqrt{2}t_S}{v_d v_u}\right)\frac{v_d v_u}{v_s} - \frac{3\kappa A_\kappa}{\sqrt{2}}v_s. \quad (15.21)
\end{aligned}
$$

The tree-level charged Higgs mass-squared matrix elements are:

$$
\begin{aligned}
(\mathcal{M}^\pm)_{11} &= \frac{v_u^2\left(g_2^2 - 2\lambda^2\right)}{4} + \left(\frac{1}{\sqrt{2}}A_\lambda\lambda s + \frac{1}{2}\lambda\kappa s^2 + \lambda t_F\right)\frac{v_u}{v_d}, \\
(\mathcal{M}^\pm)_{12} &= -\frac{v_d v_u\left(g_2^2 - 2\lambda^2\right)}{4} - \left(\frac{1}{\sqrt{2}}A_\lambda\lambda s + \frac{1}{2}\lambda\kappa s^2 + \lambda t_F\right), \\
(\mathcal{M}^\pm)_{22} &= \frac{v_d^2\left(g_2^2 - 2\lambda^2\right)}{4} + \left(\frac{1}{\sqrt{2}}A_\lambda\lambda s + \frac{1}{2}\lambda\kappa s^2 + \lambda t_F\right)\frac{v_d}{v_u}. \quad (15.22)
\end{aligned}
$$

The physical Higgs boson masses are found by diagonalising the mass-squared matrices, $\mathcal{M}_D = R\mathcal{M}R^{-1}$, where $\mathcal{M}$ also includes the radiative corrections discussed below. The rotation matrices for the diagonalisation of the CP-even and CP-odd mass-squared matrices, $R_{\pm}^{ij}$, and for the charged Higgs matrix, $\mathcal{R}^{ij}$, may then be used to construct the physical Higgs fields,

$$
\begin{aligned}
H_i &= R_+^{i1}\phi_d + R_+^{i2}\phi_u + R_+^{i3}\sigma, \\
A_i &= R_-^{i1}\varphi_d + R_-^{i2}\varphi_u + R_-^{i3}\xi, \\
H_i^{\pm} &= \mathcal{R}^{i1}H^- + \mathcal{R}^{i2}H^+,
\end{aligned}
\tag{15.23}
$$

where the physical states are ordered by their mass as $M_{H_1} \leq M_{H_2} \leq M_{H_3}$ and $M_{A_1} \leq M_{A_2}$.

15.6.1 Radiative corrections

An accurate analysis of the Higgs masses requires loop corrections. The dominant contributions at one-loop are from the top and scalar top loops due to their large Yukawa coupling. In the UMSSM, the gauge couplings are small compared to the top quark Yukawa coupling, so the one-loop gauge contributions can be dropped. Corrections unique to the NMSSM and nMSSM begin only at the two-loop level. Thus, all contributions that are model-dependent do not contribute significantly at one-loop order and the usual one-loop SUSY top and stop loops of the MSSM are universal in these extended models.

The mass squared matrix elements become

$$
\mathcal{M}_{\pm} = \mathcal{M}_{\pm}^0 + \mathcal{M}_{\pm}^1,
\tag{15.24}
$$

where the radiative corrections to the CP-even mass-squared matrix elements are given by

$$
(\mathcal{M}_+^1)_{11} = k\left[\left(\frac{(\widetilde{m}_1^2)^2}{(m_{\tilde{t}_1}^2 - m_{\tilde{t}_2}^2)^2}\mathcal{G}\right)v_d^2 + \left(\frac{\lambda h_t^2 A_t}{2\sqrt{2}}\mathcal{F}\right)\frac{v_u v_s}{v_d}\right],
$$

$$
(\mathcal{M}_+^1)_{12} = k\left[\left(\frac{\widetilde{m}_1^2\widetilde{m}_2^2}{(m_{\tilde{t}_1}^2 - m_{\tilde{t}_2}^2)^2}\mathcal{G} + \frac{h_t^2\widetilde{m}_1^2}{m_{\tilde{t}_1}^2 + m_{\tilde{t}_2}^2}(2-\mathcal{G})\right)v_d v_u - \left(\frac{\lambda h_t^2 A_t}{2\sqrt{2}}\mathcal{F}\right)v_s\right],
$$

$$
(\mathcal{M}_+^1)_{13} = k\left[\left(\frac{\widetilde{m}_1^2\widetilde{m}_s^2}{(m_{\tilde{t}_1}^2 - m_{\tilde{t}_2}^2)^2}\mathcal{G} + \frac{\lambda^2 h_t^2}{2}\mathcal{F}\right)v_d v_s - \left(\frac{\lambda h_t^2 A_t}{2\sqrt{2}}\mathcal{F}\right)v_u\right],
$$

$$
(\mathcal{M}_+^1)_{22} = k\left(\frac{(\widetilde{m}_2^2)^2}{(m_{\tilde{t}_1}^2 - m_{\tilde{t}_2}^2)^2}\mathcal{G} + \frac{2h_t^2\widetilde{m}_2^2}{m_{\tilde{t}_1}^2 + m_{\tilde{t}_2}^2}(2-\mathcal{G}) + h_t^4 \ln\frac{m_{\tilde{t}_1}^2 m_{\tilde{t}_2}^2}{m_t^4}\right)v_u^2
$$
$$
+ k\left(\frac{\lambda h_t^2 A_t}{2\sqrt{2}}\mathcal{F}\right)\frac{v_d v_s}{v_u},
$$

$$
(\mathcal{M}_+^1)_{23} = k\left[\left(\frac{\widetilde{m}_2^2\widetilde{m}_s^2}{(m_{\tilde{t}_1}^2 - m_{\tilde{t}_2}^2)^2}\mathcal{G} + \frac{h_t^2\widetilde{m}_s^2}{m_{\tilde{t}_1}^2 + m_{\tilde{t}_2}^2}(2-\mathcal{G})\right)v_u v_s - \left(\frac{\lambda h_t^2 A_t}{2\sqrt{2}}\mathcal{F}\right)v_d\right],
$$

$$
(\mathcal{M}_+^1)_{33} = k\left[\left(\frac{(\widetilde{m}_s^2)^2}{(m_{\tilde{t}_1}^2 - m_{\tilde{t}_2}^2)^2}\mathcal{G}\right)s^2 + \left(\frac{\lambda h_t^2 A_t}{2\sqrt{2}}\mathcal{F}\right)\frac{v_d v_u}{v_s}\right],
\tag{15.25}
$$

where $k = \frac{3}{(4\pi)^2}$ and the loop factors are

$$
\mathcal{G}(m_{\tilde{t}_1}^2, m_{\tilde{t}_2}^2) = 2\left[1 - \frac{m_{\tilde{t}_1}^2 + m_{\tilde{t}_2}^2}{m_{\tilde{t}_1}^2 - m_{\tilde{t}_2}^2}\log\left(\frac{m_{\tilde{t}_1}}{m_{\tilde{t}_2}}\right)\right], \quad \mathcal{F} = \log\left(\frac{m_{\tilde{t}_1}^2 m_{\tilde{t}_2}^2}{Q^4}\right) - \mathcal{G}(m_{\tilde{t}_1}^2, m_{\tilde{t}_2}^2).
\tag{15.26}
$$

Here we have defined

$$
\begin{aligned}
\widetilde{m}_1^2 &= h_t^2 \mu_{\text{eff}} \left(\mu_{\text{eff}} - A_t \tan\beta \right), \\
\widetilde{m}_2^2 &= h_t^2 A_t \left(A_t - \mu_{\text{eff}} \cot\beta \right), \\
\widetilde{m}_s^2 &= \frac{v_d^2}{s^2} h_t^2 \mu_{\text{eff}} \left(\mu_{\text{eff}} - A_t \tan\beta \right),
\end{aligned}
\tag{15.27}
$$

with Q being the $\overline{\text{DR}}$ (modified DR) scale and A_t the stop trilinear coupling.

The corrections to the CP-odd mass-squared matrices are given by [370, 374, 375, 377–380]

$$
(\mathcal{M}_-^1)_{ij} = \frac{\lambda v_d v_u v_s}{\sqrt{2} v_i v_j} \frac{k h_t^2 A_t}{2} \mathcal{F}(m_{\tilde{t}_1}^2, m_{\tilde{t}_2}^2),
\tag{15.28}
$$

where we identify $v_1 \equiv v_d$, $v_2 \equiv v_u$ and $v_3 \equiv v_s$.

The one-loop corrections to the charged Higgs mass are equivalent to those in the MSSM and can be significant for large $\tan\beta$. The charged Higgs boson in the MSSM has a tree-level mass,

$$
(M_{H^\pm}^{(0)})^2 = M_W^2 + M_Y^2,
\tag{15.29}
$$

and the extended-MSSM charged Higgs boson mass is

$$
(M_{H^\pm}^{(0)})^2 = M_W^2 + M_Y^2 - \frac{\lambda^2 v^2}{2} + \lambda \frac{\sqrt{2}(2 t_F + \kappa s^2)}{\sin 2\beta},
\tag{15.30}
$$

where $M_Y^2 = \frac{\sqrt{2}\lambda s A_\lambda}{\sin 2\beta}$ is the tree-level mass of the MSSM CP-odd Higgs boson. The case of large M_Y (or M_A in the MSSM) yields a large charged Higgs mass and is consistent with the MSSM decoupling limit, yielding a SM Higgs sector. Radiative corrections in the MSSM shift the mass by

$$
(M_{H^\pm}^{(1)})^2 = \frac{\lambda A_t s k h_t^2 \mathcal{F}}{\sqrt{2} \sin 2\beta} + \delta M_{H^\pm}^2,
\tag{15.31}
$$

where, after including $\tan\beta$ dependent terms, $\delta M_{H^\pm}^2$ is given by the leading logarithm result of the full one-loop calculation [381, 382],

$$
\delta M_{H^\pm}^2 = \frac{N_C g^2}{32\pi^2 M_W^2} \left(\frac{2 m_t^2 m_b^2}{\sin^2\beta \cos^2\beta} - M_W^2 \left(\frac{m_t^2}{\sin^2\beta} + \frac{m_b^2}{\cos^2\beta} \right) + \frac{2}{3} M_W^4 \right) \log\frac{M_{\text{SUSY}}^2}{m_t^2},
\tag{15.32}
$$

where, as usual, $N_C = 3$ is the number of colours and M_{SUSY} is the supersymmetric mass scale, taken to be 1 TeV. Model dependent terms come in at tree level, giving a charged Higgs mass after radiative corrections of

$$
M_{H^\pm}^2 = M_W^2 + M_Y^2 - \frac{\lambda^2 v^2}{2} + \lambda \frac{\sqrt{2}(2 t_F + \kappa s^2)}{\sin 2\beta} + \left(\frac{\lambda A_t s k h_t^2 \mathcal{F}}{\sqrt{2} \sin 2\beta} + \delta M_{H^\pm}^2 \right).
\tag{15.33}
$$

From the previous mass matrices, it is interesting to extract the sum rules amongst the squared masses of the neutral Higgs boson states in the various models:

$$
\sum_i M_{H_i}^2 - \sum_j M_{A_j}^2 = M_Z^2 + M_{\text{xMSSM}}^2 + \delta M^2.
\tag{15.34}
$$

The sums are over the massive Higgs bosons and M_{xMSSM} is a model-dependent mass parameter with values

$$M_{\mathrm{NMSSM}}^2 = 2\kappa(-\lambda v_d v_u + s(\sqrt{2}A_\kappa + s\kappa)), \tag{15.35}$$

$$M_{\mathrm{nMSSM}}^2 = 0, \tag{15.36}$$

$$M_{\mathrm{UMSSM}}^2 = M_{Z'}^2. \tag{15.37}$$

Here, the term δM^2 is due to the radiative corrections and takes the form

$$\delta M^2 = \mathrm{Tr}\left[\mathcal{M}_+^1 - \mathcal{M}_-^1\right]. \tag{15.38}$$

Finally, since in any supersymmetric theory that is perturbative at the GUT scale the lightest Higgs boson mass has an upper limit [383], it is important to extract a value for the latter. As the mass-squared CP-even matrix $\mathcal{M}_+$ is real and symmetric, an estimate of the upper bound on the smallest mass-squared eigenvalue may be obtained through the Rayleigh quotient:

$$M_{H_1}^2 \leq \frac{u^T \mathcal{M}_+ u}{u^T u}, \tag{15.39}$$

where u is an arbitrary non-zero vector. With the choices

$$
\begin{aligned}
u^T &= (\cos\beta, \quad \sin\beta) & \text{[MSSM]} \\
&= (\cos\beta, \quad \sin\beta, \quad 0) & \text{[NMSSM,nMSSM,UMSSM]},
\end{aligned} \tag{15.40}
$$

the well-known upper bounds of the lightest Higgs mass-squared from the mass-squared matrices of Eqs. (15.20) and (15.25) are given as follows. *(i)* MSSM [85, 86, 384]:

$$M_{H_1^0}^2 \leq M_Z^2 \cos^2 2\beta + \tilde{\mathcal{M}}^{(1)}, \tag{15.41}$$

where

$$\tilde{\mathcal{M}}^{(1)} = (\mathcal{M}_+^{(1)})_{11}\cos^2\beta + (\mathcal{M}_+^{(1)})_{22}\sin^2\beta + (\mathcal{M}_+^{(1)})_{12}\sin 2\beta. \tag{15.42}$$

(ii) NMSSM and nMSSM [350]:

$$M_{H_1^0}^2 \leq M_Z^2 \cos^2 2\beta + \frac{1}{2}\lambda^2 v^2 \sin^2 2\beta + \tilde{\mathcal{M}}^{(1)}. \tag{15.43}$$

(iii) UMSSM [369]:

$$M_{H_1^0}^2 \leq M_Z^2 \cos^2 2\beta + \frac{1}{2}\lambda^2 v^2 \sin^2 2\beta + g_{1'}^2 v^2 (Q_{H_d}\cos^2\beta + Q_{H_u}\sin^2\beta)^2 + \tilde{\mathcal{M}}^{(1)}. \tag{15.44}$$

Although the upper bounds change with the choice of the u vector, these results indicate that extended models have larger upper bounds for the lightest Higgs due to the contribution of the singlet scalar. The UMSSM can have the largest upper bound due to the quartic coupling contribution from the additional gauge coupling term, $g_{1'}$, in the $U(1)'$ extension. In the MSSM, large $\tan\beta$ values are suggested by the conflict between the experimental lower bound and the theoretical upper bound on M_{H_1}. Since the extended models contain additional terms which relax the theoretical bound, they allow smaller values for $\tan\beta$ than the MSSM.

15.6.2 Numerical results

In this subsection, we present some numerical results for the Higgs masses. Before proceeding in this respect, we ought to recap the independent parameters of the Higgs sectors of the various models. Upon enforcing EWSB and employing the minimisation conditions for the Higgs potential so as to reproduce the Z mass, one can cast the input parameters as follows. In the NMSSM case, one can choose

$$\lambda, \quad A_\lambda, \quad \kappa, \quad A_\kappa, \quad \mu_{\text{eff}}, \quad \tan\beta, \tag{15.45}$$

for the nMSSM one has

$$\lambda, \quad A_\lambda, \quad t_S, \quad \mu_{\text{eff}}, \quad \tan\beta, \tag{15.46}$$

while for the UMSSM case one can adopt

$$g_1', \quad \lambda, \quad A_\lambda, \quad \mu_{\text{eff}}, \quad \tan\beta, \tag{15.47}$$

where the effective μ parameter has a common expression in all scenarios. For the UMSSM case, we assume as an example the ESSM [24, 253, 285, 318] already introduced in the previous chapter. Fig. 15.1 shows the mass of the lightest CP-even and CP-odd Higgs bosons of the MSSM, NMSSM, nMSSM and ESSM, mapped in $\tan\beta$ and the S VEV (v_s), for a representative choice of the Higgs parameters $\lambda, A_\lambda, \kappa, A_\kappa, t_S, t_F$ and the SUSY sector inputs $M_{\text{SUSY}}, A_t, M_{\tilde{Q}} = M_{\tilde{U}}$. Two key features emerge. On the one hand, the upper limit on the lightest CP-even (and SM-like) Higgs state is larger in the extended models than in the minimal SUSY representation, thereby relieving the so-called 'little hierarchy problem' (*i.e.*, the need in the MSSM to fine-tune loop-corrections to the lightest SM-like Higgs boson mass so as to lift it well beyond M_Z and close to the measured value of 125 GeV). On the other hand, the CP-odd mass in extended models can be smaller in comparison to that of the MSSM, a key feature which we will exploit in the following section.

It is, however, instructive to scan the full Higgs and SUSY parameter space available. Upon doing so, we summarise in Fig. 15.2 the available ranges found from a random scan of the lightest CP-even, CP-odd and charged Higgs boson masses that satisfy the experimental and theoretical constraints. For each model, the values of the maximum and minimum masses are given as well as the reason for the bounds.

The lightest CP-even and CP-odd and the charged Higgs boson mass ranges differ significantly among the models. The CP-even Higgs mass range is quite restricted in the MSSM. The upper limits for the CP-even Higgs masses in the extended models saturate the theoretical bounds and are extended by 30-40 GeV compared to the MSSM, while the upper limits on the lightest CP-odd Higgs masses are artificial in the MSSM and ESSM as they change with the size of the scan parameters, such as A_λ and $\tan\beta$. The lower limits of the lightest CP-odd masses in the MSSM and ESSM reflect the LEP limits on M_{A_2}. The ESSM is similar to the MSSM since v_s is required to be large by the strict constraint on the $Z - Z'$ mixing angle (henceforth $\alpha_{ZZ'}$), decoupling the singlet state and recovering a largely MSSM Higgs sector. However, fine tuning the Higgs doublet charges under the $U(1)'$ gauge symmetry and $\tan\beta$ allows the $Z - Z'$ mixing constraint on v_s to be less severe and can result in a lower Higgs mass with respect to the MSSM. These instances along with the values $A_\lambda = A_t = 0$ GeV allow very low CP-even Higgs masses of $\mathcal{O}(1 \text{ GeV})$ and a massless CP-odd state. Since these points are distinct from the range of masses typically found in the ESSM, we do not show these points in Fig. 15.2 but simply note that they exist. However, the NMSSM and nMSSM may have a massless CP-odd state due to global $U(1)$ symmetries

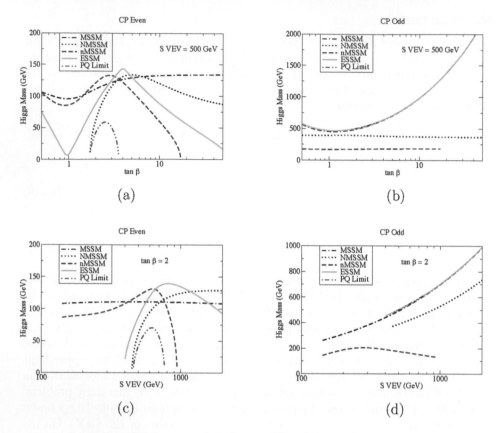

Figure 15.1 Lightest CP-even and lightest CP-odd Higgs masses versus $\tan\beta$ and v_s for the MSSM, NMSSM, nMSSM, ESSM and PQ limit. Only the theoretical constraints are applied, with $v_s = 500$ GeV (for $\tan\beta$-varying curves), $\tan\beta = 2$ (for v_s-varying curves). Input parameters of $\lambda = 0.5$, $A_\lambda = 500$ GeV, $\kappa = 0.5$, $A_\kappa = -250$ GeV, $t_F = -0.1M_{\mathrm{SUSY}}^2$, $t_S = -0.1M_{\mathrm{SUSY}}^3$, where $M_{\mathrm{SUSY}} = 500$ GeV and $Q = 300$ GeV, the renormalisation scale, are taken. The genuine SUSY parameters adopted are $A_t = 1$ TeV, $M_{\tilde{Q}} = M_{\tilde{U}} = 1$ TeV. The $U(1)_{PQ}$ limit allows one massive CP-odd Higgs whose mass is equivalent to that of the ESSM CP-odd Higgs. (The figure is taken from [25].)

discussed previously, while the upper limit on the lightest CP-odd Higgs mass depends on the specifics of the state crossing with the heavier state, A_2, which has a scan-dependent mass. In these models, the CP-odd masses extend to zero since the mixing of two CP-odd states allows one CP-odd Higgs to be completely singlet and avoid the constraints discussed above.

The charged Higgs masses are found to be as low as 79 GeV in the scans, in agreement with the imposed experimental limit of 78.6 GeV. In these cases where $M_{H^\pm} \sim 80$ GeV, the charged Higgs is often the lightest member of the Higgs spectrum. However, these cases require fine tuning to obtain values of $\mu_{\mathrm{eff}} > 100$ GeV. The upper limit of the charged Higgs mass is dependent on the range of the scan parameters as seen in Eq. (15.33). The

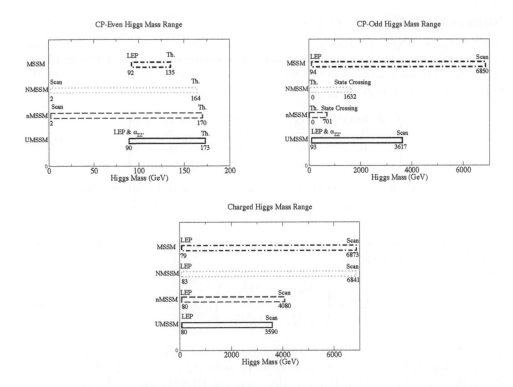

Figure 15.2 Mass ranges of the lightest CP-even and CP-odd and the charged Higgs bosons in each extended-MSSM model from the grid and random scans. Explanation of extremal bounds and their values are provided for each model. Explanations are: Th.–theoretical bound met, value not sensitive to limits of the scan parameters. Scan–value sensitive to limits of the scan parameters. State crossing–value has maximum when crossing of states occurs (specifically for A_1 and A_2 in the NMSSM and nMSSM). LEP–experimental constraints from LEP. $\alpha_{ZZ'}$–experimental constraints in the ESSM on the $Z - Z'$ mixing angle. (The figure is taken from [25].)

discrepancy in the upper limit of the charged and CP-odd Higgs mass between the MSSM and ESSM is a consequence of a lower μ_{eff} in the ESSM. Large values of μ_{eff} are more fine-tuned in the ESSM than in the MSSM since the additional gauge, $g_{1'}$, and Higgs, λ, couplings often drive $M_{H_1}^2 < 0$. Consequently, CP-odd and charged Higgs masses comparable to the higher MSSM limit are not present in the scan. The upper bound on the charged Higgs mass in the NMSSM is relaxed due to the additional parameter of the model.

15.7 THE HIGGS SECTOR OF THE BLSSM

Unlike the UMSSM scenarios described so far, the BLSSM possesses four CP-even and two CP-odd Higgs states. Hence, its whole Higgs sector deserves a separate treatment. Herein, the gauge kinetic term induces mixing at tree level between the H_u^0, H_d^0 and $\bar{\eta}, \eta$ states in the BLSSM scalar potential. Therefore, in the presence of the minimisation conditions of this potential as given in Eq. (14.44) and upon the usual redefinition of the Higgs fields, *i.e.*,

$H^0_{u,d} = \frac{1}{\sqrt{2}}(v_{u,d} + \sigma_{u,d} + i\phi_{u,d})$ and $\bar{\eta}^0, \eta^0 = \frac{1}{\sqrt{2}}(v'_{\bar{\eta},\eta} + \sigma'_{\bar{\eta},\eta} + i\phi'_{\bar{\eta},\eta})$, where $\sigma_{u,d} = \mathrm{Re}H^0_{u,d}$, $\phi_{u,d} = \mathrm{Im}H^0_{u,d}$, $\sigma'_{\bar{\eta},\eta} = \mathrm{Re}(\bar{\eta}^0, \eta^0)$ and $\phi'_{\bar{\eta},\eta} = \mathrm{Im}(\bar{\eta}^0, \eta^0)$, wherein the real parts correspond to the CP-even Higgs bosons and the imaginary parts correspond to the CP-odd Higgs bosons, the squared-mass matrix of the BLSSM CP-odd neutral Higgs fields at tree level, in the basis $(\phi_u, \phi_d, \phi'_{\bar{\eta}}, \phi'_\eta)$, is given by

$$m^2_{A,A'} = \begin{pmatrix} B_\mu \tan\beta & B_\mu & 0 & 0 \\ B_\mu & B_\mu \cot\beta & 0 & 0 \\ 0 & 0 & B_{\mu'} \tan\beta' & B_{\mu'} \\ 0 & 0 & B_{\mu'} & B_{\mu'} \cot\beta' \end{pmatrix}. \tag{15.48}$$

It is clear that the MSSM-like CP-odd Higgs A is decoupled from the BLSSM-like version A' (at tree level). However, due to the dependence of B_μ on v', one may find $m^2_A = \frac{2B_\mu}{\sin 2\beta} \sim m^2_{A'} = \frac{2B_{\mu'}}{\sin 2\beta'} \sim \mathcal{O}(1\text{ TeV})$.

The squared-mass matrix of the BLSSM CP-even neutral Higgs fields at tree level, in the basis $(\sigma_1, \sigma_2, \sigma'_1, \sigma'_2)$, is given by

$$M^2 = \begin{pmatrix} M^2_{hH} & M^2_{hh'} \\ M^{2^T}_{hh'} & M^2_{h'H'} \end{pmatrix}, \tag{15.49}$$

where M^2_{hH} is the usual MSSM CP-even neutral Higgs mass matrix, which leads to the SM-like Higgs boson with mass, at one loop level, of order 125 GeV, and a heavy Higgs boson with mass $m_H \sim m_A \sim \mathcal{O}(1\text{ TeV})$. In this case, the BLSSM matrix $M^2_{h'H'}$ is given by

$$M^2_{h'H'} = \begin{pmatrix} m^2_{A'}c^2_{\beta'} + g^2_{BL}v'^2_1 & -\frac{1}{2}m^2_{A'}s_{2\beta'} - g^2_{BL}v_{\bar{\eta}}v_\eta \\ -\frac{1}{2}m^2_{A'}s_{2\beta'} - g^2_{BL}v_{\bar{\eta}}v_\eta & m^2_{A'}s^2_{\beta'} + g^2_{BL}v'^2_2 \end{pmatrix}, \tag{15.50}$$

where $c_x = \cos(x)$ and $s_x = \sin(x)$. Therefore, the eigenvalues of this mass matrix are given by

$$m^2_{h',H'} = \frac{1}{2}\left[(m^2_{A'} + M^2_{Z'}) \mp \sqrt{(m^2_{A'} + M^2_{Z'})^2 - 4m^2_{A'}M^2_{Z'}\cos^2 2\beta'}\right]. \tag{15.51}$$

If $\cos^2 2\beta' \ll 1$, one finds that the lightest $B - L$ neutral Higgs state is given by

$$m_{h'} \simeq \left(\frac{m^2_{A'}M^2_{Z'}\cos^2 2\beta'}{m^2_{A'} + M^2_{Z'}}\right)^{\frac{1}{2}} \simeq \mathcal{O}(100\text{ GeV}). \tag{15.52}$$

The mixing matrix $M^2_{hh'}$ is proportional to $\tilde{g}$ and can be written as [21]

$$M^2_{hh'} = \frac{1}{2}\tilde{g}g_{BL}\begin{pmatrix} v_d v_{\bar{\eta}} & -v_d v_\eta \\ -v_u v_{\bar{\eta}} & v_u v_\eta \end{pmatrix}. \tag{15.53}$$

For a gauge coupling $g_{BL} \sim |\tilde{g}| \sim \mathcal{O}(0.5)$, these off-diagonal terms are about one order of magnitude smaller than the diagonal ones. However, they are still crucial for generating interaction vertices between the genuine BLSSM Higgs bosons and the MSSM-like Higgs states. Note that the mixing gauge coupling constant, $\tilde{g}$, is a free parameter that can be positive or negative [21].

A peculiar difference emerges, however, when one consider the additional (s)neutrino degrees of freedom in the BLSSM-IS, with respect to the BLSSM-I, as discussed in the previous chapter. In particular, it can be shown that the one-loop radiative corrections due to right-handed (s)neutrinos to the mass of the lightest Higgs boson when the latter is SM-like can be as large as $\mathcal{O}(100 \text{ GeV})$, thereby giving an absolute upper limit on such a mass around 170 GeV [260]. It is important to note that, unlike the squark sector, where only the third generation (stops) has a large Yukawa coupling with the Higgs boson, hence giving the relevant correction to the Higgs mass, all three generations of the (s)neutrino sector may lead to important effects, since the neutrino Yukawa couplings are generally not hierarchical. Also, due to the large mixing between the right-handed neutrinos N_i and S_{2_j} [385], all the right-handed sneutrinos $\tilde{\nu}_H$ are coupled to the Higgs boson H_2, hence they can give significant contribution to the Higgs mass correction. In this respect, it is useful to note that the stop effect is due to the running of 12 degrees of freedom (3 colours times 2 charges times 2 for left and right stops) in the Higgs mass loop corrections, just like in the case of right-handed sneutrinos for which there are also 12 degrees of freedom (3 generations times 4 eigenvalues).

To calculate the (s)neutrino corrections to the lightest Higgs mass, we computed, in a previous chapter, the explicit form of the sneutrino masses, while for the neutrino mass expressions, which are well known, we refer the reader to [260]. Due to one generation of neutrinos and sneutrinos, the one-loop radiative corrections to the effective potential are given by the relation

$$\Delta V_{\nu,\tilde{\nu}} = \frac{1}{64\pi^2}\Big[\sum_{i=1}^{6} m_{\tilde{\nu}_i}^4\Big(\log\frac{m_{\tilde{\nu}_i}^2}{Q^2} - \frac{3}{2}\Big) - 2\sum_{i=1}^{3} m_{\nu_i}^4\Big(\log\frac{m_{\nu_i}^2}{Q^2} - \frac{3}{2}\Big)\Big]. \tag{15.54}$$

The first sum runs over the sneutrino mass eigenvalues, while the second sum runs over the neutrino masses (with vanishing m_{ν_1}). In the case of degenerate diagonal Yukawa couplings, one finds that the total $\Delta V_{\nu,\tilde{\nu}}$ is given by three times the value of $\Delta V_{\nu,\tilde{\nu}}$ for one generation. This factor then compensates the colour factor of (s)top contributions.

Therefore, the genuine $B - L$ correction to the CP-even Higgs mass matrix, due to the (s)neutrinos, at the scale $\hat{Q}$ at which $\frac{\partial(\Delta V_{\nu,\tilde{\nu}})}{\partial v_k} = 0$, is given by

$$\Delta M_{ij}^2 = \frac{1}{2}\frac{\partial^2(\Delta V_{\nu,\tilde{\nu}})}{\partial v_i \partial v_j}. \tag{15.55}$$

It follows that (see [260] for details)

$$
\begin{aligned}
\frac{\partial^2(\Delta V_{\nu,\tilde{\nu}})}{\partial v_k \partial v_\ell} &= \frac{1}{32\pi^2}\sum_i (-1)^{2J_i}(2J_i+1)\frac{\partial m_i^2}{\partial v_k}\frac{\partial m_i^2}{\partial v_\ell}\log\frac{m_i^2}{\hat{Q}^2} \\
&= \frac{1}{32\pi^2}\Big[4(2Y_\nu^2 v_u)(2Y_\nu^2 v_u)\delta_{k,2}\delta_{\ell,2}\log\frac{m_{\tilde{\nu}_H}^2}{Q_0^2} \\
&\qquad -2\Big(2(2Y_\nu^2 v_u)(2Y_\nu^2 v_u)\delta_{k,2}\delta_{\ell,2}\log\frac{m_{\nu_H}^2}{Q_0^2}\Big)\Big] \\
&= \frac{m_D^4}{2\pi^2 v_u^2}\log\Big(\frac{m_{\tilde{\nu}_H}^2}{m_{\nu_H}^2}\Big)\delta_{k,2}\delta_{\ell,2}. \tag{15.56}
\end{aligned}
$$

That is, we have

$$\Delta M_{11}^2 = \Delta M_{12}^2 = \Delta M_{21}^2 = 0, \tag{15.57}$$

$$\Delta M_{22}^2 = \frac{m_D^4}{4\pi^2 v_u^2}\log\frac{m_{\tilde{\nu}_H}^2}{m_{\nu_H}^2}. \tag{15.58}$$

Therefore, the complete one-loop squared-mass matrix of CP-even Higgs bosons will be given by $M_{\text{tree}}^2 + \Delta M^2$, with

$$\Delta M^2 = \begin{pmatrix} 0 & 0 \\ 0 & \delta_t^2 + \delta_\nu^2 \end{pmatrix}, \tag{15.59}$$

where δ_t^2 refers to the (s)top contribution presented in Eq. (1) of [260] and δ_ν^2 is the (s)neutrino correction given in Eq. (15.58). In this case, the lightest Higgs bosons mass is given by

$$m_h^2 = \frac{M_A^2 + M_Z^2 + \delta_t^2 + \delta_\nu^2}{2}$$
$$\times \left[1 - \sqrt{1 - 4\frac{M_Z^2 M_A^2 \cos^2 2\beta + (\delta_t^2 + \delta_\nu^2)(M_A^2 \sin^2 \beta + M_Z^2 \cos^2 \beta)}{(M_A^2 + M_Z^2 + \delta_t^2 + \delta_\nu^2)^2}} \right]. \tag{15.60}$$

For $M_A \gg M_Z$ and $\cos 2\beta \simeq 1$, one finds that

$$m_h^2 \simeq M_Z^2 + \delta_t^2 + \delta_\nu^2. \tag{15.61}$$

If $\tilde{m} \simeq \mathcal{O}(1 \text{ TeV})$, $Y_\nu \simeq \mathcal{O}(1)$ and $M_N \simeq \mathcal{O}(500 \text{ GeV})$, one gets that $\delta_\nu^2 \simeq \mathcal{O}(100 \text{ GeV})^2$, thus the Higgs mass is of order $\sqrt{(90)^2 + \mathcal{O}(100)^2 + \mathcal{O}(100)^2}$ GeV $\simeq 170$ GeV.

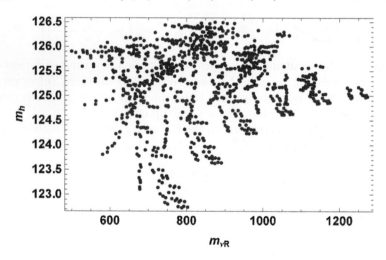

Figure 15.3 Lightest Higgs boson mass versus the lightest sneutrino mass (from [22]).

However, while it is interesting to notice that a version of the BLSSM (the inverse seesaw one) can be relieved from a fine-tuning problem, it should be recalled that an SM-like Higgs boson has now been discovered. Hence, it is worth investigating the Higgs boson spectrum in a generic BLSSM. In order to do so, in Fig. 15.3, we present the Higgs mass, m_h, as a function of the sneutrino mass, $m_{\tilde{\nu}}$, for $m_0 \in [500, 1000]$ GeV and Y_{ν_i} couplings varying from 0.1 to 0.4. As can be seen, in the BLSSM-IS the Higgs mass can be easily within the experimental limits. In particular, we have employed the Higgs mass bound as $123 \text{ GeV} \leq m_h \leq 127 \text{ GeV}$ [144,145], where we take into account about 2 GeV uncertainty in the Higgs boson mass due to the theoretical uncertainties in the calculation of the minimum of the scalar potential and the experimental uncertainties in m_t and α_s.

15.8 PHENOMENOLOGY OF EXTENDED HIGGS MODELS

Due to the additional singlet superfield $\widehat{S}$ in the NMSSM, nMSSM and UMSSM (the latter also including an additional gauge field and consequent interactions), the phenomenology of these extended Higgs models can differ strongly from the MSSM in the Higgs and neutralino sectors. Hence, below we highlight their salient features in the Higgs sector.

After the diagonalisation of the mass matrices given in the previous section, one can introduce in all generality for all the models considered the reduced couplings

$$\xi_i = \sin\beta\, S_{i2} + \cos\beta\, S_{i1} \tag{15.62}$$

of the CP-even mass eigenstates H_i ($i = 1, ...3$) to the EW gauge bosons (normalised with respect to the couplings of the SM Higgs scalar). These satisfy the following sum rule:

$$\sum_{i=1}^{3} \xi_i^2 = 1. \tag{15.63}$$

Armed with such ξ_i rescaling factors with respect to the SM Higgs boson couplings, one can set out to translate existing collider limits into bounds on the extended Higgs model parameters.

Once Higgs-to-Higgs decays are possible (or even dominant), however, Higgs searches at the LHC can become considerably more complicated: see [386–392] for discussions of possible scenarios and proposals for search channels. In principle, CP-even scalars H_2 can decay into a pair of CP-even H_1 in LEP-allowed regions of the NMSSM parameter space [386,391,393]. However, most of the studies concentrated on decays into light CP-odd scalars A motivated by an approximate PQ or R-symmetry and/or the $H \to AA$ explanation of a light excess of events seen at LEP (see [358] for a review).

Notice that all of the results mentioned above for the NMSSM are equally applicable to the case of the nMSSM. It is only with the LHC at full luminosity that one could disentangle the Higgs sectors of these two non-minimal SUSY models. Various Higgs searches have now been carried out at the LHC by ATLAS and CMS, of both SM-like channels (reinterpretable in our extended Higgs models) and in specific Higgs-to-Higgs decays. Clearly, are the latter that play a key role and are of specific importance to the NMSSM and nMSSM searches for very light pseudoscalar Higgs states down to a few GeV [394–398] (see also [399–401]), which we summarise in Fig. 15.4.

A light pseudoscalar (as previously discussed) emerges in the NMSSM and nMSSM in the PQ limit. In this limit, the LSP is the lightest neutralino $\tilde{\chi}_1^0$ and is predominantly singlino in either scenario [26, 358], but then its couplings to the SM particles are quite small, resulting in a relatively small annihilation cross section for $\tilde{\chi}_1^0 \tilde{\chi}_1^0 \to$ SM particles, which gives rise to a relic density that is typically much larger than its measured value. As a consequence, it seems to be rather problematic to find phenomenologically viable scenarios with a light pseudoscalar in the case of the NMSSM/nMSSM with approximate PQ symmetry. Nevertheless, a sufficiently light pseudoscalar can always be obtained by tuning the parameters of these two SUSY scenarios.

In contrast to the NMSSM and nMSSM, the mass of the lightest neutralino in the ESSM does not become small when the PQ symmetry-violating couplings vanish. Moreover, even when all of the latter are negligibly small, the LSP can be Higgsino-like. This allows a reasonable value for the DM relic density to be obtained if the LSP has a mass below 1 TeV (see, *e.g.*, [402,403]). Thus the approximate PQ symmetry can lead to phenomenologically viable scenarios with a light pseudoscalar in this model at the price of low fine tuning [28]. The degree of tuning required in the ESSM is illustrated in Fig. 15.5, where the region of the

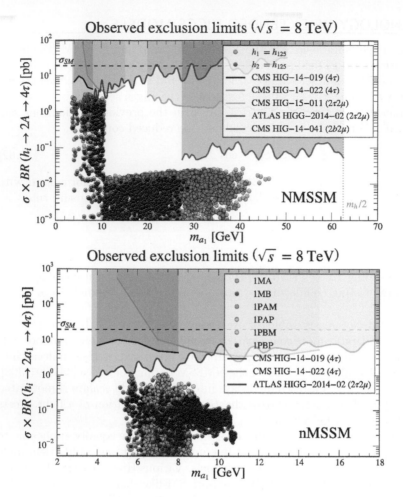

Figure 15.4 Exclusions on the mass of the lightest A boson of the NMSSM (top) and nMSSM (bottom) via the $gg \to h_{\mathrm{SM}} \to AA \to 4\tau$ cross section from a variety of ATLAS and CMS searches during Run 1. Here, $2\tau2\nu$ and $2b2\mu$ analyses have been recast in the 4τ mode. (The benchmarks for the nMSSM are those of [26].) The figure is adapted from [27].

parameter space that leads to a sufficiently small m_{A_1} for sizable A_1A_1 decays of the SM-like Higgs state (compatibly with LHC and DM data) is shown in the κ–A_κ plane. (Herein, notice that, for each value of κ, there is a lower limit on A_κ where the BR is zero because the pseudoscalar is too heavy and an upper limit above which there is a pseudoscalar tachyon.) From the plot and BR heat map, one can see that, with decreasing κ, the range of A_κ that results in a sufficiently small m_{A_1} becomes considerably wider. This corresponds to a smaller degree of fine tuning required to obtain a sufficiently light pseudoscalar.

While a very light pseudoscalar is a smoking-gun signal for the NMSSM and nMSSM, this is by no means characteristic of the BLSSM. In fact, herein, both peudoscalars are rather heavy. However, an equally distinctive mass pattern emerges, as the (BLSSM-like) h' state can be rather close in mass to the h (SM-like) state, thus generating a totally new

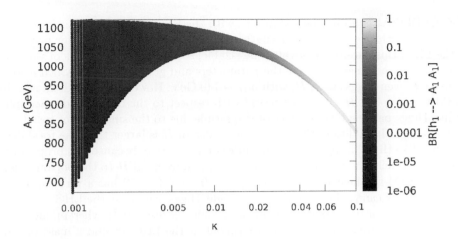

Figure 15.5 Colour contours of the BR($h_{SM} \to A_1 A_1$) in the κ–A_κ plane of the ESSM (other parameter values are specified in [28] (from which the plot is taken) for their benchmark point labelled BMA).

and interesting phenomenology that may even explain some anomalies seen in LHC data, *i.e.*, the possible presence of light Higgs boson signals in Run 1 data with mass just above the SM-like signal at 125 GeV. In order to illustrate this, in Fig. 15.6, we show the masses of the four CP-even Higgs bosons in the BLSSM for $g_{BL} = 0.4$ and $\tilde{g} = -0.4$.

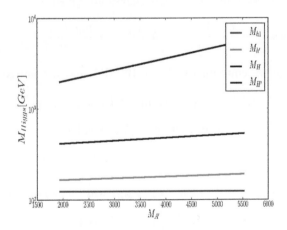

Figure 15.6 The BLSSM CP-even Higgs masses versus $m_{A'}$ for $g_{BL} = 0.4$ and $\tilde{g} = -0.4$ (from [29]).

In this plot we fix the lightest MSSM Higgs boson mass to be of order 125 GeV. As can be seen from this figure, one of the BLSSM Higgs bosons, h', can be the second lightest Higgs boson, with masses down to ~ 140 GeV. Both H and H' are instead quite heavy (both m_A and $m_{A'}$ are of order TeV). Furthermore, two remarks are in order: firstly, if $\tilde{g} = 0$, the coupling of the BLSSM lightest Higgs state, h', with the SM particles will be significantly suppressed ($\leq 10^{-5}$ relative to the SM strength), so that, in order to account for possible h' signals at the LHC, this parameter ought to be sizable; secondly, in both cases of vanishing and non-vanishing $\tilde{g}$, one may fine-tune the parameters and get a light m_A, which leads to an MSSM-like CP-even Higgs state, H, with $m_H \sim 140$ GeV. However, it is well known that in the MSSM the coupling HZZ is suppressed with respect to the corresponding coupling of the SM-like Higgs particle by one order of magnitude due to the smallness of $\cos(\beta - \alpha)$, where $\sin(\beta - \alpha) \sim 1$. In addition, the total decay width of H is larger than the total decay width of the SM-like Higgs, h, by at least one order of magnitude, because it is proportional to $(\cos\alpha/\cos\beta)^2$, which is essentially the square of the coupling of H to the bottom quark. Therefore, the MSSM-like heavy Higgs signal ($pp \to H \to ZZ \to 4l$) has a very suppressed cross section and thus cannot be a candidate for light Higgs signals at the LHC.

This sets the stage for the hypothesis made in [29,30] (see also [404]), wherein, motivated by a $\sim 2.9\sigma$ excess recorded by the CMS experiment at the LHC around a mass of order ~ 140 GeV in $ZZ \to 4l$ and $\gamma\gamma$ samples, it was shown that a double Higgs peak structure can be generated in the BLSSM, with CP-even Higgs boson masses at ~ 125 and ~ 140 GeV, a possibility instead precluded to the MSSM. In the BLSSM, this peculiar phenomenology emerges as follows.

The lightest eigenstate h is the SM-like Higgs boson, for which we will fix its mass to be exactly 125 GeV. As mentioned, numerical scans of the BLSSM parameter space performed in [29] confirm that the h' state can then be the second light Higgs boson with mass of $\mathcal{O}(140$ GeV). The other two CP-even states, H and H', are heavy (of $\mathcal{O}(1$ TeV)). The h' can be written in terms of gauge eigenstates as

$$h' = \Gamma_{31}\, \sigma_1 + \Gamma_{32}\, \sigma_2 + \Gamma_{33}\, \sigma_1' + \Gamma_{34}\, \sigma_2'. \tag{15.64}$$

Thus, the couplings of the h' with up- and down-quarks are given by

$$h'\, u\bar{u}: \; -i\, \frac{m_u}{v}\frac{\Gamma_{32}}{\sin\beta}, \qquad h'\, d\bar{d}: \; -i\, \frac{m_d}{v}\frac{\Gamma_{31}}{\cos\beta}. \tag{15.65}$$

Similarly, one can derive the h' couplings with the W^+W^- and ZZ gauge boson pairs:

$$\begin{aligned}
h'\, W^+ W^- \quad &: \quad i\, g_2 M_W \left(\Gamma_{32}\sin\beta + \Gamma_{31}\cos\beta\right), \\
h'ZZ \quad &: \quad \frac{i}{2}\Big[4g_{BL}\sin^2\theta'\,(v_{\bar{\eta}}\Gamma_{32} + v_\eta\Gamma_{31}) \\
&\quad + (v_u\Gamma_{32} + v_d\Gamma_{31})(g_z\cos\theta' - \tilde{g}\sin\theta')^2 \Big].
\end{aligned}$$

Since $\sin\theta' \ll 1$, the coupling of the h' with ZZ, $g_{h'ZZ}$, will be as follows:

$$g_{h'ZZ} \simeq i\, g_z\, M_Z\left(\Gamma_{32}\sin\beta + \Gamma_{31}\cos\beta\right), \tag{15.66}$$

where $g_z = \sqrt{g_1^2 + g_2^2}$.

The Higgs decay into $ZZ \to 4l$ is one of the golden channels, with low background, to search for Higgs boson(s). The search is performed by looking for resonant peaks in the m_{4l} spectrum, i.e., the invariant mass of the $4l$ system. In CMS [11], this decay channel shows two significant peaks at 125 GeV and around/above 140 GeV. We define by $\sigma(pp \to h')$ the total h' production cross section, dominated by gluon-gluon fusion, computed for $m_{h'} = 142$ GeV, for definiteness. (See Tab. 1 in [30] for the BLSSM parameters corresponding to this specific benchmark point, which complies well with current experimental limits.) From the

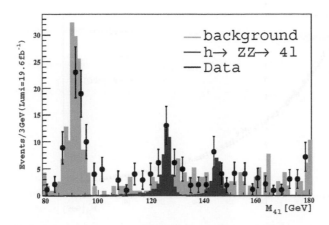

Figure 15.7 The number of events from the signal $pp \to h, h' \to ZZ \to 4l$ (red filled histogram) and from the background $pp \to Z \to 2l\gamma^* \to 4l$ (blue filled histogram) versus the invariant mass of the outgoing particles (4-leptons) against data taken from [11]. The plot is from Ref. [30].

previous discussion, it is then clear that

$$\frac{\sigma(pp \to h')}{\sigma(pp \to h)^{\text{SM}}} \simeq \left(\frac{\Gamma_{32}}{\sin \beta}\right)^2, \tag{15.67}$$

(wherein the label SM identifies the SM Higgs rates computed for a 125 GeV mass), which, for $m_{h'} \approx 140$ GeV, is of order $\mathcal{O}(0.1)$. Also, the ratio between BRs can be estimated as

$$\frac{\text{BR}(h' \to ZZ)}{\text{BR}(h \to ZZ)^{\text{SM}}} \simeq \left(1 + \frac{\Gamma_{h \to WW^*}^{\text{SM}}}{\Gamma_{h \to b\bar{b}}^{\text{SM}}}\right) \frac{F(M_Z/m_{h'})}{F(M_Z/m_h)^{\text{SM}}} \left[\left(\frac{\Gamma_{31} \sec \beta}{\Gamma_{32} \sin \beta + \Gamma_{31} \cos \beta}\right)^2 + 2F\left(\frac{M_W}{m_{h'}}\right)\right]^{-1}, \tag{15.68}$$

where

$$
\begin{aligned}
F(x) &= \frac{3(1 - 8v'^2 + 20v'^4)}{(4v'^2 - 1)^{1/2}} \arccos\left(\frac{3v'^2 - 1}{2v'^3}\right) \\
&- \frac{1 - v'^2}{2v'^2}(2 - 13v'^2 + 47v'^4) - \frac{3}{2}(1 - 6v'^2 + 4v'^4)\log v'^2. \tag{15.69}
\end{aligned}
$$

First, we analyse the kinematic search for the BLSSM Higgs boson, h', in the decay channel to $ZZ \to 4l$. In Fig. 15.7 we show the invariant mass of the 4-lepton final state from $pp \to h' \to ZZ \to 4l$ as obtained at $\sqrt{s} = 8$ TeV after 19.6 fb^{-1} of luminosity, after applying a p_T cut of 5 GeV on the four leptons. The SM model backgrounds from the Z and 125 GeV Higgs boson decays, $pp \to Z \to 2l\gamma^* \to 4l$ and $pp \to h \to ZZ \to 4l$, respectively, are taken into account, as demonstrated by the first two peaks in the plot (with the same p_T requirement). It is clear that the third peak at $m_{4l} \sim 145$ GeV, produced by the decay of the BLSSM Higgs boson h' into $ZZ \to 4l$, can reasonably well account for the events observed by CMS [11] with the 8 TeV data.

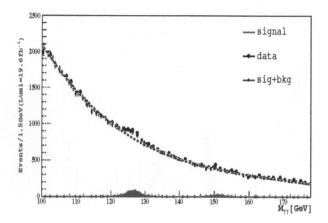

Figure 15.8 The number of events from the signal $pp \to h, h' \to \gamma\gamma$ (purple filled histogram), from the background $pp \to \gamma\gamma$ (red dotted histogram) and from the sum of these two (blue dotted histogram) versus the invariant mass of the outgoing particles (di-photons) against data taken from [31]. The plot is from Ref. [30].

Next, we turn to the di-photon Higgs decay channel, which provides the greatest sensitivity for Higgs boson discovery in the intermediate mass range (*i.e.*, for Higgs masses below $2M_W)^2$. Like the SM-like Higgs, the h' decays into two photons through a triangle-loop diagram dominated by (primarily) W and (in part) top quark exchanges. As shown above, the couplings of the h' with top quarks and W gauge bosons are proportional to some combinations of Γ_{31} and Γ_{32}, which may then lead to some suppression or enhancement in the partial width $\Gamma(h' \to \gamma\gamma)$. In the SM, BR($h \to \gamma\gamma$) $\simeq 2 \times 10^{-3}$. Similarly, in the BLSSM, we have found that, for our $m_{h'} = 142$ GeV benchmark, the BR of h' in photons also amounts to $\simeq 2 \times 10^{-3}$, hence it is within current experimental constraints.

The distribution of the di-photon invariant mass is presented in Fig. 15.8, again for a centre-of-mass energy of $\sqrt{s} = 8$ TeV and luminosity of 19.6 fb^{-1}. As previously, here too, the observed $h \to \gamma\gamma$ SM-like signal around 125 GeV is taken as background (alongside the continuum) while the (rather subtle) $Z \to \gamma\gamma$ background can now be ignored [413]. As expected, the sensitivity to the h' Higgs boson is severely reduced with respect to the presence of the already observed Higgs boson, yet a peak is clearly seen at 145 GeV or so and is very compatible with the excess seen by CMS [31].

Before closing this section, we should also mention that the $h' \to \gamma\gamma$ enhancement found in the BLSSM may be mirrored in the γZ decay channel [404] for which, at present, there exist some constraints, albeit not as severe as in the $\gamma\gamma$ case. We can anticipate (see [404]) that the BLSSM regions of parameter space studied therein are consistent with all available data.

These are tantalising hints of new physics, in particular, in the direction of a non-minimal Higgs sector, from Higgs data collected at the LHC during Run 1 that, at the time of writing, are expecting confirmation (or otherwise) from Run 2 measurements.

---

[2]The effects of light SUSY particles leading to a possible enhancement of the di-photon signal strength of the SM-like Higgs boson were studied in [13, 14, 405–412].

V

SUSY Breaking From Underlying Theory

CHAPTER 16

Gravity Mediation SUSY Breaking

16.1 MEDIATION OF SUSY BREAKING

We have seen in Chapter 3 that spontaneous SUSY breaking through a non-vanishing F- or D-term leads to a phenomenological problem, namely the existence of tree level sum rules for the mass squares:

$$\mathrm{STr}\mathcal{M}^2 = 0, \quad \Rightarrow \quad \mathrm{Tr}M^2_{\text{scalars}} = 2\mathrm{Tr}M^2_{\text{fermions}}. \tag{16.1}$$

These rules imply that the bosonic superpartner masses are of the same order as the masses of the matter fields, which are excluded experimentally. To avoid this problem, one could generate SUSY breaking in the low energy sector at the loop level. This line of thought led the way to an entirely new way of breaking SUSY. Instead of having it broken spontaneously, one can think that SUSY is broken directly in some hidden sector of the theory and then the information of such breaking is conveyed to the observable sector (ordinary particles plus their superpartners) through gauge or gravitational messengers. In this case, the theory includes an $N = 1$ globally SUSY invariant Lagrangian plus a set of terms which break SUSY explicitly in a soft manner. In this way the 'supertrace mass problem' can be overcome.

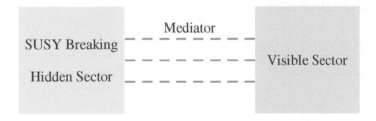

Figure 16.1 SUSY breaking in the hidden sector and transmission to the visible sector via a mediator.

Now, after many years spent in understanding the origin of the soft SUSY breaking terms, we believe that the possible mechanisms which may give rise to the SUSY soft breaking terms have three sectors, as illustrated in Fig. 16.1.

1. The observable sector, which comprises all the ordinary particles and their SUSY partners.

2. A "hidden" sector where the breaking of SUSY occurs.

3. The messengers of the SUSY breaking from the hidden to observable sector.

The two most explored alternatives that have been studied in this context are:

- SUSY breaking SUGRA where the mediators are gravitational interactions.

- SUSY is broken at a much lower scale with messengers provided by some gauge interactions.

In general, if SUSY is assumed to be broken by the non-vanishing VEV of the F-component of a single field X: $\langle F_X \rangle \neq 0$, while $\langle X \rangle = 0$, then the non-renormalisable interaction terms between this field and those in the observable sector will generate all soft SUSY breaking terms, which are characterised by a mass scale of order

$$m_{\text{SUSY}} \sim \frac{\langle F_X \rangle}{M_P}. \tag{16.2}$$

The most general Lagrangian for the interaction of X with the MSSM fields can be written as [414]

$$
\begin{aligned}
\mathcal{L}_{\text{int}} &= \int d^4\theta \left[\frac{(Z_Q)^i_j}{M_P^2} X^\dagger X Q_i^\dagger Q_j + \ldots \ldots \right. \\
&+ \left. \frac{B}{M_P} X H_u H_d + \frac{B'}{M_P^2} X^\dagger X H_u H_d + \text{h.c.} \right] \\
&+ \int d^2\theta \left[\frac{S_1}{M_P} X W_1^\alpha W_{1\alpha} + \ldots \ldots \right] + \text{h.c.} \\
&+ \int d^2\theta \left[\frac{A_{ij}}{M_P} X Q^i H_u (U^c)^j + \ldots \ldots \right].
\end{aligned}
\tag{16.3}
$$

From this Lagrangian, one can easily see that substituting the SUSY breaking VEV $\langle F_X \rangle$ induces the usual soft breaking terms of the MSSM: scalar mass-squared, μ and B-term, gaugino masses and the A-terms.

16.2 INTRODUCTION TO SUGRA

So far, we have discussed global SUSY where the superfield Φ transforms as

$$\delta\Phi = i \left(\xi Q + \bar{\xi}\bar{Q} \right) \Phi, \tag{16.4}$$

with constant ξ. If the parameter ξ depends on x_μ, i.e., $\xi = \xi(x)$, then the invariance under this kind of local SUSY transformation implies invariance under local coordinate change. Thus, local SUSY, which is called SUGRA, naturally includes gravity [415, 416]. This can be seen if one begins with a global supersymmetric Lagrangian of the Wess-Zumino model:

$$\mathcal{L}_0 = \partial^\mu \phi^* \partial_\mu \phi + \frac{i}{2} \bar{\chi} i \gamma^\mu \partial_\mu \chi, \tag{16.5}$$

where $\phi = (A+iB)/\sqrt{2}$ is a complex scalar field and χ is a Majorana spinor. This Lagrangian is invariant (up to total derivatives) under the global SUSY transformation defined by

$$\delta A = \bar{\xi}\chi, \tag{16.6}$$

$$\delta B = i\bar{\xi}\gamma_5\chi, \tag{16.7}$$

$$\delta\chi = i\gamma^\mu\partial_\mu(A + iB\gamma_5)\xi, \tag{16.8}$$

where γ_5 is given by $\gamma_5 = i\gamma_0\gamma_1\gamma_2\gamma_3$. Following the Noether procedure and assuming that $\xi \to \xi(x)$, one finds that the Lagrangian is not invariant and the following term is obtained:

$$\delta\mathcal{L} = (\partial_\mu\xi)\left[\gamma^\nu\partial_\nu(A + iB\gamma_5)\right]\gamma^\mu\chi \equiv (\partial_\mu\xi)J^\mu, \tag{16.9}$$

where J^μ is the Noether's current. Since $(\partial_\mu\xi)$ is a spinor with Lorentz index, $i.e.$, has spin-3/2, in order to cancel this term, a gauge field with spin-3/2, $i.e.$, a spinorial vector ψ_α^μ, has to be introduced,

$$\mathcal{L}' = -\frac{\kappa}{2}\psi_\mu j^\mu, \tag{16.10}$$

where κ is coupling of dimension -1 (hence the theory is non-renormalisable). Also, ψ_μ should transform as

$$\delta\psi_\mu^\alpha = \frac{2}{\kappa}\partial_\mu\xi^\alpha(x). \tag{16.11}$$

However, one can show that $\mathcal{L}+\mathcal{L}'$ is not yet invariant under the local SUSY transformation and one gets

$$\delta(\mathcal{L} + \mathcal{L}') = i\kappa\bar{\psi}_\mu^\alpha\gamma_\nu\xi_\alpha T^{\mu\nu}, \tag{16.12}$$

where $T^{\mu\nu}$ is the energy-momentum tensor of the scalar fields. This term can be canceled by introducing a new gauge field $g_{\mu\nu}$ that varies under SUSY transformation as

$$\delta g_{\mu\nu} = -i\kappa(\bar{\psi}_\mu\gamma_\nu\xi(x) + \bar{\psi}_\nu\gamma_\mu\xi(x)). \tag{16.13}$$

Note that the energy-momentum tensor $T^{\mu\nu}$ is obtained by varying the action with respect to the metric ($\delta S = \int d^4x\delta g_{\mu\nu}T^{\mu\nu}$). Therefore, it is natural to consider our gauge field $g_{\mu\nu}$ as a local metric. The corresponding spin 2-field, the graviton, is obviously the supersymmetric partner of the gravitino, a spin 3/2-field [417, 418]. In order to complete the invariance, one finds (as usual) that the normal derivative in $\delta\psi_\mu^\alpha$ should be changed to a covariant derivative: $\delta\psi_\mu^\alpha = \frac{2}{\kappa}D_\mu\xi^\alpha$, where

$$D_\mu\xi = \partial_\mu\xi + \frac{1}{2}w_\mu^{ab}\sigma_{ab}\xi, \tag{16.14}$$

with w_μ^{ab} as a spin connection (gauge field) needed to define the covariant derivative acting on the spoinors. Thus, we find the following Lagrangian for the pure gravitational part:

$$\mathcal{L} = -\frac{1}{2\kappa}\sqrt{-g}R - \frac{i}{2}\varepsilon^{\mu\nu\rho\sigma}\bar{\psi}_\mu\gamma_5\gamma_\nu D_\rho\psi_\sigma. \tag{16.15}$$

This confirms that the local version of SUSY is indeed a supersymmetric theory of gravity, which we can also obtain from the direct supersymmetrisation of the general theory of relativity. This is, in fact, an expected result, since the SUSY algebra:

$$[\alpha Q, \bar{Q}\dot{\alpha}] = 2\alpha\sigma_\mu\dot{\alpha}P^\mu \tag{16.16}$$

implies that two local SUSY transformations with $\alpha_1(x)$ and $\alpha_2(x)$ lead to a general coordinate transformation: $\alpha_1(x)\sigma_\mu\alpha_2(x)\partial^\mu$.

It is also worth noting that the gravitino $\psi_{\mu\alpha}$, which is the gauge field introduced in order to change SUSY from global to local symmetry, is the fermionic partner of boson degree of freedom (graviton), $i.e.$, $\psi_{\mu\alpha} = Q_\alpha$ (gravitino). However, it is clear from its space index that the gravity variable cannot be the metric but the vielbein e^a_μ, which is defined to manifest the Lorentz invariance locally as

$$g_{\mu\nu} = e^a_\mu e^b_\nu \eta_{ab}, \tag{16.17}$$

where a, b are the flat indices, acted upon by a local Lorentz gauge invariance. Both $g_{\mu\nu}$ and e^a_μ contain the same number of degrees of freedom: $d(d+1)/2$. The spin connection w^{ab}_μ is defined in terms of the vielbein, e^a_μ, if the free torsion condition $(T^a_{[\mu\nu]} = D_{[\mu}e^a_{\nu]} = 0)$ is imposed. In this case the gravity is described by e^a_μ only. One can also construct the field strength of w^{ab}_μ as

$$R^{ab}_{\mu\nu}(w(e)) = \partial_\mu w^{ab}_\nu - \partial_\nu w^{ab}_\mu + w^{ac}_\mu w^{cb}_\nu - w^{ac}_\nu w^{cb}_\mu. \tag{16.18}$$

It can be shown [415] that

$$R^{ab}_{\rho\sigma}(w(e)) = e^a_\mu \, (e^{-1})^{\nu b} \, R^\mu_{\nu\rho\sigma}(\Gamma(e)) \tag{16.19}$$

and also

$$R = R^{ab}_{\mu\nu} \, (e^{-1})^\mu_a \, (e^{-1})^\nu_b. \tag{16.20}$$

Therefore, Einstein-Hilbert action will be given by

$$S_{EH} = \frac{1}{16\pi G} \int d^4x \, (\det e) \, R^{ab}_{\mu\nu}(w(e)) \, (e^{-1})^\mu_a \, (e^{-1})^\nu_b. \tag{16.21}$$

Since gravity (and hence SUGRA) is not renormalisable, there is no reason to exclude non-renormalisable couplings between the other fields. In this respect, we use the most general global supersymmetric, gauge-invariant Lagrangian for chiral superfields Φ_i, which is given by

$$\mathcal{L}_{\text{global}} = \int d^4\theta K(\Phi^\dagger_i e^{2gV}, \Phi_i) + \int d^2\theta \left(W(\Phi) + f_{ab}(\Phi) W^{a\alpha} W^b_\alpha + \text{h.c.} \right), \tag{16.22}$$

where the Kähler potential K is a general real function of the superfields $\Phi^\dagger_i$ and Φ_i plus, as usual, the gauge fields are coupled with the chiral supermultiplet through $\Phi^\dagger e^{2gV}$. $W(\Phi)$ is the superpotential and W^a_α is the gauge field strength with gauge group index a. This Lagrangian can be made invariant under local SUSY transformations by including the graviton and gravitino and coupling them with other fields in an appropriate way.

In superspace formalism, the super-vielbein E^A_M and super-spin connection Ω^{AB}_M, where $M = (\mu, \alpha, \dot\alpha)$ and $A = (a, \lambda, \dot\lambda)$, are introduced. The components of the super-vielbein E^A_M are usually defined as [41]: $e^a_\mu = \delta^a_\mu$, $e^\lambda_\mu = \delta_{\mu\dot\alpha} = 0$, $e^a_\alpha = -i\sigma^a_{\alpha\dot\alpha}\bar\theta^{\dot\alpha}$, $e^\lambda_\alpha = e_{\alpha\lambda} = 0$, $e^{\dot\alpha a} = -i\theta^\alpha \sigma^a_{\alpha\dot\beta}\varepsilon^{\dot\beta\dot\alpha}$ and $e^{\dot\alpha\lambda} = e^{\dot\alpha}_{\dot\lambda} = 0$. The invariant measure in the full superspace is given by $\int d^4x d^4\theta E$, where E is defined as

$$E = \text{sdet } E^A_M, \tag{16.23}$$

with the superdeterminant (sdet) is defined as

$$\text{sdet} \begin{pmatrix} A & B \\ C & D \end{pmatrix} = \frac{\det(A - BD^{-1}C)}{\det D}. \tag{16.24}$$

Moreover, the chiral measure $\int d^4x d^2\theta$ is generalised as $\int d^4x d^4\theta \mathcal{E}$ [41], where $\mathcal{E} = \frac{1}{4}\frac{\bar{D}^2 E}{R}$. Therefore, the most general $N = 1$ invariant Lagrangian for SUGRA coupled to matter (chiral superfields and gauge superfields with canonical kinetic terms) is given by

$$
\begin{aligned}
S &= \int d^4x d^4\theta E \left[K(\Phi, \Phi^+) + \Phi^+ e^V \Phi \right] \\
&+ \int d^4x d^2\theta \mathcal{E} \left[W(\Phi) + \frac{1}{16} f_{ab} W^a W^b \right] + \text{h.c.},
\end{aligned}
\tag{16.25}
$$

where $W_\alpha = -\frac{1}{4}(\bar{D}^2 - 8R)e^{-V}D_\alpha e^V$ is the curved-space generalisation of the Yang-Mills field strength superfield. It turns out that the functions K and W of scalar fields ϕ and ϕ^* always appear in the following combination:

$$
G(\phi, \phi^*) = K(\phi, \phi^*) + \ln |W(\phi)|^2.
\tag{16.26}
$$

The function G plays the role of a Kähler potential with the Kähler metric $G_{i\bar{j}} = \partial^2 G/\partial\phi^i \partial\phi^{*j}$. We also define $G_i = \partial G/\partial\phi_i$ and $G_{\bar{j}} = \partial G/\partial\phi^{*j}$. It is also noticeable that the SUGRA Lagrangian remains invariant under the transformation

$$
\begin{aligned}
K &\rightarrow K + h(\phi) + h^*(\phi^*), \tag{16.27} \\
W &\rightarrow e^{-h} W \tag{16.28}
\end{aligned}
$$

for an arbitrary holomorphic function $h(\phi)$.

The SUGRA Lagrangian at tree level can be written as

$$
\mathcal{L}_{\text{SUGRA}} = \mathcal{L}_B + \mathcal{L}_{FK} + \mathcal{L}_F,
\tag{16.29}
$$

where $\mathcal{L}_B$ contains only bosonic fields, $\mathcal{L}_{FK}$ contains fermionic fields and covariant derivatives with respect to gravity and $\mathcal{L}_F$ contains fermionic fields but no covariant derivatives [415, 416]. In particular, the bosonic Lagrangian is given by

$$
e^{-1}\mathcal{L}_B = -\frac{1}{2}R - G_{i\bar{j}}\partial_\mu \phi^i \partial^\mu \phi^{*j} - V(\phi, \phi^*),
\tag{16.30}
$$

where the tree-evel SUGRA scalar potential, $V(\phi, \phi^*)$, is defined as

$$
V = e^G \left[G_i (G^{-1})^{i\bar{j}} G_{\bar{j}} - 3 \right] + \frac{1}{2} f_{\alpha\beta}^{-1} D^\alpha D^\beta.
\tag{16.31}
$$

In this case, the auxiliary fields of the chiral supermultiplets are given by

$$
F_i = e^G (G^{-1})_i^{\bar{j}} G_{\bar{j}} + \frac{1}{4} f_{\alpha\beta\bar{k}} (G^{-1})_i^{\bar{k}} \lambda^\alpha \lambda^\beta - (G^{-1})_i^{\bar{k}} G_{\bar{k}}^{jl} \chi_j \chi_l - \frac{1}{2}\chi_i (G_{\bar{j}}\chi^{\bar{j}}),
\tag{16.32}
$$

where λ^α refer to the gauginos associated with vector superfields and χ_i are the spinor partners of ϕ_i within chiral superfields. The auxiliary fields of the vector superfields are given by

$$
D_\alpha = i\text{Re} f_{\alpha\beta}^{-1} \left(-gG^i T_i^{\beta j}\phi_j + \frac{1}{2} i f_{\beta\gamma}^i \chi_i \chi^\gamma - \frac{1}{2} i f_i^{*\beta\gamma}\chi^i \chi_\gamma \right) - \frac{1}{2}\lambda_\alpha (G^i\chi_i).
\tag{16.33}
$$

From these expressions, it is clear that the VEV of F_i and D_α can be non-vanishing if the scalar fields involved in their first terms get non-zero VEVs. The other terms that contain fermion fields may contribute to the VEVs of F_i and D_α if there exist a strong interacting gauge force that leads to vacuum condensation of bilinear fermion-antifermion states. This case is known as 'gaugino condensation SUSY breaking'.

16.3 SPONTANEOUS SUSY BREAKING IN SUGRA

We now have a local theory (with gravitino as the gauge field) that encounters spontaneous SUSY breaking. In this case we could have a 'super-Higgs' mechanism, where the massless goldstino ($\eta = G_i \chi^i$) will be *eaten* by the spin–3/2 gravitino to render it massive [419, 420]. In SUGRA this can also be done by F–term or D–term SUSY breaking, but now the minimum of the potential is not necessarily at zero in contrast to the global case.

In the case of F-term SUSY breaking, the condition turns out to be $D_i W \neq 0$, which follows from the fact that

$$F^i = e^{G/2} (G^{-1})^{i\bar{j}} G_{\bar{j}} = e^{K/2} \sqrt{\frac{W}{\overline{W}}} (K^{-1})^{i\bar{j}} D_{\bar{j}} \bar{W}, \qquad (16.34)$$

where $D_i = \partial/\partial \phi_i + \partial K/\partial \phi_i$. We can see explicitly the mixing term of the gravitino Ψ^μ and the goldstino η (associated with the non-vanishing auxiliary field F), which can be extracted from the SUGRA action, $\sim \langle F \rangle \psi^\mu \sigma_\mu \eta$ with $\langle F \rangle \neq 0$. This term will be gauged away, while two extra degrees of freedom are obtained by the massive gravitino and its mass is given by

$$m_{3/2} = \langle e^{G/2} \rangle m_{\mathrm{Pl}} = \langle e^{K/2} \rangle \frac{|W|}{M_{\mathrm{Pl}}^2}, \qquad (16.35)$$

with G and K made dimensionless by dividing by M_{Pl}^2. As an example we give the Polonyi model [421], which has minimal Kähler potential of a singlet superfield Φ and the following superpotential:

$$W = m^2 (\beta + \Phi). \qquad (16.36)$$

It is clear that $D_\phi W \neq 0$, i.e., $F_\phi \neq 0$, and SUSY is spontaneously broken. For $\beta = 2 - \sqrt{3}$, one can show that the minimum is given by $\langle \phi \rangle = \sqrt{3} - 1$, with vanishing cosmological constant $\langle V \rangle = 0$. The gravitino will acquire a mass, which is given by

$$m_{3/2} = \frac{e^{(\sqrt{3}-1)^2}}{2} \frac{m}{M_{\mathrm{Pl}}}. \qquad (16.37)$$

Thus, the gravitino mass is of order TeV scale if the $m \sim \sqrt{\langle F \rangle} \equiv$ SUSY breaking scale is of the order of an intermediate scale $\sim 10^{11}$ GeV.

16.4 SOFT SUSY BREAKING TERMS

In the previous section we have shown that spontaneous SUSY breaking in the framework of SUGRA is possible. However, one still has to specify the mechanism that generates this breaking. SUGRA theories offer one of the simplest scenarios of SUSY breaking by F-term of a hidden superfield and gravity mediates the SUSY breaking to the observable sector of quarks, leptons, gauge fields, Higgs bosons and their super-partners, which couple to the hidden sector through gravitational interactions only. In general, the superpotential W and the Kähler potential K depend on the superfields in the observable sector, C^α and superfields in the hidden sector, h_m [422]. To the lowest order in the observable fields C^α, W and K have the following form:

$$W = \hat{W}(h_m) + \mu(h_m) H_1 H_2 + \frac{1}{6} Y_{\alpha\beta\gamma}(h_m) C^\alpha C^\beta C^\gamma, \qquad (16.38)$$

$$K = \hat{K}(h_m, h_m^*) + \tilde{K}_\alpha(h_m, h_m^*) C^{*\bar{\alpha}} C^\alpha. \qquad (16.39)$$

The bilinear term $\mu H_1 H_2$, which is allowed by gauge invariance, is introduced in the above superpotential. This term associated with the two Higgs doublets H_1 and H_2 is relevant in order to solve the μ problem in the MSSM. Also, a diagonal Kähler metric is assumed to avoid unacceptable flavour violation in the effective low-energy theory, *i.e.*,

$$\tilde{K}_{\bar{\alpha}\beta}(h_m, h_m^*) = \delta_{\bar{\alpha}\beta}\tilde{K}_\alpha. \tag{16.40}$$

The scalar fields in the hidden sector, h_m, may obtain large VEVs that induce SUSY breaking via non-vanishing VEVs of its auxiliary fields F^m. In this case, the tree-level SUGRA scalar potential is given by

$$V(h_m, h_m^*) = e^G \left(G_m K^{m\bar{n}} G_{\bar{n}} - 3\right) = \left(\bar{F}^{\bar{n}} K_{\bar{n}m} F^m - 3e^G\right), \tag{16.41}$$

where $G_m = \partial_m \equiv \frac{\partial G}{\partial h_m}$ and the matrix $K^{m\bar{n}}$ is the inverse of the Kähler metric $K_{\bar{n}m} \equiv \partial_{\bar{n}}\partial_m K$. Also $F^m = e^{G/2} K^{m\bar{n}} G_{\bar{n}}$ has been used. It is important to note that the goldstino, which is a combination of the fermionic partners of these fields, is swallowed by the gravitino via the superHiggs effect. The gravitino becomes massive and its mass is given by

$$m_{3/2} = e^{G/2} M_P = e^{K/2} \frac{|W|}{M_P^2}, \tag{16.42}$$

where we inserted the reduced Planck mass to obtain the correct unit. The gravitino sets an overall scale of the soft SUSY breaking parameters. The resulting soft breaking terms are obtained in the observable sector by replacing h_m and their auxiliary fields F^m by their VEVs in the SUGRA Lagrangian and taking the so-called flat limit, where $M_P \to \infty$, while keeping the ratio $M_S^2/M_P = m_{3/2}$ fixed (M_S denotes the scale of SUGRA breaking). In this case, the soft SUSY breaking terms are given by [422]

$$m_\alpha^2 = \left(m_{3/2}^2 + V_0\right) - \bar{F}^{\bar{m}} F^n \partial_{\bar{m}} \partial_n \log \tilde{K}_\alpha, \tag{16.43}$$

$$M_a = \frac{1}{2}(\mathrm{Re}f_a)^{-1} F^m \partial_m f_a, \tag{16.44}$$

$$A_{\alpha\beta\gamma} = F^m \left[\hat{K}_m + \partial_m \log Y_{\alpha\beta\gamma} - \partial_m \log(\tilde{K}_\alpha \tilde{K}_\beta \tilde{K}_\gamma)\right], \tag{16.45}$$

$$B = F^m \left[\hat{K}_m + \partial_m \log \mu - \partial_m \log(\tilde{K}_{H_1} \tilde{K}_{H_2})\right] - m_{3/2}. \tag{16.46}$$

It is clear that the function $\tilde{K}_\alpha$ is crucial in computing the soft terms. However, it is non-holomorphic and hard to compute. On the other hand, if one assumes that the gauge kinetic function $f_{ab} = \delta_{ab} f_a$, then one can show that the gaugino masses are given by

$$M_a = \frac{1}{2}(\mathrm{Re}f_a)^{-1} F_m \frac{\partial f_a}{\partial h_m}. \tag{16.47}$$

16.5   mSUGRA

The simplest model of SUSY breaking in the hidden sector corresponds to the minimal (canonical) kinetic terms in the SUGRA Lagrangian, namely $K = \Phi_i \Phi^{i*}$ so that

$$\tilde{K}_\alpha(h_m, h_m^*) = 1. \tag{16.48}$$

Then, one can show that the associated SUSY-breaking terms are given by

$$m_\alpha^2 = m_{3/2}^2, \tag{16.49}$$

$$A_{\alpha\beta\gamma} = F^m \hat{K}_m, \tag{16.50}$$

$$B = F^m \hat{K}_m - m_{3/2}. \tag{16.51}$$

Here we neglect the dependence of the Yukawa couplings and μ term on the hidden scalar fields. Also, $V_0 = 0$ is considered. In this model, the scalar masses and the trilinear parameters are universal. In addition, the following relation between trilinear A and bilinear B parameters is obtained:

$$B = A - m_{3/2}. \tag{16.52}$$

The gaugino masses also get a universal and simple expression in this scheme, depending on f_a, which can be given as $f_a(h_m) = c_a f(h_m)$. In a simple case like $f_a = h/M_P$, one finds that

$$M_a = m_{3/2} \frac{\sqrt{3}}{2(\sqrt{3} - 1)}. \tag{16.53}$$

However, in general the gaugino masses could be non-universal and not equal to the gravitino mass. This model of mSUGRA is quite popular because of its simplicity and for the natural explanation that it offers to the universality of the soft scalar masses. Therefore, it is considered to be a natural realisation of the cMSSM that we studied before. As we have shown, these types of models with universal soft terms face serious trouble from the new limits imposed by the Higgs mass, the dark matter searches and other new physics search experiments. As we have discussed, they are almost ruled out and a non-minimal scheme for SUSY breaking should be considered.

16.6 NO-SCALE SUGRA

The Kähler potential of what is called no-scale SUGRA [423–425] is given by

$$\hat{K} = -3 \log(h + h^*) \,, \quad \tilde{K}_\alpha = (h + h^*)^{-1}, \tag{16.54}$$

and a superpotential is independent of a hidden field, i.e.,

$$\hat{W} = \text{constant}. \tag{16.55}$$

Thus $\partial \hat{K}/\partial h = \partial \hat{K}/\partial h^* = -3(h + h^*)^{-1}$ and $\partial^2 \hat{K}/\partial h \partial h^* = 3(h + h^*)^{-2}$. Consequently, the tree-level effective potential is give by

$$V = e^{\hat{K}} \left(3 - \hat{K}_h (\hat{K}^{-1})^h_{h^*} \hat{K}^{h^*} \right) = 0, \tag{16.56}$$

for all VEVs of h. In this model, the soft parameters are given by

$$m_\alpha^2 = 0, \tag{16.57}$$
$$A_{\alpha\beta\gamma} = -m_{3/2}(h + h^*)\partial_h \log Y_{\alpha\beta\gamma}, \tag{16.58}$$
$$B = -m_{3/2}(h + h^*)\partial_h \log \mu. \tag{16.59}$$

If we assume that μ and the Yukawa couplings are hidden-field independent, then the following soft parameters are obtained:

$$m_\alpha = A_{\alpha\beta\gamma} = B = 0. \tag{16.60}$$

It is important to note that these parameters are vanishing at the high scale. However, in the evolution of these parameters from this high scale to the electroweak scale, non-vanishing values are induced by the gaugino masses. Also, as can be seen, the wino mass M_2 is the lightest gaugino masses. Therefore, both lightest neutralino (the LSP) and chargino (the NLSP) are wino-like. By radiative corrections, the charged wino becomes slightly heavier than the neutral wino. The small mass difference between these two particles leads to the long life-time of the lightest chargino. It has been shown that the lightest chargino may have a lifetime long enough to be constructed as a charged track with a detector.

16.7 AMSB

AMSB is a particular kind of gravity mediated SUSY breaking where there is no direct tree level coupling that conveys the SUSY breaking in the hidden sector to the observable sector [426,427]. As mentioned above, if there is a gauge singlet field with a non-vanishing auxiliary field, the gaugino masses in the observable sector are generated through the operator:

$$\int d^2\theta \frac{X}{M_{\text{Pl}}} \text{Tr} W^\alpha W_\alpha + \text{h.c.} \tag{16.61}$$

If there is no singlet X, then the gauginos remain massless at tree level. However, at one-loop, where the scale invariance is broken by the running of the couplings, the gaugino masses are generated. Therefore, SUSY breaking is related to the conformal anomaly and it is called AMSB [426,427].

This result can be understood by considering the one-loop renormalisation of the visible-sector global effective Lagrangian,

$$\begin{aligned}
\mathcal{L}_{\text{global}} \left(\Box, \Lambda^*, \Lambda \right) &= \frac{1}{4g^2} \int d^2\theta \left(1 - \frac{g_a^2 b}{16\pi^2} \log \frac{\Lambda^2}{\Box} \right) + \text{h.c.} \\
&+ \int d^4\theta Z_\Phi \left(\Box, \Lambda^*, \Lambda \right) \Phi^* \Phi + \text{h.c.} \\
&+ \int d^2\theta Y \Phi^3 + \text{h.c.},
\end{aligned} \tag{16.62}$$

where $\Box$ is the D'Alembert operator, Λ is the ultraviolet cut-off of the theory and W^α and Φ are the gauge field strength and chiral superfields. Z_Φ is the wavefunction renormalisation factor of Φ and Y is the Yukawa coupling constant. Finally, b is the β-function coefficient for the gauge coupling constant g. The leading SUSY breaking effect can be obtained by replacing Λ by a SUSY breaking spurion superfield $\Lambda \to \Lambda(1 + m_{3/2}\theta^2)$. Expanding the above Lagrangian, one finds the anomaly-mediated contributions to the gaugino mass M_a, scalar squared mass m_{ij}^2 and trilinear scalar coupling A are given by

$$M_a = \frac{\beta_{g_a}}{g_a} m_{3/2}, \tag{16.63}$$

$$m_i^2 = \frac{1}{2} m_{3/2}^2 \left(\beta_{g_a} \frac{\partial \gamma_i}{\partial g_a} + \beta_y \frac{\partial \gamma_i}{\partial y} \right) \gamma_{ij}, \tag{16.64}$$

$$A_{ijk} = -m_{3/2} \beta_y, \tag{16.65}$$

where $\beta_{g_a} = -\frac{1}{16\pi^2} b_a g_a^3$ and $\beta_y = \frac{dy}{dt}$ with y standing for Yuakwa coupling. The anomalous dimensions γ_i can be found in [428]. From the above expression of the soft SUSY breaking terms, it is clear that the anomaly-mediated contribution is loop-suppressed relative to the gravity-mediated contribution. However, in the case of no direct coupling between observable and hidden sectors, the anomaly-mediated terms can be the dominant contributions.

One can show that the squared masses of the sleptons are negative (tachyonic), which implies that the minimal AMSB model is not phenomelogically viable. The simplest assumption to overcome this problem is to assume a single universal scalar mass m_0^2 at GUT scale for all sfermion squared masses. In this case, the model is given in terms of two parameters only, $m_{3/2}$ and m_0. It is interesting to note that this class of models predicts rather low Higgs boson mass, namely as less than or equal to 121 GeV. The recent discovery of the Higgs boson with mass equal 125 GeV disfavours the anomaly-mediated scenario.

CHAPTER 17

Gauge Mediation SUSY Breaking

As we have learned in the previous chapter, gravity mediation, in general, would produce flavour dependent soft SUSY breaking terms. Therefore, the gravity mediation mechanism is severely constrained by the experimental limits on FCNC processes. In this chapter, we will consider an alternative mechanism of SUSY breaking at lower energy scales in which gauge interactions are the 'messengers' of SUSY breaking, as depicted in Fig. 17.1. As gauge interactions are flavour blind, squark and slepton masses are universal. This could be a natural solution to the SUSY flavour problem [429].

Figure 17.1 Gauge mediation SUSY breaking.

17.1 MINIMAL GAUGE MEDIATION SUSY BREAKING

The simplest model of gauge mediation involves an observable sector, which contains the usual quarks, leptons and Higgs boson with their SUSY partners and a hidden sector responsible for SUSY breaking. In this sector a chiral superfied X acquires a VEV along the scalar and auxiliary components,

$$\langle X \rangle = M + \theta^2 F, \tag{17.1}$$

where the parameters M and $\sqrt{F}$ can vary from tens of TeV to the GUT scale [429]. The VEVs of scalar and auxiliary fields of X can be realised through a dynamical mechanism or by inserting X into an O'Raifeartaigh-type model. As usual, we just parameterise our ignorance of the precise mechanism of SUSY breaking in terms of M and F. In addition, the model has a messenger sector formed by new chiral superfields that couple at tree level with the goldstino superfield X and that break SUSY. The interaction between chiral messenger superfields Φ and $\bar{\Phi}$ and the goldstino superfield X is given by the superpotential

$$W = \lambda X \Phi \bar{\Phi}. \tag{17.2}$$

Therefore, when X acquires a VEV, the messenger fermions get the mass λM while the mass of the scalar components can be expressed as

$$\begin{pmatrix} |\lambda M|^2 & (\lambda F)^{\dagger} \\ \lambda F & |\lambda M|^2 \end{pmatrix}. \tag{17.3}$$

Thus the scalar mass eigenvalues are given by

$$m_{\phi,\bar{\phi}} = \sqrt{|\lambda M|^2 \pm |\lambda F|}. \tag{17.4}$$

Hence, the stability condition implies

$$F < M^2. \tag{17.5}$$

However, the messenger fields (with masses much larger than the EW scale) are charged under $SU(3)_C \times SU(2)_L \times U(1)_Y$. Therefore, they coupled the observable sector through ordinary gauge interactions. In order to preserve gauge coupling unification, the messenger fields are usually assumed to form complete vector-like mutiplets of $SU(5)$. Therefore, the GMSB scenario can include n_5 pairs of $5 + \bar{5}$ [429]. These $5 + \bar{5}$ mutiplets include $SU(2)_L$ doublets l and $\bar{l}$ and $SU(3)_C$ triplets q and $\bar{q}$. In order to introduce SUSY breaking into the messenger sector and to give them very large masses, these fields may be coupled to the gauge singlet superfield X, through the superpotential

$$W = \lambda_1 X l \bar{l} + \lambda_2 X q \bar{q}. \tag{17.6}$$

Therefore, after X acquires a VEV, the spinor components of $l, \bar{l}$ and $q, \bar{q}$ form Dirac fermions with masses $\lambda_1 M$ and $\lambda_2 M$, while the scalar components, $l \pm \bar{l}/\sqrt{2}$ and $q \pm \bar{q}/\sqrt{2}$, have a squared mass $(\lambda_1 M)^2 \pm \lambda_1 F$ and $(\lambda_2 M)^2 \pm \lambda_2 F$, respectively. From this mass spectrum, it is clear that the effect of SUSY breaking through the VEV of the auxiliary field, F, induced a splitting between the fermion and scalar masses. Also, the stability of the vacuum $\langle l \rangle = \langle \bar{l} \rangle = \langle q \rangle = \langle \bar{q} \rangle = 0$ requires $F \leq |M|^2$.

Figure 17.2   One loop generation of the gaugino mass in the GMSB.

When the messenger fields are integrated out at loop level, one obtains a low energy effective theory with a set of SUSY breaking terms. The gaugino mass is generated at one loop by the diagram in Fig. 17.2, where the scalar and fermion of the appropriate messengers are running in the loop. By computing this diagram, one finds that the MSSM gaugino masses are given by [430]

$$M_a = n_5 \frac{\alpha_a}{4\pi} \frac{F}{M}. \tag{17.7}$$

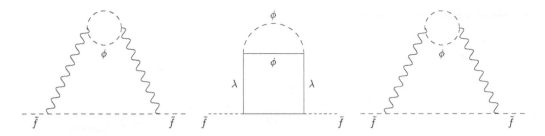

Figure 17.3 Examples of two loop generation of the scalar mass in the GMSB. Similar diagrams can be obtained from the first diagram by replacing the scalar loop by a fermion loop or by changing the three interaction vertices to the allowed four point interactions.

The leading contribution to the masses of MSSM scalars are obtained from the two loops diagram shown in Fig. 17.3. In these loops, the messenger scalars, messenger fermions, ordinary gauge boson and gaugino are exchanged. By computing these graphs, one finds that the scalar masses are given by [431]

$$\tilde{m}^2 = 2n_5\Lambda^2 \sum_{a=1}^{3} C_i \left(\frac{\alpha_a}{4\pi}\right)^2, \tag{17.8}$$

where $\Lambda = F/M$, $C_3 = 4/3$ for colour triplets and zero for singlets, $C_2 = 3/4$ for weak doublets and zero for singlets, where $C_1 = (3/5)(Y/2)^2$, Y being the ordinary hypercharge.

This equation shows that the scalar masses are functions of gauge quantum numbers only (*i.e.*, flavour universal), which is essential to avoid the stringent constraints from FCNCs. Therefore, GMSB solves the SUSY flavour problem. Also, the trilinear couplings, A-terms, are generated in GMSB at two loops and are suppressed compared to the gaugino masses. To a good approximation, one can consider that $A = 0$ at the messenger scale. They are generated from the RGE running from the messenger to the EW scale. The μ and B parameters can arise through interactions of the Higgs field with various singlets. As usual, B and the magnitude of μ will be determined by the radiative EWSB conditions. Therefore, the minimal GMSB model can be completely specified by the following parameters defined as the messenger scale:

$$M, \Lambda, \tan\beta, \text{sign } \mu, n_5. \tag{17.9}$$

From the above soft terms, we can conclude that the gaugino and sfermion masses are of the same order ($\sim F/M$) and the coloured sparticles (squarks and gluinos) are typically much heavier than the uncoloured sparticles (sleptons, charginos and neutralinos). Furthermore, it is worth noting that, in GMSB, the gravitino acquires a very light mass through the super-Higgs mechanism, namely, it is given by

$$m_{3/2} = \frac{F}{\sqrt{3}M_P} \sim \left(\frac{M\Lambda}{(100 \text{ TeV})^2}\right) \text{eV}. \tag{17.10}$$

i.e., for $\Lambda \sim 10$ TeV and $M > \Lambda$, one gets $m_{3/2} \sim \mathcal{O}(1 \text{ eV})$–$\mathcal{O}(1 \text{ keV})$. Thus, the lightest neutralino or the lightest stau is the NLSP.

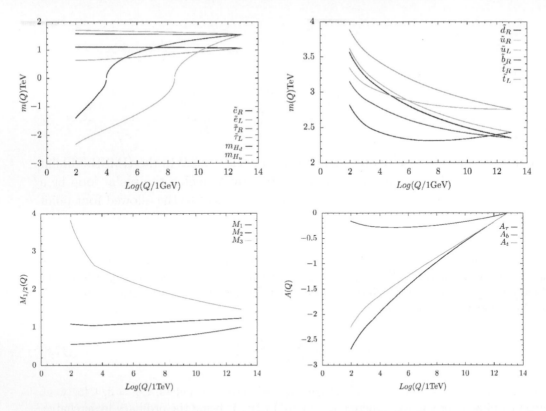

Figure 17.4 RGE running of the mass parameters in GMSB for the messenger scale with $\Lambda = 4.2 \times 10^5$, $M = 0.9 \times 10^{13}$ and $\tan\beta = 50$.

17.2 GMSB PHENOMENOLOGY

The soft SUSY breaking terms in GMSB are induced at the messenger scale M, which is typically lower than the GUT scale. The evolution of these terms with other various boundary conditions (Yukawa and gauge couplings) is performed by the RGEs from the messenger scale down to the EW scale [432]. As in the MSSM, it is crucial to prove that the running derives the squared mass of the up-sector Higgs, $m_{H_u}^2$ to a negative value at the EW scale, so that EWSB can be applied. In Fig. 17.4 we show the RGE running of the masses of squarks, sleptons, gauginos and Higgs parameters, $m_{H_u}^2$ and $m_{H_d}^2$, for $\Lambda \simeq 10^5$ GeV, $M \simeq 10^{13}$ GeV and $\tan\beta \simeq 50$. As can be seen from this plot, although $m_{H_u}^2$ starts the running from 10^{13} GeV with positive initial value, it turns quickly to a negative value at the EW scale. Therefore, the EWSB takes place and the usual symmetry breaking conditions are imposed. As in the MSSM, these conditions will be used to determine the Higgs sector mass parameters: μ and B.

We employed SARAH [283] and SPheno to carry out this running and compute the spectrum of the model. In our analysis we consider the following ranges for the GMSB parameter space:

$$10^3 \text{ GeV} \leq \Lambda \leq 10^7 \text{ GeV}, \tag{17.11}$$

$$10^8 \text{ GeV} \leq M \leq 10^{16} \text{ GeV}, \tag{17.12}$$

$$1.5 \text{ GeV} \leq \tan\beta \leq 50 \text{ GeV}, \tag{17.13}$$

$$n_5 = 1, \quad \mu > 0. \tag{17.14}$$

We also impose the following constraints

$$124 \text{ GeV} \leq m_h \leq 126 \text{ GeV}, \tag{17.15}$$

$$m_{\tilde{g}} > 1.4 \text{ TeV}, \tag{17.16}$$

$$0.8 \times 10^{-9} \leq \text{BR}(B_s \rightarrow \mu^+\mu^-) \leq 6.2 \times 10^{-9}, \tag{17.17}$$

$$2.99 \times 10^{-4} \leq \text{BR}(b \rightarrow s\gamma) \leq 3.87 \times 10^{-4}. \tag{17.18}$$

In GMSB and as in the general MSSM, the lightest CP-even Higgs mass, with one loop radiative correction proportional to the top quark mass, is given by

$$
\begin{aligned}
m_h^2 &= M_Z^2 \cos^2 2\beta \left[1 - \frac{3\sqrt{2}}{4\pi^2} G_F m_t^2 \ln\left(\frac{M_S^2}{m_t^2}\right) \right] \\
&+ \frac{3\sqrt{2}}{2\pi^2} G_F m_t^4 \left[\frac{X_t}{2} + \ln\left(\frac{M_S^2}{m_t^2}\right) \right],
\end{aligned}
\tag{17.19}
$$

where

$$X_t = \frac{2(A_t - \mu \cot\beta)^2}{M_S^2} \left(1 - \frac{(A_t - \mu \cot\beta)^2}{12 M_S^2} \right) \tag{17.20}$$

and

$$M_S = m_{\tilde{t}_1} m_{\tilde{t}_2}. \tag{17.21}$$

A comprehensive analysis for the GMSB contributions to the SM-like Higgs boson dynamics has been considered after the Higgs discovery [433–439]. It was emphasised that, in this class of models, where the A-term vanishes at the messenger (relatively low) scale, $A_t/M_{\tilde{t}}$ is always less than one. Hence $X_t \ll 1$, for any value of M and Λ, i.e., GMSB represents an example of a no-mixing scenario which leads to a low value of m_h. Therefore, the radiative corrections to the Higgs mass are dominated by the logarithm of the stop quark mass. Thus, a very heavy stop ($\sim \mathcal{O}(5 \text{ TeV})$) is required in order to account for the measured Higgs mass.

In Fig. 17.5 we plot the Higgs mass as a function of the SUSY scale M_S. We show explicitly that, in order to satisfy the Higgs mass limits: $124 \text{ GeV} \leq m_h \leq 126 \text{ GeV}$, M_S must be within the range $[3.2, 8.4]$ TeV. We also display the constraints imposed on the plane of parameter space (M, Λ) by the Higgs mass limits. It is clear that the Higgs mass imposes a stringent constraint on the parameter $\Lambda = F/M$ to be of order 10^6 GeV, while the constraints on the messenger scale $M > 10^5$ GeV are much less severe.

In Fig. 17.6 we plot the correlation between the mass of the lightest stop and that of the gluino. As can be seen, in the region where m_h is within the observed limits, the lightest stop and gluino masses are above 3 TeV. We also present the lightest stau mass versus the lightest neutralino one. As can be seen, the masses of these two particles are typically larger than 1 TeV. Also, it is possible to have regions with $m_{\tilde{\tau}} < m_{\tilde{\chi}_1^0}$, although it is more likely to have the lightest neutralino as NLSP.

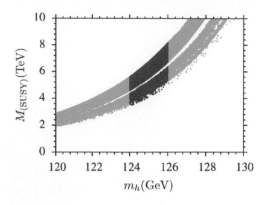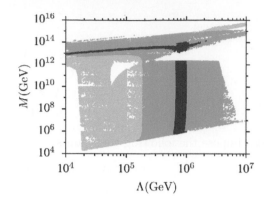

Figure 17.5 (Left) Higgs mass as a function of M_S. Red points correspond to Higgs boson mass within the experimental limits. (Right) $M - \Lambda$ plane in mGMSB ($n_5 = 1$) for $\tan \beta = 10$. Red points satisfy all constraints and lead to 124 GeV $\leq m_h \leq 126$ GeV. Gray points are consistent with the electroweak symmetry breaking, but they are excluded by the LHC. Green points satisfy all constraints except the Higgs mass.

As a concrete example, we consider the following benchmark point, which leads to $m_h \simeq 125$ GeV. In this point the initial values are given by

$$\Lambda = 4.2 \times 10^5, \qquad M = 0.9 \times 10^{13}, \qquad \tan \beta = 50. \tag{17.22}$$

We find that the corresponding spectrum is given as follows:

$$m_{\tilde{u}} = 3.9 \text{ TeV}, \qquad m_{\tilde{c}} = 3.9 \text{ TeV}, \qquad m_{\tilde{t}} = 2.8 \text{ TeV},$$
$$m_{\tilde{d}} = 3.9 \text{ TeV}, \qquad m_{\tilde{s}} = 3.9 \text{ TeV}, \qquad m_{\tilde{b}} = 3.4 \text{ TeV}, \tag{17.23}$$

$$m_{\tilde{e}} = 1.1 \text{ TeV}, \qquad m_{\tilde{\mu}} = 1.1 \text{ TeV}, \qquad m_{\tilde{\tau}} = 0.63 \text{ TeV},$$
$$m_{\tilde{\nu}_e} = 1.8 \text{ TeV}, \qquad m_{\tilde{\nu}_\mu} = 1.8 \text{ TeV}, \qquad m_{\tilde{\nu}_\tau} = 1.6 \text{ TeV}, \tag{17.24}$$

$$m_{\tilde{\chi}_1^0} = 0.5 \text{ TeV}, \qquad m_{\tilde{\chi}_1^\pm} = 1.1 \text{ TeV}, \qquad m_{\tilde{g}} = 3.9 \text{ TeV}. \tag{17.25}$$

The minimal GMSB framework yields a similar mass spectrum to the mSUGRA framework in the gaugino and Higgs sectors. If one imposes universal boundary conditions, it is possible to obtain the GMSB boundary conditions for the gaugino and Higgs masses at the messenger scale M. In contrast, the sfermion sector provides a different pattern, which cannot be obtained in the mSUGRA framework.

17.3 NEXT-TO-LIGHTEST SUSY PARTICLE

As mentioned, the phenomenolgy of GMSB is characterised by the presence of a very light gravitino $\tilde{G}$, with mass $m_{3/2} \sim 2.37$ eV, as shown in Eq. (17.10). Therefore, in this class of models, the gravitino is always the LSP. If R-parity is conserved, then it is expected that all SUSY particles decay into the NLSP, which then decays into the gravitino with a coupling suppressed by F.

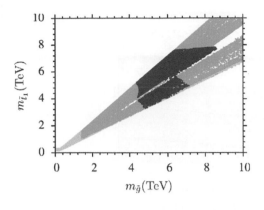
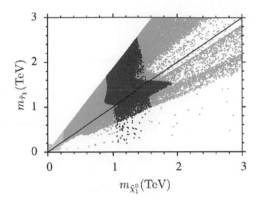

Figure 17.6 (Left) Lightest stop mass versus the gluino mass. (Right) Lightest stau mass versus the lightest neutralino mass. Red points correspond to Higgs within the experimental limits. Green and grey points as defined in the above plot.

Depending on the lifetime of the NLSP, which is determined in terms of the SUSY breaking scale F, the decay of the NLSP may occur inside or outside the detector(hence appearing to be stable). For instance, if $F > 10^{12}$ GeV, the neutralino NLSP will resemble the stable neutralino case with missing transverse energy, while the stau NLSP case gives a ionising track of a massive charged particle. For $F < 10^{12}$ GeV, the neutralino NLSP promptly decays into a photon and a gravitino (missing energy) and the stau NLSP decays into a tau and a gravitino.

In the case of stau NLSP, RH stau decays into the gravitino can potentially be probed at the LHC as missing energy. The two-body decay width is given by

$$\Gamma(\tilde{\tau}_R \to \tau_R \tilde{G}) = \frac{m_{\tilde{\tau}_R}^5}{16\pi(M\Lambda)^2}\left(1 - \frac{m_l^2}{m_{\tilde{\tau}_R}^2}\right)^4. \tag{17.26}$$

Fig. 17.7 shows the mass correlation between the gravitino and stau. The lower solid line corresponds to a ~ 1 mm free length of the stau, which can be interpreted as the prompt decay of the stau into the gravitino. The upper solid line represents the free length of 10 m, which is assumed to be the length of the detector. The results consistent with the Higgs boson data (shown in red in Fig. 17.7) yield stau solutions whose free length is longer than 10 m and, hence, it decays into a gravitino outside the detector.

17.4 THE μ PROBLEM IN GMSB

In the MSSM we have a supersymmetric mass term for the two Higgs doublets in the superpotential

$$W = \mu H_u H_d \tag{17.27}$$

and a soft SUSY breaking term $B\mu$ in the scalar potential

$$V = B\mu H_u H_d + \text{h.c.} \tag{17.28}$$

The radiative EWSB conditions imply that $B \sim \mu \sim$ TeV. However, since μ is a supersymmetric parameter, it is not connected to the scale of SUSY breaking and it can be much

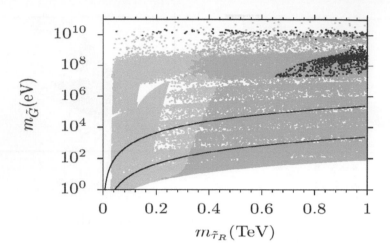

Figure 17.7 Mass correlation between gravitino and stau. The lower solid line corresponds to 1 mm free length of stau in the process $\tilde{\tau}_R \to \tau\tilde{G}$, while the upper solid line represents the free length of 10 m.

higher than TeV (up to the GUT or Planck scale). As explained in Chapter 15, this is known as the μ problem.

In GMSB, the μ parameter can be generated if one assumes a direct coupling between Higgs fields and the SUSY breaking field X [429]:

$$W = \lambda X H_u H_d. \tag{17.29}$$

Then one finds

$$\mu = \lambda M, \qquad B\mu = \lambda F. \tag{17.30}$$

Since $F \sim M^2$, it is clear that B is of order μ but very large. It is also remarkable that, if we allow for a direct coupling between the SUSY breaking hidden sector fields and the Higgs fields in the observable sector, then we can also have couplings between X and other MSSM fields and we lose any sense of SUSY breaking mediation.

In order to generate a lower value of μ and $B\mu$, a deviation from mGMSB is mandatory [440]. As an example, extra singlets S and N are assumed so that we can have the following superpotential [441]:

$$W = S(\lambda_1 H_u H_d + \lambda_2 N^2 + \lambda \Phi\bar{\Phi} - M_N^2) + X\Phi\bar{\Phi}. \tag{17.31}$$

The VEV of S can then be generated at one-loop,

$$\langle S \rangle \sim \frac{1}{16\pi^2} \frac{F^2}{M M_N^2}, \tag{17.32}$$

while F_S is only generated at the two loop level,

$$F_S \sim \left(\frac{1}{16\pi^2} \frac{F}{M}\right)^2. \tag{17.33}$$

In this case, if $F \sim M_N^2$, one obtains

$$B \sim \mu \sim \text{TeV}. \tag{17.34}$$

CHAPTER 18

SUSY Breaking in String Theory

The idea of strings was first proposed in the 1960s, in order to describe strong nuclear interactions. After it was realised in the 1970s that the massless spectrum of string theory contains a spin-2 particle, it was suggested as a consistent framework which unifies all fundemantal forces of Nature, including gravity. By replacing the notion of a point particle by a string, a consistent quantum theory of gravity emerges which reproduces Einstein's general theory of relativity and the forces of the SM. String theory is still in progress and there is no complete description of the SM based on it. Although the energy scale of the string theory effects ($\sim 10^{19}$ GeV) is much higher than the current values at the LHC ($\sim 10^3$ GeV), some predictions of the theory such as large extra dimensions and SUSY are accessible at current energies.

18.1 AN OVERVIEW OF STRING THEORY

According to string theory, the fundamental particles are not point-like but a tiny one-dimensional 'string', which can be closed or open as in Fig. 18.1 String can vibrate and different string oscillations would correspond to different physical properties, such as masses or Lorentz and gauge quantum numbers, so that they behave as different particles.

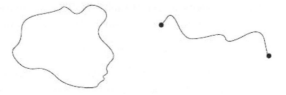

Figure 18.1 Closed string (left) and open string (right).

The string has degrees of freedom $X^\mu(\tau, \sigma), \mu = 0, 1, ...,D-1$, which specify the space-time of the world sheet point (τ, σ). The action for a string moving in flat space-time is given by

$$S = -\frac{T}{2} \int d^2\xi \sqrt{-\det g} g^{\alpha\beta} \partial_\alpha X^\mu \partial_\beta X^\nu \eta_{\mu\nu}, \tag{18.1}$$

where T is the string tension and $d^2\xi = d\tau d\sigma$. Using the symmetries of the string action (Poincaré invariance, two-dimensional reparameterisation and conformal invariance), the string equation of motion takes the simple form of a wave equation (see [415, 442–446] for more details):

$$\left(\frac{\partial^2}{\partial\tau^2} - \frac{\partial^2}{\partial\sigma^2}\right) X^\mu(\sigma, \tau) = 0. \tag{18.2}$$

The most general solution to the above equation is a superposition of left-moving and right-moving waves:

$$X^\mu(\tau, \sigma) = X_L^\mu(\tau + \sigma) + X_R^\mu(\tau - \sigma). \tag{18.3}$$

One should also take into account the proper boundary conditions, which depend on the type of string, open or closed.

Closed string

In the case of closed string, one must impose the periodicity conditions:

$$X^\mu(\tau, \sigma + \pi) = X^\mu(\tau, \sigma). \tag{18.4}$$

Thus, the general solution takes the form

$$X_R^\mu = \frac{1}{2}x^\mu + \frac{l^2}{2}p^\mu(\tau - \sigma) + \frac{i}{2}l\sum_{n\neq 0}\frac{1}{n}\alpha_n^\mu e^{-2in(\tau-\sigma)}, \tag{18.5}$$

$$X_L^\mu = \frac{1}{2}x^\mu + \frac{l^2}{2}p^\mu(\tau + \sigma) + \frac{i}{2}l\sum_{n\neq 0}\frac{1}{n}\tilde{\alpha}_n^\mu e^{-2in(\tau+\sigma)}, \tag{18.6}$$

where $l = (\pi T)^{-1/2}$. The usual canonical quantisation implies

$$[x^\mu, p^\nu] = i\eta^{\mu\nu}, \tag{18.7}$$

$$[a_m^\mu, a_n^\nu] = [\tilde{a}_m^\mu, \tilde{a}_n^\nu] = -\delta_{m,-n}\eta^{\mu\nu}, \tag{18.8}$$

where $a_n^\mu = \frac{1}{\sqrt{n}}\alpha_n^\mu$ and $a_n^{\mu^+} = \frac{1}{\sqrt{n}}\alpha_{-n}^\mu$. Similarly, $\tilde{a}_n^\mu = \frac{1}{\sqrt{n}}\tilde{\alpha}_n^\mu$ and $\tilde{a}_n^{\mu^+} = \frac{1}{\sqrt{n}}\tilde{\alpha}_{-n}^\mu$. It is convenient to work in a light-cone formalism. In this case, the mass M of physical states, in units where $l = 1$, is given by

$$M^2 = 4(N - 1), \tag{18.9}$$

with $N = \sum_{n=1}^\infty na_n^+.a_n = \tilde{N} = \sum_{n=1}^\infty n\tilde{a}_n^+.\tilde{a}_n$. Therefore, the ground state $|k, 0, 0\rangle$ has mass-squared $M^2 = -4$, which represents a spin-0 tachyon. This indicates that the closed string vacuum is unstable. The first excited state is generally of the form

$$|k, \zeta\rangle = \zeta_{\mu\nu}\left(\alpha_{-1}^\mu|k, 0\rangle \otimes \tilde{\alpha}_{-1}^\nu|k, 0\rangle\right). \tag{18.10}$$

It has mass-squared $M^2 = 0$. The polarisation tensor $\zeta_{\mu\nu}$ can be decomposed into a traceless symmetric tensor $g_{\mu\nu} = \frac{1}{2}(\zeta_{\mu\nu} + \zeta_{\nu\mu})$, which corresponds to the spin-2 graviton field, and anti-symmetric spin-2 tensor $B_{\mu\nu} = \frac{1}{2}(\zeta_{\mu\nu} - \zeta_{\nu\mu})$, in addition to the scalar field $\Phi = \text{Tr}(\zeta)$, which is the spin-0 dilaton.

Open string

In the case of open string, there are two possible boundary conditions

$$\text{Neumann}: \quad X'^\mu(\tau, \sigma)\Big|_{\sigma=0,\pi} = 0, \tag{18.11}$$

$$\text{Dirichlet}: \quad \dot{X}^\mu(\tau, \sigma)\Big|_{\sigma=0,\pi} = 0, \tag{18.12}$$

where dot and prime refer to derivative with respect to τ and σ, respectively. Dirichlet boundary conditions are equivalent to the string with fixed end points. One can show for the Neumann boundary condition that the general solution of the string wave equation is given by

$$X^\mu = x^\mu + l^2 p^\mu \tau + il \sum_{n \neq 0} \frac{1}{n} \alpha_n^\mu e^{in\tau} \cos(n\sigma). \tag{18.13}$$

Then canonical quantisation implies

$$[x^\mu, p^\nu] = i\eta^{\mu\nu}, \tag{18.14}$$

$$[a_m^\mu, a_n^\nu] = -\delta_{m,-n}\eta^{\mu\nu}, \tag{18.15}$$

where $a_n^\mu = \frac{1}{\sqrt{n}}\alpha_n^\mu$ and $a_n^{\mu^+} = \frac{1}{\sqrt{n}}\alpha_{-n}^\mu$. In light-cone formalism, the mass M of physical states, in a unit where $l = 1$, is given by

$$M^2 = N - 1, \tag{18.16}$$

with $N = \sum_{n=1}^{\infty} n a_n^\dagger a_n$. Thus, the ground state $|k,0\rangle$ is also tachyonic with mass-squared $M^2 = -1$. Hence, like the bosonic closed string, the bosonic open string does not provide a consistent quantum theory. The first excited state is given by $\alpha_{-1}^\mu |0\rangle$. It is a massless state with $(D-2)$ components, which is a vector representation of the transverse rotation group $SO(D-2)$. The higher level string states with $N \geq 2$ are all massive and will not be considered here.

Superstring

As we have seen, the bosonic (open and closed) string theory suffers from two serious problems: (i) its spectrum of quantum states contains a tachyon, which signals an unstable vacuum; (ii) it contains no fermions in its quantum spectrum and so it is not phenomenologically viable. This motivates us to construct a supersymmetric string theory by adding SUSY to the bosonic string [415, 444–449]. The RNS formalism is the standard way to introduce SUSY in string theory. It uses two-dimensional worldsheet SUSY and it requires the GSO projection to eventually realise space-time SUSY and generate a spectrum free from tachyons.

In this formalism, D-free massless Majorana spinors $\Psi_\mu(\tau, \sigma)$ are added on the string worldsheet. Thus, the worldsheet action is now given by

$$S_0 = -\frac{1}{2\pi} \int d^2\sigma(-\det h)^{1/2} \left[h^{\alpha\beta}\partial_\alpha X^\mu \partial_\beta X_\mu + i\bar{\Psi}^\mu \rho^\alpha \partial_\alpha \Psi_\mu \right], \tag{18.17}$$

where ρ^α, $\alpha = 0, 1$ are the 2×2 Dirac matrices. This action is invariant under the following global worldsheet SUSY transformation:

$$\delta X^\mu = \bar{\xi}\Psi^\mu, \tag{18.18}$$

$$\delta \Psi^\mu = -i\rho^\alpha \partial_\alpha X^\mu \xi, \tag{18.19}$$

where ξ is a two-component Majorana spinor parameter. Therefore, under local worldsheet SUSY, i.e., $\xi = \xi(\tau, \sigma)$, the action S_0 is no longer invariant. Instead, it leads to

$$\delta S_0 = \frac{2}{\pi} \int d^2\sigma(-\det h)^{1/2}\partial_\alpha\bar{\xi}J^\alpha, \tag{18.20}$$

with

$$J^\alpha = \frac{1}{2}\rho^\beta \rho^\alpha \Psi^\mu \partial_\beta X_\mu. \tag{18.21}$$

This variation can be cancelled by introducing a 2-dimensional gravitino χ_α such that $\delta\chi_\alpha = \partial_\alpha \xi$ and adding to S_0 the following S_1:

$$S_1 = -\frac{1}{\pi} \int d^2\sigma(-\det h)^{1/2} \bar{\chi}_\alpha \rho^\beta \rho^\alpha \Psi^\mu \partial_\beta X_\mu. \tag{18.22}$$

One can notice that the transformation of X_μ gives an extra term δS_1:

$$\delta S_1 = -\frac{1}{\pi} \int d^2\sigma(-\det h)^{1/2} \bar{\chi}_\alpha \rho^\beta \rho^\alpha \Psi^\mu \bar{\Psi}_\mu \partial_\beta \xi, \tag{18.23}$$

which can be cancelled by introducing

$$S_2 = -\frac{1}{4\pi} \int d^2\sigma(-\det h)^{1/2} \bar{\Psi}_\mu \Psi^\mu \bar{\chi}_\alpha \rho^\beta \rho^\alpha \chi_\beta. \tag{18.24}$$

Thus, the total action

$$S = S_0 + S_1 + S_2 \tag{18.25}$$

is invariant under the transformation:

$$\delta X^\mu = \bar{\xi}\Psi^\mu, \tag{18.26}$$
$$\delta\Psi^\mu = -i\rho^\alpha \xi \left(\partial_\alpha X^\mu - \bar{\Psi}^\mu \chi_\alpha\right), \tag{18.27}$$
$$\delta\chi_\alpha = \partial_\alpha \xi, \tag{18.28}$$
$$\delta e_\alpha^a = -2i\bar{\xi}\rho^\alpha \chi_\alpha, \tag{18.29}$$

where the velbien e_α^a is defined such that $h_{\alpha\beta} = e_\alpha^a e_\beta^b \eta_{ab}$. Now, one can show that the equations of motion are given by

$$\partial_\alpha \partial^\alpha X^\mu = 0, \tag{18.30}$$
$$i\rho^\alpha \partial_\alpha \Psi^\mu = 0. \tag{18.31}$$

Deriving these equations of motion assumed the vanishing of associated surface terms. In fact, for bosonic degrees of freedom the surface term vanishes because of the boundary conditions mentioned in the bosonic string section, while the surface term of the fermionic degrees of freedom $\Psi^\mu = \begin{pmatrix} \Psi_R^\mu \\ \Psi_L^\mu \end{pmatrix}$ would vanish in the case of a *closed string* if:

$$\Psi_R^\mu(\tau, \sigma + \pi) = \pm\Psi_R^\mu(\tau, \sigma), \tag{18.32}$$

$$\Psi_L^\mu(\tau, \sigma + \pi) = \pm\Psi_R^\mu(\tau, \sigma). \tag{18.33}$$

Note that the periodic (R) or anti-periodic (NS) boundary conditions may be chosen independently for right and left movers. For an *open string* the following conditions are required:

$$\psi_L^\mu(\tau, \sigma) = \Psi_R^\mu(\tau, \sigma), \tag{18.34}$$

$$\psi_L^\mu(\tau, \pi) = \pm\Psi_R^\mu(\tau, \pi). \tag{18.35}$$

Therefore, the mode expansions of fermionic degrees of freedom of a closed string are given by

$$\Psi_R^\mu = \sum_{n\in Z} d_n^\mu e^{-2in(\tau-\sigma)}, \qquad \Psi_L^\mu = \sum_{n\in Z} \tilde{d}_n^\mu e^{-2in(\tau+\sigma)}, \qquad (R) \tag{18.36}$$

or

$$\Psi_R^\mu = \sum_{r \in Z+1/2} b_r^\mu e^{-2ir(\tau-\sigma)}, \qquad \Psi_L^\mu = \sum_{r \in Z+1/2} \tilde{b}_r^\mu e^{-2ir(\tau+\sigma)}, \qquad \text{(NS)} \qquad (18.37)$$

while for open string:

$$\Psi_R^\mu = \frac{1}{\sqrt{2}} \sum_{n \in Z} d_n^\mu e^{-in(\tau-\sigma)}, \qquad \Psi_L^\mu = \frac{1}{\sqrt{2}} \sum_{n \in Z} d_n^\mu e^{-in(\tau+\sigma)}, \qquad \text{(R)} \qquad (18.38)$$

or

$$\Psi_R^\mu = \frac{1}{\sqrt{2}} \sum_{r \in Z+1/2} b_r^\mu e^{-ir(\tau-\sigma)}, \qquad \Psi_L^\mu = \frac{1}{\sqrt{2}} \sum_{r \in Z+1/2} b_r^\mu e^{-ir(\tau+\sigma)}. \qquad \text{(R)} \qquad (18.39)$$

In this case, the standard quantisation yields the following relations:

1. For closed string:

$$\{d_m^\mu, d_n^\nu\} = \{\tilde{d}_m^\mu, \tilde{d}_n^\nu\} = -\delta_{m+n,0}\eta^{\mu\nu} \qquad \text{(R)}, \qquad (18.40)$$

$$\{b_r^\mu, b_s^\nu\} = \{\tilde{b}_r^\mu, \tilde{b}_s^\nu\} = -\delta_{r+s,0}\eta^{\mu\nu}. \qquad \text{(NS)}. \qquad (18.41)$$

For $m = n = 0$ one finds that $\{d_0^\mu, d_0^\nu\} = -\eta^{\mu\nu}$, which are the D-dimension gamma matrices with $\Gamma^\mu = i\sqrt{2}d_0^\mu$. Therefore, the R-sector corresponds to the fermion sector in D-dimensional space-time. Further, the NS sector corresponds to the boson sector in 10D space-time.

2. Similarly for open string, one obtains

$$\{d_m^\mu, d_n^\nu\} = -\delta_{m+n,0}\eta^{\mu\nu}, \qquad (18.42)$$

$$\{b_r^\mu, d_s^\nu\} = -\delta_{r+s,0}\eta^{\mu\nu}. \qquad (18.43)$$

The mass-squared operator for closed superstring can be written as

$$M^2 = M_R^2 + M_L^2 \qquad (18.44)$$

where

$$\frac{1}{4}M_R^2 = -\sum_{n=1}^{\infty} \alpha_{-n}^\mu \alpha_{\mu n} - \sum_{r=1/2}^{\infty} r b_{-r}^\mu b_{\mu r} - a_{\text{NS}} \qquad (18.45)$$

for NS sector right-movers and

$$\frac{1}{4}M_R^2 = -\sum_{n=1}^{\infty} \alpha_{-n}^\mu \alpha_{\mu n} - \sum_{r=1/2}^{\infty} n d_{-n}^\mu d_{\mu r} - a_{\text{R}}, \qquad (18.46)$$

for R-sector right-movers. Here $a_{\text{NS,R}}$ are the normal order parameters, which are related to the dimension of space-time. In order to avoid quantum anomalies of Lorentz symmetry, one fixes the normal order parameters to some values that lead to $D = 10$. Using a light-cone gauge, one can show that $a_{\text{NS}} = 1/2$ and $a_{\text{R}} = 0$ [415, 444–449]. Similar expressions for M_L can be obtained from M_R by exchanging α, b, d with $\tilde{\alpha}, \tilde{b}, \tilde{d}$, as in bosonic string $M_L^2 = M_R^2$.

For NS, the ground state, $|k, 0\rangle_{\text{NS}}$, has a mass $M^2 = -1/2$, $i.e$, it is still tachyonic and the first excited state, $b_{-1/2}^i |k, 0\rangle_{\text{NS}}$, has mass $M^2 = 0$. Also, with a few more oscillators one gets $|\phi\rangle = b^{i_1}_{-r_1} b^{i_2}_{-r_2} \ldots b^{i_M}_{-r_M} |0\rangle$. Exchanging two of these indices yields an extra (-1) factor, which is not desirable for statistics of bosonic space-time.

In addition, the ground state, in the R-sector, $|k, 0\rangle_R$, is massless. Since $d^\mu_0 = \Gamma^\mu$ acts on $|k, 0\rangle_R$ without changing its mass, it is an irreducible representation of the Clifford algebra, *i.e.*, a spinor representation of $SO(8)$. Therefore, the ground state in the R-sector is described by a Spin(8) spinor.

It is clear that the spectrum of superstring still suffers from several problems, namely a tachyonic ground state in the NS sector and the apparent inequalities between fermions and scalar boson degrees of freedom (*e.g.*, no fermion with the same mass as the tachyon). GSO introduced the following parity which, truncating the spectrum in a very specific way to eliminate tachyons, leads to a supersymmetric theory:

$$G = (-1)^{\sum_{r=1/2}^{\infty} b^i_{-r} b^i_{-r+1}}, \qquad \text{(NS)} \qquad (18.47)$$

$$G = \Gamma_{11} (-1)^{\sum_{r=1/2}^{\infty} d^i_{-n} d^i_{-n}}. \qquad \text{(R)} \qquad (18.48)$$

The GSO projection contains only the states with a positive G-parity in the NS sector, while the states with negative G-parity should be eliminated. This means all NS sector states should have an odd number of b-oscillator excitations. In this context, the GSO projection eliminates the open-string tachyon from the spectrum, since it has negative G-parity. The first excited state, $b^i_{-1/2}|0\rangle_{NS}$, has positive G-parity and is allowed by the projection. Moreover, this massless vector boson becomes the ground state of the NS sector, which matches nicely with the fact that the ground state in the fermionic sector is a massless spinor. In the R-sector one can project on states with positive or negative G-parity depending on the chirality of the spinor ground state. The choice is purely a matter of convention.

To analyse the closed string spectrum, it is necessary to consider left-movers and right-movers. As a result, there are four possible sectors: R-R, R-NS, NS-R and NS-NS. In this case, two different theories can be obtained depending on whether the G-parity of the left- and right-moving R sectors is the same or opposite. In the Type IIB theory, the left- and right-moving R-sector ground states have the same chirality. Therefore, the two R-sectors have the same G-parity and they can be denoted as $|+\rangle_R$. The massless states in the Type IIB closed string are given by

$$|+\rangle_R \otimes |+\rangle_R, \qquad (18.49)$$

$$\tilde{b}^i_{-1/2}|0\rangle_{NS} \otimes \tilde{b}^j_{-1/2}|0\rangle_{NS}, \qquad (18.50)$$

$$\tilde{b}^i_{-1/2}|0\rangle_{NS} \otimes |+\rangle_R, \qquad (18.51)$$

$$|+\rangle_R \otimes \tilde{b}^i_{-1/2}|0\rangle_{NS}. \qquad (18.52)$$

Since $|+\rangle_R$ represents an eight-component spinor, each of the four sectors contains $8 \times 8 = 64$ physical states.

For the Type IIA theory, the left- and right-moving R-sector ground states are chosen to have opposite chirality. The massless states in the spectrum are given by

$$|-\rangle_R \otimes |+\rangle_R, \qquad (18.53)$$

$$\tilde{b}^i_{-1/2}|0\rangle_{NS} \otimes \tilde{b}^j_{-1/2}|0\rangle_{NS}, \qquad (18.54)$$

$$\tilde{b}^i_{-1/2}|0\rangle_{NS} \otimes |+\rangle_R, \qquad (18.55)$$

$$|-\rangle_R \otimes \tilde{b}^i_{-1/2}|0\rangle_{NS}. \qquad (18.56)$$

These states are very similar to the ones of the Type IIB string except that now the fermionic states come with two different chiralities. The massless spectrum of each of the Type II closed string theories contains two Majorana-Weyl gravitinos and, therefore, they form $N = 2$ SUGRA multiplets. There are 64 states in each of the four massless sectors, that we summarise below. As mentioned,

NS-NS sector: This sector is the same for the Type IIA and Type IIB cases. The spectrum contains a scalar called dilaton (one state), an antisymmetric two-form gauge field (28 states), a symmetric traceless rank-two tensor and the graviton (35 states).

NS-R and R-NS sectors: Each of these sectors contains a spin $3/2$ gravitino (56 states) and a spin $1/2$ fermion called dilatino (eight states). In the Type IIB case the two gravitinos have the same chirality while in the Type IIA case they have opposite chirality.

R-R-sector: These states are bosons obtained by tensoring a pair of Majorana-Weyl spinors. In the IIA case, the two Majorana-Weyl spinors have opposite chirality and one obtains a one-form (vector) gauge field (eight states) and a three-form gauge field (56 states). In the IIB case the two Majorana-Weyl spinors have the same chirality and one obtains a zero-form (that is, scalar) gauge field (one state), a two-form gauge field (28 states) and a four-form gauge field with a self-dual field strength (35 states).

18.2   STRING COMPACTIFICATION

In the previous section, we have considered superstrings propagating in 10D Minkowski spacetime. To connect string theory to our four dimensional spacetime, we have to assume that the extra 6 dimensions are compactified on sufficiently small scales so that they are unobservable in the current experiments [415, 444–449].

Let us start with the compactification of Type II superstring on a circle S^1 of radius R, i.e., $x^9 \sim x^9 + 2\pi R$. The associate momentum p_0^9 is given by $p_0^9 = \frac{n}{R}$ for some integer n; thus,

$$\alpha_0^9 + \tilde{\alpha}_0^9 = \frac{2n}{R}\sqrt{\frac{\alpha'}{2}}. \tag{18.57}$$

In addition, under a periodic shift $\sigma \to \sigma + 2\pi$ along the string, the string can 'wind' around the spacetime circle. This means that

$$X^9(\tau, \sigma + 2\pi) = X^9(\tau, \sigma) + 2\pi w R, \tag{18.58}$$

which leads to

$$\alpha_0^9 - \tilde{\alpha}_0^9 = w R \sqrt{\frac{2}{\alpha'}}. \tag{18.59}$$

Solving the two equations given above, one obtains

$$\alpha_0^9 = p_L \sqrt{\frac{\alpha'}{2}}, \qquad p_L = \frac{n}{R} + \frac{wR}{\alpha'}, \tag{18.60}$$

$$\tilde{\alpha}_0^9 = p_R \sqrt{\frac{\alpha'}{2}}, \qquad p_R = \frac{n}{R} - \frac{wR}{\alpha'}. \tag{18.61}$$

In this case, the mass spectrum in the remaining uncompactified $1 + 8$ dimensions is given by

$$M^2 = \frac{n^2}{R^2} + \frac{w^2 R^2}{\alpha'^2} + \frac{2}{\alpha'}(N + \tilde{N} - 2), \tag{18.62}$$

with

$$nw + N - \tilde{N} = 0. \tag{18.63}$$

A few remarks are in order. (i) There are extra terms present in both the mass formula and the level-matching condition. The winding modes are a purely stringy phenomenon, because only a string can wrap non-trivially around a circle. The usual non-compact states are obtained by setting $n = w = 0$. (ii) In the limit of $R \to \infty$ all $w \neq 0$ winding states disappear as they are energetically unfavourable, while the $w = 0$ states with all values of $n \in Z$ turn to be the momentum excitations of states of the 10 dimensional theory. (iii) In the limit of $R \to 0$ all $n \neq 0$ momentum states become infinitely massive and decouple and now the pure winding states $n = 0$, $w \neq 0$ form a continuum as it costs very little energy to wind around a small circle. So as $R \to 0$ an extra uncompactified dimension reappears. This stringy behaviour is the earmark of 'T-duality': the mass formula for the spectrum is invariant under the simultaneous exchanges

$$n \longleftrightarrow w, \qquad R \longleftrightarrow \frac{\alpha'}{R}. \tag{18.64}$$

It is clear that the choice of a compact manifold imprints itself on the four-dimensional theory and the requirement of $N = 1$ SUSY in 4D, since its phenomenological appealing singles out CY manifolds.

To compactify string theory down to 4D, one looks for vacuum solutions of the form $M_{10} = M_4 \times M_6$, where M_4 is assumed to have 4D Poincaré invariance and M_6 (or simply $\mathcal{M}$) is a compact internal 6D Euclidean space. The most general metric compatible with these requirements can be written as [450]

$$G_{MN} = \begin{pmatrix} e^{A(y)} g_{\mu\nu} & 0 \\ 0 & g_{mn}(y) \end{pmatrix}, \tag{18.65}$$

where y^m are the coordinates on $\mathcal{M}$ with the metric $g_{mn}(y)$ and the requirement of Poincaré symmetry of M_4 still allows for a warp factor which depends on $\mathcal{M}$ only. Here, we will not consider the warp factor $A(y)$. When the radius of curvature of $\mathcal{M}$ is large compared to the Planck scale, one can use the SUGRA approximation of string theory. In order to have a supersymmetric background, the SUGRA transformations of the gravitino ψ_M must vanish [450, 451]:

$$\delta_\varepsilon \psi_M = \nabla_M \varepsilon = 0. \tag{18.66}$$

The vanishing of the above variations for a given spinor ε will put some restrictions on the background fields and, in particular, on the geometry and topology of $\mathcal{M}$. This equation says that $\mathcal{M}$ admits a covariantly constant spinor. The integrability condition resulting from the above equation implies that $\mathcal{M}$ is Ricci flat:

$$R_{mn} = 0. \tag{18.67}$$

Hence the first Chern class of $\mathcal{M}$ vanishes:

$$c_1 = \frac{1}{2\pi}[\mathcal{R}]. \tag{18.68}$$

It was conjectured by Calabi and proved by Yau that Ricci-flat compact Kähler manifolds with $c_1 = 0$ admit a metric with $SU(3)$ holonomy. These metrics come in families and are parameterised by continuous parameters T_i which define the shape and size of $\mathcal{M}$. The parameters T_i appear as scalar fields (moduli) in 4D with no potential and a major goal in string theory is to generate a potential which stabilises these moduli in a way consistent with observations.

One can describe the 4D $N = 1$ models resulting from the string compactification in terms of an effective SUSY theory. This theory is characterised by a Kähler potential K, a gauge kinetic function f and a superpotential W. The tree-level superpotential W does not fix the moduli, because W is not renormalised at any order in perturbation theory due to non-renormalisation theorems [72, 452]. This means that, if SUSY is unbroken at tree level, it will remain unbroken at all orders of perturbation theory. Non-perturbative effects such as gaugino condensation [453] can correct the superpotential and fix some of the moduli.

18.3 GAUGINO CONDENSATION

Hidden sector gaugino condensation is one of the most popular schemes for the breaking of SUSY [454, 455]. This mechanism requires a hidden sector confining group that leads to a bound state of gaugino (gaugino condensate $\langle \lambda_b \lambda_b \rangle$. The chiral superfield

$$U = \delta_{ab} W^a_\alpha \epsilon^{\alpha\beta} W^b_\beta \tag{18.69}$$

is defined such that its lowest (scalar) component corresponds to the gaugino bilinear combination $\lambda_b \lambda_b$. Then U acquires a VEV if a gaugino condensation is formed. In this regard, this mechanism can be naturally realised in $E_8 \otimes E_8$ heterotic string theory where the condensate lives in one E_8, while the other forms the observable sector.

The Veneziano-Yankielowicz superpotential which describes the condensate is given by [456],

$$W^{\mathrm{np}} = \frac{1}{4} U \left(f + \frac{2}{3} \beta \ln U \right), \tag{18.70}$$

where β is the one-loop coefficient of the beta function and

$$f = S + \left(4\beta - \frac{3\delta_{GS}}{2\pi^2} \right) \ln \eta(T) \tag{18.71}$$

is the gauge kinetic function. The Dedekind function $\eta(T)$ is defined as

$$\eta(T) = q^{1/24} \prod_{n=1}^{\infty} (1 - q^n) \tag{18.72}$$

with $q = e^{2\pi i T}$. If one replaces U by its value at $\frac{\partial W}{\partial U} = 0$ [457–459], then the following superpotential is obtained [460]

$$W^{\mathrm{np}} = e^{-\frac{3}{2\beta} f} = d \frac{e^{\frac{-3S}{2\beta}}}{\eta(T)^{6 - \frac{9\delta_{GS}}{4\pi^2 \beta}}}, \tag{18.73}$$

with $d = -\beta/6e$. For $\delta_{GS} = 0$, W^{np} can be written as

$$W^{np} = \frac{\Omega(S)}{\eta(T)^6}. \tag{18.74}$$

The (standard) Kähler potential is given by [461]:

$$K = -\ln Y - 3\ln(T + \overline{T}), \tag{18.75}$$

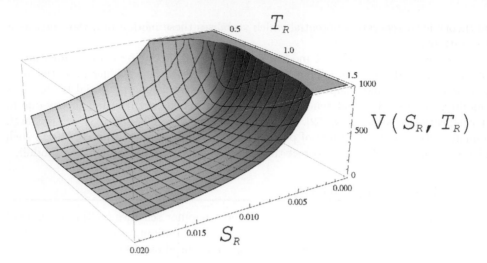

Figure 18.2 Gaugino condensation scalar potential as function of $Re[S]$ and $Re[T]$ (with $Re[T] = Im[T]$).

where $Y = S + \overline{S} + \frac{3}{4\pi^2}\delta_{GS}\ln(T + \overline{T})$. The scalar potential is expressed as

$$V = e^G \left(G_i \left(G_j^i\right)^{-1} G^j - 3\right), \tag{18.76}$$

where $G = K + \ln(|W|^2)$ and the subscripts (superscripts) denote differentiation and the sum over repeated indices runs over the (conjugate) fields in the system. One can show that the scalar potential is given by

$$V = \frac{|\eta(T)|^{-12}}{16 S_R T_R^3} \left[|2S_R\Omega_S - \Omega|^2 + \left(\frac{3T_R^2}{\pi^2}|\hat{G}_2|^2 - 3\right)|\Omega|^2 \right], \tag{18.77}$$

where $\hat{G}_2 = -(\frac{\pi}{T_R} + 4\eta^{-1}\frac{\partial\eta}{\partial T})$, $S_R = ReS$ and $T_R = ReT$. It was emphasised in [460] that the minimum of this potential in S is determined from the condition:

$$2S_R W_S - W = 0. \tag{18.78}$$

Therefore, the auxiliary field F_S, which is proportional to the above combination, vanishes identically at this minimum. Also, one can find that for any given reasonable value of S, the above potential develops a minimum around $T = 1.2$. However, this scenario does not lead to dilaton stabilisation at a reasonable value. In fact, as shown in Fig. 18.2, the scalar potential V has a minimum at $ReS \to \infty$. Since the VEV of the dilaton describes the gauge coupling constants, such a model is phenomenologically unacceptable.

Generally, the hidden sector may contain non-semi-simple gauge groups. Given the right matter content, it is plausible that gauginos condense in each of the simple group factors [462]. With a nonzero $Im\, S$, these condensates may enter the superpotential with opposite signs, thereby leading to dilaton stabilisation.

Let us consider a model with two gaugino condensates [462]. Suppose we have a gauge group $G_1 \otimes G_2$. The corresponding superpotential is given by

$$W^{np} = d_1 \frac{e^{-3S/2\beta_1}}{\eta(T)^6} + d_2 \frac{e^{-3S/2\beta_2}}{\eta(T)^6}, \tag{18.79}$$

where β_1, β_2 are the one-loop coefficients of the G_1 and G_2 beta functions, respectively. In this case, the minimisation condition, $2S_R W_S - W = 0$, leads to

$$S_R = \frac{2}{3(\beta_1^{-1} - \beta_2^{-1})} \ln \frac{d_1(3S_R\beta_1^{-1} + 1)}{d_2(3S_R\beta_2^{-1} + 1)}, \tag{18.80}$$

$$S_I = \frac{2\pi(2n + 1)}{3(\beta_1^{-1} - \beta_2^{-1})}, \quad n \in Z. \tag{18.81}$$

If $S_R = \mathcal{O}(1)$, then $3S_R\beta_a^{-1} \gg 1$ and hence

$$S_R = \frac{2}{3(\beta_1^{-1} - \beta_2^{-1})} \ln \frac{d_1\beta_2}{d_2\beta_1}. \tag{18.82}$$

The gaugino condensation in the presence of matter modifies the values of S_R and it becomes quite plausible to get a value of $S_R(= g_{\mathrm{string}}^{-2})$ in the realistic range ($S_R \sim 2$). In addition, T becomes fixed at a reasonable value ($|T| \sim 1.2$) [463]. In all cases $F_S = 0$, due to the minimisation condition, whereas F_T may be non-zero, which induces SUSY breaking.

18.4 SOFT SUSY BREAKING TERMS FROM SUPERSTRING THEORY

We now consider the soft SUSY-breaking terms obtained from 4D strings under the assumption that the dilaton S and moduli T_i, which play the role of hidden sector fields, dominate the SUSY breaking, i.e., $h_m \equiv S, T_i$, following the previous notation [422, 464, 465]. The VEV of the real part of S gives the inverse squared of the tree-level gauge coupling constant. The VEVs of the moduli parameterise the size and shape of compactification. Therefore, the effective SUGRA Kähler potential derived from superstring theory is given by

$$K(S, S^*, T_i, T_i^*, C_\alpha, C_\alpha^*) = -\log(S + S^*) + \hat{K}(T_i, T_i^*) + \tilde{K}_{\overline{\alpha}\beta}(T_i, T_i^*)C^{*\overline{\alpha}}C^\beta. \tag{18.83}$$

The forms of $\hat{K}(T_i, T_i^*)$, $\tilde{K}_{\overline{\alpha}\beta}(T_i, T_i^*)$ and $Z_{\alpha\beta}(T_i, T_i^*)$ depend on the type of compactification, while the first term is generic for any compactification. The indices α, β label the charged matter fields. To avoid flavour changing neutral current, a diagonal form for the part of the Kähler potential associated with the matter fields is assumed as $\hat{K}_{\overline{\alpha}\beta} \equiv \delta_{\overline{\alpha}\beta}\hat{K}_\alpha$.

Under the above assumptions, the scalar potential will be given by

$$V = G_{\bar{S}S}|F_S|^2 + G_{\overline{i}i}|F_{T_i}|^2 - 3e^G, \tag{18.84}$$

where $e^G = m_{3/2}^2$ is the gravitino mass-squared. In order to overcome our ignorance of the mechanism of SUSY breaking in the hidden sector, the following parameterisation for the VEVs of dilaton and moduli auxiliary fields are introduced [422]:

$$G_{\bar{S}S}^{1/2}F^S = \sqrt{3}m_{3/2}\sin\theta e^{-i\gamma_S},$$

$$G_{\overline{i}i}^{1/2}F^i = \sqrt{3}m_{3/2}\cos\theta \, e^{-i\gamma_i}\Theta_i, \tag{18.85}$$

where $\sum_i \Theta_i^2 = 1$. The angle θ and the Θ_i just parameterise the direction of the goldstino in the S, T_i field space.

Using the general expressions of soft terms in SUGRA models, it is now straightforward to compute the corresponding soft SUSY-breaking terms, which are given by

$$m_\alpha^2 = m_{3/2}^2\left[1 - 3\cos^2\theta(\hat{K}_{\overline{i}i})^{-1/2}\Theta_i e^{i\gamma_i}(\log\tilde{K}_\alpha)_{\overline{i}j}(\hat{K}_{\overline{j}j})^{-1/2}\Theta_j e^{-i\gamma_j}\right], \tag{18.86}$$

$$A_{\alpha\beta\gamma} = -\sqrt{3}m_{3/2}\left[e^{-i\gamma_S}\sin\theta - e^{-i\gamma_i}\cos\theta\Theta_i(\hat{K}_{\overline{i}i})^{-1/2}\right.$$
$$\times \left.\left(\hat{K}_i - \sum_{\delta=\alpha,\beta,\gamma}(\log\tilde{K}_\delta)_i + (\log Y_{\alpha\beta\gamma})_i\right)\right]. \tag{18.87}$$

Here the trilinear couplings are factorised as usual ($Y_{\alpha\beta\gamma}^{A} = Y_{\alpha\beta\gamma}A_{\alpha\beta\gamma}$). In addition, the gaugino masses M_a for the canonically normalised gaugino fields are given by

$$M \equiv M_a = m_{3/2}\sqrt{3}\sin\theta e^{-i\gamma_S}. \tag{18.88}$$

18.4.1 Dilaton-dominated SUSY breaking

It is worth considering the limit $\cos\theta = 0$, which corresponds to the case where the dilaton dominated the SUSY breaking. In this case, the soft SUSY breaking terms take the following simple form:

$$m_\alpha = m_{3/2}, \tag{18.89}$$
$$M_a = \pm\sqrt{3}m_{3/2}, \tag{18.90}$$
$$A_{\alpha\beta\gamma} = -M_a. \tag{18.91}$$

Here the CP phases are set to zero (to avoid the electric dipole moment bounds). These soft terms are generated at the string scale, therefore we have to run them to the EW scale and impose the EWSB conditions to analyse the corresponding spectrum. In this case the EW breaking condition reads

$$\mu^2 + \frac{M_Z^2}{2} \simeq -0.1m_0^2 + 2m_{1/2}^2 = 5.9\ m_{3/2}^2, \tag{18.92}$$

which can be satisfied easily at the TeV scale. The above set of soft terms is a special case of cMSSM. Therefore, the Higgs mass limit requires $m_{3/2}$ to be of order 1.5 TeV, as shown in Fig. 18.3, where the Higgs mass m_h is plotted versus the gravitino mass $m_{3/2}$.

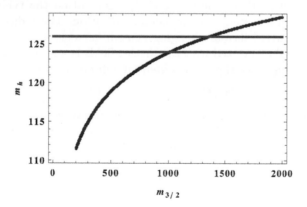

Figure 18.3 Higgs mass m_h as a function of the gravitino mass $m_{3/2}$. The area between the red (horizontal) lines shows the region for which the Higgs mass lies between 124-126 GeV.

With such heavy values of m_0 and $m_{1/2}$, the SUSY spectrum will be quite heavy and the relic abundance of the lightest neutralino will not be consistent with the latest results observed by Planck satellite.

18.4.2 Orbifold compactification

Now we consider the scenario with a single dominant modulus T (a detailed phenomenological analysis of this class of models can be found in [466–472]. In the orbifold compactification scheme, the Kähler potential is given by

$$K(S, S^*, T, T^*, C_i, C_i^*) = -\ln(S + S^*) - 3\ln(T + T^*) + \sum_i |C_i|^2 (T + T^*)^{n_i}. \quad (18.93)$$

Here, n_i are integers called 'modular weights' of the matter fields C_i. Also, the superpotential depends on S and T through the Yukawa couplings, $Y_{ijk}(T)$, the non-perturbative contributions $\hat{W}(S, T)$ and $\mu_(S, T)$. Finally, the tree-level gauge kinetic function is given by

$$f_a = k_a S, \quad (18.94)$$

where k_a is the Kac-Moody level of the gauge factor. Usually one takes $k_3 = k_2 = \frac{3}{5}k_1 = 1$. In this case, one obtains the following soft SUSY breaking terms:

$$m_i^2 = m_{3/2}^2(1 + n_i \cos^2\theta), \quad (18.95)$$

$$M_a = \sqrt{3}m_{3/2}\sin\theta e^{-i\alpha_S}, \quad (18.96)$$

$$A_{ijk} = -\sqrt{3}m_{3/2}\sin\theta e^{-i\alpha_S} - m_{3/2}\cos\theta(3 + n_i + n_j + n_k)e^{-i\alpha_T}, \quad (18.97)$$

where n_i, n_j and n_k are modular weights of fields in the corresponding Yukawa coupling Y_{ijk}. As an example for the modular weights $n_{Q_L} = n_{D_R} = -1$, $n_{u_R} = -2$, $n_{L_L} = n_{E_R} = -3$, $n_{H_1} = -2, n_{H_2} = -3$, one obtains the following non-universal scalar masses:

$$m_{Q_L}^2 = m_{D_R}^2 = m_{3/2}^2(1 - \cos^2\theta), \quad (18.98)$$

$$m_{U_R}^2 = m_{H_1}^2 = m_{3/2}^2(1 - 2\cos^2\theta), \quad (18.99)$$

$$m_{L_L}^2 = m_{E_R}^2 = m_{H_2}^2 = m_{3/2}^2(1 - 3\cos^2\theta). \quad (18.100)$$

It is clear that in order to avoid dangerous charge and colour breaking minima, the scalar squared-masses should be positive (i.e., no tachyons). The goldstino angle is thus constrained as

$$\cos^2\theta < \frac{1}{3}. \quad (18.101)$$

Furthermore, if Y_{ijk} depends on T, there appears to be another contribution to A_{ijk}. However, we do not consider such a case. In order to avoid any conflict with the experimental results on FCNC processes, we assume that the modular weights are family independent. Under this assumption, the A-parameters are given by

$$A_u = -\sqrt{3}m_{3/2}\sin\theta + 3m_{3/2}\cos\theta, \quad (18.102)$$

$$A_d = -\sqrt{3}m_{3/2}\sin\theta + m_{3/2}\cos\theta, \quad (18.103)$$

$$A_l = -\sqrt{3}m_{3/2}\sin\theta + 5m_{3/2}\cos\theta, \quad (18.104)$$

where we set the CP phases α_S and α_T equal to zero. This is an example of possible non-universal soft SUSY breaking terms that are controlled only with two parameters $m_{3/2}$ and $\cos\theta$.

In Fig. 18.4 we display the constraints from the observed limits of the Higgs mass and the relic abundance of stable neutralino LSP on the plane $(m_{3/2}, \cos\theta)$, where $\tan\beta$ varies

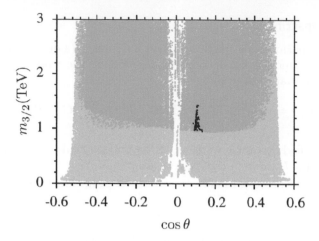

Figure 18.4 Constraints on the $(m_{3/2}, \cos\theta)$ plane in the discussed string inspired model by the Higgs mass limit and the relic abundance of the neutralino LSP. Green points correspond to Higgs boson mass within the experimental limits. Red points satisfy the relic abundance constraints. Gray points are consistent with the EWSB, but they are excluded by the LHC constraints.

from 5 to 50. Here we used the ISAJET package [473] to compute the low scale spectrum associated with the non-universal soft terms of our string inspired model at the GUT scale and also the complete relic abundance of the lightest neutralino. As can be seen from this figure, the Higgs mass limits impose a lower bound on the gravitino mass, $m_{3/2} \gtrsim 1$ TeV, and keep the parameter θ unconstrained. However, the relic abundance imposed very stringent constraints on both parameters: $m_{3/2}$ and θ. It implies that $\cos\theta \sim \mathcal{O}(0.1$ and 1 TeV$) \lesssim m_{3/2} \lesssim 1.5$ TeV. In Fig. 18.5 we show a scatter plot for the Higgs mass and lightest neutralino versus the parameter $\cos\theta$ for gravitino mass between 100 GeV and 3 TeV and $\tan\beta$ varies from 5 to 50. The Higgs mass plot shows that the relic abundance of the lightest neutralino can be satisfied by several points of parameter space but with a Higgs mass less than the observed limits. Imposing the Higgs mass bound rejects most of these points and only a few points may survive.

18.5 SUSY BREAKING AND FLUX COMPACTIFICATION

In Type IIB theory, strings can have RR and NS-NS antisymmetric 3-form field strengths (H_3 and F_3 respectively) which can wrap 3-cycles of the compactification manifold labeled by P and Q, leading to the following background fluxes:

$$\frac{1}{4\pi^2\alpha'}\int_P F_3 = L, \qquad \frac{1}{4\pi^2\alpha'}\int_Q H_3 = -K, \qquad (18.105)$$

where K and L are integers. In the effective 4D SUGRA, these geometric fluxes generate a superpotential for the CY moduli, which is of the form [474]

$$W = \int_M G_3 \wedge \Omega, \qquad (18.106)$$

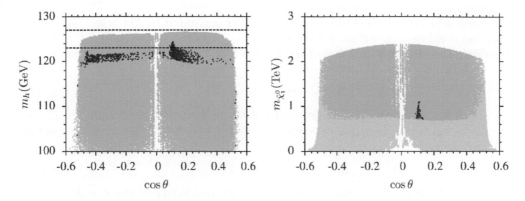

Figure 18.5 Higgs boson mass (left) and lightest neutralino mass (right) as function of $\cos\theta$ for $m_{3/2} \in [0.1, 3]$ TeV. The colour codes in the right-panel are as in Fig. 18.4.

where Ω is the holomorphic 3-form which depends on the CSM Z_i and $G_3 = F_3 - iSH_3$. The axion-dilaton field S and the overall scale modulus T are defined in Type IIB by:

$$S = \frac{e^{-\phi}}{2\pi} + ic_0, \quad T = \frac{e^{-\phi}}{2\pi}(M_{st}R)^4 + ic_4, \qquad (18.107)$$

where c_0 and c_4 are the axions from the RR 0-form and 4-form, respectively, $e^{-\phi} = g_{st}$ denotes the string coupling, $1/M_{st}^2$ is the string tension and R is the compactification radius of the CY volume $V_{CY} \equiv (2\pi R)^6$. In what follows, large T approximation will be used in the Kähle form. As can be seen from the above equation, large T corresponds to large compactification radius, which means its energy scale is low.

GKP showed that these 3-form fluxes can generically stabilise the dilaton S and all the CMS moduli Z_i [475]. However, since the Kähle modulus T does not appear in the potential, it cannot be fixed by the geometric fluxes and the potential is of no-scale type. This partial fixing of moduli in GKP framework can be understood by considering the following tree-level Kähle potential:

$$K = -3\log(T + \overline{T}) - \log(S + \overline{S}) - \log[-i\int_M \Omega \wedge \overline{\Omega}]. \qquad (18.108)$$

The superpotential (18.106) and the Kähle potential (18.108) lead to the following F-term potential:

$$V_F = e^K\left(G^{I\bar{J}}D_IW\overline{D_JW} - 3|W|^2\right) = e^K(G^{i\bar{j}}D_iW\overline{D_jW}), \qquad (18.109)$$

where I and J run over all moduli while i and j run over dilaton moduli and CSM only. The covariant derivative is defined as $D_IW = \partial_IW + (\partial_IK)W$. Here $G^{I\bar{J}} = G_{I\bar{J}}^{-1}$ and $G_{I\bar{J}}$ is given by $G_{I\bar{J}} = K_{I\bar{J}} = \partial_I\partial_{\bar{J}}K$. Note that $K^{T\bar{T}}|D_TW|^2$ cancels with $3|W|^2$, leaving the potential V_F independent of T. As can be seen from Eq.(18.109), the potential V_F is positive definite, so that its global minimum is at zero and hence the dilaton moduli and CSM are fixed by the condition $D_iW = 0$. This minimum is non-supersymmetric due to the fact that $F_T \propto D_TW \propto W \neq 0$, $i.e.$, SUSY is broken by the T field.

In order to fix T, a non-perturbative superpotential was considered by KKLT [476]. The source of these non-perturbative terms could be either generated by D3-brane instantons or by gaugino condensation within a hidden non-abelian gauge sector on the $D7$-branes. Since the dilaton moduli and CSM have been fixed at a high scale, their contribution to the superpotential is a constant W_0 and the total effective superpotential is given by

$$W = W_0 + Ae^{-aT}, \tag{18.110}$$

where A and a are constants ($1/a$ is proportional to the beta function coefficient of the gauge group in which the condensate occurs). The Kähle potential after integrating out S and Z_i is reduced to

$$K = -3\log(T + \overline{T}). \tag{18.111}$$

Here, $T = \tau + i\psi$, with τ is the volume modulus of the internal manifold and ψ is the axionic part. A supersymmetric minimum is obtained by solving the equation

$$D_T W = 0 \quad \Rightarrow \quad W_0 = -Ae^{-a\tau_0}(1 + \frac{2}{3}a\tau_0), \tag{18.112}$$

where τ_0 is the value of τ that minimizes the scalar potential. Substituting this solution in

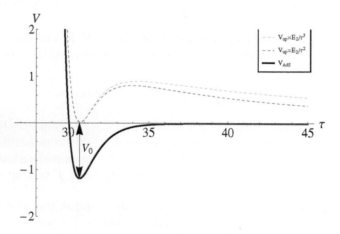

Figure 18.6 The scalar potential $V(\tau)$ (multiplied by 10^{29}) with $W_0 = -10^{-12}$, $A = 1$, $a = 1$. The blue curve shows the AdS minimum, while the green and red curves exhibit the uplifting to dS minimum via $\delta V = \frac{E_3}{\tau^3}, \frac{E_2}{\tau^2}$, respectively, with $E_3 = 3.5 \times 10^{-25}$ and $E_2 = 1.13 \times 10^{-26}$ (from [32]).

the potential

$$V = e^K \left(\frac{3}{(T + \overline{T})^2} |D_T W|^2 - 3|W|^2 \right), \tag{18.113}$$

one finds the following negative minimum

$$V_0^{AdS} = \left(-3e^K |W|^2 \right)|_{\tau_0} = -\frac{a^2 A^2 e^{-2a\tau_0}}{6\tau_0}. \tag{18.114}$$

The scalar potential as a function of $\tau = \text{Re}(T)$ is given by[1]

$$V(\tau) = \frac{aAe^{-2a\tau}\left(A(a\tau + 3) + 3W_0 e^{a\tau}\right)}{6\tau^2}. \tag{18.115}$$

---

[1]The imaginary part $\text{Im}(T) = \psi$ is frozen at zero.

It is important to uplift this AdS minimum to a Minkowski or a dS minimum in order to have realistic models. The uplift of the above AdS vacuum to a dS one will break SUSY where one needs another contribution to the potential, which usually has dependence like τ^{-n} [477]

$$\delta V(\tau) \approx |V_0^{AdS}| \frac{\tau_0^n}{\tau^n}. \tag{18.116}$$

In this case a new minimum is obtained due to shifting τ_0 to $\tau_0' = \tau_0 + \varepsilon$, where ε is given by

$$\varepsilon \simeq \frac{1}{a^2 \tau_0}. \tag{18.117}$$

Since the consistency of the KKLT requires that $\tau_0, a\tau_0 \gg 1$ [476], the shift in the minima is much smaller and we can calculate physical quantities such as masses in terms of τ_0.

There are many proposals for such uplifting, *e.g.*, adding anti-D3-branes [476], D-term uplift [478–485], F-term uplift [486, 487] and Kähler uplift [482, 488–490]. In the original KKLT scenario [476], some anti-D3-branes were added, which contribute an additional part to the scalar potential:

$$\delta V = \frac{E_3}{\tau^3}. \tag{18.118}$$

One of the drawbacks of this mechanism is that SUSY is broken explicitly due to the addition of the anti-D3-branes. In this case, the effective 4D theory cannot be recast into the standard form of 4D SUGRA and this in turn makes it very difficult to have a low energy effective theory [478]. A possible solution to overcoming this problem has been proposed in [478], where the authors used the D-term induced by the gauge fluxes on $D7$ branes. It turns out that if the matter fields charged under the gauge group on $D7$ branes are minimised at zero vevs, then the D-term gives the same contribution to the scalar potential as $\overline{D3}$. However, it is important to note that after including the non-perturbative gaugino condensation in the second step of the KKLT approach, SUSY is restored and the effective theory is fully supersymmetric. Therefore, the potential V_D is always minimised at zero, as pointed out by several authors [491], and the D-term cannot be used for uplifting the AdS SUSY vacua.

SUSY breaking in models of KKLT compactification type with phenomenological consequences has been extensively studied in [477, 481, 491, 492]. As shown in models of KKLT type, SUSY is broken by one of the uplifting mechanisms. The gravitino mass at the dS minimum is given by

$$m_{3/2} = e^{K/2} |W| \Big|_{dS} \Rightarrow m_{3/2} \simeq \frac{a\,A}{3(2\tau_0)^{1/2}} e^{-a\tau_0} \simeq \frac{W_0}{(2\tau_0)^{3/2}}. \tag{18.119}$$

Therefore, we have gravitino mass of order TeV if $(a\tau_0) \sim 32$ and, hence, $W_0 \sim 10^{-12}$. In terms of the shifts ε (18.117) of τ_0, an approximate expression for $D_T W$ near the dS minimum is given by

$$D_T W(\tau_0 + \varepsilon) = (D_T W)_\tau \varepsilon \simeq W_{T,T}\, \varepsilon = \frac{3\sqrt{2}}{a\sqrt{\tau_0}}\, m_{3/2}, \tag{18.120}$$

which is of the same order as the gravitino mass. Accordingly, the soft SUSY breaking terms are given by

$$
\begin{aligned}
m_0^2 &= \left.\frac{|W|^2}{(2\tau)^3}\right|_{dS} = m_{3/2}^2, \\
m_{1/2} &= \left.\frac{\sqrt{2\tau}}{6}D_T W(T)\frac{\partial}{\partial T}\ln(\mathrm{Re}f^*)\right|_{dS} \simeq \frac{m_{3/2}}{a\tau}, \\
A_0 &= \left.-\frac{1}{\sqrt{2\tau}}\bar{D}_{\bar{T}}\overline{W}_h\right|_{dS} = -\frac{3m_{3/2}}{a\tau},
\end{aligned}
\tag{18.121}
$$

where the gauge kinetic function f_{ab} can be chosen such that $f_{ab}(T) = f(T)\delta_{ab}$, which will lead to universal gaugino masses and is considered to have a linear dependance on the modulus field that can be derived from the reduction of Dirac-Born-Infeld action for an unmagnetised brane [493].

The soft terms indicate that SUSY breaking in KKLT is a special example of the cMSSM, where all the soft terms are given in terms of one free parameter ($m_{3/2}$) which is of order TeV. Note that $a\tau_0$ is fixed as $a\tau_0 \simeq 32$. It is well known that this type of soft term cannot account for the experimental constraints imposed by the LHC and relic abundance of the lightest SUSY particle. Even if we relax the dark matter constraints, the mass limit (~ 125 GeV) and gluino mass limit ($\gtrsim 1.4$ TeV) will imply $m_{3/2} \simeq \mathcal{O}(30\text{ TeV})$. Thus, all SUSY spectra will be quite heavy, which is beyond the LHC sensitivity. A feature of this model is the fact that gauginos are lighter than the sfermions by at least one order of magnitude. However, if one checks the parameter space for such a set of soft terms, one finds that tadpole equations at the TeV scale are not satisfied. Namely, the condition [162]

$$
\mu^2 + \frac{M_Z^2}{2} \simeq -0.1m_0^2 + 2m_{1/2}^2,
\tag{18.122}
$$

cannot account for positive μ^2. Therefore, the above set of soft terms failed to describe TeV scale phenomenology [32].

18.6 SUSY BREAKING IN A LARGE VOLUME SCENARIO

Another alternative scenario for moduli stabilisation based on a large volume scenario has been proposed by Quevedo, et al. [494, 495] in order to overcome some of the drawbacks of the KKLT model. Basically, increasing the number of Kähler moduli will worsen the situation when α' corrections are neglected. The LVS was built on the proposal that the number of complex structure moduli is bigger than the number of Kähler moduli, i.e., $h^{2,1} > h^{1,1} > 1$ as well as the inclusion of α' corrections. The $\mathcal{O}(\alpha'^3)$ contribution to the Kähler potential (after integrating out the dilaton and the complex structure moduli) is given by [496–498],

$$
K_{\alpha'} = -2\log\left[e^{-3\phi/2}\mathcal{V} + \frac{\xi}{2}\left(\frac{(S+\bar{S})}{2}\right)^{3/2}\right] + K_{cs},
\tag{18.123}
$$

with $\xi = -\frac{\zeta(3)\chi(\mathcal{M})}{2(2\pi)^3}$ and ϕ is the Type-IIB dilaton. Here $\mathcal{V}$ is defined as the classical volume of the manifold $\mathcal{M}$, which is given by

$$
\mathcal{V} = \int_{\mathcal{M}} J^3 = \frac{1}{6}\kappa_{ijk}t^i t^j t^k,
\tag{18.124}
$$

where J is the Kähler class and t_i are moduli that measure the areas of 2-cycles with $i = 1, 2, \ldots, h_{1,1}$. The complexified Kähler moduli are defined as $T_j \equiv \tau_j + i\psi_j$, while τ_j are the four-cycle moduli defined by the relation

$$\tau_j = \partial_{t_j} \mathcal{V} = \frac{1}{2} \kappa_{jkl} t^k t^l. \tag{18.125}$$

Since the superpotential is not renormalised at any order in perturbation theory, it will not receive α' corrections. But there is a possibility of non-perturbative corrections, which may depend on the Kähler moduli (as in the KKLT model) via D3-brane instantons or gaugino condensation from wrapped D7-branes. Accordingly, the superpotential is given by

$$W = W_0 + \sum_i A_i e^{-a_i T_i}, \tag{18.126}$$

where A_i is a model dependent constant and again W_0 is the value of the superpotential due to the geometric flux after stabilising the dilaton and the complex structure moduli. In this respect, the scalar potential will take the form [496, 498]

$$
\begin{aligned}
V &= e^K \left[K^{T_j \bar{T}_k} \left(a_j A_j a_k \bar{A}_k e^{(a_j T_j + a_k \bar{T}_k)} + a_j A_j e^{i a_j T_j} \bar{W} \partial_{\bar{T}_k} K - a_k \bar{A}_k e^{-i a_k \bar{T}_k} W \partial_{T_j} K \right) \right. \\
&\quad \left. + 3\xi \frac{(\xi^2 + 7\xi\mathcal{V} + \mathcal{V}^2)}{(\mathcal{V} - \xi)(2\mathcal{V} + \xi)^2} |W|^2 \right] \\
&\equiv V_{np1} + V_{np2} + V_{\alpha'},
\end{aligned}
\tag{18.127}
$$

where the α' correction is encoded in $V_{\alpha'}$. The simplest example that can realise the notion of LVS [494, 495] is the orientifold of $P^4_{[1,1,1,6,9]}$ for which $h^{1,1} = 2$ and $h^{2,1} = 272$ and therefore the volume is given by

$$\mathcal{V} = \frac{1}{9\sqrt{2}} \left(\tau_5^{\frac{3}{2}} - \tau_4^{\frac{3}{2}} \right), \tag{18.128}$$

where the volume moduli are $T_4 = \tau_4 + i\psi_4$ and $T_4 = \tau_5 + i\psi_5$ and the link to t_i is given by $\tau_4 = \frac{t_1^2}{2}$ and $\tau_5 = \frac{(t_1 + 6t_5)^2}{2}$. In this respect, the Kähler potential and the superpotential, after fixing the dilaton and complex structure moduli, are given in the string frame by

$$
\begin{aligned}
K &= K_{cs} - 2\log\left(\mathcal{V} + \frac{\xi}{2}\right), \\
W &= W_0 + A_4 e^{-\frac{a_4}{g_s} T_4} + A_5 e^{-\frac{a_5}{g_s} T_5}.
\end{aligned}
\tag{18.129}
$$

In the large volume limit, $\mathcal{V} \sim \tau_5 \gg \tau_4 > 1$, the behaviour of the scalar potential is given by [494, 495]

$$V(\mathcal{V}, \tau_4) = \frac{\sqrt{\tau_4}(a_4 A_4)^2 e^{-2a_4\tau_4/g_s}}{\mathcal{V}} - \frac{W_0 \tau_4 a_4 A_4 e^{-a_4\tau_4/g_s}}{\mathcal{V}^2} + \frac{\xi W_0^2}{\mathcal{V}^3}. \tag{18.130}$$

Minimising the potential (18.130),

$$\frac{\partial V}{\partial \mathcal{V}} = \frac{\partial V}{\partial \tau_4} = 0. \tag{18.131}$$

and solving the two equations in the two variables, $\tau_4, \mathcal{V}$, one can get one equation in τ_4

$$\left(1 \pm \sqrt{1 - \frac{3B_3 B_1}{B_2 \tau_4^{\frac{3}{2}}}}\right) \left(\frac{1}{2} - 2a_4\tau_4\right) = (1 - a_4\tau_4), \tag{18.132}$$

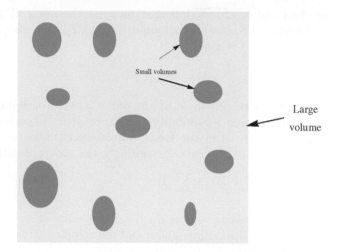

Figure 18.7 Swiss cheese structure in the large volume scenario.

where

$$B_1 \sim a_4^2 |A_4|^2, \qquad B_2 \sim a_4 |A_4 W_0|, \qquad B_3 \sim \xi |W_0|^2. \qquad (18.133)$$

Using the assumption $a_4 \tau_4 \gg 1$, which is necessary to neglect higher order instanton corrections [494], the solution is given by

$$\tau_4 = \left(\frac{4 B_3 B_1}{B_2^2} \right)^{\frac{2}{3}},$$

$$\mathcal{V} = \frac{B_2}{2 B_1} \left(\frac{4 B_3 B_1}{B_2^2} \right)^{\frac{1}{3}} e^{a_4 \left(\frac{4 B_3 B_1}{B_2^2} \right)^{\frac{2}{3}}}. \qquad (18.134)$$

Substituting for B_i by their expressions, we have

$$\tau_4 \sim (4\xi)^{\frac{2}{3}}, \qquad \mathcal{V} \sim \frac{\xi^{\frac{1}{3}} |W_0|}{a_4 A_4} e^{a_4 \tau_4 / g_s}. \qquad (18.135)$$

Therefore, potential possesses an AdS minimum at exponentially large volume $\mathcal{V}$ since it approaches zero from below in the limit $\tau_5 \to \infty$ and $\tau_4 \propto \log(\mathcal{V})$. Namely, in the latter limit, the potential has the form

$$V = \frac{W_0^2}{\mathcal{V}^3} \left(C_1 \sqrt{\ln(\mathcal{V})} - C_2 \ln(\mathcal{V}) + \xi \mathcal{V} \right). \qquad (18.136)$$

Therefore, the negativity of the potential require a very large $\ln(\mathcal{V})$. Still one has to uplift this minimum by one of the mechanisms mentioned in Sect. (18.5). This result can be generalised to more than two moduli where one of them takes a large limit while the other moduli stays small. This structure will form what is called the *Swiss-cheese* form of the CY manifold as depicted in Fig. 18.7. In this respect the volume will take the form

$$\mathcal{V} = \tau_b^{3/2} - \sum_i \tau_{s,i}^{3/2}. \qquad (18.137)$$

SUSY breaking in the LVS type has been studied in [493, 495, 499–502]. In the LVS model based on the $P^4_{[1,1,1,6,9]}$ geometry, one finds

$$K = -2 \log \left(\mathcal{V} + \frac{\xi}{2} \right),$$
$$W = W_0 + A_4 e^{-\frac{a_4}{g_s} T_4} + A_5 e^{-\frac{a_5}{g_s} T_5}. \tag{18.138}$$

Therefore, the gravitino mass will be given by

$$m_{3/2} = e^{K/2} |W| \Big|_{dS} \sim \frac{g_s^2 |W_0|}{\mathcal{V}\sqrt{4\pi}} M_p. \tag{18.139}$$

It is remarkable that depending on the large volume $\mathcal{V}$, the gravitino mass could be TeV or much larger. Considering the Kähler metric of the observable sector to have the form $\tilde{K}_{\alpha\bar{\beta}} = \tilde{K}_\alpha \delta_{\alpha\bar{\beta}}$, with $\tilde{K}_\alpha$ as constant, we have scalar soft masses of the following form:

$$m_\alpha^2 = m_{3/2}^2, \tag{18.140}$$

where we have neglected the very tiny cosmological constant value V_0. In the large volume limit, $\mathcal{V} \sim \tau_5 \equiv \tau_b \gg \tau_4 \equiv \tau_s \gg 1$, the Kähler metric and its inverse are given by

$$K_{m\bar{n}} \simeq \begin{bmatrix} \frac{1}{\mathcal{V}} & -\frac{1}{\mathcal{V}^{5/3}} \\ -\frac{1}{\mathcal{V}^{5/3}} & \frac{1}{\mathcal{V}^{4/3}} \end{bmatrix}, \qquad K^{m\bar{n}} \simeq \begin{bmatrix} \mathcal{V} & \mathcal{V}^{2/3} \\ \mathcal{V}^{2/3} & \mathcal{V}^{4/3} \end{bmatrix}. \tag{18.141}$$

The F-terms can be calculated from

$$F^m = e^{K/2} K^{m\bar{n}} D_{\bar{n}} W. \tag{18.142}$$

Hence the approximate dependence of the F-terms on the volume is given by

$$F^4 \sim \frac{1}{\mathcal{V}}, \qquad F^5 \sim \frac{1}{\mathcal{V}^{1/3}}. \tag{18.143}$$

Consider a linear dependence of the gauge kinetic function on the Kähler moduli [493], therefore the gaugino masses are given by

$$m_{1/2} \sim \frac{g_s^2 W_0 M_p}{\sqrt{4\pi}} \left(\frac{1}{\mathcal{V}} + \frac{1}{\mathcal{V}^{5/3}} \right). \tag{18.144}$$

Also, the universal A-term is given by

$$A_0 \simeq F^m K_m \sim \frac{-g_s^2 W_0 M_p}{\sqrt{4\pi}\mathcal{V}}. \tag{18.145}$$

Accordingly, at large volumes $\mathcal{V} \sim 10^{13} - 10^{15}$ the soft masses are related to the gravitino mass as follows

$$\begin{aligned} m_0 &\simeq m_{1/2} = m_{3/2}, \\ A_0 &\simeq -m_{3/2}. \end{aligned} \tag{18.146}$$

These soft terms are generated at string scale, therefore we have to run them to the EW scale and impose the EW symmetry breaking conditions to analyse the corresponding spectrum. In this case the condition (18.122) reads

$$\mu^2 + \frac{M_Z^2}{2} \simeq -0.1 m_{3/2}^2 + 2 m_{3/2}^2, \tag{18.147}$$

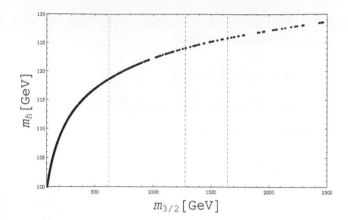

Figure 18.8 Higgs mass m_h as a function of the gravitino mass $m_{3/2}$. The region at left of the green line is disallowed by the gluino mass constraint, while the area between the red lines shows the region for which the Higgs mass lies between 124-126 GeV (from [32]).

which can be satisfied easily at the TeV scale. The above set of soft terms is a special case of the cMSSM. Therefore, the Higgs mass limit requires $m_{1/2}$ to be of order 1.5 TeV, as shown in Fig. 18.8, where the Higgs mass m_h is plotted versus the gravitino mass $m_{3/2}$ [32].

With such heavy values of m_0 and $m_{1/2}$, the SUSY spectrum will be quite heavy and the relic abundance of the lightest neutralino is not consistent with Planck's results [12].

Epilogue

What has been the discovery of the cornerstone of the SM, the Higgs boson, should now turn out to be the stepping stone into a new physics world. The latter is currently unknown to us. We know what its components ought to be: massive neutrinos, a viable DM candidate, significant CPV (above and beyond what is provided by the CKM matrix of the SM), to name but a few. We also know what its dynamics should prevent, *i.e.*, the hierarchy problem. There is one clear pathway to follow in the quest for BSM physics which would at once reconcile the above experimental and theoretical instances. This is SUSY. The search for it at the LHC is ongoing, as it has been for 40 years now, including at innumerable past facilities. However, nil results have been obtained so far. We should not despair: it took longer for the Higgs boson to be discovered! We should therefore be patient. But we should not refrain from giving ourselves all the means needed to enable us finding SUSY.

Unfortunately, we have not done so yet. We have confined ourselves to primarily testing the SUSY hypothesis in its minimal form. L.B. Slobodkin wrote (*The Role of Minimalism in Art and Science, Am. Nat.* 1986, Vol. 127, pp 257–265): "Minimalism, for my present purpose, is a process of deliberately choosing to work in the simplest possible mode that is still recognizable as part of an existing professional field. Its purpose is to encourage concern with, and comment on, the state of the field. There is usually an implication of rebellion against what the minimalist sees as uncritical acceptance of standards". It is indeed time to rebel against minimalistic assumptions about SUSY. We have tried doing so in this book. We hope to have carried the reader along and converted her or him to our cause. The reward would be tantalising. From the simplest theory of all, that matter and forces are one and the same (think here of Yves Klein monochrome paintings, supreme examples of minimalist art, Fig. E.1, left), an explanation of our cosmos (as seen by the Hubble space telescope, Fig. E.1, right) would emerge. Nothing less.

Figure E.1 Left: Yves Klein, 'Blue Monochrome' (1961). Right: Hubble Space Telescope image.

Two-Dimensional Component Spinors

WEYL SPINORS

We define a Weyl spinor as a two-component complex spinor $\psi = \begin{pmatrix} \psi_1 \\ \psi_2 \end{pmatrix}$ that transforms under $M \in SL(2,C)$, $i.e.$, $\det M = 1$, as follows:

$$\psi_\alpha \rightarrow \psi'_\alpha = M_\alpha^{\ \beta} \psi_\beta, \tag{A.1}$$

where $\alpha, \beta = 1, 2$. Unlike $SU(2)$, for $SL(2,C)$ a representation and its complex conjugate are not equivalent. Therefore, a two-component antispinor $\bar{\psi}$ is defined as

$$\bar{\psi}^{\dot\alpha} \rightarrow \bar{\psi}'^{\dot\alpha} = (M^*)^{\dot\alpha}_{\ \dot\beta} \bar{\psi}^{\dot\beta}. \tag{A.2}$$

The undotted spinors ψ_α transform under the $(\frac{1}{2}, 0)$ representation of the Lorentz group and are called *left-handed* Weyl spinors, whereas the dotted spinors $\bar{\psi}_{\dot\alpha}$ transform under the $(0, \frac{1}{2})$ representation of the Lorentz group and are called *right-handed* Weyl spinors.

The antisymmetric two index tensors

$$\varepsilon^{\alpha\beta} = \varepsilon^{\dot\alpha\dot\beta} = \begin{pmatrix} 0 & 1 \\ -1 & 0 \end{pmatrix}, \qquad \varepsilon_{\alpha\beta} = \varepsilon_{\dot\alpha\dot\beta} = \begin{pmatrix} 0 & -1 \\ 1 & 0 \end{pmatrix} \tag{A.3}$$

are used to raise and lower indices as follows:

$$\begin{aligned}
\psi^\alpha &= \varepsilon^{\alpha\beta} \psi_\beta, & \psi_\alpha &= \varepsilon_{\alpha\beta} \psi^\beta, \tag{A.4} \\
\bar{\psi}^{\dot\alpha} &= \varepsilon^{\dot\alpha\dot\beta} \bar{\psi}_{\dot\beta}, & \bar{\psi}_{\dot\alpha} &= \varepsilon_{\dot\alpha\dot\beta} \bar{\psi}^{\dot\beta}. \tag{A.5}
\end{aligned}$$

Note that $\varepsilon_{\alpha\beta}\varepsilon^{\beta\gamma} = \delta_\alpha^{\ \gamma}$ and $\varepsilon_{\dot\alpha\dot\beta}\varepsilon^{\dot\beta\dot\gamma} = \delta_{\dot\alpha}^{\ \dot\gamma}$. Since these spinors anticommute, $\psi_1\chi_2 = -\chi_2\psi_1$ and $\psi_1\bar{\chi}_{\dot2} = -\bar{\chi}_{\dot2}\psi_1$, the scalar products $\psi\chi$ and $\bar{\psi}\bar{\chi}$ are defined as

$$\psi\chi = \psi^\alpha\chi_\alpha = \varepsilon^{\alpha\beta}\psi_\beta\chi_\alpha = -\varepsilon^{\alpha\beta}\chi_\alpha\psi_\beta = \varepsilon^{\beta\alpha}\chi_\alpha\psi^\beta = \chi^\beta\psi_\beta = \chi\psi, \tag{A.6}$$

$$\bar{\psi}\bar{\chi} = \bar{\psi}_{\dot\alpha}\bar{\chi}^{\dot\alpha} = \varepsilon_{\dot\alpha\dot\beta}\bar{\psi}^{\dot\beta}\bar{\chi}^{\dot\alpha} = -\varepsilon_{\dot\alpha\dot\beta}\bar{\chi}^{\dot\alpha}\bar{\psi}^{\dot\beta} = \varepsilon_{\dot\beta\dot\alpha}\bar{\chi}^{\dot\alpha}\bar{\psi}^{\dot\beta} = \bar{\chi}_{\dot\beta}\bar{\psi}^{\dot\beta} = \bar{\chi}\bar{\psi}, \tag{A.7}$$

$$(\chi\psi)^\dagger = (\chi^\alpha\psi_\alpha)^\dagger = \bar{\psi}_{\dot\alpha}\bar{\chi}^{\dot\alpha} = \bar{\psi}\bar{\chi} = \bar{\chi}\bar{\psi}. \tag{A.8}$$

The σ_μ matrices have a dotted and an undotted index: $(\sigma^\mu)_{\alpha\dot\alpha} = (1, \sigma_i)_{\alpha\dot\alpha}$. The ε-tensor may also be used to raise the indices of the σ-matrices:

$$(\bar\sigma^\mu)^{\dot\alpha\alpha} = \varepsilon^{\dot\alpha\dot\beta}\varepsilon^{\alpha\beta}(\sigma^\mu)_{\beta\dot\beta} = (1, -\sigma_i)^{\dot\alpha\alpha}. \tag{A.9}$$

The matrices σ^μ are hermitian, so they satisfy:

$$(\sigma^\mu_{\alpha\dot\alpha})^* = (\sigma^{\mu^*})_{\dot\alpha\alpha} = (\sigma^{\mu^\dagger})_{\alpha\dot\alpha} = (\sigma^\mu)_{\alpha\dot\alpha}. \tag{A.10}$$

Therefore, the vector representation under the Lorentz group is given by $(\frac{1}{2}, \frac{1}{2})$, which can be written in terms of Weyl spinors as $V^\mu = \psi^\alpha \sigma^\mu_{\alpha\dot\alpha} \bar\psi^{\dot\alpha}$.

DIRAC SPINORS

A four-component Dirac spinor is constructed from two-component Weyl spinors:

$$\Psi = \begin{pmatrix} \psi_\alpha \\ \bar\chi^{\dot\alpha} \end{pmatrix}. \tag{A.11}$$

It transforms as the reducible $(\frac{1}{2}, 0) \oplus (0, \frac{1}{2})$ representation of the Lorentz group. Also, one introduces the Dirac matrices in the following basis, which is called the Weyl basis representation:

$$\gamma^\mu = \begin{pmatrix} 0 & \sigma^\mu \\ \bar\sigma^\mu & 0 \end{pmatrix}, \qquad \gamma_5 = i\gamma^0\gamma^1\gamma^2\gamma^3 = \begin{pmatrix} -1 & 0 \\ 0 & 1 \end{pmatrix}. \tag{A.12}$$

Dirac spinor describes a particle with both chiralities. The chirality eigenstates are:

$$\Psi_L = \frac{1 - \gamma_5}{2}\Psi = \begin{pmatrix} \psi_\alpha \\ 0 \end{pmatrix}, \qquad \Psi_R = \frac{1 + \gamma_5}{2}\Psi = \begin{pmatrix} 0 \\ \bar\chi^{\dot\alpha} \end{pmatrix}. \tag{A.13}$$

MAJORANA SPINORS

The charge conjugated spinor Ψ^c is defined as

$$\Psi^c = C\bar\Psi^T, \tag{A.14}$$

where the charge conjugation matrix C can be written as $C = i\gamma^0\gamma^2$, and Ψ^c describes the antiparticle of a given particle, with opposite internal charge. In terms of two components, Ψ^c is given by

$$\Psi^c = \begin{pmatrix} \chi_\alpha \\ \bar\psi^{\dot\alpha} \end{pmatrix}. \tag{A.15}$$

A Majorana spinor is a Dirac spinor with $\chi \equiv \psi$, i.e., it is of the form

$$\Psi_M = \begin{pmatrix} \psi_\alpha \\ \bar\psi^{\dot\alpha} \end{pmatrix}. \tag{A.16}$$

Therefore, a Majorana spinor can be defined as the four component spinor that is equal to its charge conjugated spinor: $\Psi_M^c = \Psi_M$.

Renormalisation Group Equations In Complete Matrix Form

In this appendix we provide the complete list of the RGEs for the soft SUSY breaking terms. We present it in matrix form along the lines of [208]. We also use the same notation where the scale variable is defined as $t = \log(M_{\text{GUT}}/Q^2)$. In addition, $\tilde{\alpha} = \alpha/4\pi$ and $\tilde{Y} = Y/4\pi$ are adopted. Let us start with the RGEs of gauge coupling and gaugino masses:

$$\frac{d\tilde{\alpha}_i}{dt} = -b_i \tilde{\alpha}_i^2, \tag{B.1}$$

$$\frac{dM_i}{dt} = -b_i \tilde{\alpha}_i M_i, \tag{B.2}$$

$$\tag{B.3}$$

where $b_1 = 11, b_2 = 1$, and $b_3 = -3$. The RGEs of Yukawa couplings are given by

$$2\frac{d\tilde{Y}_U}{dt} = \left(\frac{16}{3}\tilde{\alpha}_3 + 3\tilde{\alpha}_2 + \frac{13}{9}\tilde{\alpha}_1\right)\tilde{Y}_U$$
$$- \left[3\tilde{Y}_U\tilde{Y}_U^\dagger\tilde{Y}_U + 3\text{Tr}(\tilde{Y}_U\tilde{Y}_U^\dagger)\tilde{Y}_U\right] - \tilde{Y}_D\tilde{Y}_D^\dagger\tilde{Y}_U, \tag{B.4}$$

$$2\frac{d\tilde{Y}_D}{dt} = \left(\frac{16}{3}\tilde{\alpha}_3 + 3\tilde{\alpha}_2 + \frac{7}{9}\tilde{\alpha}_1\right)\tilde{Y}_D$$
$$- \left[3\tilde{Y}_D\tilde{Y}_D^\dagger\tilde{Y}_D + 3\text{Tr}(\tilde{Y}_D\tilde{Y}_D^\dagger)\tilde{Y}_D\right] - \tilde{Y}_U\tilde{Y}_U^\dagger\tilde{Y}_D - \text{Tr}(\tilde{Y}_E\tilde{Y}_E^\dagger)\tilde{Y}_D, \tag{B.5}$$

$$2\frac{d\tilde{Y}_E}{dt} = (3\tilde{\alpha}_2 + 3\tilde{\alpha}_1)\tilde{Y}_E$$
$$- \left[3\tilde{Y}_E\tilde{Y}_E^\dagger\tilde{Y}_E + \text{Tr}(\tilde{Y}_E\tilde{Y}_E^\dagger)\tilde{Y}_E\right] - 3\text{Tr}(\tilde{Y}_D\tilde{Y}_D^\dagger)\tilde{Y}_E. \tag{B.6}$$

The RGEs of the trilinear soft breaking parameters, $\tilde{Y}_F^A = (\tilde{Y}_F)_{ij}(A_F)_{ij}$ (not a matrix multiplication), are given by

$$
\begin{aligned}
2\frac{d\tilde{Y}_U^A}{dt} &= \left(\frac{16}{3}\tilde{\alpha}_3 + 3\tilde{\alpha}_2 + \frac{13}{9}\tilde{\alpha}_1\right)\tilde{Y}_U^A - 2\left(\frac{16}{3}\tilde{\alpha}_3 M_3 + 3\tilde{\alpha}_2 M_2 + \frac{13}{9}\tilde{\alpha}_1 M_1\right)\tilde{Y}_U \\
&- \left\{\left[4(\tilde{Y}_U^A\tilde{Y}_U^\dagger\tilde{Y}_U) + 6\mathrm{Tr}(\tilde{Y}_U^A\tilde{Y}_U^\dagger)\tilde{Y}_U\right] + \left[5(\tilde{Y}_U\tilde{Y}_U^\dagger\tilde{Y}_U^A) + 3\mathrm{Tr}(\tilde{Y}_U\tilde{Y}_U^\dagger)\tilde{Y}_U^A\right]\right. \\
&+ \left. 2(\tilde{Y}_D^A\tilde{Y}_D^\dagger\tilde{Y}_U) + (\tilde{Y}_D\tilde{Y}_D^\dagger\tilde{Y}_U^A)\right\},
\end{aligned}
\tag{B.7}
$$

$$
\begin{aligned}
2\frac{d\tilde{Y}_D^A}{dt} &= \left(\frac{16}{3}\tilde{\alpha}_3 + 3\tilde{\alpha}_2 + \frac{7}{9}\tilde{\alpha}_1\right)\tilde{Y}_D^A - 2\left(\frac{16}{3}\tilde{\alpha}_3 M_3 + 3\tilde{\alpha}_2 M_2 + \frac{7}{9}\tilde{\alpha}_1 M_1\right)\tilde{Y}_D \\
&- \left\{\left[4(\tilde{Y}_D^A\tilde{Y}_D^\dagger\tilde{Y}_D) + 6\mathrm{Tr}(\tilde{Y}_D^A\tilde{Y}_D^\dagger)\tilde{Y}_D\right] + \left[5(\tilde{Y}_D\tilde{Y}_D^\dagger\tilde{Y}_D^A) + 3\mathrm{Tr}(\tilde{Y}_D\tilde{Y}_D^\dagger)\tilde{Y}_D^A\right]\right. \\
&+ \left. 2(\tilde{Y}_U^A\tilde{Y}_U^\dagger\tilde{Y}_D) + (\tilde{Y}_U\tilde{Y}_U^\dagger\tilde{Y}_D^A) + 2\mathrm{Tr}(\tilde{Y}_E^A\tilde{Y}_E^\dagger)\tilde{Y}_D + \mathrm{Tr}(\tilde{Y}_E\tilde{Y}_E)\tilde{Y}_D^A\right\},
\end{aligned}
\tag{B.8}
$$

$$
\begin{aligned}
2\frac{d\tilde{Y}_E^A}{dt} &= \left(3\tilde{\alpha}_2 + 3\tilde{\alpha}_1\right)\tilde{Y}_E^A - 2\left(3\tilde{\alpha}_2 M_2 + 3\tilde{\alpha}_1 M_1\right)\tilde{Y}_E \\
&- \left\{\left[4(\tilde{Y}_E^A\tilde{Y}_E^\dagger\tilde{Y}_E) + 2\mathrm{Tr}(\tilde{Y}_E^A\tilde{Y}_E^\dagger)\tilde{Y}_E\right] + \left[5(\tilde{Y}_E\tilde{Y}_E^\dagger\tilde{Y}_E^A) + \mathrm{Tr}(\tilde{Y}_E\tilde{Y}_E^\dagger)\tilde{Y}_E^A\right]\right. \\
&+ \left. 6\mathrm{Tr}(\tilde{Y}_D^A\tilde{Y}_D)\tilde{Y}_E + 3\mathrm{Tr}(\tilde{Y}_D\tilde{Y}_D)\tilde{Y}_E^A\right\}.
\end{aligned}
\tag{B.9}
$$

The RGEs of the squark and slepton masses are given by

$$
\begin{aligned}
\frac{dm_Q^2}{dt} &= \left(\frac{16}{3}\tilde{\alpha}_3 M_3^2 + 3\tilde{\alpha}_2 M_2^2 + \frac{1}{9}\tilde{\alpha}_1 M_1^2\right)\mathbb{1} \\
&- \frac{1}{2}\left[\tilde{Y}_U\tilde{Y}_U^\dagger m_Q^2 + m_Q^2\tilde{Y}_U\tilde{Y}_U^\dagger + 2\left(\tilde{Y}_U m_U^2\tilde{Y}_U^\dagger + \bar{\mu}_2^2\tilde{Y}_U\tilde{Y}_U^\dagger + \tilde{Y}_U^A\tilde{Y}_U^{A\dagger}\right)\right] \\
&- \frac{1}{2}\left[\tilde{Y}_D\tilde{Y}_D^\dagger m_Q^2 + m_Q^2\tilde{Y}_D\tilde{Y}_D^\dagger + 2\left(\tilde{Y}_D m_D^2\tilde{Y}_D^\dagger + \bar{\mu}_1^2\tilde{Y}_D\tilde{Y}_D^\dagger + \tilde{Y}_D^A\tilde{Y}_D^{A\dagger}\right)\right]
\end{aligned}
\tag{B.10}
$$

$$
\begin{aligned}
\frac{dm_U^2}{dt} &= \left(\frac{16}{3}\tilde{\alpha}_3 M_3^2 + \frac{16}{9}\tilde{\alpha}_1 M_1^2\right)\mathbb{1} \\
&- \left[\tilde{Y}_U^\dagger\tilde{Y}_U m_U^2 + m_U^2\tilde{Y}_U^\dagger\tilde{Y}_U + 2\left(\tilde{Y}_U^\dagger m_Q^2\tilde{Y}_U + \bar{\mu}_2^2\tilde{Y}_U^\dagger\tilde{Y}_U + \tilde{Y}_U^{A\dagger}\tilde{Y}_U^A\right)\right],
\end{aligned}
\tag{B.11}
$$

$$
\begin{aligned}
\frac{dm_D^2}{dt} &= \left(\frac{16}{3}\tilde{\alpha}_3 M_3^2 + \frac{4}{9}\tilde{\alpha}_1 M_1^2\right)\mathbb{1} \\
&- \left[\tilde{Y}_D^\dagger\tilde{Y}_D m_D^2 + m_D^2\tilde{Y}_D^\dagger\tilde{Y}_D + 2\left(\tilde{Y}_D^\dagger m_Q^2\tilde{Y}_D + \bar{\mu}_1^2\tilde{Y}_D^\dagger\tilde{Y}_D + \tilde{Y}_D^{A\dagger}\tilde{Y}_D^A\right)\right],
\end{aligned}
\tag{B.12}
$$

$$
\begin{aligned}
\frac{dm_L^2}{dt} &= \left(3\tilde{\alpha}_2 M_2^2 + \tilde{\alpha}_1 M_1^2\right)\mathbb{1} \\
&- \frac{1}{2}\left[\tilde{Y}_E\tilde{Y}_E^\dagger m_L^2 + m_L^2\tilde{Y}_E\tilde{Y}_E^\dagger + 2\left(\tilde{Y}_E m_E^2\tilde{Y}_E^\dagger + \bar{\mu}_1^2\tilde{Y}_E\tilde{Y}_E^\dagger + \tilde{Y}_E^A\tilde{Y}_E^{A\dagger}\right)\right],
\end{aligned}
$$

$$
\begin{aligned}
\frac{dm_E^2}{dt} &= \left(4\tilde{\alpha}_1 M_1^2\right)\mathbb{1} \\
&- \left[\tilde{Y}_E^\dagger\tilde{Y}_E m_E^2 + m_E^2\tilde{Y}_E^\dagger\tilde{Y}_E + 2\left(\tilde{Y}_E^\dagger m_L^2\tilde{Y}_E + \bar{\mu}_1^2\tilde{Y}_E^\dagger\tilde{Y}_E + \tilde{Y}_E^{A\dagger}\tilde{Y}_E^A\right)\right].
\end{aligned}
\tag{B.13}
$$

The RGEs of the Higgs potential mass parameters $\bar{\mu}_{1,2}^2 = \mu_{1,2}^2 - \mu^2$ are given by

$$\frac{d\bar{\mu}_1^2}{dt} = \left(3\tilde{\alpha}_2 M_2^2 + \tilde{\alpha}_1 M_1^2\right) - 3\text{Tr}\left(\tilde{Y}_D(m_Q^2 + m_D^2)\tilde{Y}_D^\dagger + \bar{\mu}_1^2\tilde{Y}_D\tilde{Y}_D^\dagger + \tilde{Y}_D^A\tilde{Y}_D^{A\dagger}\right)$$
$$- \text{Tr}\left(\tilde{Y}_E(m_L^2 + m_E^2)\tilde{Y}_E^\dagger + \bar{\mu}_1^2\tilde{Y}_E\tilde{Y}_E^\dagger + \tilde{Y}_E^A\tilde{Y}_E^{A\dagger}\right), \tag{B.14}$$

$$\frac{d\bar{\mu}_2^2}{dt} = \left(3\tilde{\alpha}_2 M_2^2 + \tilde{\alpha}_1 M_1^2\right) - 3\text{Tr}\left(\tilde{Y}_U(m_Q^2 + m_U^2)\tilde{Y}_U^\dagger + \bar{\mu}_2^2\tilde{Y}_U\tilde{Y}_U^\dagger + \tilde{Y}_U^A\tilde{Y}_U^{A\dagger}\right), \tag{B.15}$$

$$\frac{d\mu^2}{dt} = \left(3\tilde{\alpha}_2 + \tilde{\alpha}_1\right)\mu^2 - \text{Tr}\left(3\tilde{Y}_U\tilde{Y}_U^\dagger + 3\tilde{Y}_D\tilde{Y}_D^\dagger + \tilde{Y}_E\tilde{Y}_E^\dagger\right)\mu^2, \tag{B.16}$$

$$\frac{dB}{dt} = -\left(3\tilde{\alpha}_2 M_2 + \tilde{\alpha}_1 M_1\right) - \text{Tr}\left(3\tilde{Y}_U\tilde{Y}_U^{A\dagger} + 3\tilde{Y}_D\tilde{Y}_D^{A\dagger} + \tilde{Y}_E\tilde{Y}_E^{A\dagger}\right)\mu^2. \tag{B.17}$$

Chargino Contributions in MIA

Here we provide the SM and SUSY contributions, at leading order in MIA, to the R_F^{AB} quantities in Eq. (12.62), where F refers to $D, E, C, B_{(u,d)}$ and $M_{\gamma,g}$ [210].

$$R_D^{LL} = \sum_{i=1,2} |V_{i1}|^2 x_{wi} P_D(x_i)$$

$$R_D^{RL} = -\sum_{i=1,2} V_{i2}^\star V_{i1} x_{wi} P_D(x_i)$$

$$R_D^{RR} = \sum_{i=1,2} |V_{i2}|^2 x_{wi} P_D(x_i)$$

$$R_D^{LR} = (R_D^{RL})^\star$$

$$R_E^{LL} = \sum_{i=1,2} |V_{i1}|^2 x_{wi} P_E(x_i)$$

$$R_E^{RL} = -\sum_{i=1,2} V_{i2}^\star V_{i1} x_{wi} P_E(x_i)$$

$$R_E^{RR} = \sum_{i=1,2} |V_{i2}|^2 x_{wi} P_E(x_i)$$

$$R_E^{LR} = (R_E^{RL})^\star$$

$$R_C^{LL} = \sum_{i=1,2} |V_{i1}|^2 P_C^{(0)}(\bar{x}_i) + \sum_{i,j=1,2} \left[U_{i1}V_{i1}U_{j1}^\star V_{j1}^\star P_C^{(2)}(x_i, x_j) \right.$$
$$+ |V_{i1}|^2|V_{j1}|^2 \left(\frac{1}{8} - P_C^{(1)}(x_i, x_j) \right) \right]$$

$$R_C^{RL} = -\frac{1}{2} \sum_{i=1,2} V_{i2}^\star V_{i1} P_C^{(0)}(\bar{x}_i) - \sum_{i,j=1,2} V_{j2}^\star V_{i1} \left(U_{i1}U_{j1}^\star P_C^{(2)}(x_i, x_j) \right.$$
$$+ V_{i1}^\star V_{j1} P_C^{(1)}(x_i, x_j) \right)$$

$$R_C^{LR} = (R_C^{RL})^\star,$$

$$R_C^{RR} = \sum_{i,j=1,2} V_{j2}^\star V_{i2} \left(U_{i1}U_{j1}^\star P_C^{(2)}(x_i, x_j) + V_{i1}^\star V_{j1} P_C^{(1)}(x_i, x_j) \right)$$

$$R_{B_u}^{LL} = 2 \sum_{i,j=1,2} V_{i1} V_{j1}^\star U_{i1} U_{j1}^\star \, x_{Wj} \sqrt{x_{ij}} \, P_B^u(\bar{x}_j, x_{ij})$$

$$R_{B_u}^{RL} = -2 \sum_{i,j=1,2} V_{i1} V_{j2}^\star U_{i1} U_{j1}^\star \, x_{Wj} \sqrt{x_{ij}} \, P_B^u(\bar{x}_j, x_{ij})$$

$$R_{B_u}^{LR} = \left(R_{B_u}^{RL}\right)^\star$$

$$R_{B_u}^{RR} = 2 \sum_{i,j=1,2} V_{i2} V_{j2}^\star U_{i1} U_{j1}^\star \, x_{Wj} \sqrt{x_{ij}} \, P_B^u(\bar{x}_j, x_{ij})$$

$$R_{B_d}^{LL} = \sum_{i,j=1,2} |V_{i1}|^2 |V_{j1}|^2 \, x_{Wj} \, P_B^d(\bar{x}_j, x_{ij})$$

$$R_{B_d}^{RL} = -\sum_{i,j=1,2} V_{i2}^\star V_{i1} |V_{j1}|^2 \, x_{Wj} \, P_B^d(\bar{x}_j, x_{ij})$$

$$R_{B_d}^{LR} = \left(R_{B_d}^{RL}\right)^\star$$

$$R_{B_d}^{RR} = \sum_{i,j=1,2} V_{i2}^\star V_{i1} V_{j1}^\star V_{j2} \, x_{Wj} \, P_B^d(\bar{x}_j, x_{ij})$$

$$R_{M_{\gamma,g}}^{LL} = \sum_i |V_{i1}|^2 \, x_{Wi} \, P_{M_{\gamma,g}}^{LL}(x_i) - Y_b \sum_i V_{i1} U_{i2} \, x_{Wi} \, \frac{m_{\chi i}}{m_b} P_{M_{\gamma,g}}^{LR}(x_i)$$

$$R_{M_{\gamma,g}}^{LR} = -\sum_i V_{i1}^\star V_{i2} \, x_{Wi} \, P_{M_{\gamma,g}}^{LL}(x_i) + Y_b \sum_i V_{i2} U_{i2} \, x_{Wi} \, \frac{m_{\chi i}}{m_b} P_{M_{\gamma,g}}^{LR}(x_i)$$

$$R_{M_{\gamma,g}}^{RL} = -\sum_i V_{i1} V_{i2}^\star \, x_{Wi} \, P_{M_{\gamma,g}}^{LL}(x_i)$$

$$R_{M_{\gamma,g}}^{RR} = \sum_i |V_{i2}|^2 \, x_{Wi} \, P_{M_{\gamma,g}}^{LL}(x_i). \tag{C.1}$$

Regarding the R_F^0 terms at the zero order in mass insertion, we have

$$R_D^0 = 2x_W \sum_{i=1,2} |V_{i2}|^2 D_\chi(x_i)$$

$$R_E^0 = 2x_W \sum_{i=1,2} |V_{i2}|^2 E_\chi(x_i)$$

$$R_C^0 = 2 \sum_{i,j=1,2} U_{i1} U_{j1}^\star V_{i2} V_{j2}^\star C_\chi^{(2)}(x_i, x_j)$$

$$R_{B_d}^0 = -\frac{1}{2} \sum_{i,j=1,2} x_{Wj} V_{i2}^\star V_{i1} V_{j1}^\star V_{j2} B_\chi^{(d)}(\bar{x}_j, \bar{x}_j, x_{ij})$$

$$R_{B_u}^0 = \sum_{i,j=1,2} x_{Wj} V_{i2} V_{j2}^\star U_{i1}^\star U_{j1}^\star B_\chi^{(u)}(\bar{x}_j, \bar{x}_j, x_{ij})$$

$$R_{M_\gamma}^0 = -x_W \sum_{i=1,2} |V_{i2}|^2 \left(F_1(x_i) + \frac{2}{3} F_2(x_i)\right)$$

$$R_{M_g}^0 = -x_W \sum_{i=1,2} |V_{i2}|^2 F_2(x_i) \tag{C.2}$$

where Y_b is the Yukawa coupling of bottom quark, $x_{Wi} \equiv m_W^2/m_{\chi i}^2$, $x_w \equiv m_W^2/\tilde{m}^2$, $x_i \equiv m_{\chi i}^2/\tilde{m}^2$, $\bar{x}_i \equiv \tilde{m}^2/m_{\chi i}^2$, and $x_{ij} \equiv m_{\chi i}^2/m_{\chi j}^2$. The chargino mixing matrices U and V are defined in Eq. (5.10). The loop functions of penguin $P_{D,E,C}$, D_χ, E_χ, C_χ, box $B^{(u,d,\tilde{g})}$,

$P_B^{(u,d,\tilde{g})}$ as well as magnetic and chromo-magnetic penguin diagrams $F_{1,2}$, $P_{M_{\gamma,g}}^{LL}$, $P_{M_{\gamma,g}}^{LR}$ are given below.

BASIC SM AND SUSY LOOP FUNCTIONS

Here we report the loop functions entering in the SM and SUSY diagrams at one-loop.

SM functions

$$
\begin{aligned}
D(x) &= \frac{x^2 \left(25 - 19x\right)}{36 \left(x - 1\right)^3} + \frac{\left(-3x^4 + 30x^3 - 54x^2 + 32x - 8\right)}{18 \left(x - 1\right)^4} \log x \\
C(x) &= \frac{x \left(x - 6\right)}{8 \left(x - 1\right)} + \frac{x \left(3x + 2\right)}{8 \left(x - 1\right)^2} \log x \\
E(x) &= \frac{x \left(x^2 + 11x - 18\right)}{12 \left(x - 1\right)^3} + \frac{\left(-9x^2 + 16x - 4\right)}{6 \left(x - 1\right)^4} \log x \\
B(x) &= -\frac{x}{4(x - 1)} + \frac{x}{4(x - 1)^2} \log x
\end{aligned}
\tag{C.3}
$$

Charged Higgs functions

$$
\begin{aligned}
D_H(x) &= \frac{x \left(47x^2 - 79x + 38\right)}{108 \left(x - 1\right)^3} + \frac{x \left(-3x^2 + 6x - 4\right)}{18 \left(x - 1\right)^4} \log x \\
C_H(x) &= -\frac{1}{2} B(x) \\
E_H(x) &= \frac{x \left(7x^2 - 29x + 16\right)}{36 \left(x - 1\right)^3} + \frac{x \left(3x - 2\right)}{6 \left(x - 1\right)^4} \cdot \log x
\end{aligned}
\tag{C.4}
$$

Chargino functions in MIA

$$
\begin{aligned}
P_D(x) &= \frac{2x \left(-22 + 60x - 45x^2 + 4x^3 + 3x^4 - 3 \left(3 - 9x^2 + 4x^3\right) \log x\right)}{27 \left(1 - x\right)^5} \\
P_E(x) &= \frac{x \left(-1 + 6x - 18x^2 + 10x^3 + 3x^4 - 12x^3 \log x\right)}{9 \left(1 - x\right)^5} \\
P_C^{(0)}(x) &= \frac{x \left(3 - 4x + x^2 + 2 \log x\right)}{8 \left(1 - x\right)^3} \\
P_C^{(1)}(x,y) &= \frac{1}{8 \left(x - y\right)} \left[\frac{x^2 \left(x - 1 - \log x\right)}{\left(x - 1\right)^2} - \frac{y^2 \left(y - 1 - \log y\right)}{\left(y - 1\right)^2}\right] \\
P_C^{(2)}(x,y) &= \frac{\sqrt{xy}}{4 \left(x - y\right)} \left[\frac{x \left(x - 1 - \log x\right)}{\left(x - 1\right)^2} - \frac{y \left(y - 1 - \log y\right)}{\left(y - 1\right)^2}\right] \\
P_B^u(x,y) &= \frac{-y - x \left(1 - 3x + y\right)}{4 \left(x - 1\right)^2 \left(x - y\right)^2} - \frac{x \left(x^3 + y - 3xy + y^2\right) \log x}{2 \left(x - 1\right)^3 \left(x - y\right)^3} \\
&\quad + \frac{xy \log y}{2 \left(x - y\right)^3 \left(y - 1\right)}
\end{aligned}
$$

$$P_B^d(x,y) = -\frac{x\,(3\,y - x\,(1 + x + y))}{4\,(x-1)^2\,(x-y)^2} - \frac{x\,\left(x^3 + (x-3)\,x^2\,y + y^2\right)\log x}{2\,(x-1)^3\,(x-y)^3}$$

$$+ \frac{x\,y^2\,\log y}{2\,(x-y)^3\,(y-1)}$$

$$P_{M_\gamma}^{LL}(x) = \frac{x\,\left(-2 - 9x + 18x^2 - 7x^3 + 3x\,(x^2-3)\log x\right)}{9\,(1-x)^5}$$

$$P_{M_\gamma}^{LR}(x) = \frac{x\,\left(13 - 20x + 7x^2 + (6 + 4x - 4x^2)\log x\right)}{6\,(1-x)^4}$$

$$P_{M_g}^{LL}(x) = \frac{x\,\left(-1 + 9x + 9x^2 - 17x^3 + 6x^2\,(3+x)\log x\right)}{12\,(1-x)^5}$$

$$P_{M_g}^{LR}(x) = \frac{x\,\left(-1 - 4x + 5x^2 - 2x\,(2+x)\log x\right)}{2\,(1-x)^4}. \tag{C.5}$$

Gluino functions in MIA

$$P_1(x) = \frac{1 - 6\,x + 18\,x^2 - 10\,x^3 - 3\,x^4 + 12\,x^3\,\ln(x)}{18\,(x-1)^5}$$

$$P_2(x) = \frac{7 - 18\,x + 9\,x^2 + 2\,x^3 + 3\,\ln(x) - 9\,x^2\,\ln(x)}{9\,(x-1)^5}$$

$$M_1(x) = \frac{1 + 4x - 5x^2 + 4x\ln(x) + 2x^2\ln(x)}{2(1-x)^4}$$

$$M_2(x) = \frac{-5 + 4x + x^2 - 2\ln(x) - 4x\ln(x)}{2(1-x)^4}$$

$$M_3(x) = \frac{-1 + 9\,x + 9\,x^2 - 17\,x^3 + 18\,x^2\,\ln(x) + 6\,x^3\,\ln(x)}{12\,(x-1)^5}$$

$$M_4(x) = \frac{-1 - 9\,x + 9\,x^2 + x^3 - 6\,x\,\ln(x) - 6\,x^2\,\ln(x)}{6\,(x-1)^5}$$

$$B_1(x) = \frac{1}{4}M_1(x)$$

$$B_2(x) = -xM_2(x). \tag{C.6}$$

QCD Factorisation Cofficients

In this appendix, we prove the cofficients of $H_i(\phi)$ and $H_i(\eta')$ that determine the amplitude of $B \to \phi K$ and $B \to \eta' K$ in QCD factorisation, along the line of Eqs. (12.86) and (12.88).

$$
\begin{aligned}
H_1(\phi) &\simeq -0.0002 - 0.0002i \\
H_2(\phi) &\simeq 0.011 + 0.009i, \\
H_3(\phi) &\simeq -1.23 + 0.089i - 0.005X_A - 0.0006X_A^2 - 0.013X_H, \\
H_4(\phi) &\simeq -1.17 + 0.13i - 0.014X_H, \\
H_5(\phi) &\simeq -1.03 + 0.053i + 0.086X_A - 0.008X_A^2 \\
H_6(\phi) &\simeq -0.29 - 0.022i + 0.028X_A - 0.024X_A^2 + 0.014X_H, \\
H_7(\phi) &\simeq 0.52 - 0.026i - 0.006X_A + 0.004X_A^2, \\
H_8(\phi) &\simeq 0.18 + 0.037i - 0.019X_A + 0.012X_A^2 - 0.007X_H, \\
H_9(\phi) &\simeq 0.62 - 0.037i + 0.003X_A + 0.0003X_A^2 + 0.007X_H, \\
H_{10}(\phi) &\simeq 0.62 - 0.037i + 0.007X_H, \\
H_{7\gamma}(\phi) &\simeq -0.0004, \\
H_{8g}(\phi) &\simeq 0.047. \quad\quad\quad\quad\quad\quad\quad\quad\quad\quad\quad\quad\text{(D.1)}
\end{aligned}
$$

Here $X_{H,A} = \left(1 + \rho_{H,A}e^{i\phi_{H,A}}\right)\ln\left(\frac{m_B}{\Lambda_{\text{QCD}}}\right)$, where $\rho_{H,A}$ are free parameters, expected to be of order of one, and $\phi_{H,A} \in [0, 2\pi]$. Note that both $H_{7\gamma}$ and H_{8g} do not have any dependence on $X_{H,A}$, since the hard scattering and weak annihilation contributions to $Q_{7\gamma}$ and Q_{8g} have been ignored [224].

Constant	Value	Constant	Value
m_{B_d}	5.279 GeV	f_{B_d}	200 ± 30 MeV
$m_d(2\text{GeV})$	7 MeV	$m_b(m_b)$	4.23 GeV
m_t	174 ± 5 GeV	$\alpha_s(M_Z)$	0.119
$|V_{cb}|$	$(40.7 \pm 1.9) \times 10^{-3}$	$|V_{ub}|$	$(3.61 \pm 0.46) \times 10^{-3}$

Table D.1 Input quantities used in the phenomenological analysis.

$$
\begin{aligned}
H_1(\eta') &\simeq 0.44 + 0.0005i, \\
H_2(\eta') &\simeq 0.076 - 0.064i + 0.006X_H, \\
H_3(\eta') &\simeq 2.23 - 0.15i + 0.009X_A + 0.0008X_A^2 + 0.014X_H, \\
H_4(\eta') &\simeq 1.76 - 0.29i + 0.026X_H, \\
H_5(\eta') &\simeq -1.52 + 0.004X_A + 0.008X_A^2 \\
H_6(\eta') &\simeq 0.54 - 0.29i + 0.006X_A + 0.027X_A^2 + 0.026X_H, \\
H_7(\eta') &\simeq 0.078 + 0.001X_A - 0.004X_A^2, \\
H_8(\eta') &\simeq -0.58 + 0.02i + 0.004X_A - 0.014X_A^2 - 0.004X_H, \\
H_9(\eta') &\simeq -0.44 + 0.054i + 0.005X_A - 0.0004X_A^2 - 0.007X_H, \\
H_{10}(\eta') &\simeq -0.80 + 0.02i - 0.004X_H, \\
H_{7\gamma}(\eta') &\simeq 0.0007, \\
H_{8g}(\eta') &\simeq -0.089.
\end{aligned}
\tag{D.2}
$$

In these expressions, the input parameters considered are shown in Tab. D.1.

Bibliography

[1] ATLAS collaboration. Search for Squarks and Gluinos With The ATLAS Detector in Final States With Jets and Missing Transverse Momentum and 20.3 fb^{-1} of $\sqrt{s} = 8$ TeV proton-proton Collision Data. *ATLAS-CONF-2013-047*, 2013.

[2] S. Chatrchyan et al. Inclusive Search for Squarks and Gluinos in pp Collisions at $\sqrt{s} = 7$ TeV. *Phys. Rev.*, D85:012004, 2012.

[3] ATLAS collaboration. Further Searches for Squarks and Gluinos in Final States With Jets and Missing Transverse Momentum at $\sqrt{s} = 13$ TeV With The ATLAS Detector. *ATLAS-CONF-2016-078*, 2016.

[4] CMS Collaboration. Search for Supersymmetry in Events With Jets and Missing Transverse Momentum in Proton-Proton Collisions at 13 TeV. *CMS-PAS-SUS-16-014*, 2016.

[5] CMS Collaboration, "CMS SUSY 2013 Stop Summary". https://twiki.cern.ch/twiki/pub/CMSPublic/SUSYSMSSummaryPlots8TeV/SUSY2013T2ttT6.pdf.

[6] ATLAS collaboration. Search for Direct Top squark Pair Production and Dark Matter Production in Final States With Two Leptons in $\sqrt{s} = 13$ TeV pp Collisions Using 13.3 fb^{-1} of ATLAS Data. *ATLAS-CONF-2016-076*, 2016.

[7] CMS Collaboration. Search for Direct Top Squark Pair Production in The Single Lepton Final State at $\sqrt{s} = 13$ TeV. *CMS-PAS-SUS-16-028*, 2016.

[8] CMS Collaboration. Search for Electroweak SUSY Production in Multilepton Final States in pp Collisions at $\sqrt{s} = 13$ TeV With 12.9/fb. *CMS-PAS-SUS-16-024*, 2016.

[9] V. Khachatryan et al. Observation of The Diphoton Decay of The Higgs boson and Measurement of Its Properties. *Eur. Phys. J.*, C74(10):3076, 2014.

[10] G. Aad et al. Measurements of Higgs Boson Production and Couplings in Diboson Final States With The ATLAS Detector at The LHC. *Phys. Lett.*, B726:88–119, 2013. [Erratum: *Phys. Lett.*, B734:406, 2014].

[11] S. Chatrchyan et al. Measurement of The properties of A Higgs Boson in The Four Lepton Final State. *Phys. Rev.*, D89(9):092007, 2014.

[12] W. Abdallah and S. Khalil. MSSM Dark Matter in Light of Higgs and LUX Results. *Adv. High Energy Phys.*, 2016:5687463, 2016.

[13] A. Belyaev, S. Khalil, S. Moretti, and M. C. Thomas. Light Sfermion Interplay in The 125 GeV MSSM Higgs Production and Decay at The LHC. *JHEP*, 05:076, 2014.

[14] M. Hemeda, S. Khalil, and S. Moretti. Light chargino effects onto $H \rightarrow \gamma\gamma$ in the MSSM. *Phys. Rev.*, D89(1):011701, 2014.

[15] E. Corbelli and P. Salucci. The Extended Rotation Curve and The Dark Matter Halo of M33. *Mon. Not. Roy. Astron. Soc.*, 311:441–447, 2000.

[16] A. A. El-Zant, S. Khalil, and H. Okada. Dark Matter Annihilation and the PAMELA, FERMI and ATIC Anomalies. *Phys. Rev.*, D81:123507, 2010.

[17] M. Bona et al. The UTfit Collaboration Report on The Status of The Unitarity Triangle Beyond The Standard Model. I. Model-independent Analysis and Minimal Flavor Violation. *JHEP*, 03:080, 2006.

[18] F. Gabbiani, E. Gabrielli, A. Masiero, and L. Silvestrini. A Complete Analysis of FCNC and CP Constraints in General SUSY Extensions of The Standard Model. *Nucl. Phys.*, B477:321–352, 1996.

[19] S. Khalil and O. Lebedev. Chargino contributions to ε and ε' . *Phys. Lett.*, B515:387–394, 2001.

[20] E. Gabrielli, K. Huitu, and S. Khalil. Comparative Study of CP Asymmetries in Supersymmetric Models. *Nucl. Phys.*, B710:139–188, 2005.

[21] B. O'Leary, W. Porod, and F. Staub. Mass Spectrum of The Minimal SUSY $B - L$ Model. *JHEP*, 05:042, 2012.

[22] S. Khalil and S. Moretti. The $B - L$ Supersymmetric Standard Model with Inverse Seesaw at the Large Hadron Collider. *Rept. Prog. Phys.*, 80(3):036201, 2017.

[23] Luigi Delle Rose, Shaaban Khalil, Simon J. D. King, Carlo Marzo, Stefano Moretti, and Cem S. Un. Naturalness and dark matter in the supersymmetric B-L extension of the standard model. *Phys. Rev.*, D96(5):055004, 2017.

[24] S. K. King, S. Moretti, and R. Nevzorov. Exceptional Supersymmetric Standard Model. *Phys. Lett.*, B634:278–284, 2006.

[25] V. Barger, P. Langacker, H. Lee, and G. Shaughnessy. Higgs Sector in Extensions of the MSSM. *Phys. Rev.*, D73:115010, 2006.

[26] D. Barducci, G. Bélanger, C. Hugonie, and A. Pukhov. Status and Prospects of The nMSSM After LHC Run-1. *JHEP*, 01:050, 2016.

[27] R. Aggleton, D. Barducci, N. Bomark, S. Moretti, and C. Shepherd-Themistocleous. Review of LHC experimental results on low mass bosons in multi Higgs models. *JHEP*, 02:035, 2017.

[28] P. Athron, M. Mühlleitner, R. Nevzorov, and A. G. Williams. Non-Standard Higgs Decays in U(1) Extensions of the MSSM. *JHEP*, 01:153, 2015.

[29] W. Abdallah, S. Khalil, and S. Moretti. Double Higgs Peak in The minimal SUSY $B - L$ Model. *Phys. Rev.*, D91(1):014001, 2015.

[30] S. Khalil and S. Moretti. Can We Have Another Light ($\sim$ 145 GeV) Higgs Boson? *arXiv:1510.05934*, 2015.

[31] CMS Collaboration. Properties of the observed Higgs-like resonance using the diphoton channel. *CMS-PAS-HIG-13-016*, 2013.

[32] S. Khalil, A. Moursy, and A. Nassar. Aspects of Moduli Stabilization in Type IIB String Theory. *Adv. High Energy Phys.*, 2016:4303752, 2016.

[33] J. F. Gunion, H. E. Haber, H. L. Kane, and S. Dawson. The Higgs Hunter's Guide. *Front. Phys.*, 80:1–448, 2000.

[34] J. F. Gunion, H. E. Haber, G. L. Kane, and S. Dawson. Errata for The Higgs Hunter's Guide. *hep-ph/9302272*, 1992.

[35] H. E. Haber and G. L. Kane. The Search for Supersymmetry: Probing Physics Beyond The Standard Model. *Phys. Rept.*, 117:75–263, 1985.

[36] https://en.wikipedia.org/wiki/Occam's_razor.

[37] William of Ockham, *Quaestiones et decisiones in quattuor libros Sententiarum Petri Lombardi.*

[38] A. Einstein, "On The Method of Theoretical Physics", The Herbert Spencer Lecture, delivered at Oxford (10 June 1933), published in Philosophy of Science, Vol. 1, No. 2 (April 1934), pp. 163-169, see p. 165.

[39] H. J. Muller-Kirsten and A. Wiedemann. *Supersymmetry: An Introduction With Conceptual and Calculational Details.* Hackensack, USA: World Scientific (2010) 439 p, 1986.

[40] I. J. Aitchison. *Supersymmetry in Particle Physics: An Elementary Introduction.* Cambridge, UK: Univ. Pr. (2007) 222 p, 2007.

[41] J. Wess and J. Bagger. *Supersymmetry and supergravity.* Princeton, USA: Univ. Pr. (1992) 259 p, 1992.

[42] M. Drees, R. Godbole, and P. Roy. *Theory and phenomenology of sparticles: An account of four-dimensional N=1 supersymmetry in high energy physics.* 2004.

[43] H. Baer and X. Tata. *Weak scale supersymmetry: From superfields to scattering events.* Cambridge University Press, 2006.

[44] P. Binetruy. *Supersymmetry: Theory, Experiment and Cosmology.* Oxford, UK: Oxford Univ. Pr. (2006) 520 p, 2006.

[45] H. K. Dreiner, H. E. Haber, and S. P. Martin. Two-component Spinor Techniques and Feynman Rules for Quantum Field Theory and Supersymmetry. *Phys. Rept.*, 494:1–196, 2010.

[46] J. Terning. *Modern supersymmetry: Dynamics and Duality.* Oxford, UK: Clarendon (2006) 324 p, 2006.

[47] S. F. Sohnius. Introducing Supersymmetry. *Phys. Rept.*, 128:39–204, 1985.

[48] S. P. Martin. A Supersymmetry primer. *hep-ph/9709356*, 1997. [Adv. Ser. Direct. High Energy Phys.18,1(1998)].

[49] D. J. Chung, L. L. Everett, G. L. Kane, S. F. King, J. D. Lykken, and L. Wang. The Soft Supersymmetry Breaking Lagrangian: Theory and Applications. *Phys. Rept.*, 407:1–203, 2005.

[50] F. Quevedo, S. Krippendorf, and O. Schlotterer. Cambridge Lectures on Supersymmetry and Extra Dimensions. *arXiv:1011.1491*, 2010.

[51] M. E. Peskin. Supersymmetry in Elementary Particle Physics. In *Proceedings of Theoretical Advanced Study Institute in Elementary Particle Physics: Exploring New Frontiers Using Colliders and Neutrinos (TASI 2006)*, pages 609–704, 2008.

[52] H. Murayama. Supersymmetry Phenomenology. In *Particle physics. Proceedings, Summer School, Trieste, Italy, June 21-July 9, 1999*, pages 296–335, 2000.

[53] J. R. Ellis. Supersymmetry for Alp Hikers. In *2001 European School of High Energy Physics, Beatenberg, Switzerland, 26 Aug-8 Sep 2001: Proceedings*, pages 157–203, 2002.

[54] J. A. Bagger. Weak Scale Supersymmetry: Theory and Practice. In *QCD and Beyond. Proceedings, Theoretical Advanced Study Institute in Elementary Particle Physics, TASI-95, Boulder, USA, June 4-30, 1995*, 1996.

[55] M. Drees. An Introduction to Supersymmetry. In *Current Topics in Physics. Proceedings, Inauguration Conference of The Asia-Pacific Center for Theoretical Physics (APCTP), Seoul, Korea, June 4-10, 1996. Vol. 1, 2*, 1996.

[56] J. D. Lykken. Introduction to Supersymmetry. In *Fields, Strings and Duality. Proceedings, Summer School, Theoretical Advanced Study Institute in Elementary Particle Physics, TASI'96, Boulder, USA, June 2-28, 1996*, pages 85–153, 1996.

[57] A. Bilal. Introduction to Supersymmetry. *hep-th/0101055*, 2001.

[58] H. E. Haber. Introductory Low-energy Supersymmetry. In *Theoretical Advanced Study Institute (TASI 92): From Black Holes and Strings to Particles Boulder, Colorado, June 3-28, 1992*, 1993.

[59] S. R. Coleman and J. Mandula. All Possible Symmetries of The S Matrix. *Phys. Rev.*, 159:1251–1256, 1967.

[60] R. Haag, J. T. Lopuszanski, and M. Sohnius. All Possible Generators of Supersymmetries of the s Matrix. *Nucl. Phys.*, B88:257, 1975.

[61] Yu. A. Golfand and E. P. Likhtman. Extension of The Algebra of Poincaré Group Generators and Violation of p-invariance. *JETP Lett.*, 13:323–326, 1971. [Pisma Zh. Eksp. Teor. Fiz.13,452(1971)].

[62] A. Salam and J. A. Strathdee. Supersymmetry and Nonabelian Gauges. *Phys. Lett.*, B51:353–355, 1974.

[63] A. Salam and J. A. Strathdee. Supergauge Transformations. *Nucl. Phys.*, B76:477–482, 1974.

[64] A. Salam and J. A. Strathdee. On Superfields and Fermi-Bose Symmetry. *Phys. Rev.*, D11:1521–1535, 1975.

[65] S. Ferrara, J. Wess, J., and B. Zumino. Supergauge Multiplets and Superfields. *Phys. Lett.*, B51:239, 1974.

[66] P. Fayet and S. Ferrara. Supersymmetry. *Phys. Rept.*, 32:249–334, 1977.

[67] I. L. Buchbinder and S. M. Kuzenko. *Ideas and Methods of Supersymmetry and Supergravity: Or A Walk Through Superspace*. Bristol, UK: IOP (1998) 656 p, 1998.

[68] J. Wess and B. Zumino. Supergauge Transformations in Four-Dimensions. *Nucl. Phys.*, B70:39–50, 1974.

[69] J. Wess and B. Zumino. Supergauge Invariant Extension of Quantum Electrodynamics. *Nucl. Phys.*, B78:1, 1974.

[70] P. Fayet and J. Iliopoulos. Spontaneously Broken Supergauge Symmetries and Goldstone Spinors. *Phys. Lett.*, B51:461–464, 1974.

[71] M.T. Grisaru, W. Siegel, and M. Rocek. Improved Methods for Supergraphs. *Nucl. Phys.*, B159:429, 1979.

[72] N. Seiberg. Naturalness Versus Supersymmetric Non-renormalization Theorems. *Phys. Lett.*, B318:469–475, 1993.

[73] J. Iliopoulos and B. Zumino. Broken Supergauge Symmetry and Renormalization. *Nucl. Phys.*, B76:310, 1974.

[74] E. Witten. Dynamical Breaking of Supersymmetry. *Nucl. Phys.*, B188:513, 1981.

[75] L. O'Raifeartaigh. Spontaneous Symmetry Breaking for Chiral Scalar Superfields. *Nucl. Phys.*, B96:331, 1975.

[76] L. Girardello and M. T. Grisaru. Soft Breaking of Supersymmetry. *Nucl. Phys.*, B194:65, 1982.

[77] H. P. Nilles. Supersymmetry, Supergravity and Particle Physics. *Phys. Rept.*, 110:1–162, 1984.

[78] G.R. Farrar and P. Fayet. Phenomenology of The Production, Decay, and Detection of New Hadronic States Associated With Supersymmetry. *Phys. Lett.*, B76:575–579, 1978.

[79] L. E. Ibanez and G. G. Ross. Low-Energy Predictions in Supersymmetric Grand Unified Theories. *Phys. Lett.*, B105:439, 1981.

[80] S. Dimopoulos, S. Raby, and F. Wilczek. Supersymmetry and The Scale of Unification. *Phys. Rev.*, D24:1681–1683, 1981.

[81] J. R. Ellis, S. Kelley, and D. V. Nanopoulos. Probing The Desert Using Gauge Coupling Unification. *Phys. Lett.*, B260:131–137, 1991.

[82] U. Amaldi, W. de Boer, and H. Furstenau. Comparison of Grand Unified Theories With Electroweak and Strong Coupling Constants Measured at LEP. *Phys. Lett.*, B260:447–455, 1991.

[83] P. Langacker and M. Luo. Implications of Precision Electroweak Experiments for M_t, ρ_0, $\sin^2 \theta_W$ and Grand Unification. *Phys. Rev.*, D44:817–822, 1991.

[84] Giunti C, C. W. Kim, and U. W. Lee. Running Coupling Constants and Grand Unification Models. *Mod. Phys. Lett.*, A6:1745–1755, 1991.

[85] H.E. Haber and R. Hempfling. Can The Mass of The Lightest Higgs Boson of The Minimal Supersymmetric Model be Larger than M_Z? *Phys. Rev. Lett.*, 66:1815–1818, 1991.

[86] J. R. Ellis, G. Ridolfi, and F. Zwirner. Radiative Corrections to The Masses of Supersymmetric Higgs Bosons. *Phys. Lett.*, B257:83–91, 1991.

[87] R. Barbieri, M. Frigeni, and F. Caravaglios. The Supersymmetric Higgs for Heavy Superpartners. *Phys. Lett.*, B258:167–170, 1991.

[88] J. M. Frere, D. R.T. Jones, and S. Raby. Fermion Masses and Induction of The Weak Scale by Supergravity. *Nucl. Phys.*, B222:11, 1983.

[89] L. Alvarez-Gaume, J. Polchinski, and M. B. Wise. Minimal Low-Energy Supergravity. *Nucl. Phys.*, B221:495, 1983.

[90] J. P. Derendinger and C. A. Savoy. Quantum Effects and $SU(2) \times U(1)$ Breaking in Supergravity Gauge Theories. *Nucl. Phys.*, B237:307, 1984.

[91] C. Kounnas, A. B. Lahanas, D. V. Nanopoulos, and M. Quiros. Low-Energy Behavior of Realistic Locally Supersymmetric Grand Unified Theories. *Nucl. Phys.*, B236:438, 1984.

[92] M. Claudson, L. J. Hall, and I. Hinchliffe. Low-Energy Supergravity: False Vacua and Vacuous Predictions. *Nucl. Phys.*, B228:501, 1983.

[93] L. E. Ibanez, C. Lopez, and C. Munoz. The Low-Energy Supersymmetric Spectrum According to N=1 Supergravity Guts. *Nucl. Phys.*, B256:218–252, 1985.

[94] S. P. Martin and M. T. Vaughn. Regularization Dependence of Running Couplings in Softly Broken Supersymmetry. *Phys. Lett.*, B318:331–337, 1993.

[95] A. Djouadi, J. Kalinowski, P. Ohmann, and P. M. Zerwas. Heavy SUSY Higgs Bosons at e^+e^- Linear Colliders. *Z. Phys.*, C74:93–111, 1997.

[96] J. Rosiek. Complete Set of Feynman Rules for The MSSM: Erratum. *hep-ph/9511250*, 1995.

[97] J. C. Romao. The Minimal Supersymmetric Standard Model. http://porthos.ist.utl.pt/romao/homepage/publications/mssm-model/mssm-model.pdf. 2001.

[98] F. Gabbiani and A. Masiero. FCNC in Generalized Supersymmetric Theories. *Nucl. Phys.*, B322:235–254, 1989.

[99] J. S. Hagelin, S. Kelley, and T. Tanaka. Supersymmetric Flavor Changing Neutral Currents: Exact Amplitudes and Phenomenological Analysis. *Nucl. Phys.*, B415:293–331, 1994.

[100] A. Djouadi. The Anatomy of Electro-weak Symmetry Breaking. II. The Higgs bosons in The Minimal Supersymmetric Model. *Phys. Rept.*, 459:1–241, 2008.

[101] S. Moretti, K. Odagiri, P. Richardson, M. H. Seymour, and B. R. Webber. Implementation of Supersymmetric Processes in the HERWIG Event Generator. *JHEP*, 04:028, 2002.

[102] Y. Okada, M. Yamaguchi, and T. Yanagida. Renormalization Group Analysis on The Higgs Mass in The Softly Broken Supersymmetric Standard Model. *Phys. Lett.*, B262:54–58, 1991.

[103] S. Heinemeyer, W. Hollik, and G. Weiglein. The Mass of The Lightest MSSM Higgs Boson: A Compact Analytical Expression at The Two Loop Level. *Phys. Lett.*, B455:179–191, 1999.

[104] M. Carena, H. E. Haber, S. Heinemeyer, W. Hollik, C.E. M. Wagner, and G. Weiglein. Reconciling The Two Loop Diagrammatic and Effective Field Theory Computations of The Mass of The Lightest CP-even Higgs Boson in The MSSM. *Nucl. Phys.*, B580:29–57, 2000.

[105] J. R. Espinosa and R. Zhang. Complete Two Loop Dominant Corrections to The Mass of The Lightest CP-even Higgs Boson in The Minimal Supersymmetric Standard Model. *Nucl. Phys.*, B586:3–38, 2000.

[106] M. Drees and M. M. Nojiri. One Loop Corrections to The Higgs Sector in Minimal Supergravity Models. *Phys. Rev.*, D45:2482–2492, 1992.

[107] D. M. Pierce, J. A. Bagger, K.T. Matchev, and R. Zhang. Precision Corrections in The Minimal Supersymmetric Standard Model. *Nucl. Phys.*, B491:3–67, 1997.

[108] A. Brignole, J. R. Ellis, G. Ridolfi, and F. Zwirner. The Supersymmetric Charged Higgs Boson Mass and LEP Phenomenology. *Phys. Lett.*, B271:123–132, 1991.

[109] A. Brignole. Radiative Corrections to The Supersymmetric Charged Higgs Boson Mass. *Phys. Lett.*, B277:313–323, 1992.

[110] E. Boos, A. Djouadi, M. Muhlleitner, and A. Vologdin. The MSSM Higgs Bosons in The Intense Coupling Regime. *Phys. Rev.*, D66:055004, 2002.

[111] G. Corcella, I. G. Knowles, G. Marchesini, S. Moretti, K. Odagiri, P. Richardson, M. H. Seymour, and B. R. Webber. HERWIG 6: An Event generator for hadron emission reactions with interfering gluons (including supersymmetric processes). *JHEP*, 01:010, 2001.

[112] K. Odagiri. Color Connection Structure of Supersymmetric QCD $(2 \to 2)$ Processes. *JHEP*, 10:006, 1998.

[113] Z. Kunszt, S. Moretti, and W. J. Stirling. Higgs Production at The LHC: An Update on Cross-sections and Branching Ratios. *Z. Phys.*, C74:479–491, 1997.

[114] M. Spira. QCD Effects in Higgs Physics. *Fortsch. Phys.*, 46:203–284, 1998.

[115] B. A. A. Barrientos and B. A. Kniehl. $W^{\pm}H^{\mp}$ Associated Production at The Large Hadron Collider. *Phys. Rev.*, D59:015009, 1999.

[116] S. Moretti and K. Odagiri. Production of Charged Higgs Bosons of The Minimal Supersymmetric Standard Model in b Quark Initiated Processes at The Large Hadron Collider. *Phys. Rev.*, D55:5627–5635, 1997.

[117] J. R. Ellis, M. K. Gaillard, and D. V. Nanopoulos. A Phenomenological Profile of The Higgs Boson. *Nucl. Phys.*, B106:292, 1976.

[118] L. Resnick, M. K. Sundaresan, and P. J. S. Watson. Is There A Light Scalar Boson? *Phys. Rev.*, D8:172–178, 1973.

[119] C. Li and R. J. Oakes. QCD Corrections to The Hadronic Decay Width of A Charged Higgs Boson. *Phys. Rev.*, D43:855–859, 1991.

[120] A. Mendez and A. Pomarol. QCD Corrections to The Charged Higgs Boson Hadronic Width. *Phys. Lett.*, B252:461–466, 1990.

[121] A. Djouadi, J. Kalinowski, and P. M. Zerwas. Two and Three-body Decay Modes of SUSY Higgs Particles. *Z. Phys.*, C70:435–448, 1996.

[122] S. Moretti and W. J. Stirling. Contributions of Below Threshold Decays to MSSM Higgs Branching Ratios. *Phys. Lett.*, B347:291–299, 1995. [Erratum: *Phys. Lett.*, B366:451, 1996].

[123] M. Spira, A. Djouadi, D. Graudenz, and P. M. Zerwas. Higgs Boson Production at The LHC. *Nucl. Phys.*, B453:17–82, 1995.

[124] S. Dawson, A. Djouadi, and M. Spira. QCD Corrections to SUSY Higgs Production: The Role of Squark Loops. *Phys. Rev. Lett.*, 77:16–19, 1996.

[125] G. Gamberini, G. F. Giudice, and G. Ridolfi. Supersymmetric Higgs Boson Production in Z Decays. *Nucl. Phys.*, B292:237, 1987.

[126] J. R. Espinosa J. R. and M. Quiros. On Higgs Boson Masses in Nonminimal Supersymmetric Standard Models. *Phys. Lett.*, B279:92–97, 1992.

[127] R. Hempfling and A. H. Hoang. Two Loop Radiative Corrections to The Upper Limit of the Lightest Higgs Boson Mass in The Minimal Supersymmetric Model. *Phys. Lett.*, B331:99–106, 1994.

[128] J. A. Casas, J. R. Espinosa, M. Quiros, and A. Riotto. The Lightest Higgs Boson Mass in The Minimal Supersymmetric Standard Model. *Nucl. Phys.*, B436:3–29, 1995. [Erratum: *Nucl. Phys.*, B439:466, 1995].

[129] M. Carena, J. R. Espinosa, M. Quiros, and C. E. W. Wagner. Analytical Expressions for Radiatively Corrected Higgs Masses and Couplings in The MSSM. *Phys. Lett.*, B355:209–221, 1995.

[130] M. Carena, M. Quiros, and C. E. W. Wagner. Effective Potential Methods and The Higgs Mass Spectrum in The MSSM. *Nucl. Phys.*, B461:407–436, 1996.

[131] N. Craig. The State of Supersymmetry After Run I of The LHC. In *Beyond The Standard Model After The First Run of The LHC Arcetri, Florence, Italy, May 20–July 12, 2013*, 2013.

[132] P. Bechtle, T. Plehn, and C. Sander. Supersymmetry. *arXiv:1506.03091*, 2015.

[133] LHC Management Board, http://wwwhlc01.cern.ch/planning.htm.

[134] CMS Collaboration, Technical proposal, LHCC/P1 (1994).

[135] ATLAS Collaboration, Technical proposal, LHCC/P2, (1994).

[136] http://home.cern/about/experiments/alice.

[137] http://lhcb-public.web.cern.ch/lhcb-public/en/detector/Detector-en.html.

[138] CMS Collaboration. Search for Supersymmetry Using Razor Variables in Events With b-jets in ppcollisions at 8 TeV. *CMS-PAS-SUS-13-004*, 2013.

[139] S. Chatrchyan et al. Search for Top-Squark Pair Production in The Single-Lepton Final State in pp Collisions at $\sqrt{s} = 8$ TeV. *Eur. Phys. J.*, C73(12):2677, 2013.

[140] ATLAS collaboration. Search for Direct Top Squark Pair Production in Final States With One Isolated Lepton, Jets, and Missing Transverse Momentum in $\sqrt{s} = 8$ TeV pp Collisions Using 21 fb^{-1} of ATLAS Data. *ATLAS-CONF-2013-037*, 2013.

[141] ATLAS collaboration. Search for Direct Third Generation Squark Pair Production in Final States With Missing Transverse Momentum and Two b-Jets in $\sqrt{s} = 8$ TeV pp Collisions With The ATLAS Detector. *ATLAS-CONF-2013-053*, 2013.

[142] CMS Collaboration. Search for Electroweak Production of Charginos, Neutralinos, and Sleptons Using Leptonic Final States in ppcollisions at 8 TeV. *CMS-PAS-SUS-13-006*, 2013.

[143] The ATLAS collaboration. Search for Direct Slepton and Direct Chargino Production in Final States With Two Opposite-sign Leptons, Missing Transverse Momentum and No Jets in 20/fb of ppcollisions at $\sqrt{(s)} = 8$ TeV With The ATLAS Detector. *ATLAS-CONF-2013-049*, 2013.

[144] S. Chatrchyan et al. Observation of A New Boson at A Mass of 125 GeV With The CMS Experiment at The LHC. *Phys. Lett.*, B716:30–61, 2012.

[145] G. Aad et al. Observation of A New Particle in The Search for The Standard Model Higgs Boson With The ATLAS Detector at The LHC. *Phys. Lett.*, B716:1–29, 2012.

[146] ATLAS Collaboration. Combined Coupling Measurements of The Higgs-like Boson With The ATLAS Detector Using up to 25 fb^{-1} of Proton-Proton Collision Data. *ATLAS-CONF-2013-034*, 2013.

[147] CMS Collaboration. Combination of Standard Model Higgs Boson Searches and Measurements of The Properties of The New Boson With a Mass Near 125 GeV. *CMS-PAS-HIG-13-005*, 2013.

[148] CMS Collaboration. A Combination of Searches for The Invisible Decays of The Higgs Boson Using The CMS Detector. *CMS-PAS-HIG-15-012*, 2015.

[149] G. Aad et al. Search for Invisible Decays of A Higgs Boson Using Vector Boson Fusion in pp Collisions at $\sqrt{s} = 8$ TeV With The ATLAS Detector. *arXiv: 1508.07869, CERN-PH-EP-2015-186*, 2015.

[150] G. Aad et al. Search for Invisible Decays of a Higgs Boson Produced in Association With a Z Boson in ATLAS. *Phys. Rev. Lett.*, 112:201802, 2014.

[151] G. Aad et al. Combined Measurement of The Higgs Boson Mass in pp Collisions at $\sqrt{s} = 7$ and 8 TeV With The ATLAS and CMS Experiments. *Phys. Rev. Lett.*, 114:191803, 2015.

[152] G. Aad et al. Search for Squarks and Gluinos in Events With Isolated Leptons, Jets and Missing Transverse Momentum At $\sqrt{s} = 8$ TeV With The ATLAS Detector. *JHEP*, 04:116, 2015.

[153] S. Chatrchyan et al. Search for Gluino Mediated Bottom and Top Squark Production in Multijet Final States in pp Collisions at 8 TeV. *Phys. Lett.*, B725:243–270, 2013.

[154] M. Carena, I. Low, and C.E.W. Wagner. Implications of a Modified Higgs to Diphoton Decay Width. *JHEP*, 08:060, 2012.

[155] G. Abbiendi et al. Search for Chargino and Neutralino Production at $s^{**}(1/2) = 192$-GeV to 209 GeV at LEP. *Eur. Phys. J.*, C35:1–20, 2004.

[156] B. Batell, S. Jung, and C. E. W. Wagner. Very Light Charginos and Higgs Decays. *JHEP*, 12:075, 2013.

[157] H. E. Haber, R. Hempfling, and A. H. Hoang. Approximating The Radiatively Corrected Higgs Mass in The Minimal Supersymmetric Model. *Z. Phys.*, C75:539–554, 1997.

[158] S. Khalil and C. Munoz. The Enigma of The Dark Matter. *Contemp. Phys.*, 43:51–62, 2002.

[159] P. A. R. Ade et al. Planck 2013 Results. XVI. Cosmological Parameters. *Astron. Astrophys.*, 571:A16, 2014.

[160] P. A. R. Ade et al. Planck 2015 Results. XX. Constraints on Inflation. 2015.

[161] G. Belanger, F. Boudjema, A. Pukhov, and A. Semenov. micrOMEGAs$_3$: A program for Calculating Dark Matter Observables. *Comput. Phys. Commun.*, 185:960–985, 2014.

[162] M. Chakraborti, U. Chattopadhyay, S. Rao, and D. P. Roy. Higgsino Dark Matter in Nonuniversal Gaugino Mass Models. *Phys. Rev.*, D91(3):035022, 2015.

[163] M. Goodman and E. Witten. Detectability of CertaIn Dark Matter Candidates. *Phys. Rev.*, D31:3059, 1985.

[164] G. Jungman, M. Kamionkowski, and K. Griest. Supersymmetric Dark Matter. *Phys. Rept.*, 267:195–373, 1996.

[165] Q. Shafi, S. H. Tanyldz, and C. S. Un. Neutralino Dark Matter and Other LHC Predictions from Quasi Yukawa Unification. *Nucl. Phys.*, B900:400–411, 2015.

[166] J. F. Navarro, C. S. Frenk, and S. D. M. White. The Structure of Cold Dark Matter Halos. *Astrophys. J.*, 462:563–575, 1996.

[167] T. Delahaye, R. Lineros, F. Donato, N. Fornengo, and P. Salati. Positrons From Dark Matter Annihilation in The Galactic Halo: Theoretical Uncertainties. *Phys. Rev.*, D77:063527, 2008.

[168] J. F. Navarro, C. S. Frenk, and S. D. M. White. A Universal Density Profile From Hierarchical Clustering. *Astrophys. J.*, 490:493–508, 1997.

[169] E. A. Baltz and J. Edsjo. Positron Propagation and Fluxes From Neutralino Annihilation in The Halo. *Phys. Rev.*, D59:023511, 1998.

[170] V. Khachatryan et al. Search for Dark Matter, Extra Dimensions, and Unparticles in Monojet Events in Proton-Proton Collisions at $\sqrt{s} = 8$ TeV. *Eur. Phys. J.*, C75(5):235, 2015.

[171] G. Aad et al. Search for New Phenomena in Final States With an Energetic Jet and Large Missing Transverse Momentum in pp Collisions at $\sqrt{s} =8$ TeV With The ATLAS Detector. *Eur. Phys. J.*, C75(7):299, 2015. [Erratum: *Eur. Phys. J.*,C75 (9):408, 2015].

[172] H. Baer, A. Mustafayev, and X. Tata. Monojets and Mono-photons From Light Higgsino Pair Production at LHC14. *Phys. Rev.*, D89(5):055007, 2014.

[173] C. Han, A. Kobakhidze, A. Saavedra N. Liu, L. Wu, and J. M. Yang. Probing Light Higgsinos in Natural SUSY From Monojet Signals at The LHC. *JHEP*, 02:049, 2014.

[174] K. Hagiwara et al. Review of Particle Physics. Particle Data Group. *Phys. Rev.*, D66:010001, 2002.

[175] L. Wolfenstein. Parametrization of The Kobayashi-Maskawa Matrix. *Phys. Rev. Lett.*, 51:1945, 1983.

[176] C. Jarlskog. Commutator of The Quark Mass Matrices in The Standard Electroweak Model and a Measure of Maximal CP Violation. *Phys. Rev. Lett.*, 55:1039, 1985.

[177] C. Jarlskog. A Basis Independent Formulation of The Connection Between Quark Mass Matrices, CP Violation and Experiment. *Z. Phys.*, C29:491–497, 1985.

[178] R. D. Peccei and H. R. Quinn. CP Conservation in The Presence of Instantons. *Phys. Rev. Lett.*, 38:1440–1443, 1977.

[179] S. Abel, S. Khalil, and O. Lebedev. EDM Constraints in Supersymmetric Theories. *Nucl. Phys.*, B606:151–182, 2001.

[180] C. A. Baker et al. An Improved Experimental Limit on The Electric Dipole Moment of The Neutron. *Phys. Rev. Lett.*, 97:131801, 2006.

[181] J. Baron et al. Order of Magnitude Smaller Limit on The Electric Dipole Moment of The Electron. *Science*, 343:269–272, 2014.

[182] W. C. Griffith, M. D. Swallows, T. H. Loftus, M. V. Romalis, B. R. Heckel, and E. N. Fortson. Improved Limit on The Permanent Electric Dipole Moment of Hg-199. *Phys. Rev. Lett.*, 102:101601, 2009.

[183] A. Manohar and H. Georg. Chiral Quarks and The Nonrelativistic Quark Model. *Nucl. Phys.*, B234:189, 1984.

[184] T. Falk, K. A. Olive, M. Pospelov, and R. Roiban. MSSM predictions for The Electric Dipole Moment of the Hg-199 Atom. *Nucl. Phys.*, B560:3–22, 1999.

[185] E. P. Shabalin. Electric Dipole Moment of Quark in a Gauge Theory with Left-Handed Currents. *Sov. J. Nucl. Phys.*, 28:75, 1978. [Yad. Fiz.28,151(1978)].

[186] J. F. Donoghue. T Violation in $SU(2) \times U(1)$ Gauge Theories of Leptons. *Phys. Rev.*, D18:1632, 1978.

[187] T. Ibrahim and P. Nath. The Neutron and The Electron Electric Dipole Moment in N=1 Supergravity Unification. *Phys. Rev.*, D57:478–488, 1998. [Erratum: *Phys. Rev.*, D60:119901, 1999].

[188] T. Ibrahim and P. Nath. The Chromoelectric and Purely Gluonic Operator Contributions to The Neutron Electric Dipole Moment in N=1 Supergravity. *Phys. Lett.*, B418:98–106, 1998.

[189] A. J. Buras, A. Romanino, and L. Silvestrini. $K \to \pi$ Neutrino Anti-Neutrino: A Model Independent Analysis and Supersymmetry. *Nucl. Phys.*, B520:3–30, 1998.

[190] J. Hisano and D. Nomura. Solar and Atmospheric Neutrino Oscillations and Lepton Flavor Violation in Supersymmetric Models With The Right-Handed Neutrinos. *Phys. Rev.*, D59:116005, 1999.

[191] S. Gamiz and E. Maria. *Kaon Physics: CP Violation and Hadronic Matrix Elements.* PhD thesis, Granada U., Theor. Phys. Astrophys., 2003.

[192] Y. Nir. CP Violation: A New Era. In *Heavy Flavor Physics: Theory and Experimental Results in Heavy Quark Physics and CP Violation. Proceedings, 55th Scottish Universities Summer School in Physics, SUSSP 2001, St. and rews, UK, August 7-23, 2001*, pages 147–200, 2001.

[193] S. Herrlich and U. Nierste. The Complete $|\Delta S| = 2$ - Hamiltonian in The Next-to-Leading Order. *Nucl. Phys.*, B476:27–88, 1996.

[194] T. Inami and C. S. Lim. Effects of Superheavy Quarks and Leptons in Low-Energy Weak Processes $k(L) \to \mu^+\mu^-$, $K^+ \to \pi^+$ Neutrino Anti-neutrino and $K^0 \leftrightarrow \bar{K}^0$. *Prog. Theor. Phys.*, 65:297, 1981. [Erratum: *Prog. Theor. Phys.*, 65:1772, 1981].

[195] S. L. Glashow, J. Iliopoulos, and L. Maiani. Weak Interactions With Lepton-Hadron Symmetry. *Phys. Rev.*, D2:1285–1292, 1970.

[196] J. R. Batley et al. A Precision Measurement of Direct CP Violation in The Decay of Neutral Kaons into Two Pions. *Phys. Lett.*, B544:97–112, 2002.

[197] G. D. Barr et al. A New Measurement of Direct CP Violation in The Neutral Kaon System. *Phys. Lett.*, B317:233–242, 1993.

[198] E. Abouzaid et al. Precise Measurements of Direct CP Violation, CPT Symmetry, and Other Parameters in The Neutral Kaon System. *Phys. Rev.*, D83:092001, 2011.

[199] L. K. Gibbons et al. Measurement of The CP Violation Parameter $\mathrm{Re}(\varepsilon'/\varepsilon)$. *Phys. Rev. Lett.*, 70:1203–1206, 1993.

[200] F. J. Gilman and M. B. Wise. Effective Hamiltonian for $\Delta s = 1$ Weak Nonleptonic Decays in The Six Quark Model. *Phys. Rev.*, D20:2392, 1979.

[201] A. J. Buras. Weak Hamiltonian, CP Violation and Rare Decays. In *Probing The Standard Model of Particle Interactions. Proceedings, Summer School in Theoretical Physics, NATO Advanced Study Institute, 68th session, Les Houches, France, July 28-September 5, 1997. Pt. 1, 2*, pages 281–539, 1998.

[202] A. J. Buras, M. Jamin, and M. E. Lautenbacher. The Anatomy of ε'/ε Beyond Leading Logarithms With Improved Hadronic Matrix Elements. *Nucl. Phys.*, B408:209–285, 1993.

[203] A. J. Buras, P. Gambino, M. Gorbahn, S. Jager, and L. Silvestrini. ε'/ε and Rare K and B Decays in The MSSM. *Nucl. Phys.*, B592:55–91, 2001.

[204] A. J. Buras. Flavor dynamics: CP Violation and Rare Decays. *Subnucl. Ser.*, 38:200–337, 2002.

[205] E. Gabrielli and S. Khalil. Constraining Supersymmetric Models From $B_d - \bar{B}_d$ Mixing and The $B_d \to J/\psi K_S$ Asymmetry. *Phys. Rev.*, D67:015008, 2003.

[206] D. Becirevic, M. Ciuchini, E. Franco, V. Gimenez, G. Martinelli, A. Masiero, M. Papinutto, J. Reyes, and L. Silvestrini. $B_d - \bar{B}_d$ Mixing and The $B_d \to J/\psi K_s$ Asymmetry in General SUSY Models. *Nucl. Phys.*, B634:105–119, 2002.

[207] G. Buchalla, A. J. Buras, and M. E. Lautenbacher. Weak Decays Beyond Leading Logarithms. *Rev. Mod. Phys.*, 68:1125–1144, 1996.

[208] S. Bertolini, F. Borzumati, A. Masiero, and G. Ridolfi. Effects of Supergravity Induced Electroweak Breaking on Rare B Decays and Mixings. *Nucl. Phys.*, B353:591–649, 1991.

[209] E. Gabrielli and G. F. Giudice. Supersymmetric Corrections to ε'/ε at The Leading Order in QCD and QED. *Nucl. Phys.*, B433:3–25, 1995. [Erratum: *Nucl. Phys.*, B507:549, 1997].

[210] D. Chakraverty, E. Gabrielli, K. K. Huitu, and S. Khalil. Chargino Contributions to The CP Asymmetry in $B \to \phi K_S$ Decay. *Phys. Rev.*, D68:095004, 2003.

[211] S. Khalil and E. Kou. On Supersymmetric Contributions to The CP Asymmetry of The $B \to \phi K_S$. *Phys. Rev.*, D67:055009, 2003.

[212] B. Dutta, C. S. Kim, and S. Oh. A Consistent Resolution of Possible Anomalies in $B^0 \to \phi K_S$ and $B^+ \to \eta' K^+$ Decays. *Phys. Rev. Lett.*, 90:011801, 2003.

[213] R. L. Arnowitt, B. Dutta, and B. Hu. $B^0 \to \phi K_S$ in Sugra Models With CP Violations. *Phys. Rev.*, D68:075008, 2003.

[214] G. L. Kane, P. Ko, H. Wang, C. Kolda, J. Park, and L. Wang. $B_d \to \phi K_S$ and Supersymmetry. *Phys. Rev.*, D70:035015, 2004.

[215] S. Khalil and E. Kou. A Possible Supersymmetric Solution to The Discrepancy Between $B \to \phi K_S$ and $B \to \eta' K_S$ CP Asymmetries. *Phys. Rev. Lett.*, 91:241602, 2003.

[216] B. Dutta, C. S. Kim, S. Oh, and G. Zhu. An Analysis of $B \to \eta' K_S$ Decays Using A Global Fit in QCD Factorization. *Eur. Phys. J.*, C37:273–284, 2004.

[217] K. Agashe and C. D. Carone. Supersymmetric Flavor Models and The $B \to \phi K_S$ Anomaly. *Phys. Rev.*, D68:035017, 2003.

[218] M. Endo, S. Mishima, and M. Yamaguchi. Recent Measurements of CP Asymmetries of $B \to \phi K_S$ and $B \to \eta' K_S$ at B-factories Suggest New CP Violation in Left-Handed Squark Mixing. *Phys. Lett.*, B609:95–101, 2005.

[219] Y. Dai, C. Huang, W. Li, and X. Wu. CP Asymmetry in $B \to \phi K_S$ in SUSY SO(10) GUT. *Phys. Rev.*, D70:116002, 2004.

[220] S. Khalil. Supersymmetric Contribution to The CP Asymmetry of $B \to J/\psi \phi$ in The Light of Recent $B_s - \bar{B}_s$ Measurements. *Phys. Rev.*, D74:035005, 2006.

[221] S. G. Kim, N. Maekawa, A. Matsuzaki, K. Sakurai, and T. Yoshikawa. CP Asymmetries of $B \to \phi K_S$ and $B \to \eta' K_S$ in SUSY GUT Model With Non-universal Sfermion Masses. *Prog. Theor. Phys.*, 121:49–72, 2009.

[222] A. Ali, G. Kramer, and C. Lu. Experimental Tests of Factorization in Charmless Nonleptonic Two-Body B Decays. *Phys. Rev.*, D58:094009, 1998.

[223] M. Beneke, G. Buchalla, M. Neubert, and C. T. Sachrajda. QCD Factorization for $B \to \pi\pi$ Decays: Strong Phases and CP Violation in The Heavy Quark Limit. *Phys. Rev. Lett.*, 83:1914–1917, 1999.

[224] M. Beneke and M. Neubert. QCD Factorization for $B \to PP$ and $B \to PV$ Decays. *Nucl. Phys.*, B675:333–415, 2003.

[225] Y. Amhis et al. Averages of b-Hadron, c-Hadron, and τ-Lepton Properties as of Summer 2014. 2014.

[226] A. Kagan. The Phenomenology of Enhanced $b \to sg$. In *B physics and CP violation. Proceedings, 2nd International Conference, Honolulu, USA, March 24-27, 1997*, 1997.

[227] G. L. Fogli, E. Lisi, A. Marrone, D. Montanino, A. Palazzo, and A. M. Rotunno. Global Analysis of Neutrino Masses, Mixings and Phases: Entering The Era of Leptonic CP Violation Searches. *Phys. Rev.*, D86:013012, 2012.

[228] L. Basso, A. Belyaev, D. Chowdhury, M. Hirsch, S. Khalil, S. Moretti, B. O'Leary, W. Porod, and F. Staub. Proposal for Generalised Supersymmetry Les Houches Accord for See-Saw Models and PDG Numbering Scheme. *Comput. Phys. Commun.*, 184:698–719, 2013.

[229] S. Weinberg. Baryon and Lepton Nonconserving Processes. *Phys. Rev. Lett.*, 43:1566–1570, 1979.

[230] E. Ma. Pathways to Naturally Small Neutrino Masses. *Phys. Rev. Lett.*, 81:1171–1174, 1998.

[231] P. Minkowski. $\mu \to e\gamma$ at A Rate of One Out of 10^9 Muon Decays? *Phys. Lett.*, B67:421–428, 1977.

[232] M. Gell-Mann, P. Ramond, and R. Slansky. Complex Spinors and Unified Theories. *Conf. Proc.*, C790927:315–321, 1979.

[233] T. Yanagida. Horizontal gauge symmetry and masses of neutrinos. In Proceedings of The Workshop on The Baryon Number of The Universe and Unified Theories, Tsukuba, Japan, 13-14 Feb 1979.

[234] S.L. Glashow, in *Quarks and Leptons*, eds. M.Lèvy et al. (Plenum, New York 1980), p. 707.

[235] R. N. Mohapatra and G. Senjanovic. Neutrino Mass and Spontaneous Parity Non-conservation. *Phys. Rev. Lett.*, 44:912, 1980.

[236] J. Schechter and J. W. F. Valle. Neutrino Masses in $SU(2) \times U(1)$ Theories. *Phys.Rev.*, D22:2227, 1980.

[237] T. P. Cheng and L. Li. Neutrino Masses, Mixings and Oscillations in $SU(2) \times U(1)$ Models of Electroweak Interactions. *Phys.Rev.*, D22:2860, 1980.

[238] R. Foot, H. Lew, X. G. He, and G. C. Joshi. Seesaw Neutrino Masses Induced by A Triplet of Leptons. *Z.Phys.*, C44:441, 1989.

[239] E. K. Akhmedov, M. Lindner, E. Schnapka, and J. W. F. Valle. Dynamical Left-Right Symmetry Breaking. *Phys.Rev.*, D53:2752–2780, 1996.

[240] R. N. Mohapatra and J. W. F. Valle. Neutrino Mass and Baryon Number Nonconservation in Superstring Models. *Phys.Rev.*, D34:1642, 1986.

[241] L. J. Hall and M. Suzuki. Explicit R-Parity Breaking in Supersymmetric Models. *Nucl. Phys.*, B231:419, 1984.

[242] J. W. F. Valle. Supergravity Unification With Bilinear R-Parity Violation. In *Proceedings, 6th International Symposium on Particles, strings and cosmology (PASCOS 1998)*, pages 502–512, 1998.

[243] J. A. Casas and A. Ibarra. Oscillating Neutrinos and $\mu \to e\gamma$. *Nucl. Phys.*, B618:171–204, 2001.

[244] J. Adam et al. New Constraint on The Existence of The $\mu^+ \to e^+\gamma$ Decay. *Phys. Rev. Lett.*, 110:201801, 2013.

[245] B. Aubert et al. Searches for Lepton Flavor Violation in The Decays $\tau^\pm \to e^\pm\gamma$ and $\tau^\pm \to \mu^\pm\gamma$. *Phys. Rev. Lett.*, 104:021802, 2010.

[246] U. Bellgardt et al. Search for The Decay $\mu^+ \to e^+e^+e^-$. *Nucl. Phys.*, B299:1, 1988.

[247] K. Hayasaka et al. Search for Lepton Flavor Violating τ Decays into Three Leptons With 719 MILLION Produced $\tau^+\tau^-$ Pairs. *Phys. Lett.*, B687:139–143, 2010.

[248] J. Hisano, T. Moroi, K. Tobe, and M. Yamaguchi. Lepton Flavor Violation Via Right-Handed Neutrino Yukawa Couplings in Supersymmetric Standard Model. *Phys. Rev.*, D53:2442–2459, 1996.

[249] J. R. Ellis, J. Hisano, S. Lola, and M. Raidal. CP Violation in The Minimal Supersymmetric Seesaw Model. *Nucl. Phys.*, B621:208–234, 2002.

[250] D. F. Carvalho, M. E. Gomez, and S. Khalil. Lepton Flavor Violation With Non-Universal Soft Terms. *JHEP*, 07:001, 2001.

[251] L. Calibbi, A. Faccia, A. Masiero, and S. K. Vempati. Lepton Flavour Violation From SUSY-GUTs: Where Do We Stand for MEG, PRISM/PRIME and A Super Flavour Factory. *Phys. Rev.*, D74:116002, 2006.

[252] F. Deppisch, H. Päs, A. Redelbach, and R. Rückl. Lepton Flavor Violation in The SUSY Seesaw Model: An Update. *Springer Proc. Phys.*, 98:27–38, 2005.

[253] S. F. King, S. Moretti, and R. Nevzorov. Theory and Phenomenology of An Exceptional Supersymmetric Standard Model. *Phys. Rev.*, D73:035009, 2006.

[254] M. Malinsky, J. C. Romao, and J. W. F. Valle. Novel Supersymmetric $SO(10)$ Seesaw Mechanism. *Phys. Rev. Lett.*, 95:161801, 2005.

[255] M. Hirsch, M. Malinsky, W. Porod, L. Reichert, and F. Staub. Hefty MSSM-Like Light Higgs in Extended Gauge Models. *JHEP*, 02:084, 2012.

[256] M. Hirsch, W. Porod, L. Reichert, and F. Staub. Phenomenology of the minimal supersymmetric $U(1)_{B-L} \times U(1)_R$ extension of the standard model. *Phys. Rev.*, D86:093018, 2012.

[257] V. De Romeri, M. Hirsch, and M. Malinsky. Soft Masses in SUSY SO(10) GUTs With Low Intermediate Scales. *Phys. Rev.*, D84:053012, 2011.

[258] S. Khalil and A. Masiero. Radiative $B - L$ Symmetry Breaking in Supersymmetric Models. *Phys. Lett.*, B665:374–377, 2008.

[259] P. P. Fileviez and S. Spinner. The Fate of R-Parity. *Phys. Rev.*, D83:035004, 2011.

[260] A. Elsayed, S. Khalil, and S. Moretti. Higgs Mass Corrections in The SUSY $B - L$ Model with Inverse Seesaw. *Phys. Lett.*, B715:208–213, 2012.

[261] A. Elsayed, S. Khalil, S. Moretti, and A. Moursy. Right-Handed Sneutrino-Antisneutrino Oscillations in a TeV Scale Supersymmetric $B - L$ Model. *Phys. Rev.*, D87(5):053010, 2013.

[262] L. Basso, B. O'Leary, W. Porod, and F. Staub. Dark Matter Scenarios in The Minimal SUSY $B - L$ Model. *JHEP*, 09:054, 2012.

[263] B. Holdom. Two $U(1)$'s and Epsilon Charge Shifts. *Phys. Lett.*, B166:196, 1986.

[264] K. S. Babu, C. F. Kolda, and J. March-Russell. Implications of Generalized $Z - Z'$ Mixing. *Phys. Rev.*, D57:6788–6792, 1998.

[265] F. del Aguila, G. D. Coughlan, and M. Quiros. Gauge Coupling Renormalization With Several $U(1)$ Factors. *Nucl. Phys.*, B307:633, 1988. [Erratum: *Nucl. Phys.*, B312:751,1989].

[266] F. del Aguila, J. A. Gonzalez, and M. Quiros. Renormalization Group Analysis of Extended Electroweak Models From The Heterotic String. *Nucl. Phys.*, B307:571, 1988.

[267] W. Porod R. Fonseca, M. Malinsky and F. Staub. Running Soft Parameters in SUSY Models With Multiple $U(1)$ Gauge Factors. *Nucl. Phys.*, B854:28–53, 2012.

[268] P. H. Chankowski, S. Pokorski, and J. Wagner. Z' and The Appelquist-Carrazzone Decoupling. *Eur. Phys. J.*, C47:187–205, 2006.

[269] D. Suematsu. Vacuum Structure of The μ Problem Solvable Extra $U(1)$ Models. *Phys. Rev.*, D59:055017, 1999.

[270] F. Braam and J. Reuter. A Simplified Scheme for GUT-Inspired Theories With Multiple Abelian Factors. *Eur. Phys. J.*, C72:1885, 2012.

[271] L. Basso, S. Moretti, and G. M. Pruna. A Renormalisation Group Equation Study of The Scalar Sector of The Minimal $B - L$ Extension of The Standard Model. *Phys. Rev.*, D82:055018, 2010.

[272] L. Basso, S. Moretti, and G. M. Pruna. Constraining The g_1' Coupling in The Minimal $B - L$ Model. *J. Phys.*, G39:025004, 2012.

[273] J. Alcaraz et al. A Combination of Preliminary Electroweak Measurements and Constraints on The Standard Model. *hep-ex/0612034*, 2006.

[274] J. Erler, P. Langacker, S. Munir, and E. Rojas. Improved Constraints on Z' Bosons From Electroweak Precision Data. *JHEP*, 08:017, 2009.

[275] K. Nakamura et al. Review of Particle Physics. *J. Phys.*, G37:075021, 2010.

[276] G. Cacciapaglia, C. Csaki, G. Marandella, and A. Strumia. The Minimal Set of Electroweak Precision Parameters. *Phys. Rev.*, D74:033011, 2006.

[277] D. Adams, talk given at Rencontres de Moriond, EW Interactions and Unified Theories, 3-10 March 2012.

[278] A. El-Zant, S. Khalil, and A. Sil. Warm Dark Matter in A $B - L$ Inverse Seesaw Scenario. *Phys. Rev.*, D91(3):035030, 2015.

[279] S. Khalil. Low Scale $B - L$ Extension of The Standard Model at The LHC. *J. Phys.*, G35:055001, 2008.

[280] S. Khalil and S. Moretti. Heavy Neutrinos, Z' and Higgs Bosons at The LHC: New Particles From An Old Symmetry. *J. Mod. Phys.*, 4(1):7–10, 2013.

[281] S. Khalil and S. Moretti. A Simple Symmetry As A Guide Toward New Physics Beyond The Standard Model. *Front.In Phys.*, 1:10, 2013.

[282] W. Abdallah, A. Awad, S. Khalil, and H. Okada. Muon Anomalous Magnetic Moment and $\mu \to e\gamma$ in $B - L$ Model With Inverse Seesaw. *Eur. Phys. J.*, C72:2108, 2012.

[283] F. Staub. SARAH 4 : A Tool for (Not Only SUSY) Model Builders. *Comput. Phys. Commun.*, 185:1773–1790, 2014.

[284] S. Khalil. Radiative Symmetry Breaking in Supersymmetric $B - L$ Models With An Inverse Seesaw Mechanism. *Phys. Rev.*, D94(7):075003, 2016.

[285] J. L. Hewett and T. G. Rizzo. Low-Energy Phenomenology of Superstring Inspired E_6 Models. *Phys. Rept.*, 183:193, 1989.

[286] A. Leike. The Phenomenology of Extra Neutral Gauge Bosons. *Phys. Rept.*, 317:143–250, 1999.

[287] P. Langacker. The Physics of Heavy Z' Gauge Bosons. *Rev. Mod. Phys.*, 81:1199–1228, 2009.

[288] L. Basso, A. Belyaev, S. Moretti, and C. H. Shepherd-Themistocleous. Phenomenology of The Minimal $B - L$ Extension of the Standard Model: Z' and Neutrinos. *Phys. Rev.*, D80:055030, 2009.

[289] L. Basso, A. Belyaev, S. Moretti, G. M. Pruna, and C. H. Shepherd-Themistocleous. Z' Discovery Potential at The LHC in The Minimal $B - L$ Extension of The Standard Model. *Eur. Phys. J.*, C71:1613, 2011.

[290] L. Basso, S. Moretti, and G. M. Pruna. Phenomenology of The Minimal $B - L$ Extension of the Standard Model: The Higgs Sector. *Phys. Rev.*, D83:055014, 2011.

[291] L. Basso, S. Moretti, and G. M. Pruna. Theoretical Constraints on The Couplings of Non-Exotic Minimal Z' Bosons. *JHEP*, 08:122, 2011.

[292] W. Emam and S. Khalil. Higgs and Z' Phenomenology in $B - L$ Extension of The Standard Model at LHC. *Eur. Phys. J.*, C52:625–633, 2007.

[293] M. Abbas, W. Emam, S. Khalil, and M. Shalaby. TeV scale $B - L$ phenomenology at LHC. *Int. J. Mod. Phys.*, A22:5889–5908, 2007.

[294] K. Huitu, S. Khalil, H. Okada, and S. K. Rai. Signatures for Right-Handed Neutrinos at The Large Hadron Collider. *Phys. Rev. Lett.*, 101:181802, 2008.

[295] A. A. Abdelalim, A. Hammad, and S. Khalil. $B - L$ Heavy Neutrinos and Neutral Gauge Boson Z' at The LHC. *Phys. Rev.*, D90(11):115015, 2014.

[296] G. Arcadi, Y. Mambrini, M. H. Tytgat, and B. Zaldivar. Invisible Z' and Dark Matter: LHC vs LUX Constraints. *JHEP*, 03:134, 2014.

[297] ATLAS Collaboration. Search for high-mass dilepton resonances in 20 fb^{-1} of pp collisions at $\sqrt{s} = 8$ TeV with the ATLAS experiment. *ATLAS-CONF-2013-017*, 2013.

[298] S. Chatrchyan et al. Search for Heavy Narrow Dilepton Resonances in pp Collisions at $\sqrt{s} = 7$ TeV and $\sqrt{s} = 8$ TeV. *Phys. Lett.*, B720:63–82, 2013.

[299] L. Basso, A. Belyaev, S. Moretti, and G. M. Pruna. Probing The Z' Sector of The Minimal $B - L$ Model at Future Linear Colliders in The $e^+e^- \to \mu^+\mu^-$ Process. *JHEP*, 10:006, 2009.

[300] L. Basso, A. Belyaev, S. Moretti, G. Pruna, and C. H. Shepherd-Themistocleous. Phenomenology of The Minimal $B - L$ Extension of The Standard Model. *PoS*, EPS-HEP2009:242, 2009.

[301] M. Abbas and S. Khalil. Neutrino Masses, Mixing and Leptogenesis in TeV Scale $B - L$ Extension of The Standard Model. *JHEP*, 04:056, 2008.

[302] T. Sjostrand, S. Mrenna, and P. Z. Skands. A Brief Introduction to PYTHIA 8.1. *Comput. Phys. Commun.*, 178:852–867, 2008.

[303] J. de Favereau, C. Delaere, P. Demin, A. Giammanco, V. Lemaitre, A. Mertens, and M. Selvaggi. DELPHES 3, A Modular Framework for Fast Simulation of A Generic Collider Experiment. *JHEP*, 02:057, 2014.

[304] F. Gianotti et al. Physics Potential and Experimental Challenges of The LHC Luminosity Upgrade. *Eur. Phys. J.*, C39:293–333, 2005.

[305] V. D. Barger and T. Han. Triple Gauge Boson Production at e^+e^- and pp Supercolliders. *Phys. Lett.*, B212:117–122, 1988.

[306] V. Hankele and D. Zeppenfeld. QCD Corrections to Hadronic WWZ Production With Leptonic Decays. *Phys. Lett.*, B661:103–108, 2008.

[307] W. Abdallah, J. Fiaschi, S. Khalil, and S. Moretti. Mono-Jet, -Photon and -Z Signals of A Supersymmetric $(B - L)$ Model at The Large Hadron Collider. *JHEP*, 02:157, 2016.

[308] W. Abdallah, J. Fiaschi, S. Khalil, and S. Moretti. Z'-Induced Invisible Right-Handed Sneutrino Decays at The LHC. *Phys. Rev.*, D92:055029, 2015.

[309] ATLAS Collaboration, Expected Performance of The ATLAS Experiment—Detector, Trigger and Physics. *arXiv:0901.0512*, 2015.

[310] B. C. Allanach, S. Grab, and H. E. Haber. Supersymmetric Monojets at The Large Hadron Collider. *JHEP*, 01:138, 2011. [Erratum: *JHEP*, 09:027, 2011].

[311] M. Drees, M. Hanussek, and J.S. Kim. Light Stop Searches at The LHC With Monojet Events. *Phys. Rev.*, D86:035024, 2012.

[312] S. Khalil and H. Okada. Dark Matter in $B - L$ Extended MSSM Models. *Phys. Rev.*, D79:083510, 2009.

[313] S. K. King, S. Moretti, and R. Nevzorov. Gauge Coupling Unification in The Exceptional Supersymmetric Standard Model. *Phys. Lett.*, B650:57–64, 2007.

[314] F. del Aguila, G. Blair, M. Daniel, and G. G. Ross. Superstring Inspired Models. *Nucl. Phys.*, B272:413, 1986.

[315] Y. Hosotani. Dynamical Gauge Symmetry Breaking as The Casimir Effect. *Phys. Lett.*, B129:193, 1983.

[316] M. Cvetic and P. Langacker. Implications of Abelian Extended Gauge Structures From String Models. *Phys. Rev.*, D54:3570–3579, 1996.

[317] M. Cvetic and P. Langacker. New Gauge Bosons From String Models. *Mod. Phys. Lett.*, A11:1247–1262, 1996.

[318] P. Langacker and J. Wang. $U(1)'$ Symmetry Breaking in Supersymmetric E_6 Models. *Phys. Rev.*, D58:115010, 1998.

[319] S. F. King. Neutrino Mass Models. *Rept. Prog. Phys.*, 67:107–158, 2004.

[320] E. Keith and E. Ma. Generic Consequences of A Supersymmetric U(1) Gauge Factor at The TeV Scale. *Phys. Rev.*, D56:7155–7165, 1997.

[321] J. Rich, O. D. Lloyd, and M. Spiro. Experimental Particle Physics Without Accelerators. *Phys. Rept.*, 151:239–364, 1987.

[322] P. F. Smith. Terrestrial Searches for New Stable Particles. *Contemp. Phys.*, 29:159–186, 1988.

[323] T. K. Hemmick et al. A Search for Anomalously Heavy Isotopes of Low Z Nuclei. *Phys. Rev.*, D41:2074–2080, 1990.

[324] L. E. Ibanez and G. G. Ross. $SU(2)_L \times U(1)_Y$ Symmetry Breaking as a Radiative Effect of Supersymmetry Breaking in Guts. *Phys. Lett.*, B110:215–220, 1982.

[325] J. R. Ellis, D. V. Nanopoulos, and K. Tamvakis. Grand Unification in Simple Supergravity. *Phys. Lett.*, B121:123, 1983.

[326] J. R. Ellis, J. S. Hagelin, D. V. Nanopoulos, and K. Tamvakis. Weak Symmetry Breaking by Radiative Corrections in Broken Supergravity. *Phys. Lett.*, B125:275, 1983.

[327] T. Hambye, E. Ma, M. Raidal, and U. Sarkar. Allowable Low-Energy E_6 Subgroups From Leptogenesis. *Phys. Lett.*, B512:373–378, 2001.

[328] E. Ma and M. Raidal. Three Active and Two Sterile Neutrinos in An E_6 Model of Diquark Baryogenesis. *J. Phys.*, G28:95–102, 2002.

[329] E. Ma. Neutrino Masses in An Extended Gauge Model With E_6 Particle Content. *Phys. Lett.*, B380:286–290, 1996.

[330] E. Keith and E. Ma. Efficacious Extra U(1) Factor for The Supersymmetric Standard Model. *Phys. Rev.*, D54:3587–3593, 1996.

[331] D. Suematsu. Neutralino Decay in The μ-Problem Solvable Extra U(1) Models. *Phys. Rev.*, D57:1738–1754, 1998.

[332] Y. Daikoku and D. Suematsu. Mass Bound of The Lightest Neutral Higgs Scalar in The Extra $U(1)$ Models. *Phys. Rev.*, D62:095006, 2000.

[333] M. Cvetic and S. Godfrey. Discovery and Identification of Extra Gauge Bosons. *hep-ph/9504216*, 1995.

[334] J. Kang and P. Langacker. Z' Discovery Limits for Supersymmetric E_6 Models. *Phys. Rev.*, D71:035014, 2005.

[335] M. Dittmar, A. Nicollerat, and A. Djouadi. Z' Studies at The LHC: An Update. *Phys. Lett.*, B583:111–120, 2004.

[336] J. D. Lykken L. J. Hall and S. Weinberg. Supergravity As The Messenger of Supersymmetry Breaking. *Phys. Rev.*, D27:2359–2378, 1983.

[337] U. Ellwanger. Nonrenormalizable Interactions From Supergravity, Quantum Corrections and Effective Low-Energy Theories. *Phys. Lett.*, B133:187–191, 1983.

[338] J. E. Kim and H. P. Nilles. The μ Problem and The Strong CP Problem. *Phys. Lett.*, B138:150, 1984.

[339] K. Inoue, A. Kakuto, and H. Takano. Higgs as (Pseudo)Goldstone Particles. *Prog. Theor. Phys.*, 75:664, 1986.

[340] A. A. Anselm and A. A. Johansen. SUSY GUT With Automatic Doublet - Triplet Hierarchy. *Phys. Lett.*, B200:331–334, 1988.

[341] G/ F. Giudice and A. Masiero. A Natural Solution to The μ Problem in Supergravity Theories. *Phys. Lett.*, B206:480–484, 1988.

[342] E. Accomando et al. Workshop on CP Studies and Non-Standard Higgs Physics. 2006.

[343] V. Barger, H. E. Logan, and G. Shaughnessy. Identifying extended Higgs models at the LHC. *Phys. Rev.*, D79:115018, 2009.

[344] V. Barger, P. Langacker, M. McCaskey, M. Ramsey-Musolf, and G. Shaughnessy. Complex Singlet Extension of the Standard Model. *Phys. Rev.*, D79:015018, 2009.

[345] V. Barger, P. Langacker, M. McCaskey, and and G. Shaughnessy M. Ramsey-Musolf. LHC Phenomenology of an Extended Standard Model with a Real Scalar Singlet. *Phys. Rev.*, D77:035005, 2008.

[346] V. Barger, P. Langacker, and G. Shaughnessy. Collider Signatures of Singlet Extended Higgs Sectors. *Phys. Rev.*, D75:055013, 2007.

[347] V. Barger, P. Langacker, and G. Shaughnessy. Singlet extensions of the MSSM. *AIP Conf. Proc.*, 903:32–39, 2007. [,32(2006)].

[348] H. P. Nilles, M. Srednicki, and D. Wyler. Weak Interaction Breakdown Induced by Supergravity. *Phys. Lett.*, B120:346, 1983.

[349] J. R. Ellis, J. F. Gunion, H. E. Haber, L. Roszkowski, and F. Zwirner. Higgs Bosons in a Nonminimal Supersymmetric Model. *Phys. Rev.*, D39:844, 1989.

[350] M. Drees. Supersymmetric Models With Extended Higgs Sector. *Int. J. Mod. Phys.*, A4:3635, 1989.

[351] L. Durand and J. L. Lopez. Upper Bounds on Higgs and Top Quark Masses in the Flipped $SU(5) \times U(1)$ Superstring Model. *Phys. Lett.*, B217:463, 1989.

[352] U. Ellwanger, M. Rausch de Traubenberg, and C. A. Savoy. Particle Spectrum in Supersymmetric Models With A Gauge Singlet. *Phys. Lett.*, B315:331–337, 1993.

[353] U. Ellwanger, M. Rausch de Traubenberg, and C. A. Savoy. Higgs Phenomenology of The Supersymmetric Model With A Gauge Singlet. *Z. Phys.*, C67:665–670, 1995.

[354] U. Ellwanger, M. Rausch de Traubenberg, and C. A. Savoy. Phenomenology of Supersymmetric Models With A Singlet. *Nucl. Phys.*, B492:21–50, 1997.

[355] T. Elliott, S. F. King, and P. L. White. Unification Constraints in The Next-to-Minimal Supersymmetric Standard Model. *Phys. Lett.*, B351:213–219, 1995.

[356] S. F. King and P. L. White. Resolving The Constrained Minimal and Next-to-Minimal Supersymmetric Standard Models. *Phys. Rev.*, D52:4183–4216, 1995.

[357] D. A. Demir. Dynamical Relaxation of The CP Phases in Next-to-Minimal Supersymmetry. *Phys. Rev.*, D62:075003, 2000.

[358] U. Ellwanger, C. Hugonie, and A. M. Teixeira. The Next-to-Minimal Supersymmetric Standard Model. *Phys. Rept.*, 496:1–77, 2010.

[359] Ya. B. Zeldovich, I. Yu. Kobzarev, and L. B. Okun. Cosmological Consequences of The Spontaneous Breakdown of Discrete Symmetry. *Zh. Eksp. Teor. Fiz.*, 67:3–11, 1974. [Sov. Phys. JETP40,1(1974)].

[360] H. P. Nilles, M. Srednicki, and D. Wyler. Constraints on The Stability of Mass Hierarchies in Supergravity. *Phys. Lett.*, B124:337, 1983.

[361] A. B. Lahanas. Light Singlet, Gauge Hierarchy and Supergravity. *Phys. Lett.*, B124:341, 1983.

[362] H. P. Nilles and N. Polonsky. Gravitational Divergences As A Mediator of Supersymmetry Breaking. *Phys. Lett.*, B412:69–76, 1997.

[363] J. Bagger and E. Poppitz. Destabilizing Divergences in Supergravity Coupled Supersymmetric Theories. *Phys. Rev. Lett.*, 71:2380–2382, 1993.

[364] J. Bagger, E. Poppitz, and L. Randall. Destabilizing Divergences in Supergravity Theories at Two Loops. *Nucl. Phys.*, B455:59–82, 1995.

[365] Y. Jain. On Destabilizing Divergencies in Supergravity Models. *Phys. Lett.*, B351:481–486, 1995.

[366] S. A. Abel. Destabilizing Divergences in The NMSSM. *Nucl. Phys.*, B480:55–72, 1996.

[367] C. F. Kolda, S. Pokorski, and N. Polonsky. Stabilized Singlets in Supergravity As A Source of The μ-Parameter. *Phys. Rev. Lett.*, 80:5263–5266, 1998.

[368] S. A. Abel, S. Sarkar, and P. L. White. On The Cosmological DomaIn Wall Problem for The Minimally Extended Supersymmetric Standard Model. *Nucl. Phys.*, B454:663–684, 1995.

[369] M. Cvetic, D. A. Demir, J. R. Espinosa, L. L. Everett, and P. Langacker. Electroweak Breaking and The μ Problem in Supergravity Models With An Additional $U(1)$. *Phys. Rev.*, D56:2861, 1997. [Erratum: *Phys. Rev.*, D58:119905, 1998].

[370] D. A. Demir and L. L. Everett. CP violation in Supersymmetric $U(1)'$ models. *Phys. Rev.*, D69:015008, 2004.

[371] T. Han, P. Langacker, and B. McElrath. The Higgs Sector in A $U(1)'$ Extension of The MSSM. *Phys. Rev.*, D70:115006, 2004.

[372] C. Panagiotakopoulos and K. Tamvakis. Stabilized NMSSM Without DomaIn Walls. *Phys. Lett.*, B446:224–227, 1999.

[373] C. Panagiotakopoulos and K. Tamvakis. New Minimal Extension of MSSM. *Phys. Lett.*, B469:145–148, 1999.

[374] C. Panagiotakopoulos and A. Pilaftsis. Higgs Scalars in The Minimal Nonminimal Supersymmetric Standard Model. *Phys. Rev.*, D63:055003, 2001.

[375] A. Dedes, C. Hugonie, S. Moretti, and K. Tamvakis. Phenomenology of A New Minimal Supersymmetric Extension of The Standard Model. *Phys. Rev.*, D63:055009, 2001.

[376] C. Panagiotakopoulos and A. Pilaftsis. Light Charged Higgs Boson and Supersymmetry. *Phys. Lett.*, B505:184–190, 2001.

[377] S. Ham, S. K. Oh, and B. R. Kim. Absolute Upper Bound on The One Loop Corrected Mass of S_1 in The NMSSM. *J. Phys.*, G22:1575–1584, 1996.

[378] D. J. Miller, R. Nevzorov, and P. M. Zerwas. The Higgs Sector of The Next-to-Minimal Supersymmetric Standard Model. *Nucl. Phys.*, B681:3–30, 2004.

[379] D. A. Demir and N. K. Pak. One Loop Effects in Supergravity Models With An Additional $U(1)$. *Phys. Rev.*, D57:6609–6617, 1998.

[380] H. Amini. Radiative Corrections to Higgs Masses in Z' Models. *New J. Phys.*, 5:49, 2003.

[381] M. A. Diaz and H. E. Haber. One Loop Radiative Corrections to The Charged Higgs Mass of The Minimal Supersymmetric Model. *Phys. Rev.*, D45:4246–4260, 1992.

[382] H. E. Haber and R. Hempfling. The Renormalization Group Improved Higgs Sector of The Minimal Supersymmetric Model. *Phys. Rev.*, D48:4280–4309, 1993.

[383] G. L. Kane, C. F. Kolda, and J. D. Wells. Calculable Upper Limit on The Mass of The Lightest Higgs Boson in Any Perturbatively Valid Supersymmetric Theory. *Phys. Rev. Lett.*, 70:2686–2689, 1993.

[384] Y. Okada, M. Yamaguchi, and T. Yanagida. Upper Bound of The Lightest Higgs Boson Mass in The Minimal Supersymmetric Standard Model. *Prog. Theor. Phys.*, 85:1–6, 1991.

[385] S. Khalil, H. Okada, and T. Toma. Right-Handed Sneutrino Dark Matter in Supersymmetric $B - L$ Model. *JHEP*, 07:026, 2011.

[386] U. Ellwanger, J. F. Gunion, and C. Hugonie. Difficult Scenarios for NMSSM Higgs Discovery at The LHC. *JHEP*, 07:041, 2005.

[387] S. Chang, P. J. Fox, and N. Weiner. Naturalness and Higgs Decays in The MSSM With A Singlet. *JHEP*, 08:068, 2006.

[388] S. Chang, R. Dermisek, J. F. Gunion, and N. Weiner. Nonstandard Higgs Boson Decays. *Ann. Rev. Nucl. Part. Sci.*, 58:75–98, 2008.

[389] U. Aglietti et al. Tevatron for LHC Report: Higgs. *hep-ph/0612172*, 2006.

[390] R. Barbieri, L. J. Hall, A. Y. Papaioannou, D. Pappadopulo, and V. S. Rychkov. An Alternative NMSSM Phenomenology With Manifest Perturbative Unification. *JHEP*, 03:005, 2008.

[391] A. Djouadi et al. Benchmark Scenarios for The NMSSM. *JHEP*, 07:002, 2008.

[392] I. Rottlander and M. Schumacher. Investigation of The LHC Discovery Potential for Higgs Bosons in The NMSSM. 2009.

[393] A. Djouadi, U. Ellwanger, and A. M. Teixeira. Phenomenology of The Constrained NMSSM. *JHEP*, 04:031, 2009.

[394] S. Munir N. Bomark, S. Moretti and L. Roszkowski. LHC Phenomenology of Light Pseudoscalars in The NMSSM. *PoS*, EPS-HEP2015:162, 2015.

[395] N. Bomark, S. Moretti, and L. Roszkowski. Detection Prospects of Light NMSSM Higgs Pseudoscalar via Cascades of Heavier Scalars From Vector Boson Fusion and Higgs-strahlung. *arXiv:1503.04228*, 2015.

[396] N. Bomark, S. Moretti, S. Munir, and L. Roszkowski. A light NMSSM Pseudoscalar Higgs Boson at The LHC Run 2. In *2nd Toyama International Workshop on Higgs as a Probe of New Physics (HPNP2015) Toyama, Japan, February 11-15, 2015*, 2015.

[397] N. Bomark, S. Moretti, S. Munir, and L. Roszkowski. Revisiting A Light NMSSM Pseudoscalar at The LHC. *PoS*, Charged2014:029, 2015.

[398] N. Bomark, S. Moretti, S. Munir, and L. Roszkowski. A light NMSSM Pseudoscalar Higgs Boson at The LHC Redux. *JHEP*, 02:044, 2015.

[399] S. F. King, M. Muhlleitner, R. Nevzorov, and K. Walz. Discovery Prospects for NMSSM Higgs Bosons at the High-Energy Large Hadron Collider. *Phys. Rev.*, D90(9):095014, 2014.

[400] S. F. King, M. Muhlleitner, R. Nevzorov, and K. Walz. Natural NMSSM Higgs Bosons. *Nucl. Phys.*, B870:323–352, 2013.

[401] S. F. King, M. Muhlleitner, and R. Nevzorov. NMSSM Higgs Benchmarks Near 125 GeV. *Nucl. Phys.*, B860:207–244, 2012.

[402] N. Arkani-Hamed, A. Delgado, and G. F. Giudice. The Well-Tempered Neutralino. *Nucl. Phys.*, B741:108–130, 2006.

[403] G. Chalons, M. J. Dolan, and C. McCabe. Neutralino Dark Matter and The Fermi Gamma-Ray Lines. *JCAP*, 1302:016, 2013.

[404] A. Hammad, S. Khalil, and S. Moretti. Higgs Boson Decays into $\gamma\gamma$ and $Z\gamma$ in The MSSM and The $B - L$ Supersymmetric SM. *Phys. Rev.*, D92(9):095008, 2015.

[405] J. A. Casas, J. M. Moreno, K. Rolbiecki, and B. Zaldivar. Implications of Light Charginos for Higgs Observables, LHC Searches and Dark Matter. *JHEP*, 09:099, 2013.

[406] A. Djouadi, V. Driesen, W. Hollik, and J. L. Illana. The Coupling of The lightest SUSY Higgs Boson to Two Photons in The Decoupling Regime. *Eur. Phys. J.*, C1:149–162, 1998.

[407] A. Arbey, M. Battaglia, A. Djouadi, F. Mahmoudi, and J. Quevillon. Implications of a 125 GeV Higgs for Supersymmetric Models. *Phys. Lett.*, B708:162–169, 2012.

[408] S. Heinemeyer, O. Stal, and G. Weiglein. Interpreting The LHC Higgs Search Results in The MSSM. *Phys. Lett.*, B710:201–206, 2012.

[409] L. J. Hall, D. Pinner, and J. T. Ruderman. A Natural SUSY Higgs Near 126 GeV. *JHEP*, 04:131, 2012.

[410] M. Carena, S. Gori, N. R. Shah, and C. E. M. Wagner. A 125 GeV SM-like Higgs in The MSSM and The $\gamma\gamma$ Rate. *JHEP*, 03:014, 2012.

[411] K. Schmidt-Hoberg and F. Staub. Enhanced $h \to \gamma\gamma$ Rate in MSSM Singlet Extensions. *JHEP*, 10:195, 2012.

[412] L. Basso and F. Staub. Enhancing $h \to \gamma\gamma$ With Staus in SUSY Models With Extended Gauge Sector. *Phys. Rev.*, D87(1):015011, 2013.

[413] S. Moretti. Variations on a Higgs theme. *Phys. Rev.*, D91(1):014012, 2015.

[414] M. A. Luty. 2004 TASI Lectures on Supersymmetry Breaking. In *Physics in D ≥ 4. Proceedings, Theoretical Advanced Study Institute in elementary particle physics, TASI 2004, Boulder, USA, June 6-July 2, 2004*, pages 495–582, 2005.

[415] D. Bailin and A. Love, *Supersymmetric Gauge Field Theory and String Theory*, Taylor and Francis press, 1994.

[416] D. G. Cerdeno and C. Munoz. An Introduction to Supergravity. 1998. [PoScorfu98,011(1998)].

[417] D. Z. Freedman, P. Van Nieuwenhuizen, and S. Ferrara. Progress Toward a Theory of Supergravity. *Phys. Rev.*, D13:3214–3218, 1976.

[418] S. Deser and B. Zumino. Consistent Supergravity. *Phys. Lett.*, B62:335, 1976.

[419] E. Cremmer, B. Julia, J. Scherk, P. Van Nieuwenhuizen, S. Ferrara, and L. Girardello. Super-higgs effect in supergravity With general scalar interactions. *Phys. Lett.*, B79:231, 1978.

[420] E. Cremmer, S. Ferrara, L. Girardello, and A. Van Proeyen. Yang-Mills Theories With Local Supersymmetry: Lagrangian, Transformation Laws and SuperHiggs Effect. *Nucl. Phys.*, B212:413, 1983.

[421] J. Polonyi. Generalization of The Massive Scalar Multiplet Coupling to The Supergravity. *KFKI-77-93*, 1977.

[422] A. Brignole, L. E. Ibanez, and C. Munoz. Towards A Theory of Soft Terms for The Supersymmetric Standard Model. *Nucl. Phys.*, B422:125–171, 1994. [Erratum: *Nucl. Phys.*, B436:747, 1995].

[423] E. Cremmer, S. Ferrara, C. Kounnas, and D. V. Nanopoulos. Naturally Vanishing Cosmological Constant in $N = 1$ Supergravity. *Phys. Lett.*, B133:61, 1983.

[424] J. R. Ellis, A. B. Lahanas, D. V. Nanopoulos, and K. Tamvakis. No-Scale Supersymmetric Standard Model. *Phys. Lett.*, B134:429, 1984.

[425] R. Barbieri, E. Cremmer, and S. Ferrara. Flat and Positive Potentials in $N = 1$ Supergravity. *Phys. Lett.*, B163:143, 1985.

[426] G. F. Giudice, M. A. Luty, H. Murayama, and R. Rattazzi. Gaugino Mass Without Singlets. *JHEP*, 12:027, 1998.

[427] L. Randall and R. Sundrum. Out of This World Supersymmetry Breaking. *Nucl. Phys.*, B557:79–118, 1999.

[428] C. Pica and F. Sannino. Beta Function and Anomalous Dimensions. *Phys. Rev.*, D83:116001, 2011.

[429] G. F. Giudice and R. Rattazzi. Theories With Gauge Mediated Supersymmetry Breaking. *Phys. Rept.*, 322:419–499, 1999.

[430] L. Alvarez-Gaume, M. Claudson, and M. B. Wise. Low-Energy Supersymmetry. *Nucl. Phys.*, B207:96, 1982.

[431] G. F. Giudice and R. Rattazzi. Extracting Supersymmetry Breaking Effects From Wave Function Renormalization. *Nucl. Phys.*, B511:25–44, 1998.

[432] S. Dimopoulos, S. D. Thomas, and J. D. Wells. Sparticle Spectroscopy and Electroweak Symmetry Breaking With Gauge Mediated Supersymmetry Breaking. *Nucl. Phys.*, B488:39–91, 1997.

[433] I. Gogoladze, Q. Shafi, and C. S. Un. Reconciling The Muon $g - 2$, a 125 GeV Higgs Boson, and Dark Matter in Gauge Mediation Models. *Phys. Rev.*, D92(11):115014, 2015.

[434] M. Ibe, S. Matsumoto, T. T. Yanagida, and N. Yokozaki. Heavy Squarks and Light Sleptons in Gauge Mediation From The viewpoint of 125 GeV Higgs Boson and Muon $g - 2$. *JHEP*, 03:078, 2013.

[435] P. Grajek, A. Mariotti, and D. Redigolo. Phenomenology of General Gauge Mediation in light of a 125 GeV Higgs. *JHEP*, 07:109, 2013.

[436] J. L. Feng, Z. Surujon, and H. Yu. Confluence of Constraints in Gauge Mediation: The 125 GeV Higgs Boson and Goldilocks Cosmology. *Phys. Rev.*, D86:035003, 2012.

[437] P. Byakti and D. Ghosh. Magic Messengers in Gauge Mediation and Signal for 125 GeV Boosted Higgs Boson. *Phys. Rev.*, D86:095027, 2012.

[438] J. L. Evans, M. Ibe, S. Shirai, and T.T. Yanagida. A 125GeV Higgs Boson and Muon $g - 2$ in More Generic Gauge Mediation. *Phys. Rev.*, D85:095004, 2012.

[439] M. A. Ajaib, I. Gogoladze, F. Nasir, and Q. Shafi. Revisiting mGMSB in Light of a 125 GeV Higgs. *Phys. Lett.*, B713:462–468, 2012.

[440] Z. Komargodski and N. Seiberg. μ and General Gauge Mediation. *JHEP*, 03:072, 2009.

[441] G. R. Dvali, G. F. Giudice, and A. Pomarol. The μ Problem in Theories With Gauge Mediated Supersymmetry Breaking. *Nucl. Phys.*, B478:31–45, 1996.

[442] M. B. Green, J. H. Schwarz and E. Witten, *Superstring Theory: Volume 1, Introduction*, Cambridge University Press, 1987.

[443] J. Polchinski, *String Theory: Volume 1*, Cambridge University Press, 1998.

[444] L. Ibanez and A. M. URANGA, *String Theory and Particle Physics : An Introduction to String Phenomenology*, Cambridge University Press, 2012.

[445] R. Blumenhagen, D. Lust and S. Theisen, *Basic Concepts of String Theory*, Springer-Verlag Berlln Heidelberg, 2013.

[446] K. Becker, M. Becker and J. Schwarz, *String Theory and M-Theory: A Modern Introduction*, Cambridge University Press, 2007.

[447] M. B. Green, J. H. Schwarz, E. Witten, *Superstring Theory: Volume 2, Loop amplitudes, Anomalies and Phenomenology* , Cambridge University Press, 1987.

[448] J. Polchinski, *String Theory: Volume 2*, Cambridge University Press, 1998.

[449] J. H. Schwarz. Superstring Theory. *Phys. Rept.*, 89:223–322, 1982.

[450] P. Candelas, G. T. Horowitz, A. Strominger, and E. Witten. Vacuum Configurations for Superstrings. *Nucl. Phys.*, B258:46–74, 1985.

[451] A. Strominger. Superstrings With Torsion. *Nucl. Phys.*, B274:253, 1986.

[452] M. Dine and N. Seiberg. Nonrenormalization Theorems in Superstring Theory. *Phys. Rev. Lett.*, 57:2625, 1986.

[453] M. Dine, R. Rohm, N. Seiberg, and E. Witten. Gluino Condensation in Superstring Models. *Phys. Lett.*, B156:55, 1985.

[454] H. P. Nilles. Dynamically Broken Supergravity and The Hierarchy Problem. *Phys. Lett.*, B115:193, 1982.

[455] S. Ferrara, L. Girardello, and H. P. Nilles. Breakdown of Local Supersymmetry Through Gauge Fermion Condensates. *Phys. Lett.*, B125:457, 1983.

[456] G. Veneziano and S. Yankielowicz. An Effective Lagrangian for The Pure N=1 Supersymmetric Yang-Mills Theory. *Phys. Lett.*, B113:231, 1982.

[457] B. de Carlos, J. A. Casas, and C. Munoz. Massive Hidden Matter and Gaugino Condensation. *Phys. Lett.*, B263:248–254, 1991.

[458] J. A. Casas, Z. Lalak, C. Munoz, and G. G. Ross. Hierarchical Supersymmetry Breaking and Dynamical Determination of Compactification Parameters by Nonperturbative Effects. *Nucl. Phys.*, B347:243–269, 1990.

[459] D. Lust and T. R. Taylor. Hidden Sectors With Hidden Matter. *Phys. Lett.*, B253:335–341, 1991.

[460] S. Khalil, O. Lebedev, and S. Morris. CP violation and Dilaton Stabilization in Heterotic String Models. *Phys. Rev.*, D65:115014, 2002.

[461] E. Witten. Dimensional Reduction of Superstring Models. *Phys. Lett.*, B155:151, 1985.

[462] B. de Carlos, J. A. Casas, and C. Munoz. Supersymmetry Breaking and Determination of The Unification Gauge Coupling Constant in String Theories. *Nucl. Phys.*, B399:623–653, 1993.

[463] T. Barreiro, B. de Carlos, and E. J. Copeland. Stabilizing The Dilaton in Superstring Cosmology. *Phys. Rev.*, D58:083513, 1998.

[464] A. Brignole, L. E. Ibanez, C. Munoz, and C. Scheich. Some Issues in Soft SUSY Breaking Terms From Dilaton / Moduli Sectors. *Z. Phys.*, C74:157–170, 1997.

[465] A. Brignole, L. E. Ibanez, and C. Munoz. Soft Supersymmetry Breaking Terms From Supergravity and Superstring Models. *Adv. Ser. Direct. High Energy Phys.*, 21:244–268, 2010.

[466] S. Khalil, A. Masiero, and F. Vissani. Low-Energy Implications of Minimal Superstring Unification. *Phys. Lett.*, B375:154–162, 1996.

[467] S. Khalil, A. Masiero, and Q. Shafi. Sparticle and Higgs Masses WithIn Minimal String Unification. *Phys. Lett.*, B397:197–203, 1997.

[468] S. Khalil and A. Masiero Q. Shafi. From $b \to s\gamma$ to The LSP Detection Rates in Minimal String Unification Models. *Phys. Rev.*, D56:5754–5760, 1997.

[469] Y. Kawamura, S. Khalil, and T. Kobayashi. Phenomenological Implications of Moduli Dominant SUSY Breaking. *Nucl. Phys.*, B502:37–58, 1997.

[470] S. Khalil and T. Kobayashi. Yukawa Unification in Moduli-Dominant SUSY Breaking. *Nucl. Phys.*, B526:99–114, 1998.

[471] S. Khalil, T. Kobayashi, and A. Masiero. CP violation in Supersymmetric Model With Nondegenerate A-terms. *Phys. Rev.*, D60:075003, 1999.

[472] S. Khalil and T. Kobayashi. Supersymmetric CP Violation ε'/ε Due to Asymmetric A-matrix. *Phys. Lett.*, B460:341–347, 1999.

[473] F. E. Paige, S. D. Protopopescu, H. Baer, and X. Tata. ISAJET 7.69: A Monte Carlo Event Generator for pp, $\bar{p}p$, and e^+e^- Reactions. *hep-ph/0312045*, 2003.

[474] S. Gukov, C. Vafa, and E. Witten. CFT's From Calabi-Yau Four Folds. *Nucl. Phys.*, B584:69–108, 2000. [Erratum: *Nucl. Phys.*, B608:477, 2001].

[475] S. B. Giddings, S. Kachru, and J. Polchinski. Hierarchies From Fluxes in String Compactifications. *Phys. Rev.*, D66:106006, 2002.

[476] S. Kachru, R. Kallosh, A. D. Linde, and S. P. Trivedi. De Sitter Vacua in String Theory. *Phys. Rev.*, D68:046005, 2003.

[477] A. Linde, Y. Mambrini, and K. A. Olive. Supersymmetry Breaking Due to Moduli Stabilization in String Theory. *Phys. Rev.*, D85:066005, 2012.

[478] C. P. Burgess, R. Kallosh, and F. Quevedo. De Sitter String Vacua From Supersymmetric D-terms. *JHEP*, 10:056, 2003.

[479] G. Villadoro and F. Zwirner. De-Sitter Vacua via Consistent D-terms. *Phys. Rev. Lett.*, 95:231602, 2005.

[480] A. Achucarro, B. de Carlos, J. A. Casas, and L. Doplicher. De Sitter Vacua From Uplifting D-terms in Effective Supergravities From Realistic Strings. *JHEP*, 06:014, 2006.

[481] K. Choi and K. S. Jeong. Supersymmetry Breaking and Moduli Stabilization With Anomalous $U(1)$ Gauge Symmetry. *JHEP*, 08:007, 2006.

[482] E. Dudas and Y. Mambrini. Moduli Stabilization With Positive Vacuum Energy. *JHEP*, 10:044, 2006.

[483] M. Haack, D. Krefl, D. Lust, A. Van Proeyen, and M. Zagermann. Gaugino Condensates and D-terms From D7-branes. *JHEP*, 01:078, 2007.

[484] C. P. Burgess, J. M. Cline, K. Dasgupta, and H. Firouzjahi. Uplifting and Inflation With D3 Branes. *JHEP*, 03:027, 2007.

[485] A. Awad, N. Chamoun, and S. Khalil. On Flux Compactification and Moduli Stabilization. *Phys. Lett.*, B635:136–140, 2006.

[486] R. Kallosh and A. D. Linde. O'KKLT. *JHEP*, 02:002, 2007.

[487] O. Lebedev, H. P. Nilles, and M. Ratz. De Sitter Vacua From Matter Superpotentials. *Phys. Lett.*, B636:126–131, 2006.

[488] A. Westphal. De Sitter String Vacua From Kähler Uplifting. *JHEP*, 03:102, 2007.

[489] M. Rummel and A. Westphal. A Sufficient Condition for De Sitter Vacua in Type IIB String Theory. *JHEP*, 01:020, 2012.

[490] J. Louis, M. Rummel, R. Valandro, and A. Westphal. Building An Explicit De Sitter. *JHEP*, 10:163, 2012.

[491] K. Choi, A. Falkowski, H. P. Nilles, and M. Olechowski. Soft Supersymmetry Breaking in KKLT Flux compactification. *Nucl. Phys.*, B718:113–133, 2005.

[492] M. Endo, M. Yamaguchi, and K. Yoshioka. A Bottom-up Approach to Moduli Dynamics in Heavy Gravitino Scenario: Superpotential, Soft Terms and Sparticle Mass Spectrum. *Phys. Rev.*, D72:015004, 2005.

[493] J. P. Conlon, S. S. Abdussalam, F. Quevedo, and K. Suruliz. Soft SUSY Breaking Terms for Chiral Matter in IIB String Compactifications. *JHEP*, 01:032, 2007.

[494] V. Balasubramanian, P. Berglund, J. P. Conlon, and F. Quevedo. Systematics of Moduli Stabilisation in Calabi-Yau Flux Compactifications. *JHEP*, 03:007, 2005.

[495] J. P. Conlon, F. Quevedo, and K. Suruliz. Large-Volume Flux Compactifications: Moduli Spectrum and D3/D7 Soft Supersymmetry Breaking. *JHEP*, 08:007, 2005.

[496] K. Becker, M. Becker, M. Haack, and J. Louis. Supersymmetry Breaking and α' Corrections to Flux Induced Potentials. *JHEP*, 06:060, 2002.

[497] T. W. Grimm and J. Louis. The Effective Action of N = 1 Calabi-Yau Orientifolds. *Nucl. Phys.*, B699:387–426, 2004.

[498] V. Balasubramanian and P. Berglund. Stringy Corrections to Kähler Potentials, SUSY Breaking, and The Cosmological Constant Problem. *JHEP*, 11:085, 2004.

[499] B. C. Allanach, F. Quevedo, and K. Suruliz. Low-Energy Supersymmetry Breaking From String Flux Compactifications: Benchmark Scenarios. *JHEP*, 04:040, 2006.

[500] J. P. Conlon and F. Quevedo. Gaugino and Scalar Masses in The Landscape. *JHEP*, 06:029, 2006.

[501] S. Angus and J. P. Conlon. Soft Supersymmetry Breaking in Anisotropic LARGE Volume Compactifications. *JHEP*, 03:071, 2013.

[502] L. Aparicio, M. Cicoli, S. Krippendorf, A. Maharana, F. Muia, and F. Quevedo. Sequestered de Sitter String Scenarios: Soft-terms. *JHEP*, 11:071, 2014.

Subject index

Printed and bound by CPI Group (UK) Ltd, Croydon, CR0 4YY
01/11/2024
01782603-0007